DESERTIFICATION: ITS CAUSES AND CONSEQUENCES

Professor Mensching
with best regards

1/11/79.

OTHER PERGAMON TITLES OF INTEREST:

MEDVEDKOV

Amelioration of the Human Environment

PACEY

Water for the Thousand Millions

UNITED NATIONS

Water Supply and Management

WORTHINGTON

Arid Land Irrigation in Developing Countries

DESERTIFICATION: ITS CAUSES AND CONSEQUENCES

Compiled and Edited by
the Secretariat of the United Nations Conference on Desertification
Nairobi, Kenya, 29 August to 9 September, 1977

PERGAMON PRESS
OXFORD · NEW YORK · TORONTO · SYDNEY · PARIS · FRANKFURT

U. K. | Pergamon Press Ltd., Headington Hill Hall, Oxford OX3 0BW, England

U. S. A | Pergamon Press Inc., Maxwell House, Fairview Park, Elmsford New York 10523, U.S.A.

CANADA | Pergamon of Canada Ltd. 75 The East Mall, Toronto, Ontario, Canada

AUSTRALIA | Pergamon Press (Aust.) Pty. Ltd., 19a Boundary Street, Rushcutters Bay, N.S.W. 2011, Australia

FRANCE | Pergamon Press SARL, 24 rue des Ecoles, 75240 Paris Cedex 05, France

WEST GERMANY | Pergamon Press GmbH, 6242 Kronberg-Taunus, Pferdstrasse 1, West Germany

First edition 1977

Library of Congress Catalog Card No. 77-81423

Published as a supplement to the international journal MAZINGIRA

Printed in Great Britain by Netherwood & Dalton Co. Ltd.

ISBN 0 08 022033 9 flexi
ISBN 0 08 022023 1 hard

CONTENTS

PREFACE

It has been the practice of the United Nations, in preparing the documentation for world conferences, to commission a range of papers, each covering one aspect of the subject under consideration. Desertification, the subject of a global conference to be held in Nairobi, Kenya, from 29 August to 9 September, 1977, seemed to demand a different approach, one that would cover the subject while at the same time organizing it. The General Assembly had charged the Conference Secretariat with the task of assessing "all available data and information on desertification and its consequences on the development process of the countries affected". In carrying out this directive, the Secretariat encountered a curious situation. The causes of desertification are understood, it was unanimously agreed by the expert consultants to the Conference, and the techniques for arresting the process are known. Yet this knowledge is fragmented among a variety of other disciplines—climatology, agronomy, range management, veterinary science, geography, ecology, biology—among whose practitioners the Secretariat had sought guidance.

As an attempt was made to organize the subject of desertification, it seemed to fall naturally under four headings—climate, ecological change, technology (embracing the techniques for combating the process), and its social and behavioural aspects. These components of the subject gave rise to the notion that it would best be covered by four fairly extended studies, which we called Component Reviews, rather than the long set of briefer papers characteristic of world conferences. At the same time, and because the Component Reviews would be somewhat technical and sectoral in character, it was decided that the four reviews would be summed up in an easily readable, integrated exposition which we would title an Overview. While the Component Reviews would be presented to the Conference as background documents, the Overview would be offered as a principal document which would not only elaborate the causes and consequences of desertification but would also provide scientific justification for its cure, as embodied in the Plan of Action to Combat Desertification, consideration of which would be the Conference's principal business.

It is our hope that we have been successful in organizing what had been a diffuse subject, and will facilitate its consideration by the world conference.

Mostafa K. Tolba
Executive Director
The United Nations Environment Programme
and
Secretary-General
The United Nations Conference On Desertification

DESERTIFICATION: AN OVERVIEW

CONTENTS

PREFACE

1. In December 1974 the United Nations General Assembly passed a resolution, 3337 (XXIX), calling for an international conference on desertification, to be held in 1977. The General Assembly specified that to prepare for this Conference a world map should be developed showing areas vulnerable to desertification, all available information on desertification and its consequences for development should be gathered and assessed, and a plan of action to combat desertification should be prepared with emphasis on the development of indigenous science and technology. In a subsequent resolution, 3511 (XXX), the General Assembly stressed the need "for additional research to clarify a number of fundamental problems of desertification".

2. This additional research took the form of case studies directed toward key aspects of the desertification process. Six such case studies were financed and carried out by the specialized agencies of the United Nations systems. They analysed the process of desertification in (1) Chile and (2) Tunisia, both with predominantly cold-season rainfall, in (3) India and (4) Niger, both with predominantly warm-season rainfall, and in (5) the Indus Valley and (6) the Tigris–Euphrates Valley, both irrigated areas subject to waterlogging and salinization. In addition, a number of governments co-operated by developing associated case studies focused on desertification problems within their borders. These governments include Australia, China, Iran, Israel, the Soviet Union and the United States.

3. Another set of studies relates to the possibility of co-operative, transnational efforts to combat desertification. These so-called feasibility studies, prepared by specialists, concern the construction of green belts on the northern and southern rims of the Sahara, the management of groundwater aquifers in north-east Africa and the Arabian peninsula, the monitoring of desertification processes in South America and the Middle East, and livestock and rangeland management in fragile, dryland ecosystems.

4. The accumulation of the currently available information on desertification may be said to include the requested desertification maps. A world map of areas vulnerable to degradation was prepared by UNESCO, FAO and WMO at a scale of 1 to 25 million and maps of the desert areas of North Africa and South America have been prepared at a scale of 1 to 5 million. In addition, some of the case studies were accompanied by more detailed maps of the areas under consideration.

5. To present the available information in a coherent way, the broad subject of desertification was broken down into four major elements, and recognized specialists were commissioned to write a review of each element. The authors of these four component reviews received the advice and assistance of an international panel of experts. The reviews are entitled *Climate and Desertification, Ecological Change and Desertification, Population, Society and Desertification*, and *Technology and Desertification*.

6. This Overview seeks to provide a brief account of the main findings of the four component reviews. To do so properly, it has sometimes gone beyond the component reviews, as, for example, in making reference to the case studies and the feasibility studies. Limited in length, the Overview cannot be regarded as a summary of all aspects of desertification. Its viewpoint is more specifically directed to showing how the elements of the Plan of Action to Combat Desertification, as submitted to the Conference, emerge

directly from the information presently available and from past efforts to combat desertification, as carried out in many parts of the world. In the same sense, the Plan of Action has served as the focal point around which all preparations for the conference have been organized.

7. The four component reviews, together with the Overview, represent an attempt to provide the delegates to the conference on desertification with a more organized kind of documentation. It is the hope of the Conference Secretariat that the delegates will find this system more useful and convenient than the more customary procedure of providing an extended catalogue of documents each of which covers a limited aspect of the subject under consideration.

I. THE PROBLEM OF DESERTIFICATION

The Sahel – a Region at Risk

8. In the Sahel, as in other dry lands, rainfall is scanty and highly variable. For longer than history remembers, nomads have pastured their herds in the arid grasslands that rim the southern edge of the Sahara desert; such lands have seemed suitable for little else, since only nomadism can take full advantage of the fact that the rain will fall in one place and not in another. In the less arid southern Sahel, villagers have evolved cropping systems adapted to a wet season that becomes shorter and increasingly unreliable northward. Both pastoralists and farmers of the Sahel live under the constant threat that the rains, such as they are, will fail, and that the land will be affected by drought. Drought struck the Sahel in 1911 and again in 1940.

The Great Drought, 1968–73

9. In 1968 drought returned. At Rosso in the Mauritanian Sahel, which received an average (1935–72) of 284 mm of rain a year, only 122 mm fell in 1968. At the time, this seemed no more than a quirk in the as yet unpredictable weather patterns, since the rainfall returned to normal in 1969 with 295 mm. But in 1970, the rains failed again with a mere 149 mm, then again in 1971 (126 mm) and worst of all in 1972 (54 mm). This was typical of the whole broad stretch of the Sahel, where by 1973 the situation was catastrophic. A spectacle of vast migrations and refugee camps, brought to the world's attention, served as the immediate stimulus for the 1974 call by the United Nations General Assembly for international co-operation to combat desertification, including a world conference on desertification to which a plan of action would be submitted.

10. What was the scene in the Sahel by 1973, the fifth year of drought? Lake Chad had shrunk to one-third its normal size and was no longer one body of water. In the preceding winter, the Niger and Senegal rivers had failed to flood, leaving much of the best cropland in five countries (Niger, Mali, Upper Volta, Senegal and Mauritania) unwatered and barren. Failures of the rains meant the loss of valuable annual pastures in the Northern Sahel, as shown in the Niger Case Study, and as the drought continued and the water store became more depleted, there was widespread death of sustaining shrubs and trees. With the drying out of shallow and seasonal wells over much of the Sahel, the grazing range of the pastoralists became critically restricted. Famished and weakened livestock concentrated around larger watering points, where vegetation and soil were totally destroyed, or were driven southward in an often fruitless search for pasture. They left behind a stripped landscape, baking in the sun, where patches of newly-created desert seemed to grow and link up, producing an impression that the great Sahara desert was "marching southward". Further south, they found the farmers in a similar drought-stricken state after repeated crop failures, and the traditional grazing of stubble was no longer available. Regional food stores were exhausted, with complete breakdown of the Sahelian livelihood systems.

11. By 1973, the last year of the drought, a large programme of international assistance had been mounted for the distressed countries of the Sahel. Contributions in

cash, but mainly in food, by governments, the United Nations system and private individuals approached a value of $200 million by 1974. This was emergency relief, primarily intended to prevent starvation. It could do little about the collapse of the agricultural base of five countries (Mauritania, Upper Volta, Mali, Niger and Chad), already among the poorest nations in the world, and severe damage to the agricultural base of two other (Senegal and Gambia). For these countries, the failure of agriculture meant a loss of their tax base and a situation close to bankruptcy. In the absence of reliable statistics, it is not easy to say how many people died as a result of the drought, but estimates have ranged between 100,000 and 250,000. Yet this was less than in the terrible year 1913, remembered by the old people as the driest in this century. The amount of drought-induced disease is also impossible to calculate precisely. Outbreaks of measles and spinal meningitis took on epidemic proportions among children.

12. The Sahelian drought meant a drastic slash in productivity and income in the six countries most affected. Two million nomadic pastoralists lost as much as half of their livestock – in the worst local situations, losses exceeded 90 per cent. For almost 15 million villagers, harvests yielded less than half of the usual crop during most of the years between 1968 and 1973. The repercussions were felt not only by individuals and families but resounded as disastrous cuts in national incomes.

13. The consequences of the Sahelian drought cannot be grasped solely in terms of its climatic severity. It must be appreciated that even before 1968 the six Sahelian countries and the people who live in them were already among the poorest in the world. In the list of thirteen least developed countries, four of them are in the Sahel. These countries have gross national products that amount to less than $100 a year per person. Chronic poverty and lack of capital are characteristic of desert and dryland communities in developing countries. This is a major reason why they are so vulnerable to drought disaster.

14. As the Sahelian drought advanced, food stocks dwindled to extinction, and famine, prevalent by 1971, was general throughout the area by 1972. In the pastoral areas worst affected, there was a complete breakdown of livelihood systems and a mass exodus to towns and refugee camps located in the less affected south. Some parts of Upper Volta lost 80 per cent of their inhabitants. Many died on these migrations. To a large number of nomadic refugees the journey brought survival, since they found food, medical care and eventually wage employment in the towns toward which they headed, but with the loss of their animals, the traditional basis of their existence had gone. Surveys have since shown that a sizable proportion of the refugees may never return to their former homelands. In such ways, severe droughts leave a permanent legacy.

15. Such an exodus can be regarded as something between a minor social revolution and a temporary magnification of a regular and pre-existing out-migration, both seasonal and permanent, of dryland dwellers to towns in and beyond the region. Although these livelihoods systems benefit from the remittance of money sent from wage-earners in towns, still, even before the drought, out-migration had represented a constant and even essential adaptation. Such movements, persisting through good years as well as bad, may be necessary to survival, but they also suggest the land's lack of capacity to support its populations. They also suggest the prevalence of local perceptions that a better life might be possible elsewhere.

16. But in a way, the untold story of the Sahelian drought is that of human survival. That death and disease were much less than might have been expected is a testimonial partly to the relief effort but more importantly to the response of dryland peoples when

faced with an ancient enemy. It is a response based on mobility and facilitated by complicated networks of interrelationships, such as the loan of stock within the community, exchanges between pastoralists and farmers, and connections with relatives or friends in the towns.

17. Climatologists have asked if the recent drought in the Sahel signifies a long-term climatic shift to more arid conditions across this immense territory, where 25 million people live in six Sahelian countries. A review of rainfall records in the area has led to a conclusion that the Sahelian drought falls within the range of expectable climatic phenomena, something to be anticipated at long intervals, perhaps three or four times a century. Such anticipations are engraved in the attitudes of the people who live in the Sahel. They know that sooner or later drought will strike again, and that their livelihood systems must be adapted to cope with it.

18. The great drought in the Sahel also gave rise to a number of other questions. Can such occurrences be predicted so that people can prepare for them? What should be done to see people through such successions of lean years, in the form both of emergency relief and longer-term rehabilitative measures to promote recovery. What are the best rehabilitative measures to be applied after such an occurrence? These questions are the more pertinent because the Sahel drought occurred at a time of rainfall failures, in other parts of the world's dry lands, including particularly East Africa and the deserts of Pakistan and India, where crops and livestock were also lost.

Drought and Desertification

19. The drought ended by 1974 when favourable rainfall returned to the Sahel, but at least a decade will be required to restock the pastures and at least as long again before the land returns to something like its former state. Some of the land may take centuries to recover. In short, a legacy of the drought was an intensification of desert-like conditions in a land where rainfall is normally scanty. How long the land takes to recover is a measure of its resilience. If not carefully treated, the land may never recover completely and more desert-like conditions will persist, involving a permanent loss of biological productivity. It is this long-term decline in biological productivity that we call desertification. Drought, through its short-lived but recurrent stress, can advance the process of desertification in this way, especially when man fails to respond properly, accentuating its effects and interfering with the land's natural powers of recovery.

20. Desertification is far from a novel experience for mankind. It has played a part in the decline of civilizations from the earliest historical times. For example, because of improper drainage, salts concentrated in the lands irrigated by the Sumerians and Babylonians, thus destroying their agricultural productivity. Prolonged desiccation damaged the agricultural basis of the Harappans, who had constructed an early civilization in what is now Pakistan. The Mediterranean littoral of Africa was more productive in Roman times than it is today. No one can say how much land has been lost to man's use or severely degraded since the practice of agriculture began, but it is certainly a great deal – some have estimated it as high as the total amount of land now in cultivation. There is agreement that the rate of land degradation has increased significantly during the last decades, and some experts have suggested that it has reached at least 50,000 km^2

per year. This is a startling figure for a world likely to be faced with food shortages.

21. Man's realization that desertification is accelerating is an important factor in his increasing attention to the problem. Experience of desertification may not be new, but there is something novel in the recognition of it as a serious global problem. It is so recent, indeed, that the word "desertification" has not yet found a place in the dictionaries, with the result that there is some dispute as to how precisely it should be defined. Paving a road destroys the biological productivity of the ground covered by pavement, but few would call that desertification.

22. Although deserts are not without life, they can be viewed as areas with extremely limited agricultural potential. Deserts occur in a variety of types, hot and cold, stony and sandy, but all are characterized by rainfall deficiencies so marked that cultivation or stock-rearing are possible only with special adaptations, as, for example, by the development of irrigation. Desertification, as the extension or intensification of desert conditions, involves a decline in the productivity of the land, and it is this which make it fundamentally a human problem. Desertification anywhere affects the whole global community; for example, lowered wheat yields in one country will have worldwide repercussions on the price of wheat. However, the human impact of desertification is far greater on the people who live where it is happening and who depend on arid lands for their livelihoods. This is especially true of the developing countries. There, where capital is lacking to buy relief from "outside", desertification worsens conditions of poverty, brings malnutrition and disease, erodes the basis of the national economy, and then brings deterioration of social services already hampered by remoteness and lack of funds. All this affects the ability of dryland communities to respond to succeeding droughts, each of which would then tend to advance further the deterioration of living standards, which represents the human aspect of desertification.

23. Vulnerability to desertification and the severity of its impact are partly governed by climate, in that the lower and more uncertain the rainfall, the greater the potential for desertification. Other natural factors also come into play, such as the seasonal occurrence of rainfall, as between hot season, when it is quickly evaporated, and cool season. Also important are non-climatic factors such as the structure and texture of the soil and the topography and the types of vegetation encountered. Above all, liability to desertification is a function of pressure of land use, as reflected in density of population or livestock or in the extent to which agriculture is mechanized.

24. Areas regarded as subject to desertification are displayed on the World Map of Desertification, prepared by the United Nations Food Agriculture Organization with the assistance of UNESCO and the World Meteorological Organization. It shows that areas assessed as being at high or very high risk occupy most of the arid and semi-arid regions and extend into adjacent subhumid regions. Neglecting the very cold dry lands and the extreme deserts themselves, the latter not subject to further degradation, there remains an area of potentially productive but threatened dry lands covering 45 million km^2 or 30 per cent of the world's land surface. These occur so widely that two-thirds of the 150 nations of the world are affected. Through its sheer extent, therefore, desertification is a global problem.

25. It must also be borne in mind that desertification has an impact beyond the lands immediately affected. Dust storms can move soil great distances, and increased flooding may occur far downstream due to overly rapid run-off from lands denuded of trees and plants in upstream catchments undergoing desertification.

Risk to People, Production and Environment

26. If desertification were allowed to develop or proceed further on the geographical scale suggested by the World Map of Desertification, almost the entire population of the earth's dry lands could be said to face eventual risk. They contain between 600 and 700 million people, and in terms of broad areas and livelihood systems their numbers are as shown in Table 1.

TABLE 1. *Estimates of drylands[a] populations by region[b] and livelihood group (in thousands)*

			Livelihood populations in dry lands	
Region	Dry lands total population[c]	Urban based	Cropping based	Animal based
Mediterranean Basin	106,800	42,000 (39%)	60,000 (57%)	4,200 (4%)
Sub-Saharan Africa	75,500	11,700 (15%)	46,800 (62%)	17,000 (23%)
Asia and the Pacific	378,000	106,800 (28%)	260,400 (69%)	10,300 (3%)
Americans	68,100	33,700 (50%)	29,300 (43%)	5,100 (7%)
	628,400	194,200 (31%)	397,100 (63%)	37,100 (6%)

[a]Meigs classification (1953) including extremely arid, arid, and semi-arid area.
[b]Groupings as designated by UNEP Governing Council for regional meetings.
[c]Total world population was estimated to be 3.85 billion in 1974.

27. The dry lands under threat must be seen for what they are, the home of one-sixth of the world's population and although they are regions of low productivity per unit of area, their total production is enormous, especially of meat, cereals, fibres and hides. Under proper management, the dry lands have an even greater potential for production. The problem of supplying everyone with adequate food, shelter and clothing is now recognized as urgent and a problem that will increase in difficulty as the world's population continues to expand. Efforts to resolve this problem with anything less than the enormous total output of the dry lands would have little prospect of success. The agricultural potential of the dry lands is essential to the welfare of mankind, and their conservation is accordingly of global concern.

28. The dry lands also serve as reserves sheltering an important range of plant life, including the genetic fore-runners of many mankind's staple grains – wheat, barley, sorghum and maize. They undoubtedly contain other potential domesticates. The Green Revolution has focused new attention on the critical importance of this botanical heritage, particularly as a resource which can be used to keep highly cultured strains, such as the so-called "miracle wheat", resistant to destruction by disease. As ecotypal reserves with a

variety of interesting and useful natural settings, the dry lands constitute a precious human heritage. In recent years they have come increasingly to serve as areas to which people go – and where they often remain – in quest of health and recreation.

29. More than a mere threat, desertification is actively at work. A great many people live in dry lands that are now undergoing the process, and their livelihoods are already affected. Estimates of present losses give rise to a pessimistic outlook, suggesting that the world will lose close to one-third of its arable lands by the end of the century. Such losses will take place while the food requirements of the human race are rising at least as rapidly as the human population is growing. This demands an increase in food by at least one-third before the end of the century merely to maintain present dietary standards, as inadequate as they may be in some parts of the world.

30. Estimates of the present rate at which land is undergoing degradation and how much it would cost each year to rehabilitate these losses was agreed on by a group of expert consultants to the Conference Secretariat. These estimates appear as Table 2. Precise figures on losses through degradation and the cost of rehabilitation are not known, and these educated estimates must be taken as indicating a rough order of magnitude of the elements under consideration. The table shows that the costs of rehabilitation are high, but are nevertheless justified by the benefits. Prevention is less costly, however, and the estimated rate of land degradation provides a strong argument that preventive measures, embodied in proper land management, should be developed on a massive scale and without delay.

TABLE 2. *Preliminary estimates of orders of magnitude of costs and benefits of corrective measures*

(1)	(2)	(3)	(4)	(5)	(6)	(7)	(8)	(9)	(10)
						Estimated cost of salvage[c]			
		Estimated value		Gain					
Type of land[a]	Annual rate of land degradation (000 ha)[b]	If not salvaged ($ per ha)[b]	If salvaged ($ per ha)[b]	(4)–(3) per ha $	(2) × (5) Total million $	per ha $	total (2 × 7) million $	Net gain per ha (5)–(7) $	Total net benefits (2) × (9) million $
Irrigated	125[d]	250	2000	1,750	219	850 (250–2000)	106	900	112.5
Range	3,600	25	125	100	360	50[e] (10–200)	180[e]	50	180.0
Rainfed crop	1,700	100	400	300	510	100 (50–150)	170	200	340.0
Total	5,425				1,089		456		632.5

[a]Arid and semi-arid lands only; subhumid lands omitted. Totals are estimated roughly at 4000 million hectares, of which 250 million are integrated, 3600 million are rangeland and 170 million are rainfed cropland.

[b]Annual rate of land degradation is based on annual rate of change of classes of land to more degraded conditions. The degree of degradation from higher to lower classes of land has been converted to more limited areas assumed to be deteriorating from land yielding highest net return (if salvaged) to land at the point of going out of production (if not salvaged). The total areas subject to desertification, in varying degrees of severity, are estimated as follows, by type of problem, in million hectares: waterlogging (25), salinization (20), range deterioration (3600), dryland deterioration (170).

[c]Figures within parentheses give ranges of salvage costs. It follows from footnote b that cost of salvage is the maximum, equivalent to the cost of reclamation or restoration of practically completely desertified land. Because desertification is a continuous process, the more prudent course of action would seem to be that corrective investment begin as soon as practicable, and that it initially apply to saving lands which offer the highest returns to ensure continued maximum production.

[d]Due to waterlogging, salinization and, to a lesser extent, alkalinization.

[e]Excludes certain auxiliary social costs and benefits.

TABLE 3. *Estimates of populations and livelihoods resident in areas recently undergoing severe desertification (in thousands)*

Region	Total population	Urban based	Cropping based	Animal based	Area (km^2)
Mediterranean basin	9,820	2,995 (31%)	5,900 (60%)	925 (9%)	1,320,000
Sub-Saharan Africa	16,165	3,072 (19%)	6,014 (37%)	7,079 (44%)	6,850,000
Asia and the Pacific	28,482	7,740 (27%)	14,311 (54%)	6,431 (19%)	4,361,000
Americas	24,079	7,683 (32%)	13,417 (56%)	2,979 (12%)	17,545,000
	78,546	21,490 (27%)	39,642 (51%)	17,414 (22%)	30,076,000

31. The numbers of people immediately threatened, their general location and livelihood systems, are shown in Table 3. Of the 78 million people threatened, about a third may be in a position, because of high income or other advantages, to avoid the worst consequences of desertification. This still leaves about 50 million people who are immediately menaced through the destruction of their livelihoods and who are faced by the grim prospect of uprooting themselves from everything familiar and of migrating to other areas frequently ill-equipped to receive them.

Urgency of the Problem

32. The need for action to combat desertification is all the more urgent because the process is a dynamic one. Desertification can feed on itself and become self-accelerating. With delay, rehabilitation becomes increasingly lengthy and expensive, and degradation may reach a threshold beyond which it is irreversible in practical and economic terms. Fundamental preventive measures should be introduced as soon as possible in the form of land-use practices which are both socio-economically and environmentally appropriate, which ameliorate microclimates and soils and which prevent desertification from making further encroachments.

II. PROCESSES OF DESERTIFICATION

Water and Energy Balances

33. To see precisely what happens when desertification occurs, attention should be focused on that shallow meeting place between soil and atmosphere, where plants thrive and where a balance is maintained between incoming and outgoing energy and between water received and lost.

34. When rain fails, some of the water is taken up directly by plants, some filters into the soil, where it may remain in storage, and the rest evaporates or runs off. Some soil moisture, that intercepted by plants, is put back into the atmosphere by the plants in transpiration. Some of the moisture may seep into deeper layers to collect in underground reservoirs or aquifers, where it may remain for thousands of years, or may migrate slowly from pleateau to depression or back to the ocean itself.

35. The soil–air meeting place participates in an energy balance activated by the rays of the sun or through atmospheric heating. Some energy is reflected by the surface layer back into the atmosphere and into space. Some is held by the soil in storage, thereby warming and lighting the earth, and it is this energy and that from the sun directly that is used by plants to carry out the processes of photosynthesis and growth. Some of the plants are eaten by grazers or browsers, and these animals in turn may be eaten by carnivores, with all animals returning energy and moisture to the atmosphere in respiration and to soil in the form of humus. The excreta of animals, their decomposing carcasses and the decomposition of plants supply the soil with organic nutrients, most densely in the topmost layers and thinning out below. These relations are illustrated in Fig. 1.

Adaptation to the Arid Environments

36. In arid situations the cycling of water and energy takes on special characteristiics because of deficient and variable rainfall and abundant solar energy from cloudless skies. Vegetation is generally sparser than in humid areas, provides less cover to the ground surface and returns less organic matter to the topsoil. During occasional intense rainfall runoff may occur in spate, but water at the surface tends rapidly to be lost through evaporation, and in the long intervening dry spells the soil is parched and heated by the powerful sun.

37. However scanty it may be, the dryland vegetation constitutes a fundamental resource which transforms solar energy into food and which protects and stabilizes the surface of the ground. This vegetation survives by adapting to water deficit in ways which are important because they determine seasonal differences in the usefulness of dryland pastures.

38. Part of the plant population consists of short-lived ephemerals which germinate and complete their life cycles rapidly after rain, remaining as seed through intervening dry periods. Such plants are commonly fleshy and palatable and are preferred by grazing animals. Other plants, such as perennial grasses, wither and die back to the root stock or to bulbs in dry spells and shoot anew with fresh rains. These plants form more durable

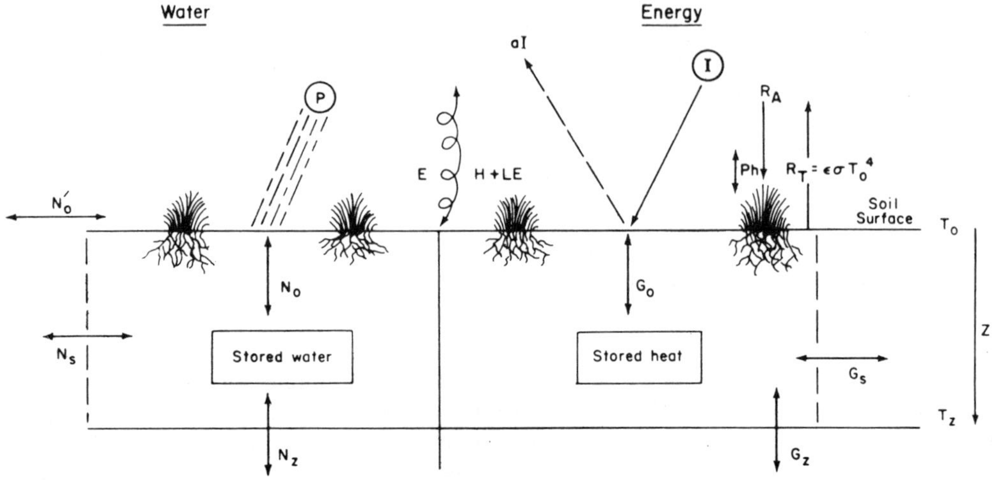

Fig. 1.

Water exchanges are on the left, energy on the right. The diagram refers to a partly vegetated surface whose temperature (for radiative emission and convective forcing) is T_O. Most natural surfaces in dry climates have extensive areas of bare soil, as shown.

The water exchanges are forced by precipitation, P, a very intermittent input. The water reaching the surface may percolate, N_O, or run-off at the surface, N_O'. The soil layer, of depth z, stores water – typically of order 10 cm of precipitation when wet. If storage is full, the remaining rainfall may outflow laterally, N_S, or percolate to ground water. With high water tables on sloping sites, N_O', N_S and N_Z may all reverse, so that there is run-on. Total water surplus $N = N_O' + N_O + N_Z$. If there is no change in storage, $P = N + E$, where E is the evapotranspiration.

The energy exchanges are forced by global solar radiation, I, direct (shadow-casting) plus diffuse, a fraction (a, the albedo) of which is reflected, aI. The surface temperature is mainly a function of the absorbed solar radiation $I(1 - a)$. There is a diffuse infrared radiation received from the atmosphere, R_A, and infrared is emitted from plants and soil, $R_T=\epsilon\sigma T_O{}^4$, where ϵ is the emissivity (usually > 0.9), σ is the Stéfan-Boltzman constant, and T_O is surface temperature. $-R_T + R_A$ is the net longwave heating, generally negative. The net radiation is $R = I(1 - a) - R_T + R_A$. A small part of R, G_O is conducted to the soil, where it may be stored. If level z is the depth at which there is no annual temperature cycle, $G_Z = 0$, and G_S is negligible. Hence G_O virtually vanishes over the year. The remaining heat, $R - G_O$, feeds convective fluxes, H of heat, and LE of latent heat. If $T_O < T_A$ (air temperature), H is negative, though small, and LE may also be negative (dew formation). In this case the convective fluxes force the net radiation, rather than the reverse – the typical nocturnal condition. Ph is the net photosynthesis, generally 2 orders of magnitude smaller than R (see component review on ecological change). If there is no change in storage, and no net synthesis, $R = I(1-a) - R_T + R_A - G_O = H + LE$. Note that *a, the albedo, is a function of the state of the vegetation and soil*, tending to rise from values near 0.17 in fresh green savanna to ~0.35 in sandy deserts (Oguntoyinbo, 1974). Degradation of vegetation in general raises the albedo, as does drying the surface. Absorbed solar radiation, and hence surface temperature T_O, are functions of a, as are net longwave cooling ($R_A - R_T$), the soil heat flux G_O, the convective fluxes of heat and latent heat, H + LE, and net photosynthesis, Ph. Hence changes in albedo drastically alter the entire energy and water balances.

Increases in soil compaction due to overstocking, use of vehicles, drought contraction or pan formation decrease the infiltration rate, and hence increase N_O' at the expense of N_O, so increasing the chance of sheet erosion and gullying. They also alter the thermal conductivity, and hence G_O. These processes are discussed in the component review on ecological change.

T_O is low by day when the soil is moist, and higher when it is dry – by 20 to 30C in tropical conditions. With moist soil and low surface temperature, R is large, and the Bowen ratio H/LE is very low – of order 0.10. With dry soil and high surface temperature, R is lower, and the Bowen ratio rises to very high values.

pastures, are attractive and palatable to stock when green, and may provide valuable hay, but are of little pastoral value when thoroughly dried out. Nevertheless, their extensive fine root systems remain to bind the topsoil and contribute importantly to its organic content. Lastly, there are the longer-lived perennial plants which resist water loss by such adaptations as woody stems and leathery leaves. These include the larger plants such as shrubs and trees, which remain nutritious during the dry periods, when they can provide an important food source to browsing animals, although their adaptations may reduce their palatability and attractiveness for some stock. They have the additional role of protecting the ground surface, providing shade and preserving an environment which favours the response of important shorter-lived plants.

Impact of Land Use on Equilibrium on Dry Land

39. Under natural conditions and through appropriate strategies, the dryland ecosystems maintain a balanced exchange of water and energy, but a favourable equilibrium is readily disturbed when man makes use of the land. For example, where meagre vegetation is further reduced to expose the ground surface, humus will be mineralized and soil structure lost. Rain will fall directly on the soil and break it down, and the sun will bake a thin crust which prevents additional water from sinking in. As the water budget deteriorates in the soil beneath, the level of groundwater in nearby wells may fall. The water lost to the soil store now contributes to over-rapid run-off. Where the surface has been loosened or disturbed as by the trampling of animals, the topmost soil layer, that with the best structure and containing the bulk of plant food, may be washed away, or blown away in dust storms. The denuded soil is essentially infertile, with poor structure and water relations. All these changes constitute a shift towards a more hostile environment for plants, with the result that the vegetation responds less well to rain and produces less biomass, and many plants tend to die off at an increasingly early stage of drought. Such changes are typical of desertification.

40. In its initial stages desertification may merely involve a shift to a more desert-like and less productive ecosystem, with water, energy and nutritional balances less favourable to plant growth than before. But land use in arid regions poses problems which continually menace the prevailing equilibrium. This is at least partly because of fluctuations in rainfall between drought years and good years which, not yet predictable, are difficult for the land user to respond to effectively.

41. For example, in dryland pastoral economies, large numbers of stock tend to build up during runs of good years, too many to be supported through the inevitably ensuing drought. There is a natural reluctance to cut back on stock numbers in the first dry year, and a tendency to hang on until drought is seen to be established. But by that time dryland pastures are probably being overgrazed toward a state that threatens eventual regeneration. By this time, too, prices for surplus stock will probably have shrunk because the market is glutted, and destocking through sale of surplus numbers will be opposed by economic forces. For the same reasons, destocking may be prevented during the periods critical to the regeneration of pastures following the rains that end drought.

42. Dryland farmers, too, show a tendency during runs of good years to extend their cropping onto ever more marginal lands, into areas of higher climatic risks, pushing back the pastoralists in the process. This is particularly the case when pressure on the land is increased through population growth or restrictive systems of land tenure or the

short-sighted introduction of mechanization. The expectable but unpredictable onset of drought will find the marginal land prepared for planting, stripped of its protective natural vegetation and vulnerable to erosion. Such land enters a run of dry years without defences and may emerge in too degraded a condition to support even livestock.

43. Thus, the delayed response of the land user through cycles of good and insufficient rainfall may convert periodic drought into an engine of long-term desertification. But this need not be so. Land-use practices should combine with efficient marketing systems to make possible an appropriate response to drought, a natural and inevitable factor of dryland agriculture, and a recurring event that must be taken into account. Land should also take advantage equally of the rainier years, employing them to replenish the ultimate agricultural resources, the fertility of the soil and the production of vegetation.

44. The main processes and stages of desertification can be summarized as follows. In pastoral rangelands, there is an initial deterioration in the composition of pastures subject to excessive grazing in dry periods, particularly a reduction in the proportion of edible perennial plants and an increase in the proportion of annual and inedible species. The thinning and death of vegetation in dry seasons increases the extent of bare ground, and this is followed in turn by a deterioration in the conditions at the surface of the soil that are vital to plant growth, particularly in impoverishment of plant–water relations. The response of ephemerals to rain suffers accordingly. With consequent increase in run-off, sheet and fully erosion set in on sloping ground, and the topsoil and its store are lost. All these changes mean a decrease in plant productivity and a decline in the palatability and durability of the native pastures. With continuing erosion, formerly productive lands may be lost through soil stripping and gully extension. These changes are even more drastic where devegetation occurs in strategic areas, as on watershed uplands, and the processes are advanced where soils are exposed and disturbed in dryland cultivation.

45. In areas of rainfed farming, desertification often originates on land cleared for cultivation or left fallow. Removal of the original vegetation cover exposes the soil to accelerated wind and water erosion. The beating action of rain on naked soil puddles the surface which crusts when the sun comes out, reducing infiltration and further increasing run-off. This in turn, leads to increased soil erosion which ultimately, unless halted by protective measures, strips away the fertile surface soil and exposes infertile subsoils. Gullies may form on the lower parts of slopes and impede farming operations, or prevent them entirely. Sediment deposited at the foot of slopes covers plants, fills waterways and aggravates flooding in low-lying areas which follows increased run-off from the slopes above.

46. Water and wind erosion work together, as redeposited silts from surfaces stripped by water erosion are particularly vulnerable to wind transport. Wind erosion starts with the movement of coarse soil particles in one part of a field, then progresses downwind with increasing severity as bouncing soil particles knock other particles into the air in a kind of snowballing effect. Finer materials are lifted into the air and carried away over long distances as dust; coarser sandy materials drift over the surface until they are trapped by plants in accumulations as hummocks and small dunes. Removal of fine topsoil materials means the loss of the most productive and nutritious portions of the soil complex, while sterile sand accumulations cover plants and good soil. A further harmful effect of high-velocity sand drift is the destruction of young crops by the blasting impact of moving sand. Fine airborne particles may carry soil-borne diseases, irritate respiratory tracts of humans and animals, cause wear on machinery parts and reduce visibility.

47. The principal manifestations of desertification on irrigated lands are the salinization and alkalinization of soils, due to inadequate leaching of salts contained in the soil or added in irrigation water. Salinization and waterlogging commonly occur together. Where the soil is waterlogged, the upward movement of saline groundwater leaves salts on the surface where water evaporates. On soils that are not waterlogged, salinization can still occur when water containing soluble salts moves from irrigation furrows into the ridges where crops are planted or to high spots in poorly levelled land. Under-irrigation of weakly permeable soils can also lead to salinization if the irrigation water is salty.

Desertification can Feed on Itself

48. When left alone, dryland ecosystems disturbed by land use or stressed by drought, will usually return to what they were. Recovery tends to advance at a slow pace because of the low productivity of dry lands and is usually episodic, with more rapid recovery in years of above-average rainfall. Eventually, former water and energy balances will be restored, with the recovery of the original vegetation. This is a measure of the natural resilience of the dry lands.

49. Where pressure of land use persists through drought, these same ecosystems are shown to be fragile, and processes can be set in motion whereby desertification becomes self-accelerating. This can occur where sand dunes are stripped of vegetation, as near watering points or other places where stock tend to congregate, and drifting sand destroys more vegetation and mobilizes extending surfaces, and dunes slowly advance and engulf less damaged sites. It can occur where destruction of vegetation initiates accelerating erosion, removing sediment which, in turn, buries fields or pastures downstream, or in denuded areas, where hot, drying winds become increasingly prevalent. In irrigated systems, lack of drainage allows watertables to rise and waterlog and salinize fields to the point where they must be abandoned. Because self-acceleration can occur through a variety of circumstances, desertification will often advance inexorably unless preventive measures are undertaken. As it advances, it becomes even more difficult and more expensive to treat, with the costs of reclamation continually rising until the stark equilibrium of extreme desert is reached and the land has for all practical purposes passed beyond hope of rehabilitation.

50. Deserts themselves are not the sources from which desertification springs. Except for hot winds, the deserts themselves supply none of the essential impetus for the processes described. Desertification breaks out, usually at times of drought stress, in areas of naturally vulnerable land subject to pressures of land use. These degraded patches, like a skin disease, link up to carry the process over extended areas. It is generally incorrect to envision the process as an advance of the desert frontier engulfing usable land on its perimeter: the advancing sand dune is in fact a very special and localized case. Desertification, as a patchy destruction that may be far removed from any nebulous front line, is a more subtle and insidious process.

The Advance of Desertification

51. Yet the drier lands are the most vulnerable to desertification, and the dry lands, for climatic reasons, tend to ring the desert cores in belts which are progressively more humid outwards, of which the Sahel-Sudan-Guinea sequence in West Africa is one

example. As the patchwork of desertification grows and links up, it can eventually join with the climatic desert, and the final result will look exactly as if the desert itself had spread. The deserts extended in this way form wastelands over areas that from a strictly climatic point of view should not be entirely desert. Thus one observer reported in 1975 that the desert had advanced southward in the Sudan by 90–100 km over the preceding 17 years. The vast extent of such extended desert lands provides a basis for estimating how much of the land surface has been degraded through man's use of it in past times. To the extent that desertification contains self-reinforcing elements, its effects may extend outward through a whole system of climatic and land-use belts. As arid pastureland turns into complete desert, neighbouring semi-arid land, once suitable for rainfed cropping, may deteriorate into arid pasture. But since the process does not spring from the desert cores, it need not always work outwards, and the semi-arid or subhumid cropland may be the first to deteriorate to desert status.

III. THE CAUSES OF DESERTIFICATION

52. Desertification arises from the interaction between a difficult, unreliable and sensitive dryland environment and man's use and occupation of it in his efforts to make a living. Some understanding of the controls of dryland climates helps towards an appreciation of climatic factors in desertification.

The Desert Belts

53. Although their boundaries have shifted over time, deserts must always have characterized the earth's subtropical zones. Global patterns of air circulation dictate that the subtropics can be regions of subsiding air. When air subsides it warms up and its capacity to hold moisture increases, so inhibiting the formation of rain. This accounts for a prevalence of dry climates between latitudes 15 and 30 degrees north and south of the equator. However, the dry climates are extended into other latitudes and their patterns complicated by additional factors, such as distance from the rain-supplying oceans, the seasonal high-pressure zones of large continental areas linked with monsoon systems or the presence of mountain barriers down which air spills on their lee sides, creating rain shadows.

54. The play of these factors is evident in the distribution of deserts, as shown on the World Desertification Map. There are five main desert belts: (1) the Sonoran desert of north-western Mexico and its continuation in the desert basins of the south-western United States; (2) the Atacama desert, a thin coastal strip running west of the Andes from southern Ecuador to central Chile, whence dry climates extend eastwards into Patagonia; (3) a vast belt running from the Atlantic Ocean to China and including the Sahara, the Arabian desert, the deserts of Iran and the USSR, the Rajasthan desert of Pakistan and India, and the Takla-Makan and Gobi deserts in China and Mongolia, (4) the Kalahari and its surrounding arid lands in southern Africa, (5) most of the continent of Australia. Outside these principal desert regions there are isolated areas of arid lands in many parts of the world, such as the Guajira peninsula in Colombia, south western Madagascar and part of north eastern Brazil. Within the desert belts there are many climatic contrasts resulting from difference in temperature, in the season when rain (if any) falls and in degree of aridity. At one extreme are the cold deserts such as the Tibetan plateau, where settlement is precluded by low temperatures and where environmental degradation has accordingly been small. On the other are hot deserts such as the inner Sahara, where plant growth and land use are absent outside of irrigable cases because of extreme aridity. These extreme deserts do not concern us; they are not subject to further desertification and they remain unclassified on the World Desertification Map.

55. More extensive than the extreme deserts are the world's arid lands, with not enough rain to support cropping but with sufficient vegetation to support pastoralism below about 200–250 mm, depending on temperature. Outside these are the semi-arid lands, depending on temperature and season, where cultivation of drought-resistant crops is generally possible with the use of moisture-conserving practices (outer limits between about 200–600 mm depending on temperature). Finally, the greener margins of the dryland belts comprise the drier parts of the subhumid zones where land use and

settlement become more intensive, but which must be considered as liable to deterioration to desert conditions. This can include areas with applicable rainfall up to 800 mm annually. The classification of land, and its condition, are not determined exclusively by the amount of rainfall since the effect of rain also depends on how readily it evaporates, and evaporation will take place more rapidly in lower latitudes and when rain falls in the hot season. Arid and semi-arid lands, together with their subhumid margins, comprise what are here called dry lands. It is in the dry lands, where desertification is taking place, that its causes must be sought.

Shifting Limits of Dryland Climates

56. It is evident that the boundaries of the dryland belts are not fixed. Much of the Sahel, for example, contains old sand ridges, now quite fossil under a cover of vegetation, which indicate an extension southwards some 20,000 years ago of Saharan climates and moving sands, some 400 km southward of their present limit. In the same region, Lake Chad was much more extensive 10,000 years ago, indicating semi-arid or subhumid conditions had encroached northward into what is now arid. Dry conditions have also shown shifts in an east–west direction; some 8000 years ago, Rajasthan in India was perhaps 1500 km east of the arid zone. These climatic changes have been shown to be part of global shifts of the climate belts, related to changes in the earth's atmospheric circulation and energy budget. They are linked to the great changes of the Ice Age and of subsequent millennia, during which temperature shifts are known to have been accompanied by changes in rainfall patterns.

57. Such changes, with durations of centuries or millennia, have continued and have directly affected the possibilities of man's occupation and use of the dry lands in the past. For instance, some of the presently hyperarid Sahara was open to pastoralists and hunters under semi-arid conditions between about 8000 and 6000 years ago. The definite records of later changes are mainly from higher northern latitudes and give temperatures rather than rainfall. From 1600 to 1850, for example, the northern hemisphere underwent a cooling known as "The Little Ice Age"; then followed a warming which continued into the 1940s, since when temperatures have again declined in the north while they are reported to be rising in the southern hemisphere. Some have wondered whether this new cooling may not mark the resumption of another "Little Ice Age" in the northern hemisphere.

Climate Change as a Cause of Desertification

58. The fact of continual climatic change gives rise to the question: "Do recent droughts in the Sahel and elsewhere form part of a change to a more arid climate, expressed as an equatorward shift of the dryland belt?" Those replying yes would note that over the same period deficiency of rainfall in the dry lands was accompanied by greater rainfall in the wet equatorial belt. The consequences would be that the dryland inhabitants would face a long period of increased aridity in their part of the world following a century or more of relatively favourable climate.

59. Unfortunately the question cannot be answered with confidence partly because the period is still too short for climatic records to be used with certainty and partly because a firm basis for long-term prediction of climate has not yet been established. Our present

understanding of the global mechanism of atmospheric circulation remains inadequate. Tantalizing hints of a basis for prediction have served to spur research. These lie in the establishment of "teleconnections", long-distance relationships among climatic phenomena as, for example, between the high-pressure zone in the North Atlantic and the climate of the Sahel, and in the development of continually improved models of the global atmospheric system. The latest Sahelian drought was far from unprecedented, even in the relatively short historical record, and it cannot alone form evidence of a change of climate. At the same time, it would be unwise to rule out the possibility of change, as recorded, for example, in increasing aridity in Sudan, and the consequences for change should be looked at; particularly in areas of strong rainfall gradient such as the Sahel. Change might mean that droughts would become more frequent and more severe, and that plans for land management should take into account the possibility of an even less reliable climate in the future. This should not, however, be taken to imply that man is the victim of recently accelerated desertification rather than its active agent. Clearly the answer to the question has great significance for strategies to combat desertification.

Man-made Climatic Change

60. A related suggestion is that man himself may have contributed to recent increased climatic aridity through modifications in the energy exchange that have followed his degradation of desert ecosystems. These modifications include an increase of dust and aerosols in the lower atmosphere, noted particularly in association with the recent African and Asian droughts, and an increased reflection of solar energy from denuded dryland surfaces. It has been argued that these events could have diminished convection in the atmosphere above the desert, with an ensuing reduction in the frequency of rainstorms.

61. At this time, however, the significance of such processes and the resulting direction of climatic change remain in doubt. It is possible that man has accentuated climatic stresses in this way, but it is most unlikely that his activities have been prime causes of any *general* deterioration of dryland climates, which are, after all, expressions of the fundamental circulation of the atmosphere. What is certain is that the direct physical consequences of the man-made changes on the local, effective climate, such as the adverse effects of surface denudation on the soil–water balance, are many times more important than any indirect large-scale climatic effects.

Climatic Fluctuations contribute to Desertification

62. Climatic boundaries in the dry lands are subject to short-term shifts, corresponding to sequences of lean years and fat years. In general, the drier the climate, the greater the rainfall variability and the higher the drought risk. Such fluctuations are expressed geographically in expansions and contractions of the dryland belts, such that a semi-arid region may experience arid conditions at one time, subhumid conditions at another.

63. These fluctuations, although not so regular as to be predictable, can be divided into short-period, say 2–4 years, which introduce periodic stress into livelihood systems, and those of greater amplitude and duration which can lead to significant changes in the patterns and structure of land use, such as the extension of cultivation in good years or large build-ups in stock numbers in runs of good years. It may not be possible to adjust these expansions promptly when drought inevitably follows. It is when drought strikes

land-use systems that are stretched beyond their usual limits that its consequences can be disastrous and maximum and long-lasting degradation can occur. Recovery from such degradation will be slow at best, and if land-use pressures continue unabated, recovery may be partial only, to a lower plane of productivity than formerly. Desertification will then have occurred.

64. To the climatic factors in desertification must be added the sensitivity of dryland ecosystems and their fragility under land use. Because of lack of water, they generally support only scanty amounts of plant and animal life, and because life is thinly spread the soils are poor in organic matter and nutrients. This small but essential reserve of fertility is concentrated in the upper few centimetres of the topsoil, while the soil below is commonly salt-affected through lack of leaching, and poorly structured in consequence, with adverse plant–water relations. These ecosystems are delicately balanced, especially during drought. Their necessary adaptation to water deficiency results in life forms which are highly specialized in their adaptation to climate and soil. Undue destruction of the plant cover can be followed rapidly and in accelerating fashion by physical degradation through relatively extreme action by wind and water, and regeneration of the original vegetation may be long-delayed or even prevented entirely.

65. Any use of dry lands that does not take account of their limitations, and of the patchwork contrasts in productivity and vulnerability which accompany them, will constitute misuse and lead to desertification. The problem is compounded by fluctuating climate and land condition, and a tendency toward overoptimistic assessments of the potential of dry lands for sustained production, often made on the basis of remembered best years. Such optimism is often a response to external pressures to which the users of dry lands are increasingly subjected – from commercial markets, their own rising expectations, and from population growth. It is often associated with the introduction of inappropriate technology in the hope of short-term gains, for instance deep ploughing of croplands or undue pulverization of topsoils through mechanical tillage.

66. Traditional systems of land use have met these environmental challenges in various ways so as to maintain flexibility and spread the risks. For instance, pastoralists may herd several kinds of animals, each capable of profiting from different parts of the ecosystem, or they may range widely as nomads to spread and lighten the grazing load. But with increasing technological inputs arising from the push towards higher productivity, there is a trend towards diminished flexibility. For instance, as dryland farmers extend towards the climatic limits of cropping, they must depend increasingly on the hardiest cereals, wheat and barley, with fallowing to conserve soil moisture. Lack of flexibility is particularly evident in large-scale commercial ranching and farming, which tend to be highly specialized, but it may also exist in traditional cropping systems, as shown in the Chilean case study, through inequitable systems of land ownership and land tenure.

67. Technological innovations are often brought in from more humid environments without full regard for the particular equilibrium of the dryland ecosystems. For instance, the Niger case study demonstrates how the introduction of deep wells has indeed improved the availability of water but at the same time has had the effect of increasing herd size while decreasing herd mobility, leading to local overgrazing and excessive trampling around the new watering points. Technological changes can give rise to desertification through excessive demands on limited natural resources.

68. The concept of proper land use must be extended beyond agricultural practices. Modern man is threading the dry lands with road and highways; he is exploring them for

mineral resources, opening mines, sinking oil wells, constructing pipelines and canals, establishing factories, and building cities in them. Increasingly, he is intruding into arid lands for purposes of health and recreation. These new activities are important because they diversify the bases of human occupation of the dry lands and provide an important additional source of revenue. However, they must be undertaken with a full regard for the delicate natural balance that prevails there. For example, soil surfaces disturbed by earth-moving machinery will take longer to revegetate and stabilize in these conditions, and may become subject to wind drifting. Many of the new activities in the dry lands have been made possible by advanced technology, but it is this same ever-growing technological capacity that so strengthens man's ability to disrupt and damage a sensitive environment.

Rapidity of Feed-backs in Drylands Situations

69. The sparsity of life in dry lands and the tenuous linkages among their life forms mean that small changes can trigger profound physical effects. Where the equilibrium is delicate, as under water stress, even a small change in one component can radiate effects through the entire ecosystem. Dry lands are sensitive to minor shifts in their water and energy balances, as reflected in changes in plant cover. Changes brought on by seemingly minor disturbances can follow with startling rapidity, sometimes sufficient to throw the system beyond the critical threshold whence natural recovery will not normally occur.

Desertification: a Product of the Interaction between Men and a Difficult Environment

70. Failure in resilience usually arises from sudden and severe disturbance, and such disturbances, in the present world, are almost always the work of man. Human activities that may have a less pronounced impact in more richly endowed environments, where a greater richness of land and life aids the process of restoration, may in the dry lands have more drastic consequences, with desertification a common outcome.

71. If man is the chief instrument of desertification, the process should not be viewed exclusively from the human side. Desertification results from the interaction between man and a difficult and changing environment. It occurs when man penetrates such environments and acts there – often out of his need for survival – without an understanding of or proper regard for their sensitivities and limitations.

IV. DESERTIFICATION IN ACTION

Desertification and Livelihood Systems

72. The interaction between man and a difficult environment, which results in desertification when land is misused, can well be examined in terms of what man does in the dry lands, how he wrests a living from an environment which confronts him with great difficulties.

73. Dryland agriculture takes three main forms: extensive pastoralism or the herding of livestock, rainfed cropping and irrigation agriculture. These range from highly commercialized or specialized systems, dependent on external markets, to traditional systems which have evolved strategies and acquired skills to cope with the stresses and risks imposed by dryland environments. Such tested practices should not be lightly dismissed; indeed, they should be regarded as a basis for further development.

74. Yet it cannot be said that any land-use practice has been so self-regulating, so perfectly adapted to arid conditions or so far-seeing that it has incurred no cost to the environment. Desertification has accompanied all land-use systems, although its effects may have been less destructive in the past when people were fewer and the land less densely occupied. The accelerated desertification of recent years may in part be attributed to breakdowns in traditional practices. The old ways have come under intense pressures which have eroded ancient social, economic and political constraints. Such pressures have come from population growth, the spread of the use of money, aspirations toward higher standards of living, the introduction of technological innovations and the incorporation of dryland agriculture into often remote and powerful commercial systems, where prices seem to fluctuate without relation to the needs or interests of the prime producer.

Dryland Pastoral Systems: General Aspects

75. The pastoral systems characteristic of the dry lands use grazing or browsing animals to harvest a thin crop of natural vegetation. In semi-arid lands, stock raising is increasingly linked with crop production, but in arid regions beyond the reach of the farm, pastoralism is dominant, except where irrigation is possible, and on his remote natural pastures the herdsman is subject to the extremes of climatic hazards.

76. Herdsmen have found many ways of coping with the climatic stress that typifies their arid lanscapes. Ordinarily, they spread their stock thinly over large areas so that grazing pressure is lightened and they can take advantage of the patchwork ecosystem of arid lands where topographical variety yields good pasture only here and there. They are mobile, often traversing great distances to reach seasonal pastures. Pastoralists, including nomads, must not be thought of as "wandering". Familiar with the land and what it provides, they have a clear idea of where they are headed.

77. They usually employ some measures to modify the ecosystems in which they

function. They will limit the size of their herds, if necessary by selling off surplus animals. They will exercise some control of pastures by deferred or rotational grazing or by letting certain rangelands rest to allow them to accumulate moisture over several years. They will develop additional watering points, extending the area and duration of grazing and diluting the pressure on older pastures. They sometimes burn pastures to facilitate the growth of more palatable plants. Sometimes, too, they will provide supplementary feed by cutting hay or growing forage crops under irrigation.

78. Some herdsmen have access to alternative sources of income. They sometimes develop handicraft industries. They might engage in hunting and gathering, or perhaps commerce, a natural adjunct to their mobility, but one increasingly denied them as railroads and truck routes penetrate the dry lands. The picturesque herdsman-caravaneer, who supplemented his income by attacking caravans as often as he escorted them, has become a figure out of the past.

79. Pastoralism can range from traditional subsistence systems, often nomadic, through more sedentary systems closely linked with cropping, to the great commercial ranches which mainly serve as exporters from the arid zones. All tend to have links with the outside where the chief markets are to be found – for hides, wool and stock on the hoof. In the more commercial pastoral systems, stock will be bred in the arid zones and fattened in areas closer to the market. Whether this practice can be extended to less commercial operations in the Sudano-Sahelian region is the subject of one of the feasibility studies. As breeding areas, arid zones have certain advantages, such as freedom from disease, long outdoor range periods and high protein levels in pastures.

80. Under the pressures to which they are increasingly subject, pastoralists sometimes behave as if they regarded their breeding stock, rather than the land and its vegetation, as their ultimate resource. This leads to an apparent disregard for the ecology of the plant communities on which the stock feed. Comparatively little attention may be paid to the performance of pastures under stress, to the requirements for successful germination or to the impact of selective grazing on the whole plant community.

81. Pastoral systems are generally afflicted with what might be called a time-lag problem. With the onset of drought there is a natural reluctance to reduce stock numbers built up in preceding favourable years, and overgrazing and damage to drought-stressed pastures can result. Similarly, when the drought ends, there is a temptation to restock too rapidly those rangelands that are in need of regeneration. Conversely, herds reduced by destocking or death in dry years may confront the return of better grazing with numbers too scanty to take full advantage of it. To introduce the kind of flexibility that would make a fully opportunistic use of the land presents problems to which careful attention should be given.

82. When deterioration comes to grazing land, it is particularly important to observe its first stages as expressed in the condition of vegetation, not only because plants are the basic grazing resource but also because of the part they play in the stability of dryland ecosystems. First to be removed are the more palatable plant species leaving the terrain to less desirable types. During drought stress, when they constitute the only feed, valuable, soil-holding perennials can be grazed to extinction. Exposure can have harmful effects on bare ground, which diminishes the response of ephemeral green feed following the next rains and hinder the slower return of desired perennials. Destruction of vegetation and pulverization of the soil become particularly noticeable at points where stock congregate, such as at watering places, where trampling is usually severe.

Nomadic Pastoralism

83. Stock-herding nomads have found ways of using land too arid for any other agricultural purposes. With a diet obtained from their herds and supplemented by food gathering, mobile pastoralists have achieved standards of health and nutrition often superior to their more sedentary neighbours.

84. It is, of course, mobility that provides the herdsman with his principal weapon against a harsh environment. His movements may be continuous, or he may move back and forth between fixed seasonal pastures, a practice known as transhumance. Flocks and herds are generally owned by families, but other resources – pasture, watering places, fuel – are often communally-owned and their use regulated by custom. Some pastoralists have mutually beneficial relationships with the farmers on their periphery. Such links may include ownership of cropland, provision of seasonal labour to oasis settlements and the right to graze stubble in exchange for the manure dropped by grazing animals.

85. If the flexibility of pastoralism resides principally in mobility, which spreads pressure on the land and dilutes risk, the system also contains other important adjustments. Pastoralists may do some rainfed cropping at the edge of the migratory range. Some of them emigrate to outside employment, relieving pressure on local resources and sending home cash remittances which have become an established part of many pastoral economies.

86. Increasingly over the past 50–100 years, pastoral nomadism has found itself at bay. The political status of the nomads had declined, and with it their control of grazing rights, their relations *vis-á-vis* adjacent crop-based systems, and their role in desert transport and commerce. Their essential mobility has come into disfavour for political and administrative reasons, and has proved to be an obstacle in providing them with education, health care and other essential social services.

87. In addition, imbalances have arisen in systems of pastoral nomadism which have heightened their potential for desertification. Among nomads, too, modern health care has reduced mortality and given rise to marked population growth even if at a lesser rate than among their settled neighbours. Population growth, combined with improved veterinary care and reinvested income from outside activity, has led to an explosion in livestock numbers. Expansion of the systems, together with breakdowns in traditional authority, have affected rational management and made improvements difficult to achieve within the structure of existing practices. Traditional subsistence activities have increasingly fallen into a neglect hastened by the use of money. Sedentarization, whether voluntary or enforced, has resulted in severe degradation around permanent settlements where former nomads continue to herd livestock without their former mobility. Technology has been introduced without concern for all factors in the environment, which is to say in disregard of a proper ecological approach. The use of off-the-road vehicles for hunting or fuel gathering has been particularly destructive. The provision of large watering points, out of harmony with traditional migration patterns, has led to unusual concentration of stock and extreme local degradation. Grazing ranges have shrunk because of invasion by crop-based systems or political restrictions on the movements of animals, a particularly important problem in Africa, where colonial boundaries, drawn with little regard for geographical realities, have congealed into national frontiers. Although African countries have taken a tolerant view of the movements of herds and the needs of pastoralists, customs posts now stand athwart ancient stock routes, and occasionally borders are closed to all such movements.

88. That nomadic pastoralism is in trouble is evident in increasing desertification associated with such systems. Pastures have widely deteriorated, showing increased surface instability especially in their more vulnerable elements such as once-vegetated sand dunes. Advanced physical degradation around watering points, over-exploitation of shallow groundwater has dried up wells and rendered pastures inaccessible. Degraded pastures, dependent ever more on annual species, have shown themselves to be increasingly vulnerable to drought with all that this implies – destruction of livestock, enforced abandonment of grazing lands, deterioration in the diet and health of the people involved. Increasingly, former nomads are leaving the rangelands, and there is evidence that the emigrants are largely the young and the more able bodied.

Traditional more Sedentary Pastoral Systems

89. Although nomadic herdsmen occupy enormous areas, a greater number of livestock is maintained in more sedentary pastoral systems, involving village farmers in somewhat less arid environments. In these settings, animal husbandry is usually linked with rainfed cropping. The greater concentration of animals, and their very marked concentration around fixed points, such as watering places and village perimeters, leads to more intensive pasture degradation than in nomadic systems. Sedentary pastoralism is linked with other forms of desertification, such as the removal of woody plants for fuel and increased animal and human traffic.

90. Desertification also arises from the cropping elements in these systems, usually practised on marginal farmland, where the herdsman-farmer may constitute an impoverished part of a commercial system. The farming, a secondary activity, may suffer from a shortage of labour, sometimes due to out-migration.

91. Unfortunately, the animal element and the cropping element in these systems fail to buttress each other in times of drought stress, when both elements will tend to collapse together. When crop yields are low and plant residues used for forage are in short supply, the farmer-herdsman must utilize whatever forage may be available within his own limited land holdings or in communal grazing areas. This frequently leads to denuding the land of all plant cover, exposing both grazing land and cropland to accelerated water and wind erosion. That, in turn, reduces the likelihood that adequate forage will be available the next year even if moisture conditions then are favourable.

Commercial Ranching Systems

92. Commercial ranching tends to specialize in one kind of animal or breed, selected for the value of its product. Such systems compensate for environmental risk and the low productivity of pastures by adopting very low stocking rates (lower than those of nomadic pastoralists, for example). They are rarely sited in the most arid, remote or infertile regions. Yet since land is a low-cost item, commercial ranches tend to be large units which can take advantage of economies of scale.

93. Such ranches tend to minimize labour costs, especially in high-wage economies. Stock is set to graze in large, fenced enclosures and controlled with a minimum of handling, although winter stalling and feeding may be required in temperate to cold dry lands. The comparatively small labour force is highly mobile, whether on horseback,

driving off-the-road vehicles or piloting aircraft. For special tasks, such as fencing or shearing, or at times of seasonal demand, contract labour will often be employed.

94. These systems have inherent weaknesses, some of them stemming from animal specialization, which increases environmental risk and commercial vulnerability and results in inefficient use of the pasture complex.

95. A lack of close control of grazing is often combined in commercial ranching with an inability to manipulate the impact of grazing on pastures. Despite the expansion of agricultural extension services, it may not be possible to organize selective grazing in accordance with the relative performance, as known, of pasture species under grazing stress and to the requirements of soil and plant life. Nevertheless, as shown in the Australian Case Study, there is a failure to exercise the broad management controls available; for example, deferred grazing for the successful germination of desirable perennials after good rains. Since land and vegetation are the low-cost elements, they are not always viewed as the ultimate resource base, the livestock itself being so regarded.

96. Commercial ranching is dependent on external markets whose forces may or may not be in harmony with wise stocking policies and practices as called for by the local environment. The tendency to maximize profits can readily lead to poor ecological management and overstocking within a short-term perspective. Such ranching is often controlled by corporate managers or absentee landlords who tend to be less immediately concerned with the state of the range, or by leaseholders whose tenancy is too short to give the necessary perspective for rangeland improvement.

97. These systems increase their vulnerability already heightened by specialization by ignoring subsistence elements. Food and supplies are purchased at the market.

98. The growth of large units and the labour economies imposed on them lead to a progressive decline in the population they support with consequent out-migration, particularly of the young and landless. The populations of ranching areas are generally on the decrease, with an accompanying decline in secondary service centres.

99. Capitalization and technical improvements tend to buffer commercial ranching against the immediate consequences of overgrazing, and high prices for its products will yield cash returns which may further delay a proper response to the degradation of pastures. Since a determination of range trends is difficult in any case, delay can lead to irreversible deterioration of pasture long before economic collapse occurs or even before the situation is truly appreciated.

100. Unlike more traditional systems, commercial ranching makes large-scale use of heavy machinery for construction and road building. Such machinery can disturb the environment, producing localized degradation. The generally greater capacity for ecological manipulation in these technically advanced systems may have drastic feedback consequences. For example, the control of brush fires has resulted in invasion by undesirable scrub in such places as southern Australia and the south-western United States.

101. Rainfed cropping systems, referred to in general as "dry farming", are typical of the semi-arid lands, which include those regions in which agriculture was first practised by man. Farming is possible there only through the adoption of special techniques whose primary objective is the collection, storage, protection and utilization of every drop of water; in particular, cropping and fallowing in alternate years allows the conservation of moisture over seasons. Drought-resistant crops are selected for planting, notably the cereals – wheat, barley, rye, sorghum, millet – so typical of dry farming.

102. The greater productivity of crop farming, when compared with extensive

pastoralism, has encouraged its extension to climatic limits mainly set by the amount, seasonal incidence, duration and variability of rainfall, and in high latitudes, additionally by light and temperature. This extension, often fostered by pressure of population and aided by technological improvements, such as new, drought-resistant or rapidly maturing crop varieties, has reached into areas that were once exclusively pastoral. The Niger Case Study illustrates such movement, despite legislation against it, in the decade or so of good rainfall that preceded the recent Sahelian drought. At the same time, the entry of cash crops, such as cotton and groundnuts in the Sahel, into these systems has led to a breakdown of earlier arrangements under which nomadic herds graze stubble lands after harvesting.

103. Commercial dry farming compensates for climatic risks by producing crops of high quality, hard wheats for instance, suitable for transportation and storage and commanding good prices. Many semi-arid regions are comparatively free from crop diseases, notably rust. They provide extensive stretches of sparsely settled land suitable for tillage by large-scale mechanized agriculture. The cereal crops produced are easily handled, transported and stored.

104. Clearing for agriculture involves a much more drastic transformation of natural ecosystems than does pastoralism. Dry farming exposes and disturbs the soil, increasing the risk of erosion. Certain dry-farming techniques enhance this risk. Tillage for the preparation of seedbeds can set the stage for erosion if the surface soil is left in a finely divided condition. Bare-fallowing is also a common practice. Here the land is left stripped of vegetation to allow the infiltration of an additional season's rainfall and to minimize losses through transpiration. Such fields are also fine-tilled to prevent capillary loss of moisture and to promote the aerobic nitrification of nitrogen compounds.

105. Many of these systems spread across open plains which were already subject to wind erosion of silty soils, creating a dust nuisance and sometimes dust storms. Sand drift and dune formation are also common on sandier alluvial soils near old river channels. All such effects are worsened by the removal of trees and high-standing vegetation over extensive areas prepared for dry farming.

106. Dry farming tends to specialize in both crops and techniques, and it does so at the expense of mixed farming, which would include crop rotations involving legumes and the raising of animals. This results in an undue removal of nutrient materials transported away in the off-farm sale of grain and in the burning of straw and litter after harvesting with combines. Decades of producing one specialized crop have resulted in the depletion and breakdown of many semi-arid soils that once possessed excellent structure and fertility. When this happens, yields decline and erosion increases, especially on finer textured soils such as those formed in wind-blown parent material.

107. These systems support much denser and more settled populations than do pastoral systems, and man and his works exert a much stronger impact in them. Many such systems have been worked for millennia and provide a history of land use – and of land deterioration – extending over thousands of years.

Regional Problems in Rainfed Cropping

108. Rainfed cropping systems comprise several types as determined by climate and other environmental conditions. Each is marked by its characteristic crops, technology and

cultural setting. Each is vulnerable to desertification, which takes on distinctive forms in each setting and calls on distinctive measures to combat it.

The Mediterranean Region

109. The Mediterranean semi-arid regions have winter cyclonic rainfall and warm to subtropical temperature régimes. It is a region of mountains and plains with degradable calcareous or leached soils in which cultivation has been extended into areas of very low rainfall (locally less than 200 mm annually). Winter cereals may alternate with summer crops. Farming is often combined with animal husbandry, especially of sheep and goats. Tree farming is particularly important. Connections between rainfed and irrigated cropping are particularly close in the Mediterranean, with the former deeply involved in water management, as in the terracing of slopes whose upkeep requires not only much labour but social stability as well.

110. The Mediterranean has a long history of land use by fairly dense populations. Its history is also one of cities, many of them large and important. Man has thus had a profound impact on the Mediterranean ecosystems, which provide, in fact, the longest historical record of desertification. Land degradation appears to have been associated with the early spread of sedentary agriculture and its related settlements. In some parts of the Mediterranean, desertification has reached advanced stages. It appears there is the deforestation of once-wooded uplands. Forests have given way to dwarf, leather-leaved dryland shrubs, or to bare earth, with soil sometimes stripped completely from slopes to uncover calcareous crusts or naked rock.

111. This stripping of upslope watersheds has damaged downslope water régimes. In many places, run-off has become ephemeral and spasmodic, sometimes giving rise to catastrophic flooding in the lowlands and to increased deposition of silt in valley bottoms. Siltation, a problem in ancient days, continues to pose a threat, as for example to the useful life of large reservoirs in modern water-control schemes. And yet siltation can be used to advantage, as among the ancient Nebataeans, who used the fertile blankets of soil in downstream valleys.

112. The Mediterranean region shows broad deterioration in groundwater reserves accompanied by a lowering of groundwater tables and a decline in water quality. Seawater intrusion became a problem in coastal areas of Israel after 10–20 years of pumping out fresh water and stopping groundwater recharge from run-off. The lowering of groundwater levels in pumped areas where recharge is restricted can occur in as short a time as 2–3 years depending on aquifer characteristics and the amount of pumping.

113. Cultivated footslopes often show marked gullying, particularly where cultivation has been unwisely extended over the past 50 years because of rising population pressure. The loss of soil has been considerable and with it the loss of potentially cultivable land.

114. Despite growing population, labour shortages have developed as a result of heavy migration to cities and towns, and water management has been affected. Terraces and *qanat* systems, the irrigation tunnels so characteristic of Iran, have suffered from lack of proper maintenance.

115. The removal of trees and shrubs has accelerated wind erosion of light soils, stripping them of nutrients and humus. In many places, as in southern Tunisia, wind

erosion has led to the formation of coppice dunes and made the land unsuitable for cultivation.

116. Some lowland soils, particularly in basins of interior drainage, have been affected by salinization.

Mediterranean-type Regions

117. Other areas with a Mediterranean-type semi-arid climate are distributed around the world. Such regions include, for example, semi-arid portions of southern Australia, south-western Cape Province in South Africa and parts of California.

118. These are typically regions of highly mechanized agriculture producing cereals for export. Their devotion to monoculture has resulted in a lack of leguminous rotation crops and a virtual absence of animal husbandry, thus limiting the return of organic matter to the soil. This, associated with the export of the crop and the removal of the vegetable litter produced by mechanical harvesting, has depleted the soil of nutrients. Light-textured grey-brown or black-earth soils have increasingly been subject to wind erosion. Deterioration following continuous cropping was reflected in the inter-war years by falling yields. Such failings are not, of course, universal in these regions. Parts of South Australia, for example, provide models of good dry-farming practices, rotating cereals with legumes and applying fertilizers to maintain a good content and structure of nutrients in the soil. Wheat yields in the U.S. have increased over the past 30 years despite a certain amount of continuing erosion. This is due to improved varieties, the increased use of fertilizer and better land management.

119. Depending on the setting, desertification in various forms has made its appearance in all Mediterranean-type regions.

120. Extensive gullying of slopes, as in Cape Province, has become a particular handicap to mechanized agriculture. Tilled but unvegetated surfaces that occur where bare fallowing is practised have been extensively subject to general sheet erosion by water. The almost complete clearance of vegetation associated with large-scale mechanized agriculture has resulted in wind erosion of light soils and once-stabilized dunes, causing sand drift and the mobilization of fresh dunes as in, for example, the Mallee region of South Australia.

121. The clearance of deep-rooted shrubs and their substitution by crops or fallow has reduced transpiration in favour of evaporation and increased run-off from cleared slopes. This has altered the water balance in valley soils and brought on salinization. Such effects are particularly noticeable in areas of sluggish natural drainage, as in northern Victoria and south-western Western Australia where saline groundwater has come to the surface with increased effluent seepage from slopes.

Subtropical to Warm-temperate Regions

122. In subtropical to warm-temperate regions of dry farming, transitional rainfall régimes are characterized by two seasons of precipitation, both winter and summer. Under such conditions, winter cereals can be combined with a variety of spring-sown crops, such as cotton in the south-west of the United States or fodder grass, millet and nuts in areas between the Ukraine and the Caspian Sea, resulting in a favourable mixed agriculture with a more continuous cover. Such farming systems are generally young, having been established over the past century in such rich soils as the black earths of the southern

USSR. In these systems, soil depletion with falling yields has only recently become evident. The application of mineral fertilizers and the replacement of nutrient stores are increasingly required. The plains topography of these regions, characterized by an absence of trees, has promoted the degradation of soil through wind and water erosion.

Cool Temperature Semi-arid Regions

123. Cool-temperature semi-arid regions typically have rain in spring and early summer. They include, for example, a broad strip from south central Asia into North China and the dry prairies of Canada, where much of the surface is exposed and where long, cold and dark winters result in a short growing season restricted to spring cereals and great difficulty in introducing cover crops other than grass. Under such conditions, it is difficult to combine animal husbandry with cropping systems.

124. Erosion is the characteristic form of desertification in such regions, which like warmer dry-farming regions, largely consist of open plains. Their structureless and light-textured soils, often lying on carbonate or hardpan layers, are most affected during dry winter or late summer.

Tropical Semi-arid Summer Monsoon Regions

125. Tropical semi-arid summer monsoon regions are typified by the Sudanian belt, with its 300 mm to 600 mm annual rainfall, to the south of the African Sahel. They also include the margin of the Rajasthan desert in the north-west India. They tend to grade into subhumid savannas, lands which must also be considered as vulnerable to desertification.

126. In these regions, opan savanna woodland is cleared, usually by burning, to provide a seedbed, although clearance is not complete and many trees may be left standing. The pattern is generally that of shifting agriculture, known as the *swidden* system. Four to five years of continuous cropping are followed by abandonment, when successional regrowth may be harvested, gum arabic for example, or grazed by cattle, with the growth of grass encouraged by burning.

127. This is mainly subsistence farming by peasants, who grow grain crops such as sorghum or millet. The warm climate may allow a second crop, such as groundnuts or cotton in Africa, increasingly grown for cash. Most swidden farmers maintain animals, and adjacent pastoral peoples may introduce an element of animal husbandry into these systems, with the rights to graze on stubble obtained through various types of exchange or through cash payments.

128. Although burning can be important in promoting new growth of grass and in providing a seedbed, it can be harmful where excessive and ill-timed. It can then lead to the destruction of humus and the loss of fertility, stability and good water relations in the topsoil, together with the seed store. It can also result in the disappearance of shade trees and valuable shrubs.

129. During periods of above-average rainfall, these systems have tended to encroach on neighbouring animal-based systems because of population increase or for the extension of cash cropping. Such encroachments are successful until the rains fail, as they inevitably do. The severe imbalances which then appear can act as a major accelerator of

desertification, as they did in the recent Sahalian drought, affecting not only the farmland itself but also the pastoral areas which farming had invaded.

130. Desertification in these systems often appears as a marked decline in fertility following the consumption of soil nutrients and a deterioration in the structure of the subtropical soils. This often comes about because population pressures and a resulting land hunger act to speed up the agricultural cycle, bring the slash-and-burn farmer back to the same piece of land in 15 years, say, instead of 20. The introduction of equipment unsuited to the particular conditions of these regions has resulted in deeper tillage and a consequent pulverization of the soil. The rise of cash cropping accelerates the removal from the soil of its humus and nutrients.

131. As fertility declines, crop yields are less, and an adverse impact becomes self-accelerating. To make up the difference, the land is worked even more intensively, diminishing its fertility even more. Rainfall in these regions, while localized, is often intense, causing pluvial erosion of cultivated surfaces. Soil surfaces become puddled and soil structure severely damaged. The dry spells that alternate with the onset of rains bake a crust on the surface, hindering the germination and development of seedlings. During the dry winters, wind erosion lifts clouds of dust from these lands, sometimes transporting it over enormous distances. Soils in the Caribbean islands have been added to by what has been lost from distant Africa, and that only a minute fraction of what has rained into the Atlantic Ocean.

Irrigated Cropping Systems

132. Irrigation provides the most productive basis for agriculture in arid regions and serves as a vital supplement to crop production in semi-arid regions. About 13 per cent of the world's cultivated lands are irrigated. Although not all of these 250 million hectares are located in deserts and dry lands, most of them are, and irrigation techniques are closely linked to the prevalence of arid conditions.

133. If present nutritional deficits are to be corrected and an expanding world population adequately nourished, world food production must rise dramatically. A 30 per cent increase in cereal/production alone has been projected as essential between 1970 and 1985. If that seems primarily a problem relating to rainfed agriculture, some of the increase will have to come from the further development of irrigation, which will be called on to supply other elements of the diet as well.

134. Compared with rainfed agriculture, irrigation can lead to a six-fold increase in yields of cereals and a four-to-five-fold increase in root crops. The importance of irrigation to agricultural development is revealed by the fact that the irrigated harvest area in developing countries is expanding at a rate of 2.9 per year compared with an annual expansion of 0.7 per cent for rainfed crops. Irrigation in arid lands can therefore be expected to play a critical role in satisfying the world's food requirements. Measures to combat desertification in such systems are accordingly of the utmost urgency.

135. Its remarkable productivity is one aspect of the importance of irrigated agriculture in arid lands. The productivity of rainfed cropping as carried out in areas with 250–600 mm of annual rainfall is much lower because of this limitation in available

moisture. Not only does irrigation increase yields but it also allows the replacement of fallowing systems by annual cropping.

136. The increased stability of crop systems with removal of drought risk and uncertainty is another advantage of irrigation. Irrigation increases the efficiency of cropping systems. For instance, the application of fertilizer and the planting of higher-yield crop varieties are greatly facilitated wherever productivity is not limited by the availability of water. Animal-based systems are made more stable and efficient when they are carried out adjacent to irrigation, which can provide them with forage crops as supplementary feed and can store reserves against the threat of drought. Irrigation diminishes the risk of desertification in cropping systems. The planting of trees and a more consistent vegetation cover replace fallowing and the open and exposed landscapes characteristic of other dryland systems. Irrigation provides water which can be used to reclaim desert lands, whether by supporting a plant cover or by the leaching of salinized soils.

137. As rich producers of cash crops, irrigation systems serve as important economic resources for arid lands. They provide a basis for dense settlement and its related social amenities in regions that once supported only sparse populations. As such, irrigated lands can be used for the resettlement programmes that desertification elsewhere sometimes makes necessary.

138. It is not merely because they are short of rainfall that arid lands are particularly suited to irrigation. Situated as many arid regions are in the cloudless sub-tropical zones of atmospheric subsidence, they are favoured with long hours of sunshine. This makes irrigated lands suitable for multicropping and the growing of early maturing, warmth-demanding crops that command high prices in regions that are not so sunny. Algeria or Israel, for instance, produce winter and spring flowers and tropical fruits that are shipped off for sale in Europe.

139. Again, under conditions of low rainfall, carefully irrigated soils suffer only limited leaching of nutrients. Plants grown in low atmospheric humidity are relatively free of diseases, such as rust in cereals, that flourish in moisture conditions.

140. Many arid lands are rich in land forms and soils, such as sloping piedmont plains and well-drained terraces of interior river systems, that are favourable to irrigation and remarkably productive when water is brought to them. Many such places still remain to be exploited by intensive cropping.

141. Irrigation, however, is often a costly, technically complex procedure that requires skilled management and sound experience if its full advantages are to be realized. Furthermore, it gives rise to changes in all the major ecosystem régimes – soil, water and atmosphere – that may introduce unwanted effects leading to desertification unless appropriate precautions are incorporated into the system.

142. A failure to apply efficient principles of water management will lead to water wastage and hence loss of productivity. Such wastage can occur at any point in the system – through seepage and evaporation during storage, conveyance or distribution or as a result of bad timing in water application, by over-watering or poor techniques of field application.

143. Seepage, over-watering and inadequate drainage can result in waterlogging of soils, which reduces productivity through inadequate aeration and its associated salinity, eventually leading to the loss of cultivable lands. This is a problem locally associated with low-lying tracts and areas of heavy soils.

144. When soils are inadequately leached of the soluble-salts contained in irrigation water, then excess evaporation and transpiration will result in salinization and alkalization of soils. When drainage is inadequate, whether natural or artificial, salts accumulate. The process commonly begins where natural seepage occurs, as along canals or the margins of irrigated land commanded by higher ground, in an irrigated terrace, for instance, or where there is seepage from a network of channels. It spreads where irrigation has been carried into areas of unsuitable soil, such as alkaline clays, or into unsuitable terrain, such as flood-plain sumps or the higher parts of poorly levelled lands. In such situations, when leaching is inadequate, salt crystals will appear on the surface in abundance.

145. Salinization and alkalization become general problems wherever artificially raised watertables, associated with waterlogging, capillary rise or pollution from salinized outflow, prevent the proper leaching of salts. Salinization also occurs when the irrigation water is too salty. It has been estimated that half of all irrigated soils in arid lands are affected by salinization to some degree. The eventual result can be found in lowered yields, restrictions in the choice of crops and the final loss of irrigable lands which can only be reclaimed at great expense. In monetary terms, no type of desertification is more costly to man.

146. Improper watering, inappropriate, tillage of moist soils and the leaching of soils containing gypsum can lead to a deterioration of soil structure and compaction. This results in poor aeration, reduced transmission of irrigation water and finally to lowered yields. When watertables are lowered through excess pumping, the ground can subside. In newly irrigated fields, subsidence also occurs on loess soils or as the soil becomes compacted under cropping, particularly if it was relatively loose in its natural condition or high in gypsum content. Further irrigation then becomes impossible without expensive relevelling.

147. Irrigation calls for particular skills in the application of water and the tillage of watered soils if its great potential for increased productivity is to be developed and sustained. The efficiency of irrigation schemes rests in the last analysis on the individual cultivator. When cultivators lack the appropriate agricultural experience, irrigation systems and the lands they water can suffer great damage. The case Study in Iraq showed the critical importance of agricultural extension services if large irrigation projects, such as that of the Greater Mussayeb, are to function without deterioration.

148. Potentially beneficial to health through improved nutrition and water supplies, irrigation can also bring with it a distinctive set of health problems, particularly where mismanagement occurs. Malaria has been identified as a problem in the early irrigation civilizations of the Nile and the Tigris-Euphrates. The transmission of water-related diseases, such as bilharzia, malaria and typhoid fever, is facilitated by water mismanagement which results in the formation of stagnant pools. Lack of water supplies and sanitation under conditions of dense irrigation settlement are associated with typhoid fever and intestinal parasites that give rise to chronic ill health and labour inefficiency. Such conditions are illustrated in the Iraq and Pakistan Case Studies.

149. Laws and traditions can present obstacles to efficient irrigation by establishing curious restrictions on water use, illogical subdivisions of the land or contractual limitations on tenant activities.

150. Irrigation gives birth to cities in arid lands and to the social stresses that arise when people of diverse backgrounds come into contact with one another in a new social and economic environment. Dense settlements have a profound impact on the surrounding

desert environment, and this impact can be very damaging where populations have no tradition of close settlement.

151. Irrigation systems can be based on surface waters or on groundwater. Each type brings with it its own characteristic problems.

152. Systems involving the use of surface waters range from flood farming or floor-recession agriculture in floodplains, through annual basin irrigation using flood banks, to perennial irrigation using man-made storage reservoirs and canals. Systems of the last type, based on rivers flowing through or on large upland sources of run-off, support the largest populations and the most intensive agricultural production in the arid lands. Such systems call for advanced, large-scale management.

153. Devegetation, surface deterioration or gullying due to overgrazing or the extension of croplands on watershed areas and the breakdown of works such as terracing intended to control run-off, all give rise to problems in the management of surface water. All such deterioration promotes increasingly spasmodic and violent local flooding and sediment movement which complicates water management and gives rise to flood damage and siltation in storage reservoirs and on irrigated lands.

154. Systems employing surface waters are confronted with problems of storage. Reservoir capacity can be lost to siltation, while seepage can involve the loss of salinization of stored irrigation water, and evaporation, too, can enhance salinity. Such systems are also confronted with problems relating to water conveyance. Water losses in travel, averaging 50 per cent, arise from seepage and evaporation in networks of channels, and such problems can become acute in very large systems containing long distribution channels. Irrigation water can become increasingly saline as it is recycled in surface run-off or subsurface flow, when it becomes contaminated by saline soils, particularly where discharge declines as the water moves down-valley.

155. When surface-water systems are adjacent to large rivers, they are subject to a risk of flooding. When they are located in large desert river basins, they always involve problems of water resources and water rights at local, regional and international levels.

156. Systems involving the use of groundwater suffer from their own distinctive set of problems. These are usually smaller schemes than the often elaborate systems that draw on surface water. All told, they probably amount to less than 10 per cent of the extent of surface-water systems, but they are particularly important within oasis settings, including those within extreme deserts and in plains overlying large aquifers. Such systems sometimes exploit shallow subsurface water by means of hand-dug wells. In other settings, they go deeper and require the use of pumps. They may tap artesian supplies or they may "mine" non-renewable deep-water sources. The efficient use of deep, non-renewable aquifers is the subject of a feasibility study set in the Arabian Peninsula and North-east Africa.

157. Groundwater is commonly more saline than surface water. Limitations in supplies and difficulties in terrain may present great difficulties in obtaining effective leaching and drainage. Salinization of soils, often working through the mechanisms described in connection with surface-water use, is a frequent problem in groundwater systems.

158. Problems also arise from over-exploitation of limited water supplies. As water is used up, shallow sources may be abandoned, and the pumping becomes increasingly expensive as draw-down affects marginal wells. Where the irrigated area is too extensive, well discharge may be inadequate for both irrigation and leaching. The watertable may fall. Less favourably sited wells may become so drained that marginal lands will be abandoned. Heavily exploited water may suffer from an increasing accumulation of salts

through recharge by salinized water, thus aggravating the problem of soil salinization. Seawater may encroach on aquifers that are intensely exploited in coastal dry lands.

159. Problems can arise in groundwater systems due to the raising of the watertable as supplies are brought up from depth. When drainage is hindered this can lead to the pollution of acquifers by saline soil water. Conversely, watertables may be lowered in acquifers as supplies are drawn from them. This can cause land subsidence on a major scale, as in California's Sacramento Delta or in Mexico City.

160. In groundwater irrigation too, problems can arise in connection with water management and water rights leading to conflicts and resulting in inefficiencies. The Case Study set in Pakistan shows the importance of allocating the correct amount of land to the appropriate number of cultivators.

161. Many irrigation schemes in developing countries are faced with social and economic problems arising from the skilled farming, high investment of labour and capital and dense populations called for by irrigation development. These are well exemplified in the Iraq and Pakistan Case Studies. The needs include agricultural extension services and demonstration farms, co-operative schemes to assist with purchasing, marketing and capital loans, ancillary agricultural activities such as poultry or livestock farming to increase income, and amenities and services appropriate to close settlement, including schools, health services and housing. Alternative employment, such as in agriculture-based industries, are also required as population increases.

Fishing, Hunting and Gathering

162. There are communities that still gain their subsistence by traditional methods of hunting, fishing or gathering or some combination of these activities. More often, however, such pursuits are supplementary to agricultural systems of livelihood, and when compared with the latter, their environmental impact is generally local and slight. When these activities are affected by desertification, it is likely that they have been damaged by adjacent agricultural systems with their much more pronounced impact.

163. Since diminished biological productivity is the hallmark of desertification, that label can be applied to circumstances leading to reductions in wildlife populations or to the loss of their habitats.

164. Wildlife, particularly game animals, are a source of food for humans and of income to landowners who maintain herds for hunting. Game ranching is a well-established business in southern Africa and is becoming important elsewhere. Wild herbivores forage on a variety of plants: elephants feed on trees, wild goats on thorny shrubs, zebras on grass. In mixed communities of wild animals, the carrying capacity of the land is greater than in a one-species community because different parts of the ecosystem are exploited by different animals. There is some evidence that wild animals, such as oryx, are more efficient converters of woody plants into meat than are domestic animals. The domestication of potentially valuable wild animals remains a largely unexplored opportunity to increase food supplies.

165. In addition to its value as a food supplement, dryland wildlife, an intrinsic part of such ecosystems, constitutes a vital element in the natural environmental balance. Its presence may therefore be essential in restorative measures to combat desertification. As

part of the world's ecotypal heritage, wildlife is intrinsically worthy of preservation. As a tourist attraction, dryland animals may also serve as an economic resource.

166. The large native herbivores in dryland ranges have become reduced in numbers almost everywhere. Some species are threatened with extinction. This is in part due to heightened hunting pressure as man increasingly intrudes into the more remote dryland refuge in search of oil, minerals or pleasure. Such intrusions are increasingly made in off-the-road cross-country vehicles by men carrying sophisticated weapons. It is also due to deterioration of animal habitats as described in connection with animal-based livelihood systems. As their situations become ever more precarious, dryland animals decline in both vigour and numbers. Competition with domestic animals, whether real or merely perceived, is a factor leading to deliberate reductions in the populations of larger animals or to their exclusion from habitats they formerly occupied. Under these worsened conditions, the impact of supplementary hunting by pastoral peoples has become increasingly severe, especially when accelerated by the general increase in human populations.

167. Dryland fishing communities are also subject to desertification. Desert lakes, coastal lagoons and perennial dryland rivers support fishing industries which contribute important amounts of protein-rich supplements to local diets. For example, the Lake Chad fisheries reportedly produce an annual yield of 100,000 tonnes. Lake Chad, however, was much reduced during the recent Sahelian drought, since drought reduces discharge by rivers and shrinks the lakes which they feed. Shallow bodies, such as Lake Manchar in Pakistan, have suffered shrinkage when the waters that feed them were diverted for other purposes such as irrigation.

168. Lakes and lagoons can be salinized by excessive evaporation, by the increased salinity of entering waters in irrigated regions or by the encroachment of seawater. The Aswan dam, for example, has prevented Nile sediments from reaching the beach barriers that once protected the lakes in the delta and has allowed seawater to penetrate them.

169. The degradation of river catchments increases siltation and generates turbidity in the lakes and lagoons fed by such rivers. This acts to kill aquatic vegetation and causes a decline in fish catches. Conversely, the sardine fisheries in the Eastern Mediterranean failed after the Aswan dam cut off the Nile-borne nutrients that once supported them. At the same time, however, the dam generated a new fresh-water fishery in Lake Nasser, the long reservoir backed up behind it.

170. The desertification of watersheds brings on adverse changes in river régimes leading to increasingly spasmodic discharge, to siltation or flood scour in river channels and to the destruction of aquatic ecosystems including fisheries.

171. Mining and tourism are livelihood systems practised in all types of climate and environment, but they take on special importance in the drylands as alternative resources in conditions of relative scarcity. Oil revenues, for instance, have fundamentally improved the prospects of a number of developing desert nations and indeed have given them the means to combat desertification. Mining and tourism provide an impetus to the establishment or development of dryland settlements and communications. Yet they have not been carried out without environment impact, especially in the ecosystems that make up the dry lands.

172. Mining and mineral based industries, including the extraction and processing of oil, cause direct disturbance of vegetation, soil and terrain, not only in the actual mining

operation itself but also in such ancillary activities as the construction of roads and pipelines and the development of heavy vehicular traffic.

173. Disturbed and denuded soil is subject to wind erosion with increasing dust nuisance and sand drift. Disturbed ground is also vulnerable to accelerated water erosion, with consequent siltation and obstruction of surface drainage. These problems are exacerbated in dry lands where water for irrigation is in short supply and where denuded ground is recolonized by vegetation only at a slow natural rate.

174. Airborne or waterborne wastes from mining and industry cause pollution of soils and groundwater, a problem difficult to deal with in the dry lands where there is rarely enough water to remove pollutants through leaching or surface drainage.

175. The patterns of air circulation over dry lands are characterized by limited atmospheric mixing and frequent temperature inversions which held the lower layers of air firmly in place. In such conditions, atmospheric pollution tends to hang on instead of being carried away. Held in place under brilliant sunlight, pollutants are then subject to photochemical synthesis which can transform them into the noxious smog that afflicts such dryland cities as Los Angeles or Mexico City. In the San Bernardino mountains, downwind from Los Angeles, smog has decimated pine forests and has damaged nearby vegetable and citrus crops.

176. Mineral industries can bring about desertification for scarce resources such as water, wood (for fuel and construction materials), energy sources and labour. Such competition is often detrimental to local agricultural livelihood systems, an example of sustainable systems being damaged by short-term non-sustainable activities.

177. Problems arise from the impact of mining settlements in dry lands. Apart from the physical impact and demands common to settlements anywhere, dryland mining towns bring with them an array of social problems related to their often temporary character, their remoteness, and to the unusual and changing composition of population which may be wholly or partly foreign to the dryland setting. One usual effect is an utter disregard on the part of newly arrived mining-town inhabitants for what they take to be useless vegetation.

178. Tourism and recreation have been drawn to deserts and dry lands by warm sunny climates, a dry healthy atmosphere and natural landscapes with distinctive life forms where parks and reserves can be easily established. Dry lands contain archaeological and folklore attractions and they provide ideal settings for certain kinds of sanatoria, particularly those that treat respiratory ailments. Their popularity as tourist and recreation areas has been aided by the development of communications to and within them and has risen steeply with the increasing leisure and affluence of industrialized societies, especially those that experience cold winters. The tourist industry is an increasingly important source of revenue and employment in dry lands, although complaints are often heard that the control and benefits of tourism remain outside the dryland communities. Tourism and recreation can also serve as active agents of desertification.

179. The construction of tourist roads and camps and the resulting increase of traffic, particularly by cross-country vehicles, is disturbing to and destructive of vegetation and soil cover in the usually sensitive landscapes that constitute scenic attractions. This leads to accelerated erosion. Attractive plant or animal species can be rediced or even wiped out by the uncontrolled gathering of wild flowers or by disturbing animals at critical periods in their life cycles. Tourist settlements give rise to problems of health and sanitation, and these can be exacerbated through contacts between tourists and local populations.

180. Commercial tourism can have an uncontrolled impact on traditional communities resulting in interactive social complications, including the resentment of local populations at being regarded merely as objects of interest. Seasonal labour requirements have great effects on local livelihood systems as does an increased demand for the products of local craft industries. Labour may be drawn away from agricultural systems and land management may deteriorate.

Dryland Settlements

181. Aridity stimulates the formation of nucleated settlements since their necessary supports, such as water and agricultural land, tend to be localized in deserts and dry lands. Depending on how "urban" is defined, between 20 and 30 per cent of the 680 million people living in dry lands are urban. Dryland cities include some of the oldest in the world. Today they function as irrigation centres (including oasis settlements), garrisons, communications and caravan centres, political, administrative and regional services centres, or they may be focused on tourism, sanatoria, mining or other industries.

182. Deserts and dry lands have been subject to an accelerated urbanization over the past 50 years, often superimposed on general population increases. In Iran, for example, where the population has more than tripled since 1900 to a present total of about 35 million, the percentage of urban population has increased from 20 per cent to 40 per cent. Today, the world's dry lands contain nine metropolitan centres with more than one million people each. Expanding dryland cities share many of the problems of cities in more humid lands. But situated where they are, they bring with them additional problems as agents of desertification.

183. Dryland communities have a direct and often adverse impact on the lands surrounding them. As concentrations of people and traffic, including livestock traffic in agricultural settlements, they are often surrounded by naked perimeters of bare ground subject to constant disturbance. Movement on such perimeters is rarely confined to established roads or tracks. The result is intensified dust nuisance and localized sand drift. After rains, these bare surfaces become muddy and filled with stagnant pools which can constitute a health hazard. Such conditions may extend right into settlements built on an open grid pattern of large blocks, like most Australian outback towns, which contain extensive uncontrolled surface which cannot be grassed because of water shortages.

184. Waste disposal in dryland settlements is confronted with particular difficulties. The water-borne disposal of domestic or industrial wastes is hampered by a lack of water for flushing or leaching, by slow rates of bio-degradation and by problems in revegetating waste dumps. This leads to chemical and bacterial pollution of soils and groundwater with attendant health hazards, particularly in more primitive conditions. Included in this problem is the impact of feedlots and slaughterhouses located in town perimeters.

185. Rubbish is often dumped on the outskirts of dryland towns in sparsely settled areas difficult to supervise. Dumping is encouraged by a widespread attitude that desert land is inexhaustible and otherwise worthless. Atmospheric pollution from vehicles or the burning of fuels in cities is aggravated by the same dryland conditions that effect atmospheric pollution from mining or industry – low rates of atmospheric mixing, temperature inversions and a high level of photochemical synthesis.

186. Like all towns, dryland settlements make demands on their hinterlands, and more so in developing nations where communications may be poor. Desert towns in developed

188. In developing countries especially, demand for wood and charcoal tends to devegetate an expanding area around the city with the usual adverse consequences. As time passes, supplies must be brought in from farther and farther away at continually rising cost to the consumer.

189. The expanding settlement may engulf the cultivated land that supported its earlier growth. Conversely, unrealistic land-boom sales in the United States have caused lots and roadways to be scraped out at distances often remote from towns, where they lie stagnant without further development, constituting a source of accelerated erosion. Although the growth of cities in deserts or dry lands may entail smaller losses in agricultural productivity than in humid areas, such losses occur in environments that are very sensitive to disturbance and they may be very important locally.

190. The demand for labour by urban services and industries, reinforced by higher wages, may draw workers from adjoining livelihood systems, whose works then deteriorate. Forms of agriculture requiring intensive upkeep, such as rainfed terraced agriculture or *qanat*-fed irrigated cropping, have suffered particularly from labour shifts of this kind.

191. Just as settlements have an impact on their surroundings, so desertification of a region has an impact on the cities and settlements located within or near it.

192. During drought, rural peoples migrate in large numbers to nearby towns. This happened in the recent Sahelian drought, where urban population growth rates, already very high at 10 per cent per year, briefly doubled. Although towns provide a successful escape for the migrant in terms of wages and welfare, such movements impose severe burdens on urban housing and services and tend to intensify the adverse environmental impacts that cities and towns already exert.

193. The accelerated growth of cities, so characteristic of the contemporary world, places continuous stress on urban water resources, a stress that is aggravated in periods of low rainfall. When desertification affects the hinterland or surrounding regions, the city's water supply can be further stressed by increasing siltation in surface water storages, reducing their useful life, and by lowered groundwater tables and a deterioration in water quality.

194. Desertification of surrounding lands will heighten environmental stress within the settlement. The town may experience hot winds and more frequent dust storms, particularly in periods of summer drought when local shade and shelter will also have diminished. These impacts will be most strongly felt when settlements and houses are unsuitably designed and particularly in the temporary dwellings of newly arrived urban immigrants.

Woodcutting and Wood Gathering

195. Among most dryland populations, the main source of energy for cooking and heating will continue to be woody plant material for at least some decades to come. The cutting or gathering of woody plants for this purpose has been a major cause of desertification, and the labour of collecting such fuel or the cost of buying it has become a major factor in the lives of dryland dwellers in many developing countries. In many dryland areas, wood cutting and charcoal burning are industries carried out by professionals. To prevent deterioration and to meet energy needs, woodlots and forests should be established near settlements and preserved from indiscriminate cutting and browsing with their use regulated so as to achieve sustained yield. For example, rotational

economies may import many of their necessities from far away. Under any circumstances, however, the impact of the modern city on its surroundings is considerable.

187. Below certain limits, *per capita* consumption of water increases with urbanization, and to meet its domestic, industrial and power generating needs the town may compete for water with adjoining agricultural systems, as Mexico City does. Where a city is dependent on groundwater, its rising needs may lead to a lowering of regional watertables, as in the Tucson basin of Arizona.

systems of fuel gathering have been proposed. Plantings of trees to protect environments in and around settlements should be completely excluded from exploitation.

196. At the same time, alternative energy sources must be developed and exploited wherever possible to lighten the load on vegetation resources. The arid lands are particularly favoured with solar and wind energy. The development and distribution of cheap solar cookers, with adequate instructions for use and arrangements for maintenance, is of paramount importance. Small wind generators have been used with success in many dryland settings. Additionally, the use of cheap cooking stoves, lighting systems and power generators using mineral fuels is already under development, and such developments should be pursued.

Desertification Resulting from War

197. A new and more sombre cause of desertification can be found in the chemical and biological weapons that have recently been added to the arsenals of warfare. The effects of chemical deforestation can last for generations and can, indeed, devastate tropical forests with their thin and fragile soils, permanently changing the character of the landscape. The consequence of the use of such weapons would be particularly devastating in dry lands where biological recovery is relatively slow and where profound physical changes can ensue which are relatively permanent.

198. On 18 May 1977 a convention was opened for signature at Geneva through which countries would agree to avoid the military use of techniques for environmental modification whose effects would be widespread, long-lasting or severe. By 1 June thirty-four countries had signed the convention.

V. THE HUMAN CONSEQUENCES OF DESERTIFICATION

The Impact of Desertification on man

199. Desertification is a human problem. Its most important aspect lies in its impact on man himself – on the individual, the family, the community and the nation. The environmental degradation and the biological and physical stress described as desertification in the different dryland livelihood systems have their direct counterparts in physical, emotional, economic and social consequences for man.

200. As with the environmental manifestations, the impact of desertification on human beings shows a corresponding vulnerability, chronic or progressive, upon which are superimposed those critical periodic stresses that result in human disaster. The effects of desertification on man appear most dramatically in the mass exodus that accompanies drought crisis. A survey of such events might give rise to an impression that the impact of desertification on man proceeds from crisis to crisis. But this would ignore the chronic effects of land degradation as reflected in persistent out-migration in good years as well as bad. The abandonment of land and expensive installations in deteriorating irrigation projects occurs without waiting for the onset of drought. The degradation of rainfed farmland will strike at productivity and income even in years of good rainfall. Yet it is the crisis of drought that most effectively draws attention to persisting problems of desertification.

201. The recent Sahelian drought generated an international relief effort which eventually reached substantial size. Nonetheless, the effort was less than a total success, and this is indicative of underlying weakness which existed before the drought crisis and which have undoubtedly survived it. That food and medicine could not be brought in time to those who needed them most underlines the remoteness of the drought-afflicted regions and their lack of transport facilities. Administrative failures and bureaucratic obstacles hampered relief operations locally.

202. A traditional antagonism between nomad and peasant in the Sahel, which led to inequities in the distribution of food to refugees, draws attention to the kinds of cultural conflicts and political bias which also face anti-desertification programmes. A lack of effective communication between refugees and those who had come to help them serves as reminder that literacy rates in affected countries are sometimes very low, below 10 per cent in most countries of the Sahel, and that programmes of education must accompany measures for improving land use, programmes that should involve more than just reading and writing.

203. As dryland ecosystems react with greater or lesser sensitivity to climatic stress and the pressures of land use, suggesting an order of priorities for measures directed toward physical improvement, so differences in the inherent vulnerability of dryland communities, as reflected in the greater or lesser sufferings of their inhabitants, suggest priorities on human grounds. Full consideration should be given to whether or not international action should be directed toward the most vulnerable nations, and national action to the most vulnerable communities, rather than to areas of the greatest disturbance, although, of course, the two categories might on occasion coincide.

204. However they are ordered, remedial programmes should have as their perspective the treatment of long-term disabilities, not merely relief from temporary hardship.

Human and Social Manifestations of Desertification

205. The persistence of such disabilities provides evidence of chronic difficulties in many marginal dryland communities. As such, they may not be specific to desertification but broadly common to families and livelihoods on the margin of the modern world, particularly in remote and hazardous environments and where tradition, social inequality or political indifference further isolate people and livelihood systems form the resources and capacity needed to improve their lives.

206. Among these adverse consequences are hunger, disease, emigration and premature death brought on by continued crop failure or the massive destruction of livestock, particularly in marginal subsistence societies where malnutrition increases vulnerability to epidemic diseases such as measles. Few diseases are specific to desertification, eye disease such as trachoma marginally so, and certain diseases, such as bilharzia, are linked with inefficiencies in irrigation systems. Debility leads to further inefficiencies in another kind of cycle that may become self-accelerating.

207. In developed nations, such as the United States or Australia, the first noticeable sign of desertification may be the farmer's loss of income. In subsistence societies, it will more likely appear as a loss of assets – pasture or crop – which can endanger the physical well-being of those affected. As drought continues, livelihood systems will approach collapse. In nomadic societies, this stage may be marked by self-enforced sedentarization of pastoralists, on marginal cropland. Out-migration to the towns, both seasonal and permanent, may show a marked increase among nomads and among those who work cropping systems as well. When the migrants find employment, relief appears back home in the form of remitted wages, but migration can easily reach the point at which the local livelihood systems suffer from labour shortages, thus weakening them further. Although traditional societies have various ways of coping with such problems, some are particularly vulnerable to them, those especially which have become commercially or technologically more specialized or those in which traditional social bonds have broken down.

208. When the stage of incipient breakdown was reached in the Sahelian drought, nomadic pastoral communities generally fared sedentary agriculturalists afflicted with crop failure. In developed countries, buffering of the livelihood system at this stage may take the form of government drought relief and loans, as it did in the USSR, for example, in 1972 and 1975. It is likely that indebtedness will increase and the less viable holdings will be abandoned. Selective depopulation will occur in rural settings, while the local towns based on rural services and industries will suffer economic depression.

209. If drought lasts long enough, livelihood systems will collapse utterly. When this stage is reached, the greatest hardships fall on those communities that are most exposed environmentally and least equipped to transfer to a new livelihood system. At the breakdown point in the Sahelian drought, the exposed nomadic pastoralists suffered most, with many areas experiencing a mass exodus of people ill fitted for life in towns.

210. Apart from the physical hardships involved, such an upheaval brings with it severe emotional stress. Those most hard hit may succumb to an apathy stemming from their felt loss of status. Social disasters of this magnitude are more characteristic of marginal

societies in marginal lands than they are of developed nations. Industrialized societies have access to resources that can blunt the impact of disaster. Some societies have worked to incorporate their most remote and vulnerable populations into the larger community. Such actions are most effective when they support dryland peoples in their efforts to maintain sustained productivity in their harsh and demanding environments.

VI. MEASURES TO COMBAT DESERTIFICATION

Principles which Should Guide All Measures Against Desertification

211. Any measures undertaken to combat desertification must be informed by the scientific and humanitarian principles long recognized by the member states which compose the United Nations. Desertification is a human problem, and measures to combat it must ultimately be directed toward people, toward sustaining and improving their livelihoods. Thus, measures to combat desertification must be seen as having human and social objectives. They must be inspired by an acknowledgement of the right of people living in dry lands to acceptable standards of health, nutrition, education, livelihood and social well-being, consistent with human dignity.

212. Account must be taken of traditional social values and an appropriate respect shown for life styles and ancient knowledge developed through long adaptation to the dryland environment.

213. Priorities in programmes to combat desertification should be influenced by the severity of its impact on the populations concerned, and by the degree of their vulnerability, rather than by the severity of its impact on the land alone.

214. The approach should be an integrated one, in which proposals involving technological or environmental change are linked with social and economic measures undertaken to advance development.

215. Measures to combat desertification will not succeed without the willing participation of local communities. The need must be recognized to work through existing livelihood systems and established social patterns. The involvement of the community must be sought, as by enlisting the example of community leaders. It may be necessary to create incentives toward community participation. The practicality and advantages of proposed measures should be demonstrated at the earliest stage through realistic pilot projects. The enlistment of community participation should not be thought of exclusively in terms of outside experts persuading people to do what the experts think is good for them. Local knowledge should also be enlisted. Sometimes the best procedure would be the elimination of obstacles to the good land use practices that local people would otherwise prefer to carry out.

216. From the outset, programmes should contain some measures selected because they relate to immediate local problems, because they demonstrate prompt action within the community, because they are possible with existing resources, and because they promise convincing results within a reasonable time.

217. Advantage should be taken of crisis situations when social structures and livelihood systems have been disrupted and people are more prepared to consider change and to carry out whatever restructuring of dryland livelihood systems that are required for sustained production.

218. The ideal objective is the recovery and maintenance of ecological balance in the dry lands in the interests of sustained productivity, but this must be reconciled with the needs of local populations. Some degree of environmental disturbance, as determined by pressing human needs, must be tolerated in land management.

219. On the other hand, it must be recognized that land use pressures have been the major factor in the advance of desertification. Accordingly, changes in land use will be required, and these bring with them a need for corresponding social changes. Some element of control may be required, but it will not succeed without a sympathetic community response.

220. Where limits are set by rainfall, the productivity of dry lands per unit area will generally be low. Such lands can command only modest investments in keeping with their productivity. Reclamation and preservatory measures should be realistically designed so as not to ask for investments that the land will never repay.

221. Since rainfall will remain variable in the dry lands, they will continue to be high-risk areas for most land use systems, and this should be reflected in development plans. However, measures to stabilize their livelihood systems and buffer them against periodic drought should not deprive them of their flexibility and the risk-spreading strategies characteristic of traditional dryland practices.

222. Apart from limitations set by climate, dryland ecosystems will remain sensitive to land use pressure because their soils and dynamics are delicately balanced. The best designed dryland livelihood system will still require constant surveillance if balance is to be sustained. It is therefore essential that campaigns against desertification should not be presented as sets of single episodes. Development plans must incorporate systems of monitoring that will indicate how campaigns are proceeding and when they should be altered. This requirement strongly underlines the need to develop indigenous science and technology, so that assessment, monitoring and planning will not be added to the list of imported items.

223. A geographical spread as immense as the dry lands comprises a vast variety of biophysical, economic and social settings. Desertification processes and problems are correspondingly varied and complex. Any plan of action to combat desertification will recognize this, and with it that there can be no single set of remedies. Recommendations must take account of different situations and be flexible enough to encompass a wide range of conditions.

224. A review of the desertification problem strongly supports the contention that past failures to maintain balanced livelihood systems in dry lands are the outcome of an inability to apply existing knowledge of physical processes rather than from any lack of understanding of what those processes are. The same is true of the design of measures to combat desertification. Accordingly, plans of action should address themselves first of all to the removal of obstacles to the application of existing knowledge, to the adaptation of existing knowledge to local situations in the social as well as in the physical sphere, and to problems of acceptance and participation among local communities. Plans of action should stress action rather than future research.

225. It should not be taken for granted that action to combat desertification will take first place among national commitments. A plan of action to combat desertification should not appear to preempt already established national priorities. Nevertheless, it should be kept in mind that action on the ground will largely be carried out by national organizations, and presentation of the plan should accordingly aim to influence governmental attitudes toward the problem of desertification and should seek to secure the active commitment of governments. This is most likely to occur when combative measures are linked to broad national plans for development and appear to be consistent with national goals.

226. Because dryland ecosystems are fragile, they are particularly vulnerable to misapplied technology. Techniques and equipment tried with success in more humid regions have contributed to desertification in drier environments. When innovations are suggested, attention should be paid to the impact they will have on the dryland environment and to their adaptability to local livelihood systems. In developing countries, attention must also be paid to low cost, simplicity of operation and acceptability by the local community. It follows that modifications of existing technology and practice are likely to prove more effective than radical innovation.

227. The best defence against misapplied technology is the ecological approach, which should govern all actions concerning land – including dry lands. According to ecological principles, ecosystems are regarded as complexes of interdependent elements – plants, animals, soils, hydrology and energy balance – a change in any of which has repercussions on all the others. Secondary as well as primary effects must be recognized. Successful land use practices collaborate with ecosystems instead of fighting them. The use of dry lands must be carefully timed to conform with their extreme climatic patterns, and this involves resting and fallowing. The complementarity of different environments is vital to the tactics of land use, and this will give rise to exchanges of various kinds. Use must be carefully adjusted to the land's capability.

228. Beyond these guiding principles, measures to combat desertification must take as their point of departure the identification of the process when it occurs and the assessment of its nature and severity. This is not a call for further research. It is rather a recognition of the fact that desertification is a complex phenomenon that must be met by measures specifically adapted to different situations.

229. Measures to combat desertification will take on distinctive characteristics depending on the nature of the land and the livelihood systems practised there. They must also reflect an understanding of the specific situation encountered. An example of the kind of careful study that is often needed appears in the component review, *Ecological Change and Desertification*, which refers to gullying on Arizona rangeland at the turn of the century originally attributed to overgrazing.

> "A close examination of the evidence led Cooke and Reeves (the researchers) to a much more complicated explanation. They found that there had been little gullying during the period of heaviest stocking levels; and the gullies had only begun to develop well after the numbers of cattle had fallen. In search of another explanation, they first looked to the rainfall and found, with earlier workers, that although there had been little sustained change in annual amounts during the period of gullying, there had been a consistent increase in intensity at some seasons. Nevertheless this alone had probably not been enough to form the gullies, for the history of each particular case showed that there had almost always been some local disturbance of the surface such as the cutting of a roadside ditch or the insertion of a culvert. Driven by enthusiasm and a sense of impending crisis, observers of dryland conditions seldom have time for these niceties . . . ."

Identification and Assessment of Desertification

230. Experiences indicate that the long-term progressive deterioration that constitutes desertification may not be readily identifiable against the background of short-term

environmental fluctuations that spring from periodic shifts in rainfall. There is a consequent need for regular monitoring of the status of dryland ecosystems to provide early warning of trends, to identify areas in which change is taking place, and to provide a basis for the investigation of causes and processes. It is in terms of such information that measures for prevention or reclamation will ultimately be designed.

231. Because the problem is global, calling for international effort and a worldwide exchange of information, monitoring should be established in the form of worldwide surveillance carried out on a uniform basis in which data can be intelligibly exchanged. Such an arrangement might usefully be identified and co-ordinated as a "world desert watch".

232. Global surveillance of the status of dryland ecosystems and of land use can be achieved most economically through the remote sensing powers of specialized orbiting satellites. The so-called LANDSAT system, already in operation, has this capacity. LANDSAT now provides imagery with a resolution of at least 50 metres, with prospects that higher resolutions will soon be achieved.

233. The first step in the use of a satellite such as LANDSAT is to employ it for the identification and mapping of distinct units on the ground. This can be carried out in false colour imagery (LANDSAT bands 4, 5 and 7) or in black and white. The information can be obtained in even more detail on a digital basis, using computer-compatible tapes. Much of the world's dry lands, perhaps 85 per cent is already covered by LANDSAT. The mapping would define functional environment types as determined by their geology, landform and surface drainage, each type characterized by certain soils and vegetation cover. The characteristics of each unit would be established from imagery or tapes and supported by already existing information on geology, soils and vegetation. The findings would be validated on the ground, this so-called "ground truth" being determined by field sampling and traverses.

234. Initial demarcation of the topographical and soils units by a skilled photo-interpreter would be inexpensive, costing on the order of a few dollars per thousand hectares. The building up of ground truth would be a separate and continuing operation. Different combinations of boundaries would allow the information so obtained to be expressed in terms of a variety of references, such as pasture, land vegetation or salinity. The achievable resolution is adequate for general surveys of land status, for planning for extensive land use such as pastoralism, or as a first stage in the identification of likely areas for more intensive kinds of land use. The maps obtained could provide a framework for the interchange of experience among comparable environments.

235. To fix trends in dryland ecosystems, repeated monitoring on a uniform basis is required. This can be obtained from the LANDSAT system via remote sensing satellite-to-ground receiving stations. Each of these has an effective radius of about 2700 km but at the present time, only a part of the dry lands is properly covered. Access to a ground receiving station provides an opportunity for manipulating the data output to conform with local needs.

236. The storage, handling and reproduction of the data from the ground receiving station, and its integration with data from other sources, call for linked computer-based data systems which will generally form part of a national land-data system. Information can be related to a given topographical unit or to another geographical subdivision by the use of a standard system, and experience suggests that a 1–2 km grid provides adequate definition for the general surveillance of dry lands.

237. Feasibility studies set in South America and south-west Asia are proposing to validate a transnational approach to monitoring desertification on the above basis. The annual cost of establishing a ground receiving station and linked data system would be $2 million, which amounts, as suggested, to dollars per thousands of hectares covered. The establishment and validation of such a system using pilot test areas would take about 3 years, and such systems are not likely to be generally established before 5 years. It looks as if world coverage would be most economically achieved through regional groupings of countries.

238. In addition to ground truth, satellite data can also be supplemented by airborne imagery, which, although more expensive, is capable of higher resolution. The further investigation of areas revealed by LANDSAT as undergoing desertification or as potentially suitable for more intensive land use will call for mapping and monitoring at the finer scales provided by conventional air photography.

239. The evidence obtained from satellite data and other sources, linked through the data bank and supplemented with the information already available, should provide the basis for a number of important activities.

240. The first of these should be the construction of a map showing types of desertification present and the relative vulnerability of the demarked ground units to further desertification. Then regional plans can be formulated for measures to combat desertification, linked with plans for improved land use, for the re-establishment of disturbed populations or for whatever else conditions call for. Following the regional plan, specific combative measures can be designed and sites selected for demonstration or pilot projects.

The Conditions of Dryland People must also be Assessed

241. If measures to combat desertification require continuous assessment of vulnerable lands, they demand also a comparable understanding of the people who live in such places. Experience with existing programmes has indicated that physical problems associated with desertification are commonly more amenable to solution than the typically human problems.

242. Assessments of physical conditions should therefore be accompanied by efforts to obtain a more precise understanding of the state of dryland peoples. Surveys should be undertaken, perhaps through a strengthening of census services and techniques, of their demographic characteristics, of the state of their health, and of their social and economic circumstances. On the basis of what the surveys reveal, measures can be designed to combat malnutrition, ill health, poverty, illiteracy, excessive population growth and other social and economic disadvantages experienced by people living in the dry lands. Social and economic changes, such as resettlement or alternative livelihood systems, should be presented and proposed as integral parts of a plan to improve conditions and not as mere afterthoughts to environmental measures.

243. Social, economic and technological acceptance are likely to be as important as environmental compatibility in determining the effectiveness of what is proposed. Studies should be directed toward uncovering the social obstacles to good land use practice. The Case Studies have shown that systems of land tenure and tenancy are of critical importance in determining whether or not the land will be properly managed and maintained.

244. Whether they concern land or people, plans must be flexible. They should incorporate periodic checks on the progress of measures put into operation and should allow for concurrent reassessment of the problem in human as well as in physical terms.

Measures to Combat Desertification in Extensive Pastoral Systems

245. To combat desertification in pastoral systems means to adopt grazing practices that will allow vegetation to recuperate. In areas too dry for rainfed cropping, the natural vegetation usually forms the most efficient pasture in terms of upkeep, grazing returns and protection of the soil surface. The maintenance of a plant cover that will sustain the pastoral system under most conditions is the obvious goal of combative efforts. Anything more – intensive reclamation, for example, by planting programmes or mechanical controls – will be feasible only in restricted areas where the physical processes of desertification threaten installations, communications, settlements or valuable cropland.

246. The experience of the Sahelian drought indicated that the death of livestock was chiefly due to the failure of pastures rather than of water supplies. Accordingly, conservation measures should be introduced for the control of grazing access to dryland ranges where such measures do not exist, including fencing where necessary and economically feasible.

247. As a first step, surveys should be initiated to determine the useful productivity of the main varieties of dryland pasture under differing seasonal conditions, the requirements of pasture plants for successful regeneration under grazing, and the dimensions of the grazing impact of a proposed system composed of certain animals in certain numbers. Surveys must take into account the dual role of perennials as surface protectors and as fodder during drought. A logical first step in the assessment of dryland pastures is to map them, indicating their distinct topographic, soil and water conditions.

248. Surveys lead to preliminary assessments of carrying capacity under a variety of conditions and these, in turn, form the basis for appropriate grazing strategies. "Carrying capacity" is an overworked but poorly understood term, and research might properly be employed to clarify and quantify it. Meanwhile, reasonable estimates must be used to support grazing strategies incorporating a number of elements.

249. They should incorporate possibilities for deferred or rotational grazing and for the establishment of protected reserves as seed reservoirs, grazing reserves in the event of drought, and plant and wildlife refuges in which genetic variety can be conserved. As far as possible, they should preserve the mobility, flexibility, diversity and low stocking rates traditional in dryland grazing systems. Consideration should be given to fencing those parts of the rangeland subject to concentrated stock movements, those made up of particularly vulnerable pasture type because of soil or the formation of the land, or such sensitive areas as town perimeters.

250. Opportunity should be taken to enrich natural pastures locally by developing simple water-harvesting schemes, such as through the use of trenches and flood banks in areas of natural flooding. These areas should generally be treated as controlled reserves, available for the breeding of animals, as a resource against drought and for the harvesting of forage. Controlled reserves can also be established in the form of green belts as suggested in the feasibility studies concerned with the establishment of such protected reserves on the northern and southern borders of the Sahara. Consideration should be given to using such areas for subsistence cropping. They should be fenced off from the

open range and their use integrated into the general grazing scheme. Range conditions should be periodically surveyed to determine what grazing pressures are doing to the land and vegetation and with a view to adjusting the grazing system when required and to guide herds to ephemeral vegetation.

251. While grazing strategies refer to average stocking rates, attention must also be given to localized concentrations, as on long tracks and around salt licks, watering points and settlements, and measures should be taken to avoid intensive local grazing and trampling. Consideration might be given to the establishment of watering points of moderate size in a network that gives adequate access to all pastures being grazed. Measures should be introduced for the controlled and responsible use of such watering points, including the levying of charges on graziers who use communal supplies.

252. When the trend of rangeland conditions indicates that grazing pressure should be reduced, a number of measures can be taken for which foresight and advance preparation are called for. Such preparations would include the improvement of transport facilities, assistance with breeding programmes to improve productivity per animal and measures to reduce the risk of losses from breeding herds. Marketing outlets should have been established for the efficient disposal of surplus animals, for example, in stratified management programmes as suggested by the feasibility study on the stratification of Sahelian rangelands, with subsidies and price supports where necessary.

253. Pastoral systems will need help in coping with recurrent drought stress. A number of measures can be taken to provide this help, such as the setting aside of grazing and forage reserves, the provision of transport facilities for the movement of stock, financial assistance to restore herd numbers following drought, and insurance against drought losses.

254. Mutual support from adjoining crop-based systems has traditionally provided pastoralists with an important safeguard. Such arrangements should be maintained and strengthened where possible. They have included market exchanges, arrangements for stubble or fallow grazing (in exchange for natural fertilizer), and the introduction of forage crops into crop-based systems. Arrangements vary widely, from the incorporation of seasonal nomadic pastoral systems into schemes for irrigated agriculture, as in Soviet Central Asia, to the integration of animal-based and rainfed crop systems in zones of controlled land use, as in green belts around the Sahara.

Measures to Combat Desertification Nomadic Pastoral Systems

255. Recent years have shown an increased tendency for pastoral nomads to settle down in fixed habitations. This happens because of changing personal goals or attitudes, because of drought disaster, or as a result of government programmes. Nomadic herding is then left to part of the former community, which comes increasingly to resemble more settled pastoral systems. These changes will continue, and assistance should be given to accommodate them.

256. Such assistance might take several forms. Consideration might be given to the establishment of properly designed settlements equipped with water supplies and community services. Nomads can be aided to develop ancillary farming, whether irrigated or rainfed, particularly for subsistence or forage crops. Wherever nomads have settled down, measures should be taken to reduce the environmental impact of stock concentrations or fuel-gathering activities among people unaccustomed to living in

permanent settlements. Woodland or range reserves can be established near settlement perimeters.

257. Over recent years, nomadic pastoralists have been increasingly at a disadvantage relative to adjacent farmers, particularly during periods of above-average rainfall when cropping tends to encroach on pasture lands. Care should be taken to preserve the traditional access by pastoralists to rangelands and watering points, by legislation or taxation policies if necessary.

258. Little has been done to strengthen nomadic pastoralism by using traditional practices, with all their adaptations, as a base. Measures could be taken to improve livestock quality through breeding programmes in an effort to increase yields from smaller herds and decrease losses through disease. Control of grazing can be effected through technical advice, preferably directed toward the reinforcement of traditional practice and authority. Breeding and marketing schemes can be developed in harmony with traditional systems. Additional watering points can be provided that are moderate in size, cheap to construct and easy to maintain. Here the use of wind pumps should be introduced. The use of such waters should be controlled to conform with broad grazing programmes and should aim to bring all pastures into effective use.

259. These animal-based systems are often at the extreme edge of environmental productivity and are therefore vulnerable to periodic extreme of protracted drought. This situation should be recognized by setting aside food reserves and in the advance planning of emergency measures.

260. The seasonal or permanent out-migration characteristic of these communities has long provided them with supplementary income in the form or remittances from once-nomadic wage earners. Plans to combat desertification should seek to accommodate and assist such population movements through appropriate schemes for the settlement of persons wishing to move. Alternative sources of livelihood might be provided in the local setting, as through employment in tourism, craft industries and services or through the establishment of new industry or agricultural activities. Attempts should be made to reduce the selective out-migration of the most able workers whose loss tends to impoverish the local community.

Measures to Combat Desertification in Commercial Ranching Systems

261. Commercial ranching suffers from the greater likelihood of disturbance around fixed installations. Attention should be given to arrangements for moving and yarding stock, and some installations, such as yards, paddock gates and troughs, may have to be shifted periodically to avoid extreme effects.

262. The use of mechanized transport and other equipment characterizes commercial ranching. Care should be taken in routing and grading tracks and roadways, especially where protective stone covers are involved. Attention should be paid to possibilities of stabilizing surfaces as an alternative to grading fresh routes. Particular case is required where run-off is channeled along tracks or their margins.

263. High labour costs in commercial ranching mean a minimum use of manpower, which causes difficulties when labour-intensive measures, such as planting, are called for. Proposals for pastoral development should include an assessment of environmental impact and an estimate of the likely costs of reclamation measures, an expense that may be tax-deductible.

264. Vulnerability to price fluctuations, including those on distant, international markets, introduces an additional hazard into commercial ranching systems, reinforcing the hazard of climatic variability. Depressed markets may lead to the abandonment of properties and the loss of installations and may discourage appropriate long-term investment. Government-assisted marketing and price-stabilization schemes should be introduced when necessary.

265. The advantage of commercial ranching include lower stocking rates, better control of stock movements and watering points, and improved facilities for the transport of stock and forage by road or rail. Such advantages point to the need for more imaginative stocking policies, particularly those that avoid extreme grazing pressure. Enlightened policies would include the maintenance of public rangelands within the pastoral district. Provision should be made in advance for the transport of stock to such reserves when circumstances require it, for destocking in times of drought and for restocking when the rains return, and for access to forage when drought occurs. Such provisions may call for outside assistance.

266. Mechanized equipment can be used to counter extreme desertification in local situations. Encouragement should be given to research on improved methods of revegetation, including soil treatment, pitting or furrowing, seeding and fertilizing. Assistance might include technical advice, loan of plant stocks, the provision of seeds and fertilizers, and financial subsidies for approved measures.

267. In operation, commercial ranching can take advantage of economies of scale. It is also more subject to governmental regulation, either directly through lease provisions or indirectly through financial or taxation policies. Recommended stocking practices can be enforced through these means, which can also be used to achieve the subdivision or amalgamation of holdings so as to favour operations on recommended lines.

Combating Desertification in Rainfed Cropping Systems

268. Studies of the relation between agriculture and climate, such as those carried out by the World Meteorological Organization in western Asia and Saharan Africa, have done much to determine the connections between climate and the water needs of cereal crops, thus fixing the probability of the occurrence of effective seasons on the basis of climatic records. These studies should be extended and improved through additional meteorological recording and investigations into the water requirements of crops at different stages of growth and under a range of soil conditions. By providing good estimates of climatic risk, such studies will support policies of land zoning, and measures should be taken to discourage the extension of cropping beyond certain environmental limits.

269. At the same time and because cropping represents a more productive use of the land, attempts should be made to expand the safe-cropping area through the introduction of crops that are more resistant to extreme conditions and through improved methods of cultivation and water conservation. Such actions should be supported by demonstration projects and extension services. In selecting drought-resistant or salt-resistant crops, attention should be paid to the genetic qualities of local plants whether they have already been domesticated or not.

270. Research should be encouraged that will lead to improved weather forecasting

with accompanying warning systems, particularly for such critical periods as seeding, germination and harvesting.

271. It is important that plans for the reclamation and improved use of rainfed croplands should be a part of integrated schemes for the use of functional areas such as drainage catchments and which recognize the interdependence of upland, piedmont and valley with their associated land use.

272. A first step in formulating a plan is to map land types and land use at a scale appropriate to cropping (1:50,000 to 1:250,000, depending on conditions). The land units mapped should be classified according to potential use as determined by the existence of hazards, such as steepness and length of slope, the presence of stones or rocks, the risk of flooding, the quality of the drainage and vulnerability to wind erosion.

273. Recommendations as to how the various parts of the land should be used will constitute the plan, which must recognize appropriate limits to rainfed cropping, as determined by rainfall, terrain, soils and relationships with adjacent land uses such as forestry or grazing. The marginal lands outside these limits should be removed from cropping by acquisition, by financial inducements or by the establishment of forest, grazing or water-catchment reserves. When such measures involve the disruption of traditional livelihood systems, they are unlikely to succeed unless they form part of larger schemes of rural reconstruction involving appropriate changes in land tenure, the consolidation of holdings, or resettlement programmes offering alternative livelihoods.

274. Clean fallowing, or allowing a field to rest while stripped of vegetation, provides a way of conserving the moisture in the soil. Like several such techniques, clean fallowing happens to increase the land's vulnerability to erosion. Safeguards can be erected by improved methods of rainfed cropping and by measures which maintain ground cover and improve soil structure. The things that can be done to counter risks and improve productivity vary considerably among different situations and different systems of rainfed agriculture.

275. In regions of a Mediterranean type, traditional combinations of tree and field crops should be encouraged. An element of livestock husbandry should be retained, increasing the diversity of these systems, their resilience, and hence their resistance to climatic stress.

276. In some Mediterranean regions, decay and disuse have affected certain traditional methods for the conservation of soil and water, such as terracing and water-spreading systems. These old systems should be brought back into service, maintained and improved, and assistance should be provided for such purposes. Tree planting should be encouraged, whether in shelter belts or in coppice groves for firewood. Tillage should avoid powdering light topsoils, and farm machinery, some of which may have to be freshly designed, should be suitable for working such situations as terraced slopes. Strip cropping should be introduced as a counter to wind erosion. More use should be made of crop rotations, including legumes, at the expense of fallow.

277. Crop rotation, including cover crops to be ploughed back into the soil, should also be introduced into mechanized systems of rainfed monocropping. Such systems should restrict the burning or removal of litter, and livestock should be introduced to graze on feed crops or crop residues. Strip cropping should be encouraged, with inducements on occasion, as well as the planting of shelter belts on open plains. To combat salinization on valley floors, deep-rooted varieties or salt-tolerant pasture can be planted.

278. In the swidden system, the slash-and-burn agriculture so typical of rainfed cropping in dry lands with summer rain, the farmer will return to a particular plot after its vigour has been restored by extended fallow, often as long as 20 years. Shortening the cycle, coming back too soon, can have adverse effects on plant recovery and regrowth and on soil fertility. When this happens, measures should be taken to restore the cycle to its older rhythm, perhaps by expanding the area available to cultivation or by removing population pressures through resettlement or the development of alternative livelihoods.

279. In these systems, valuable substances, such as gum arabic, can sometimes be extracted from the natural regrowth during the fallow part of the cycle. Steps can be taken to increase the value of regrowth by introducing new trees or by adopting good forestry practices.

280. Traditional crops and ancient tillage practices have sometimes become fixed in these systems where new varieties and alternative techniques would work better to maintain the fertility and structure of tropical soils and to diminish the effects of pluvial erosion and soil crusting. Swidden agriculture should be closely scrutinized everywhere with a view to reducing its impact on the land.

281. Once rainfed cropland has been degraded, efforts to rehabilitate it should form part of larger actions directed toward water management, improved land use and the control of erosion. Within broader plans, the specific actions to be taken will depend on the course that degradation has followed.

282. Gullying, a particularly unsightly form of erosion, can be arrested by planting trees in upper catchments and along gully margins and by grassing areas that feed the gullies with flows. Also helpful are the construction of diversion banks and furrows across gully heads, the installation of check dams and silt traps along gully courses. Under favourable conditions, gullies can simply be filled in and their banks regrassed.

283. Sheet erosion, which scours topsoil from wide areas, can be countered with contour banks and ditches, with grassed contours strips and by means of terraces.

284. Wind erosion, which blows soil away from rainfed cropland and which causes sand drift and dune encroachment, can be countered by planting shrubs and trees in shelter belts (at a spacing four times as far apart as their eventual height). Fences can be constructed or lines of resistant shrubs and trees planted as barriers against oncoming sand, upwind of threatened areas. Bare sand can be covered with matting, bituminous coating or mulches of vegetation litter.

285. Sand surfaces can be stabilized by seeding and planting proper successions of vegetation, including plants which thrive in sand in association with shrubs and trees. Finally, dunes can be levelled or reshaped to remove slip faces in conjunction with action to prevent their reappearance.

Combating Desertification in Irrigated Cropping Systems

286. The amount of irrigated land lost annually to desertification (some hundreds of thousands of hectares) is about equal to the amount of land newly brought under irrigation each year. Large investments are involved in the breakdown and abandonment of such intensive, highly capitalized agricultural projects. Irrigable land is scarce, and new enterprises are enormously expensive. Such considerations stress the importance of

maintaining existing irrigation schemes by countering desertification with preventive action whenever it affects them. That desertification can be successfully countered in affected irrigation systems is the message contained in the Pakistan and Iraq Case Studies.

287. The most prevalent form of desertification in irrigated cropping systems occurs when waterlogging causes salts and alkali to infect soils, particularly where drainage is poor and leaching inadequate. It is a problem that emphasizes the importance of preliminary surveys and testing of proposed irrigation projects to assure adequate design. Most salinization problems arise from design deficiencies.

288. Good design should be based on an understanding of how much water is available for irrigation and its silt and salt loads, including seasonal variations. A close study must be made of the soils in the area embraced by the project, their texture and salinity, and especially of their water properties, as these will determine drainage requirements and how much water will be available to crops. Water requirements should be determined for proposed cropping systems. The position and salt content of the groundwater table should also be determined as seasonal fluctuations in both. This will require some understanding of the hydraulic properties of the soil's lower layers, of how these layers store and transmit water.

289. These investigations should yield a map showing salt hazards and how they might restrict the proposed cropping system. On the basis of the map and the surveys, design work can continue with particular emphasis on the distribution of the water and on effective drainage systems and the subdivisions of the system as determined by estimated water needs. Finally, design should take account of the services and communications the system will require and the settlements that serve and are served by it.

290. Whether under development or in operation, irrigation schemes should be run by operating authorities equipped with professional staff, adequate funding and the powers to control land use. As a way of proceeding, especially with new schemes, the authority should undertake pilot projects which can be expanded into research and demonstration projects as they prove their worth.

291. Irrigation schemes require constant maintenance. The main distribution canals should be properly banked and lined, as with concrete, to reduce seepage. Canals and drainage ditches should be kept clear of silt and weeds and pools of stagnant water eliminated. Take-offs or turn-outs, where water is drawn into the system, should be designed and maintained to keep silt loads to a minimum. The plots to be irrigated should be leveled to ensure even watering and leaching, and where local subsidence occurs, leveling should be carried out periodically. While provisions and requirements for adequate leaching should be maintained, there should also be checks against over-irrigation.

292. Irrigation schemes are sometimes established where farmers are unfamiliar with this type of agriculture. The tillage of heavy soils under irrigation calls for particular skills, as does the application of irrigation water at prescribed stages in the development of the crops. Extension services must be provided if irrigation schemes are to work successfully. Land-holders should also be given assistance in the form of credit, purchasing and marketing plans, and where suitable and desirable, in the development of agricultural co-operatives. Improved land use should be encouraged through such measures as economic incentives and tax concessions.

293. When irrigation schemes are designed, individual or family holdings should be shaped to ensure an appropriate level of intensive use, without being too large to preclude effective maintenance, a point emphasized in the case studies. Encouragement should be

given to an appropriate balance of subsistence and cash crops, tree and field crops. Forage crops may be included if circumstances favour a livestock component. Great care should be taken in the allocation of holdings and in the formulation and administration of regulations for their proper management.

294. When successful, irrigation schemes inevitably give rise to close settlement, to towns, often inhabited by people unaccustomed to congestion and its attendant problems. Housing should be planned for and provided at the same time that land-holdings are allocated. Houses should be equipped with potable water and sanitation services, and all the more so, where diseases can be transmitted through the irrigation system itself. Indeed, new communities should be provided with all the standard community services, including health, education, welfare and cultural centres, and these should be sited as part of the land settlement plan. Transport services should be established.

295. Irrigation projects based on groundwater supplies encounter special difficulties because groundwater quality is usually lower than that of surface waters and the threat of salinization is generally higher. Limitations in groundwater supplies may hinder proper leaching. Groundwater supplies must be kept in balance with the requirements of land use, and enough water must be provided for both irrigation and leaching. Generally, discipline applied to water use must be stricter when irrigation is based on groundwater rather than surface water.

296. Such discipline may include central control over the siting of bores and wells and the installation of pumping equipment. Monitoring must be constant of such factors as groundwater levels, draw-down and salinity, and the proper staff must be on hand to conduct such monitoring or any other investigations as required.

297. When based on groundwater supplies, irrigation schemes often suffer from poor drainage, with increased chances that the groundwater supplies will be contaminated by saline irrigation run-off. Such schemes are often characterized by networks of small distributing channels under individual control affected by wastage through seepage and higher risks of salinization.

298. Many such problems arise because older groundwater-based irrigation projects often grew up without any planning, and their operations remain hampered because of entrenched rights to land and water. Such old projects should be rationalized, with compensation when necessary. Groundwater assessment, together with the mapping and classification of land types – the information used to plan a new system – would provide a basis for rationalizing older systems and for their continuous reassessment.

299. When irrigated lands have suffered salinization or other forms of desertification, they should be surveyed as a first step to reclamation. By determining what topographic changes have occurred and the degree of salinization of soil and groundwater levels, an estimate can be made of what is needed to leach and drain affected lands and what else might be required to restore the system – by relevelling of ground surfaces, for example, or renewal of irrigation channels. How drainage will be carried out – whether by tubewells, tile drainage or open ditches – will depend on groundwater conditions, soil properties and costs of land and labour. To decide among alternatives, a cost-benefit analysis may be needed.

300. When the situation has been made clear, decisions can be made on priorities, which might include abandonment of lands most severely affected, and a reclamation programme designed in terms of the availability of water, labour and capital. After the programme has been implemented, reclaimed lands can be re-allocated, but not without

clear regulations on what can be done with them. Reclamation provides an occasion for the enforcement of practices that will prevent desertification from recurring.

Combating Desertification in Mining

301. The dry lands have always held vast treasure in mineral resources, including much of the modern world's petroleum, and it can be expected that new discoveries will be exploited there in a now familiar pattern: revenues will be large compared with other local sources of income; direction and financing will come from outside the region, and almost all financial benefits will be exported away.

302. In the past, or so it has generally seemed, such resources would have been exploited whatever the local human consequences and environmental impact. Nowadays, it is agreed that the region and the local community should be protected from the worst consequences of such exploitation, which is indeed expected to make a proper contribution to regional development and welfare. To assure this, mining proposals must contain an assessment of their environmental impact, and the proprietors of the mines will be expected to meet the full costs of environmental protection and reclamation. Their operations must be so conducted that they contribute to the general development of the region.

303. It may be difficult to maintain principles when great riches are involved, but in any competition for scarce resources, such as water or land, the rights and needs of the local community should receive priority. When mining or drilling operations are about to be introduced, the local community should participate fully in planning and in all other decisions that concern them, and arrangements should be made for continuing consultation.

304. The dry lands should be favoured with the same standards of environmental protection that are applied in more humid areas. Indeed, dry lands may require additional precautions because of the special sensitivity of the arid environment, its susceptibility to air pollution, groundwater pollution, dust nuisance and surface disturbance. As an example, restrictions should be placed on the grading of unsealed roads in dry lands and on their use by heavy vehicles. Precautions should be taken to prevent ground pollution by oil, particularly in pumping operations.

305. The activity of mining or drilling and the people who carry it out, many of them brought in from outside, will have all sorts of effects on the surrounding region. Plant and animal reserves may have to be established on the perimeter of the activity, with restrictions on hunting or plant removal over a wider surrounding area. Employees brought in from outside should be placed in suitably designed settlements equipped with proper services.

306. Mining or drilling ventures will view local communities as a source of labour and a supplier of food and materials, and fulfilling these roles can affect a community adversely. It sometimes happens that a once-isolated, traditional society is brought into sudden contact with people of a very different kind, often rootless, sometimes violent, accustomed to a transient, unstable society. It will be difficult to maintain the principle that the rights and needs of the local community should be protected and that local

people are given every opportunity to participate in and benefit from the new development.

Combating Desertification Associated with Tourism

307. Many of the considerations relating to mining and drilling have equal application to tourist activities and installations in deserts and dry lands. Local communities should share in the benefits of tourism. It should provide them with opportunities for employment, improved communications and access to other support services and improved markets for local products, including those of craft industries. But before local communities can share in the benefits of tourism, they may have to be protected from it.

308. For example, local livelihood systems, such as pastoralism, may have to be protected from interference by tourist activities. The information tourists are provided should include comments on the local people, their customs and way of life and how they have adapted to a harsh environment, to help ensure respect for their practices and for themselves as persons. Protection may have to be given to sites and objects of traditional cultural importance. In the competition for scarce resources such as water, land and pasture, the needs of local communities should be assigned first priority. This viewpoint and the protection required may best be achieved when local communities participate in the planning and management of tourist activities.

309. The natural environment will also require protection against tourist activities. Great care must be taken in the siting, design and maintenance of tourist roads, camps and rest areas. Traffic restrictions will be needed, particularly on the use of cross-country vehicles, and roads subject to heavy traffic will have to be paved. Lodges and camps will have to be served with proper facilities, for water, sanitation, rubbish disposal and the control of local traffic. Penalties should be applied to combat littering. Plants and animals will require protection, particularly of endangered or attractive species. Archaeological and scientific sites, interesting geological formations and natural monuments will all require special protection.

310. The concept of environmental management, so important to sustaining productivity in agriculture, should be extended to the tourist industry. This might involve the establishment of reserves or wilderness areas from which tourists would be excluded and which would serve as refuges and a source of regeneration for plants and animals. Or it might embrace the concept of natural parks for controlled tourism in which the tourist could view an interesting and typical range of natural ecosystems without causing them damage. The management of such parks should incorporate the concept of "recreational carrying capacity" with "deferred" or "rotational" uses to allow for the seasonal vulnerability of species and to spread the impact of tourism. It is obvious that such parks must be adequately staffed with professionals capable of providing tourists with expert guidance.

311. The development of tourism should be generally controlled in the interest of environmental protection. Such control can be exercised by Tourism Ministries or tourist boards on which local communities and land users are represented or can be heard. Each tourism proposal should be required to incorporate an environmental impact study, and approval of the proposal should be subject to the provision of adequate environmental protection. The costs of such protection and of reclamation, if subsequently needed, should be borne by the project.

Combating Desertification Associated with Human Settlements

312. Dryland settlements can range all the way from the one-family homestead with its thorn-tree fencing to great, modern cities with millions of inhabitants. The usual dryland settlement, however, will be a village or small town that has grown up to serve the needs of the livelihood systems practised in an arid setting. A number of measures can be taken to improve conditions in such settlements and reduce their adverse impact on the environment.

313. Reserves should be established surrounding settlements and extending for a few kilometres out from their limits and within which grazing, farming and fuel gathering are restricted. Such reserves must be well fenced on their boundaries and wherever they are traversed by roads. They should be regarded as areas affording regeneration of natural vegetation, but they may be subject to land treatment and planting where degradation is advanced.

314. Special measures will be required to check active physical degradation around settlements when it threatens urban land and gardens. It may be necessary, for example, to stabilize moving sands and to check gullies or fill them in. Roads in and near settlements should be paved or otherwise improved. Traffic should be confined to roads by fencing. Open areas inside settlements which form sources of dust nuisance or which retain stagnant water after rain should be brought under control. Grassing and planting of shelter belts may be required, but attention should be given to types of wind-stable ground cover which require little maintenance and consume little water, as for example gravel surfaces relieved by the planting of local trees and shrubs. Adequate storm drainage should be provided to handle the run-off from rains which if infrequent are often intense when they come.

315. Services such as water supply, sanitation, waste disposal and street maintenance should not only meet general standards but should be reinforced to cope with the special problems arising from the dryland environment.

316. Assistance and encouragement should be given to residents to improve conditions in and around their own homes. Insulating or screening materials might be provided or help given in the reconstruction of homes or in the establishment of gardens, shelters and shade belts. If much can be done to improve the conditions of existing settlements, control must be exercised over their further growth.

317. Proposals to expand settlements or to establish new towns should incorporate environmental impact assessments which take into account the possibilities for desertification that such activities bring with them. The assessments should include estimates of future demand for water and energy and for land presently used for other purposes, and of the consequences of these projected demands. They must include estimates of requirements for waste disposal, sanitation and other services.

318. New housing and settlements should be designed to reduce stresses imposed by the desert environment; for example, by the layout and orientation of houses, by screening, insulation and cooling devices and the provision of outdoor living areas, all planned to be compatible with local life styles. Roofs should be designed to catch and store storm water and should be adaptable to the use of solar heaters. Settlements should incorporate shelters and the control of open spaces to reduce the threat of wind, dust and moving sand. Perimeter reserves and controlled recreation areas should be included as a normal part of urban plans.

319. Research should be encouraged into architectural and living problems in desert regions. Studies should be made of the use of solar energy at various scales for domestic needs and industry, of the use of wind energy in small installations, and of other alternative energy sources which can reduce the use of wood as fuel. Local materials should be studied for their use in construction. Progress can be made in improving insulation and cooling systems, including those employing solar power. Trees and shrubs should be examined for their suitability as protection and ornaments in desert settlements. Research should continue on techniques for the desalination of water, on recycling water, and on the use of brackish water in sanitation and industry. Studies can result in improvements in subsurface water storage and the purification of water supplies. Methods of waste disposal can be more compatible with the arid environment.

320. Some control needs to be exercised over the relationships between settlements and their hinterlands. In recent decades, urban growth in and near deserts has been linked to out-migration from nearby rural areas. Since such migration will continue, it should be anticipated in plans for housing and community services. Urban development plans should form an integral part of regional development and resettlement schemes.

321. Urban development, with its demands for water, fuel, construction materials, land and labour, should not be carried out to the detriment of adjacent livelihood systems. The prior needs and rights of those systems should be protected from the environmental impact of planned settlement growth, and the siting and design of settlements should be influenced by such considerations. At the same time, rural people should be made aware of the possible advantage to them of nearby settlements, and they should be involved in planning new communities and preparing for the growth of established settlements.

IN CONCLUSION

322. This survey of desertification contains many suggestions, both explicit and implicit, for combating the process and the reclaiming of land that has suffered the ravages of degradation. Many of these suggestions appear as recommendations in the Plan of Action to Combat Desertification that will be submitted to the United Nations Conference on Desertification to be held in the late summer of 1977.

323. Some suggestions call for additional research and an improved understanding of ways in which desertification operates and of methods for combating it. This is all to the good, as is any proposal that would make the task of land maintenance and reclamation easier. But the fact is that most instances of desertification can be dealt with through knowledge and experience that are available right now. The Romans applied terracing to convert the sea coast plains of North Africa into the breadbasket of the Mediterranean. Good land use practices transformed the Great Plains of the United States and Canada into the wheat empire that they are today. The SCARP project in Pakistan has begun to reclaim one million waterlogged acres of once-productive irrigated land. In Iran and Turkmenia millions of hectares of stabilized sand dunes are now coming into production and great improvement has been achieved on other millions of hectares of protected rangeland.

324. The immense changes affecting the contemporary world have brought the problem of desertification into sharper focus than ever before, just as pressures on the sensitive dryland ecosystems are more intense than ever before. Desertification can be halted and ravaged land reclaimed in terms of what is known now. All that remains is the political will and determination to do it.

CLIMATE AND DESERTIFICATION

F. KENNETH HARE

Institute for Environmental Studies

CONTENTS

[*N.B.* Recommendations for action are shown in **bold type.**]

ACKNOWLEDGEMENTS

The senior author received help from so many quarters that it is impossible to list all the individuals who provided it. He is especially indebted to his direct UNEP and WMO advisers, namely H. Flohn, R. A. Bryson, O. Ashford, J. Adem, C. C. Wallén, J. A. Mabbutt, G. White, M. Kassas, H. Dregne and the authors of the other component reviews, R. W. Kates, D. Johnson, A. Warren and M. Anaya.

The work done by D. Henning of the University of Bonn was fundamental to the present document, in that it provided the essential quantitative classification of the dry climates in energy-based terms. Figure 2 and Appendix I describe this work.

At the University of Toronto the author was greatly helped by Tzvi Galchen (formerly of Israel), who contributed Appendix III and valuable criticism; Keith Hendrie (from Australia), who prepared Appendix II; and Emanual Oladipo (from Nigeria), who helped in an analysis of the water balance of West Africa. All three had previous dry climate experience.

Much valuable criticism was also received from the senior staff at UNEP headquarters, from Dr. A. T. Grove, from Dr. Marion Clawson, and from the author's colleagues at the Institute for Environmental Studies. Very useful material was provided by A. Gilchrist and P. Rowntree at the Meteorological Office, Bracknell, United Kingdom. Much help was also received from staff members of the National Center for Atmospheric Research, Boulder, Colorado (see Appendix III).

The typescript and illustrations were prepared by the author's colleagues M. Klausen, T. Davey, G. Matthews, J. Wilcox and J. Davie. Their work is greatly appreciated.

The author and the secretariat of the United Nations Conference on Desertification wish to express their appreciation to Encyclopaedia Britannica, Inc., for permission to reproduce Figs. 4 and 6.

ABSTRACT

1. The report seeks answers to four questions: (i) What is our present knowledge about secular or long-term shifts of climate? (ii) To what extent can climatic variations be attributed to man's actions? (iii) What are the prospects for longer-term (season and longer) forecasts? and (iv) What is the likelihood of significant human amelioration of present conditions?

2. Natural climatic variation has occurred in the dry world on many time-scales. Recent desertification is the result of the interaction of naturally-recurring drought with unwise land-use practices. There is interaction between the large-scale dynamic climate of the earth, reflecting the general circulation of atmosphere and oceans, and the physical climate of the earth's surface (which depends also on human use and misuse).

3. The present-day dry climates, in which desertification is most wide-spread, are defined by means of the Budyko–Lettau dryness ratio, defined as the ratio of the annual net radiation at the surface to the energy required to evaporate a year's normal rainfall. A world map of this ratio, prepared by D. Henning, is presented. The areas subject to desertification (dry lands under pastoral and rainfed agricultural use) lie mainly between values of 2 and 7.

4. Aridity arises from persistent, widespread atmospheric subsidence, or from more localized subsidence in the lee of mountains. It may also arise from the absence of humid airstreams and of rain-inducing disturbances. Clear skies and low humidities in most regions give the dry climates very high solar radiation incomes (averaging over 200 watts per square metre), which lead to high soil temperatures. The light colour and high reflectivity (albedo) of many dry surfaces cause large reflectional losses, and long-wave cooling is also severe. Hence net radiation incomes are relatively low – of order 80 to 90 watts per square metre.

5. The world distribution of these climates depends mainly on the subsidence associated with the sub-tropical high pressure belts, which migrate poleward in summer, and equatorward in winter. This gives a threefold structure to the arid zone – a Mediterranean fringe, with rains only in winter; a desert core (in about 20–30 degrees latitude) with little or no rain; and a tropical fringe with rains mainly in the high-sun season. A few tropical areas have two brief rainy periods. Throughout the zone rainfall variability is high. Not all sub-tropical areas are, however, dry, because the sub-tropical high pressure belt is broken up into cells that allow some regions to receive abundant rain.

6. Climatic variation is reviewed on both long and short time-scales. The world's deserts and semi-deserts are of high antiquity, though they have shifted in latitude and varied in extent during geological history. The modern phase of climate began with a major change about 10,000 B.P. (before present), when a rapid warming removed most of the continental ice-sheets. The Sahara and the Indus valley were at first moist, but severe natural desiccation took place about 4000 years B.P., since when aridity has been profound. Inland Australia was dry throughout. Less is known of other regions.

7. Recent climatic variations, such as the Sahelian drought, are natural in origin, and are not unprecedented. Statistical analysis of rainfall shows a distinct tendency for abnormal wetness or drought to persist from year to year, especially in the Sahel.

Prolonged desiccation, lasting a decade or more, is common in the records, and often ends abruptly with excessive rainfall. This persistence suggests that feedback mechanisms may be at work, whereby drought feeds drought, rain feeds rain.

8. These feedbacks are among the processes investigated by a variety of modelling exercises. General circulation models (GCMs) aim to simulate the statistics of the earth's dynamic climate, and they may also be used to test feedback hypotheses. When dry surfaces are degraded by overstocking or unwise cultivation, albedo increases. Several GCM experiments, supported by others of a more restricted kind, suggest that this increase in albedo should also increase subsidence, and hence aridity. The modelling also suggests connections between drought over the Sahel and anomalies of sea surface temperature over the Atlantic. Other teleconnections for which there is modelling support link sea surface temperature anomalies over the Arabian Sea to rainfall over dry north-west India and Pakistan.

9. Natural living communities in these latitudes are well adapted to drought, and in general protect the soil against wind and water erosion. As the desert is approached, however, the vegetation becomes more open-structured, and areas of bare soil become extensive. Deflation of fine materials from these surfaces has been in progress for an immense period. Bottom deposits from the Atlantic, for example, contain a record of Saharan and Sahelian losses far back into the Pleistocene. Desertification accelerates this natural deflation.

10. Several avenues leading towards solutions are suggested. A realistic statistical appraisal of drought incidence, or of prolonged adverse or favourable weather, should be available to decision-makers in all affected countries, and The World Meteorological Association (WMO) should take the lead in recommending how this should be done. Such an appraisal requires skill and experience, since it is important to derive the statistics from the right population, and to understand the limitations of learning from recent climatic history.

11. Rainfall enhancement by carefully controlled cloud-seeding programmes, backed up by the use of satellite imagery in tracing the path of rain-storms, offers some prospect of success, chiefly where rainfall is already sufficient for cultivation. The proposed West African WMO-sponsored experiment should be strongly supported, but there is no guarantee of success.

12. Flooding (or irrigation) of desert basins is not seen as an effective way of modifying the large-scale climate of dry regions. Calculations for the Chad basin suggest that the diversion of 5000 cubic metres per second into an irrigation scheme in the vicinity of the present lake floor might increase rainfall in a belt over West Africa by about 7 per cent. Numerical modelling of other lake schemes suggest an even smaller potential yield.

13. Controlled land use, so as to ameliorate the surface microclimates, offers the most certain return. The maintenance of an adequate plant cover is the key to such control. The capacity of a region to survive a drought, and then to recover its productivity, depends on its soil's capacity to retain nutrients, organic substances, and a high infiltration capacity. A favourable microclimate also requires these qualities.

14. Solutions to the problem demand further research into climate-desertification links. It is desirable, for example, that the monsoon experiment of the Global Atmospheric Research Programme (sponsored by WMO and the International Council for Scientific Unions) be sufficient in scope as to cover the entire African–Asian monsoonal province.

In the modelling area generally it is desirable that, in addition to general circulation modelling, models of the climatonomic variety devised by Lettau, plus modelling that tends to bridge the gap between climatology, ecosystem simulation and management techniques, should be encouraged.

15. The conclusions of the review take the form of answers to the four questions posed at the outset:

(a) *What is the present state of knowledge about secular or long-term shifts of climate?*

16. Rainfall in the dry regions is very variable in time and space. There is a weak 2–3-year rhythm in some areas, and a suggestion of a 10–20 or 30-year recurrence interval in others. These weak modes are overridden by apparently non-periodic variations of large amplitude. Within the life-time of an individual there may be several droughts up to 4 years in length, and once or twice they may last longer, as did the Sahelian drought. The same is true of wet periods. These variations are of natural origin, but may be accentuated by feedback mechanisms. Historic and prehistoric records show many much longer fluctuations. In several regions there is evidence of drought or humid phases lasting one or two centuries. The present mainly arid phase in Africa, north-west India and Pakistan is four millennia old, whereas in Australia it is much older.

(b) *To what extent can climatic changes be attributed to man's influences?*

17. Overstocking or unwise cultivation of dry land surfaces during periods of drought is a direct cause of deteriorating surface microclimates, and of increases in surface albedo. Modelling experiments suggest that large increases in albedo lead to further decreases in rainfall, and hence further stress on livestock and crops. If this is so widespread as to create significant regional increases of albedo – for example, across the entire Sahel – such feedbacks may modify the large-scale climate. On this scale it is now believed that the addition of carbon dioxide (from fossil fuels) and halocarbons to the atmosphere must tend to raise surface temperatures, and hence to modify the position of the sub-tropical high pressure belt. The preponderance of opinion is that the atmosphere's increasing particle load, much of it of human origin, is of less importance.

(c) *What are the prospects for longer-term (seasonal and longer) forecasts?*

18. There is no present method available for such forecasts, and progress towards it will be slow. This will result from further development of general circulation models. Some progress may also be derived from the discovery of meaningful periodicities, and in particular teleconnections. Significant longitudinal correlations of rainfall, streamflow, pressure and sea surface temperature are already known from tropical latitudes.

(d) *What is the likelihood of significant human amelioration of present conditions?*

19. Better land-use methods will improve local microclimates, and the proposed green belts will be of use here. Planned experiments in rainfall-enhancement also offer some hope in the moister parts of the dry belt. Wider use of satellite imagery will also aid maximum use of each year's rainfall. Schemes for artificial lakes, or major irrigation systems, are unlikely to have significant effects on large-scale climate. Realistic use of existing statistics will improve decision-making, even if the advice given can only be crude.

I. INTRODUCTION

1. There must always have been deserts on the earth, for they are due to atmospheric subsidence, and on an unequally heated, rotating earth such subsidence must always occur in sub-tropical latitudes (Hare, 1961). Such climates occur at sea as well as over land, and it is not unrealistic to talk of marine deserts. But it is on land that the stark facts of desert climate strike home, and impinge most harshly on man. It is there that lack of rain and excess of heat can render the surface largely or completely empty of life.

2. The geological record contains many sedimentary rocks that are unmistakably of desert origin. But the desert zones have shifted significantly during the long history of the earth, and have expanded and contracted many times. Within the past 4000 years we have historical evidence that such changes have affected North Africa, the Middle East and India–Pakistan. The older changes, on the geological time-scale, were clearly of natural origin. The later changes may well have been influenced or exacerbated by man.

3. The present United Nations Conference concerns itself with the *process* of desertification, rather than with the deserts themselves. It poses the questions: why have the world's deserts appeared to expand over the centuries of human history, and is this expansion accelerating? Is the entire process natural, or is man's work at fault? Can we reverse the decay? If so, how?

4. In this study the objective is to try to answer these questions from the standpoint of the climatologist. It is not easy to disentangle natural from man-made climatic change. A man-made change is physically real, and may be irreversible. Natural systems are typically self-regulating, and hence stable; but they are not always so. They may contain potential instabilities that can be triggered by man's actions.

5. We shall look for answers to the four questions listed below.

(a) What is the present state of knowledge about secular or long-term shifts of climate?
(b) To what extent can climatic variations be attributed to man's actions?
(c) What are the prospects for longer-term (seasonal and longer) forecasts?
(d) What is the likelihood of significant human amelioration of present conditions?

6. In general the present study confirms the widely held view that desertification is at present a man-made effect, arising from heavy stocking, unwise cultivation of semi-arid land areas during periods of drought, or the collapse of established conservation systems such as terraced cultivation. The major stress in this review lies on the specific role of climate itself, and on its interactions with the use of land and renewable resources by mankind.

7. The analysis of climate is a quantitative and statistical discipline. Much of it depends on physical laws that are expressible in differential equations capable of numerical solution. The large-scale climate of the earth is governed mainly by the global distribution of radiative energy, by the inequalities of land and sea, and by the general circulation of the atmosphere and oceans. Sometimes we talk of the *dynamic climate* of the earth when we are focusing on this large-scale, atmospheric behaviour, because the motion of the atmosphere is so all-important. The world's drier climates, desert, semi-arid and sub-humid alike, are dynamically caused, and hence need to be studied through the general circulation of the atmosphere.

8. But there is another aspect of climate that is much closer to the phenomenon of desertification. This is the *physical climate* of the earth's surface, by which we mean the system of exchanges and equilibria that link atmosphere to water surfaces, soil and biota. In principle this physical climate is not distinct from the dynamics of the atmospheric circulation; in practice, however, its study involves a different set of working methods, and particularly choice of scale. The fundamental causes of dry climates are primarily dynamical, but their details depend on the intricate web of physical and biological processes at work near the earth's surface.

9. The physical climate of a locality is transformed when man alters the nature of the surface. Such alterations may also affect the dynamic climate by feedback processes acting on regional, continental or even planetary scales. But the primary impact of faulty land use is on physical climate on the local scale, which we usually call the *microclimate*. Hence an understanding of the desertification process depends on our ability to weigh the influence of large-scale dynamical processes in the atmosphere (governing the *macro-climate*) against the intricately local effects of human interference with physical processes working at the earth's surface (the microclimate).

10. Obviously the search for this understanding brings the climatologist into the closest possible contact with the ecologist, for ecological change is also microclimatic change – i.e. change in the associated patchwork of small-scale climate. If one degrades a natural ecosystem by over-grazing or cultivation, one simultaneously alters the surface albedo (reflectivity)[1], roughness and conductivity for heat, water and trace gases. Ecosystem change is inevitably climatic change. Between a bare soil surface and complete vegetation cover, the thermal microclimate may differ by an amount equivalent to many degrees of latitude in macroclimate. Desertification can be seen, moreover, as a reduction in potential productivity, i.e. in the capacity of the biota to convert solar energy and rainfall to their own purposes. Unfortunately the border territory between the physical climatologist and the ecologist has been little explored in dry areas: but it holds the key to an understanding of the desertification process.

11. Another vital interface in studies of this kind brings into contact climatology, archaeology and history. The spread of desert climates in North Africa, the Middle East and northern India–Pakistan has been closely involved with the prehistory and history of man. Important climatological lessons can be derived from the record of cultural innovation and decay. Climatology can in reverse shed a substantial light on cultural change.

12. Desertification is a process that cannot readily be tied down to a specific region of the earth. Around the nickel smelters of Sudbury in the humid climate of central Ontario, Canada, is a desert landscape created by acid rainfall due to the long-term emission of sulphur dioxide from the smelters. The dust-bowl years of the 1930's in the Great Plains of North America removed the topsoil from vast areas of a mid-latitude continental surface. These were episodes of desertification in developed countries – and they have not yet been fully contained. But today it is along the desert margin of Africa and Asia that the process seems to be working maximum damage.

[1] The reflectivity of a surface is the reflected fraction of incident radiation of a specific wavelength. The total reflectivity of a surface with respect to solar radiation of all wavelengths is called its *albedo*.

II. THE WORLD'S DRY CLIMATES

13. In this section we shall discuss the definition and location of the dry climates, together with their characteristics and mode of origin. We shall also discuss briefly their zonality and seasonality.

A. THE DRY CLIMATES: DEFINITION AND LOCATION

14. The best-known and most widely accepted classification of the dry climates is that of Meigs (1961), prepared in connection with the arid zone programme of the United Nations Educational, Scientific and Cultural Organization (UNESCO). This classification attempts to relate available rainfall to potential evapotranspiration[2] by means of the moisture index (I_m) of Thornthwaite (1948). In the present review the index preferred is Budyko's (1958) radiational index of dryness, D, also called the *dryness* ratio by Lettau (1969). The two indices are related in a simple fashion summarized in Appendix I. Isolines of one index are roughly isolines of the other in tropical and sub-tropical latitudes.

15. The dryness ratio is

$$D = R/LP \tag{1}$$

where R = mean annual net radiation (i.e. radiation balance),
P = mean annual precipitation,
L = latent heat of vaporization for water (see also Fig. 1 for details).

16. In the warm, dry climates energy supply is excessive and rainfall deficient. Hence we might expect an index of aridity to depend on their relative magnitudes. This is the objective of the dryness ratio defined by equation (1). In this equation the net radiation is the global solar radiation, *plus* atmospheric radiation received at the surface, *less* solar radiation reflected back, and *less* radiation emitted by the surface (see Fig. 1). It is given the value it attains under *actual* conditions, rather than under the conditions of a hypothetically moistened surface, as originally proposed by Budyko. The dryness ratio at a given site indicates the number of times the net radiative energy income of the surface could evaporate the mean annual precipitation.

17. In Budyko's studies (1958, 1974) the usage is as follows:

Dryness ratio	Vegetational response
> 3.4	desert
2.3–3.4	semi-desert
1.1–2.3	steppe or savanna

[2] I.e. evapotranspiration off a moist surface.

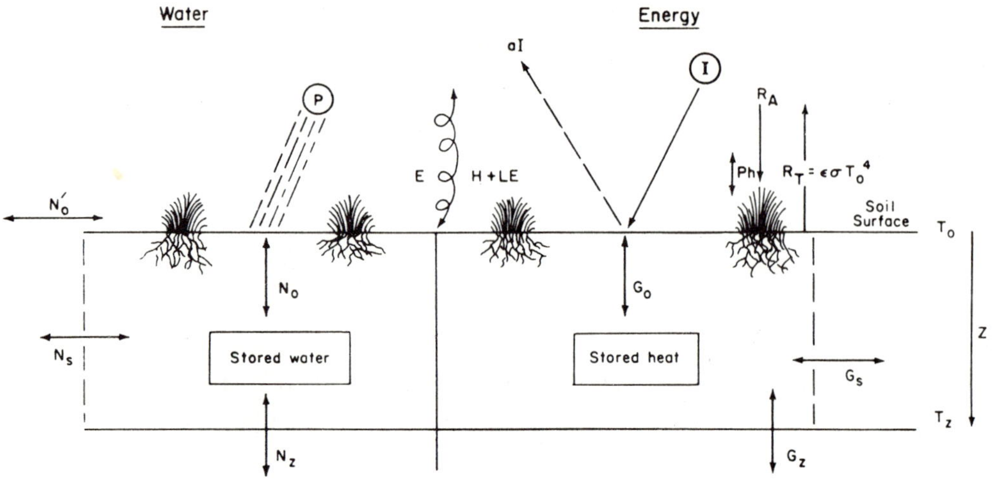

Fig. 1. Water exchanges are on the left, energy on the right. The diagram refers to a partly vegetated surface whose temperature (for radiative emission and convective forcing) is T_0. Most natural surfaces in dry climates have extensive areas of bar soil, as shown.

The water exchanges are forced by precipitation, P, a very intermittent input. The water reaching the surface may percolate, N_0, or run-off at the surface, N_0'. The soil layer, of depth z, stores water – typically or order 10 cm of precipitation when wet. If storage is full, the remaining rainfall may outflow laterally, N_S, or percolate to ground water. With high water tables on sloping sites, N_0, N_S and N_Z may all reverse, so that there is run-on. Total water surplus $N = N_0' + N_0 + Ns,z$. If there is no change in storage, $P = N + E$, where E is the evapotranspiration.

The energy exchanges are forced by global solar radiation, I, direct (shadow-casting) plus diffuse, a fraction (a, the albedo) of which is reflected, aI. The surface temperature is mainly a function of the absorbed solar radiation $I(1 - a)$. There is diffuse infrared radiation received from the atmosphere, R_A, and infrared is emitted from plants and soil, $R_T = \epsilon\sigma T_0{}^4$, where ϵ is the emissivity (usually > 0.9), σ is the Stéfan-Boltzmann constant, and T_0 is surface temperature. $-R_T + R_A$ is the net longwave heating, generally negative. The net radiation is $R = I(1 - a) - R_T + R_A$. A small part of R, G_0, is conducted to the soil, where it may be stored. If level z is the depth at which there is no annual temperature cycle, $G_Z = 0$, and G_S is negligible. Hence G_0 virtually vanishes over the year. The remaining heat, $R - G_0$, feeds convective fluxes, H of heat, and LE of latent heat. If $T_0 < T_A$ (air temperature), H is negative, though small, and LE may also be negative (dew formation). In this case the convective fluxes force the net radiation, rather than the reverse – the typical nocturnal condition. *Ph* is the net photosynthesis, generally 2 orders of magnitude smaller than R (see component review on ecological change). If there is no change in storage, and no net photosynthesis, $R = I(1 - a) - R_T + R_A - G_0 = H + LE$. Note that *a, the albedo, is a function of the state of the vegetation and soil*, tending to rise from values near 0.17 in fresh green savanna to ~ 0.35 in sandy deserts (Oguntoyinbo, 1974). Degradation of vegetation in general raises the albedo, as does drying the surface. Absorbed solar radiation, and hence surface temperature T_0, are functions of a, as are net longwave cooling ($R_A - R_T$), the soil heat flux G_0, the convective fluxes of heat and latent heat, $H + LE$, and net photosynthesis, *Ph*. Hence changes in albedo drastically alter the entire energy and water balances.

Increases in soil compaction due to overstocking, use of vehicles, drought contraction or pan formation decrease the infiltration rate, and hence increase N_0' at the expense of N_0, so increasing the chance of sheet erosion and gullying. They also alter the thermal conductivity, and hence G_0. These processes are discussed in the component review on ecological change.

T_0 is low by day when the soil is moist, and higher when it is dry – by 20° to 30°C in tropical conditions. With moist soil and low surface temperature, R is large, and the Bowen ratio H/LE is very low – of order 0.10. With dry soil and high surface temperature, R is lower, and the Bowen ratio rises to very high values.

It is clear that Budyko calls "desert" many arid ecosystems with significant biomass. In this study we shall regard as deserts all land areas from which perennial life has been essentially excluded by lack of rainfall. This corresponds in our experience to dryness ratios of at least 10, and resembles the usage of Le Houérou (1970). Settlement of such areas is essentially confined to riverine or ground-water oases. We shall concentrate most attention on the land areas having dryness ratios between 2 and 7, because it is in this broad zone that desertification phenomena seem most widespread.

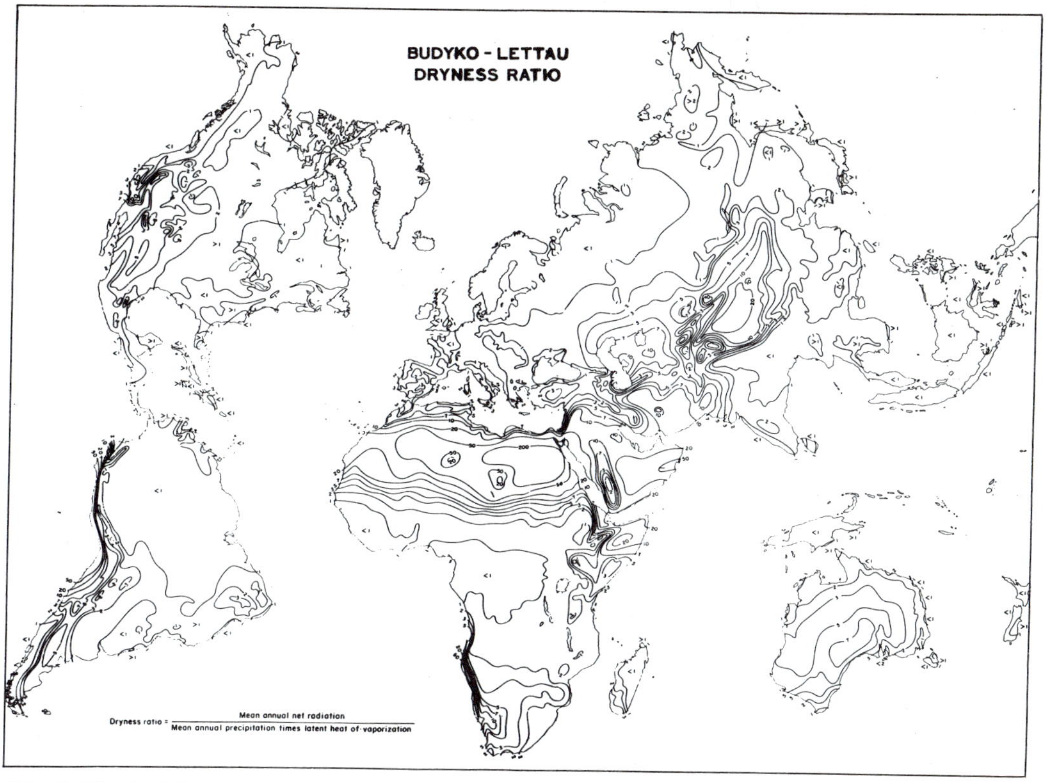

Fig. 2. The Budyko–Lettau dryness ratio over the land areas of the earth (reduced and simplified from original computations by Henning, see Appendix I). The ratio shows the number of times the mean annual net radiation income could evaporate the local mean annual rainfall. All detail has been omitted in humid areas, i.e. where the ratio is less than unity.

18. Figure 2 gives the world distribution of the dryness ratio, as calculated by Henning (see Appendix I). It is clear that the arid and semi-arid regions of the earth ($D > 2$) include the following:

(a) an area in Mexico north of 16°N, merging with the dry plateaus, basins and plains of the western U.S. The dry cores are in inland California, Arizona and adjacent parts of Mexico;

(b) a narrow strip of coastal South America from the Equator to 35°S, west of the Andean slopes, plus a broader strip east of the Andes from 18°S to southern Patagonia, together with small areas in eastern Brazil, Colombia and Venezuela;

(c) the vast North African–Asian belt extending from the Atlantic African coast north of 16°N eastwards to the Nile valley, the Arabian peninsula and thence into the

Asian mainland. The belt includes the Caspian–Aral sea plains, and most of the plateau and basins of Sinkiang and Mongolia, plus the lower Indus–Rajasthan belt of India–Pakistan. Dry terrain extends to the northern districts of China;

(d) a smaller belt in southern Africa south of 6°S, and north of Cape Province and Natal. This includes not only the Kalahari desert, but much of the plateaus of the interior;

(e) the entire Australian continent, with the exception of south-western Western Australia and South Australia, and the hilly eastern and northern areas from Victoria and Tasmania to Queensland and the great tropical peninsulas.

19. In all five of these major regions the process of desertification has been at work. But it has been the plight of countries in region (c), especially in Africa north of the Sahara, that has focused world attention on the problem. Moreover, it has been in tropical and sub-tropical latitudes where the problem has been most acute, and it is to those belts that most attention has been given in the review.

B. ORIGIN AND CHARACTERISTICS OF DRY CLIMATES

20. Significant rainfall is always caused by the uplift of humid air, and the absence of rainfall is good evidence of the lack of such uplift. Even in humid airstreams rain does not fall unless something disturbs the stability enough to cause uplift. There are many parts of the earth where prolonged drought occurs at times of high humidities. The most remarkable example is the extensive semi-arid and desert area of north-west India and Pakistan.

21. In most parts of the dry world, however, a combination of factors keeps the atmosphere stable for long periods, and hence rainless. The atmosphere is stable (i.e. discourages upward motion) when temperature decreases only slowly with height, or even increases (inversion of temperature). Clearly stability is favoured if the moist lower layers of the atmosphere are warmed aloft or cooled below. This happens when the air moves downwards, or subsides, which leads to dynamical warming aloft. It also happens if warm air passes over cold surfaces. In low latitudes this happens whenever warm airstreams pass over cold ocean surfaces, as, for example, off the west coasts of north and southern Africa, off northern Chile and Peru, and off southern California, the so-called *cold-water coasts.*

22. In more detail we can recognize the following main causes of aridity:

(i) *Widespread, persistent atmospheric subsidence*, which results from the general circulation of the atmosphere. The present-day circulation tends to create such subsidence in the sub-tropical latitudes of both hemispheres (Fig. 3). The major desert belts of the earth lie beneath these zones of subsidence – the Sonoran desert of Mexico and the south-western United States; the Saharan–southwestern Asian belt; the Kalahari of south-west Africa; and the Australian desert.

(ii) *Localized subsidence* induced by mountain barriers or other special physiographic features. West of the Andes, for example, there is almost continuous subsidence in a narrow strip over coastal north Chile and Peru (Dirección Meteorológica de Chile, 1976) which leads to the remarkable Atacama desert (Miller, 1976; Johnson, 1976). But most such areas lie in the lee of mountains across the westerly belts, and are hence in mid-latitudes. The dryness is caused by the descent of the westerly currents east of the

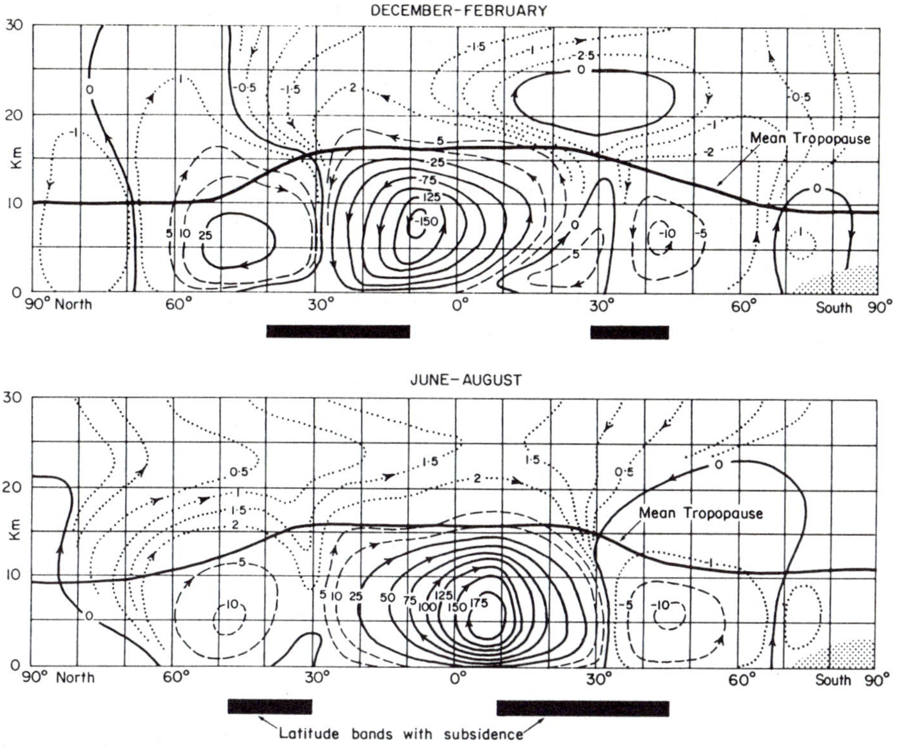

Fig. 3. Computed atmospheric motion in a longitudinally averaged meridional plane (mainly after J. F. Louis, 1974). The diagram shows for the entire earth the longitudinally averaged south–north and upward–downward motions. The latitudes of average subsidence (descending motion) are indicated by the heavy horizontal bars. Such motion tends to inhibit rainfall and cloudiness.

summits. The extensive dry areas of western North America, western and southern Argentina (Prohaska, 1976) and parts of inner Asia are mainly caused in this way.

(iii) *Absence of rain-inducing disturbances* causes dry weather even in areas of moist air. The dry grasslands of west Texas and Oklahoma may be rainless for weeks while being traversed by a broad, humid airstream from the Gulf of Mexico. Most rain is caused by the travel of organized disturbances across a region – i.e. systems that involve actual uplift of humid air. Thus the aridity of the Mediterranean summer, though in part due to subsidence, arises mainly from the absence of the cyclonic disturbances that bring the rains of winter. There is plenty of water in the air, but nothing to bring it down as rain.

(iv) *Absence of humid airstreams.* The relationship between the water available for precipitation (*precipitable water*) and the precipitation that actually falls is by no means simple. As we have just seen, dry weather may be prolonged in areas of high humidity. But abundant rainfall occurs only in regions that have access to humid airstreams, which are necessary though not sufficient for it. Some inner-continental regions are quite remote from such sources. Thus the desert and steppe surfaces of central Asia are cut off from the moist monsoonal currents of the south by the Himalayas and the Tibetan plateau. The dry season of West Africa arises from the almost unbroken dominance of the weather by dry Saharan airstreams, which are partly of mid-latitude origin.

23. These four controls are interdependent, but their relative effect depends on season and locality. We can broadly distinguish between:

(a) almost continuously dry climates, leading to desert surface conditions, in which there is no season of appreciable rainfall;
(b) semi-arid or sub-humid climates with a short wet season of varying intensity in which humid airstreams and rain-inducing disturbances penetrate; and
(c) the rare sub-humid areas in which rainfall is infrequent, but not confined to a special season.

24. We can further distinguish under (b) between winter-dry climates of the monsoonal or Sahelian–Sudanian variety, in which rain falls during the months of highest sun; and Mediterranean-style climates in which rain falls mainly from winter-season disturbances. The latter are widespread in the Middle East and North Africa, and occur also in California, Chile, South Africa, Western Australia and South Australia.

25. For reasons that will appear later the major impact of desertification is being felt today primarily in the winter-dry tropical climates, which are most widespread in Africa and monsoon Asia. These are not only areas of high interannual climatic variability, they are also among the oldest habitats of mankind. It is one of the ironies of human history that climates that offered stimulus to early man have proved inimical in modern times.

26. Throughout the tropical and sub-tropical continental dry climates the skies tend to be clear much of the time. Even during the rainy season bright sunshine amounts usually average over 50 per cent of the possible. Prolonged overcast conditions are rare. Accordingly very high solar radiation incomes are achieved almost everywhere. Global solar radiation (i.e. direct sunlight plus diffuse sunlight from sky or clouds) averages over 200 watts per square metre (W m^{-2}) in most places, and reaches 275 W m^{-2} in the heart of the Sahara. These are the highest solar energy incomes on earth.

27. The high soil temperatures induced by bright sunshine and the dryness of the atmosphere combine, however, to produce strong long-wave radiative cooling of the surface. The latter tends to be light in colour, and to have a high albedo (between 0.20 and 0.35 over typical dry surfaces); hence between a fifth and a third of the solar radiation is lost by reflection. The net radiation (see Fig. 1), which is the sum of the various radiative heating and cooling processes, is hence surprisingly low (Tetzlaff, 1974). Henning (1970, and personal communication) has recently completed calculations of the annual average net radiation for the entire earth. He finds values of less than 80 W m^{-2} for the central Sahara and Kalahari, increasing outwards to near 90 W m^{-2} in the Sahel and the corresponding margin of the Kalahari. He gives similar figures for north-west India, Pakistan and much of the Middle East, but slightly higher values for Australia, Mexico and southern Saudi Arabia.

C. NATURAL CLIMATIC ZONALITY AND SEASONALITY

28. Later in this study we give detailed attention to the natural variability of dry climates on a variety of time-scales. But we must immediately stress the seasonal and interannual changes that are natural to such climates, and which have to be coped with by the economies of indigenous peoples.

29. Most dry climates have dry and wet seasons, which reproduce themselves annually (though by no means identically). Thus in dry north-west India and Pakistan rainfall is

mainly confined to the months July–September, when the warm, humid south-west monsoon covers the region. The same is true of Sahelian and Sudanian North Africa. Rainfall in the semi-arid or sub-humid regions of northern, eastern and central Australia falls mainly in the equivalent southern hemisphere period, i.e. January–March. In the northern part of the Argentinian dry belt it comes earlier in the summer. By contrast, rainfall in Africa north of the Sahara, across much of the Middle East, in north-east Brazil and in scattered areas of South Africa and southern Australia, is almost restricted to the local winter season.

30. Such seasonal contrasts are of dynamical origin. They are caused by changes in the general circulation of the atmosphere. Through much of the dry world, in fact, certain months see the penetration of humid air-streams, often accompanied by disturbances favouring the outbreak of showers and thunderstorms. In the remainder of the year subsidence, stability and (as a rule) lower humidities curtail or prevent rainfall. It is airstream changes, plus variations in the tendency towards subsidence and stability, that produce the familiar dry and wet seasons of the drier tropical and warm temperate climates.

31. In most of these regions the key to the climate lies in the behaviour and position of the sub-tropical high pressure belt in the troposphere (i.e. the basal 15 km of the atmosphere). We show in Fig. 4 the world topography of the 300-millibar pressure surface (approximately 9 km above sea-level). The wind on such charts blows parallel to the contours of equal geopotential (roughly equal to the height above sea-level in the units chosen) such that lower geopotential is to the left of the wind-vector in the northern hemisphere, and to the right in the southern. Hence the contours are roughly streamlines of the average wind. Figure 5 shows the shift of the axis of the sub-tropical high pressure belt between January and July in both hemispheres. It clearly shifts poleward in the summer hemisphere, and equatorward in the winter hemisphere. This suggests that the tropical and sub-tropical zones should have three idealized régimes:

(a) A central régime, centered on the mean annual latitude of the sub-tropical high pressure belt, which is traversed twice by the highest pressure, and which is likely in consequence to be dominated throughout the year by the associated subsidence, clear skies and rainlessness. This central régime includes the true sub-tropical desert regions.
(b) A poleward margin that is covered by high pressure and subsidence in the summer season only, but by circumpolar westerlies in winter, with their travelling disturbance. This is the régime with winter rainfall, or Mediterranean-type conditions; and
(c) An equatorward margin dominated by subsidence and high pressure only in winter, and characteristically invaded by easterlies in the summer months (i.e. the high sun season). This is the summer rain type typical of savanna lands.

32. In practice this symmetrical structure is complicated by two circumstances. The first is that the sub-tropical high pressure belt is markedly cellular at low altitudes, and the second is the development of the great monsoonal anomalies of Africa, Asia and (to a minor extent) North America.

33. The lower tropospheric sub-tropical high pressure belt (below about 4 km) is shifted slightly poleward from the 300-millibar locus of Fig. 4, and is broken up into separate cells. These permit the systematic migration of equatorial airstreams poleward

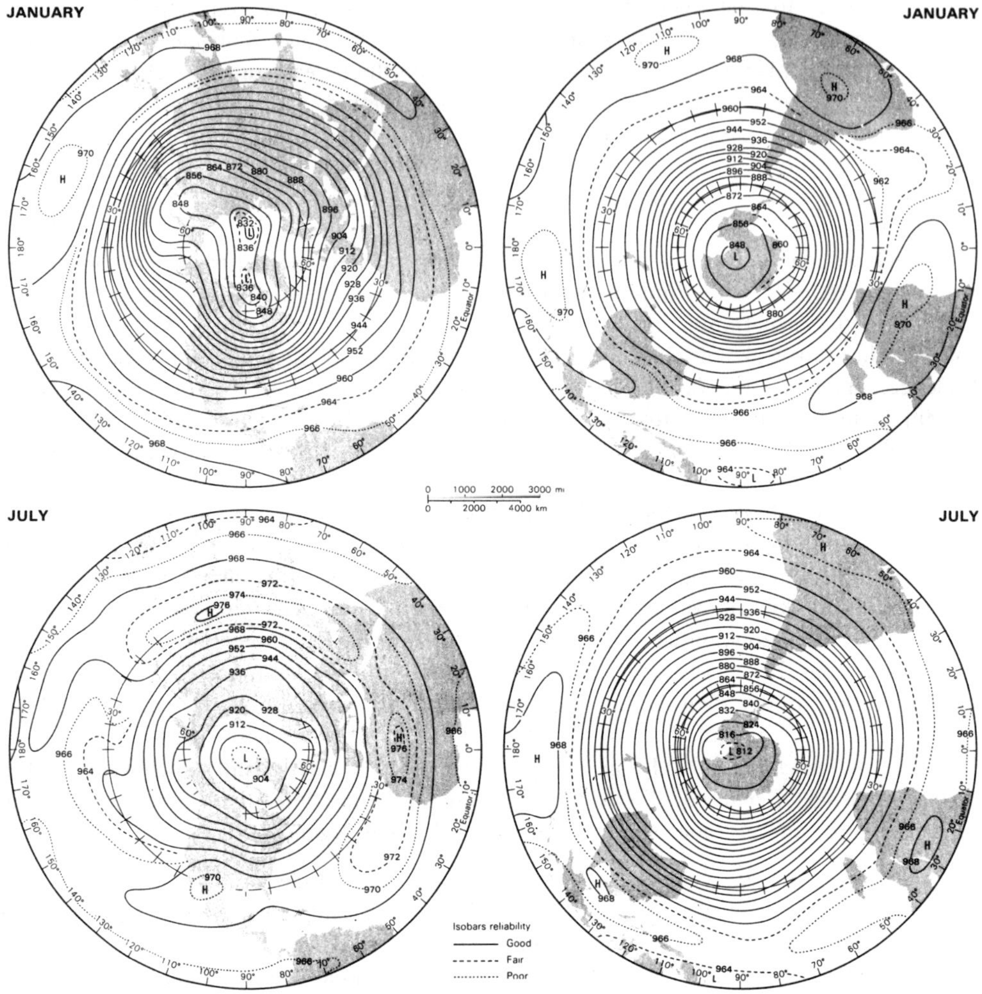

Fig. 4. Contours (tens of geopotential metres) of the 300 mb pressure surface (near 9 km above sea-level) for January and July, showing sub-tropical high pressure belts and circumpolar westerly vortices at this level (after Crutcher, Jenne, Van Loon and Taljaard 1969). Average wind flows parallel to the contours at a speed proportional to the gradient, with low pressure to the left of the wind in the northern hemisphere, and to the right in the southern (Courtesy Encyclopedia Britannica Inc.).

round the western ends of the cells, and of mid-latitude airstreams equatorward round the eastern ends. In most longitudes of the southern hemisphere the cells migrate eastwards, so that the effect produces alternations of northerly and southerly airstreams. But in the northern hemisphere the cells tend to remain anchored over the oceans, thus providing certain preferred longitudes for persistent poleward or equatorward airstream movement.

34. Thus in the summer season the persistence of high pressure between the Azores and Bermuda (Fig. 6) produces almost continuous northerly or north-easterly airstream over the westernmost Sahara, cool and moist relative to the air over the interior. A similar current flows southwards over coastal California and Mexico. In the southern

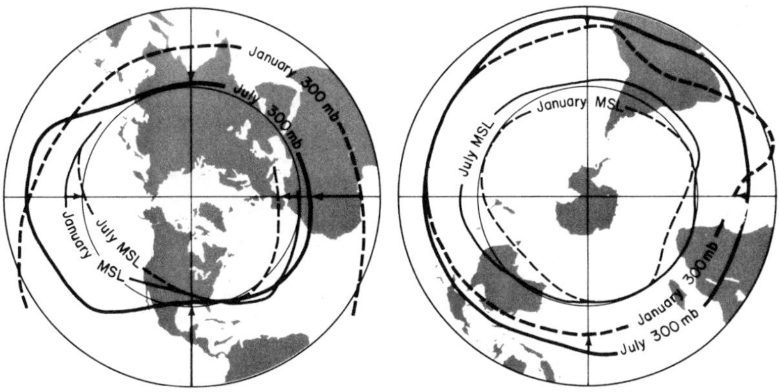

Fig. 5. January to July shifts of the axis of the sub-tropical high pressure belt at 300 mb (roughly 9 km above sea-level) and at sea-level. Note that the belt generally shifts poleward in the summer hemisphere and equatorward in the winter. The axes of high pressure can be compared with the latitudes of average subsidence on Fig. 3.

hemisphere cool southerlies occur along the west coasts of Chile, Peru and southern Africa. All these currents are strongly diffluent[3] and divergent, and hence encourage subsidence. All have the effect of intensifying the aridity but of mitigating the summer heat.

35. The sub-tropical deserts and semi-deserts do not, however, form a continuous round-the-world ring in either hemisphere. The North Atlantic and Pacific cells of high pressure promote the north-westward and northward efflux of warm humid air from the tropical oceans. This effectively banishes the aridity usual to the latitudes from South-east Asia, south-east U.S.A. and the Caribbean countries. Similar moist regions occur on eastern sub-tropical coasts of South America, southern Africa and Australia.

36. The monsoonal anomalies are characteristic of three main areas: (i) west and central Africa, (ii) India–Pakistan and much of South-east Asia, and (iii) tropical Australia. In each area the trade wind circulation typical of tropical latitudes is replaced for part of the year by very warm, moist westerlies – the south-west monsoons of West Africa and India–Pakistan in June–September, and the north-west monsoon of northern Australia in December–March. These monsoonal currents bring a season of humid heat with scattered thunderstorms and shower activity over the dry belts, though the falls are usually infrequent and small in Rajasthan, the Indus lowlands and central Australia.

37. All the régimes just discussed have an unusually high interannual variability in rainfall and cloudiness, and it is important to distinguish between interseasonal and interannual changes. There are some desert areas where rain has never fallen since records began, and the area of the world that has experienced at least one wholly rainless year is large, extending well into the vegetated semi-arid surrounds.

38. The causes of this variation from year to year, as for the interseasonal changes, are mainly dynamical. There may, for example, be a persistent shift in the latitude of the

[3] *Diffluence* occurs when the streamlines of the wind become wider apart, with diminished speed. *Divergence* occurs when the flow is such that there is a net export of mass per unit area of the airstream. Confluence and convergence are the equivalent negative states.

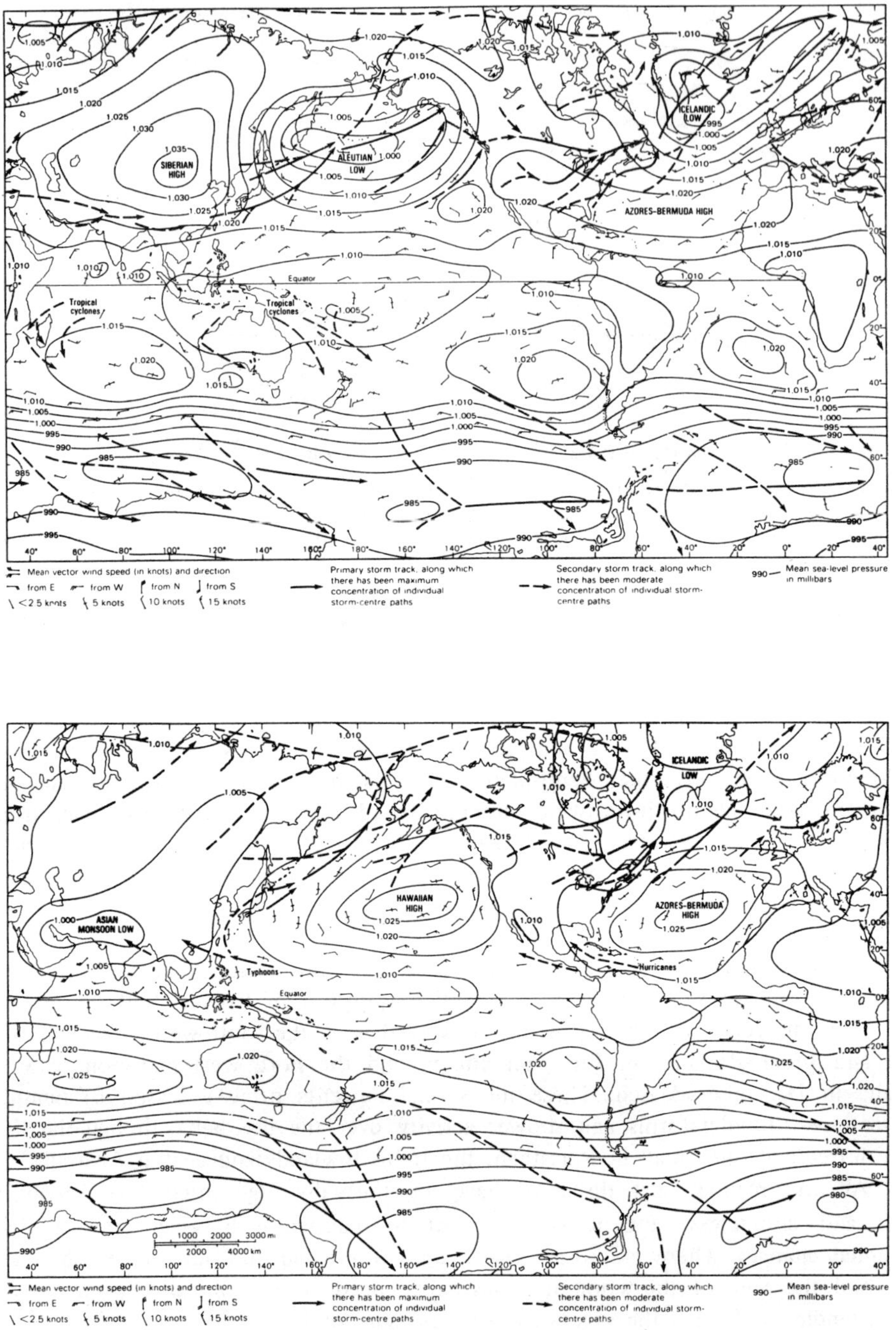

Fig. 6. Mean sea-level pressure charts for January and July (after Crutcher and Davis, 1969). Note the cellular character of sub-tropical high pressure belt at this low level, especially in the northern hemisphere (Courtesy Encyclopedia Britannica Inc.).

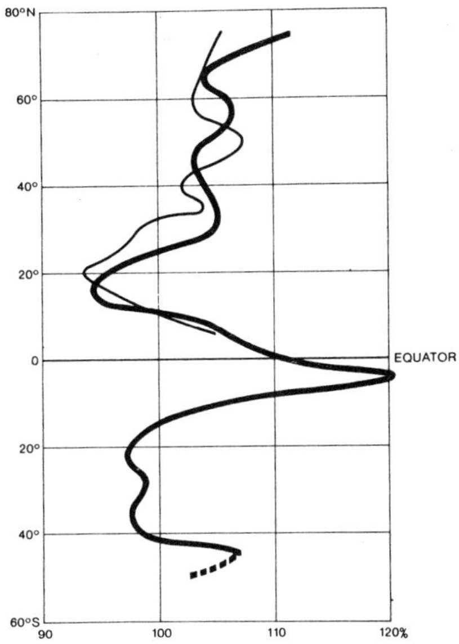

Fig. 7. Deviations from 1931–60 normal values of 1970–2 precipitation, averaged over longitude, after Lamb (1976). Thin curve shows the 1960–9 deviation from the same normals. There is a strong suggestion of a southward shift of rainfall belts during the 1970–2 period.

sub-tropical high pressure belt, and of the airstreams that flank it. As Lamb (1976) points out, the 1970–2 period shows world-wide precipitation anomalies (Fig. 7) that strongly suggest such a shift in both hemispheres. To quote him:

> "The equatorial rains performed [in 1970–2] a more restricted seasonal migration north and south, and their dominant position – indicated by a very sharp positive anomaly of rainfall at about 5° S – seems to have become further south than before. . . . Owing to the restricted seasonal migration of the equatorial rain system, places in latitude 10–20°N and 12–20°S have experienced droughts or monsoon failures, repeated or continued over a number of years."

39. Other rainfall anomalies in the sub-tropics, however, can be traced to changes in the east–west cellular structure of the sub-tropical high pressure belts, and hence to altered airstream movement. The excessive rainfalls over inland Australia since 1972 have been due in part to a southward shift of the sub-tropical high pressure belt in summer, but also to unusually deep penetration of the continent by moist monsoon airstreams from the north – a characteristic cellular effect discussed at greater length below. Paradoxically the rains have recently (1976) been accompanied in parts of south-eastern Australia by intense drought.

40. It has been shown that rainfall anomalies over the Caribbean, tropical Atlantic and tropical North Africa are interconnected (Hastenrath, 1976), which again suggests cellular variations along the sub-tropical high pressure belt. Annual precipitation over the Central American–Caribbean area is significantly and positively correlated with rainfall in the Sahel, with the discharge of several major African rivers, including the Senegal and Nile, and with the level of Lake Chad. In fact it is in such correlations over longitude that the Sahelian rainfall shows the most significant links with other, distant climatic régimes.

41. In summary, both the characteristic distribution of the continental rainfall belts and the anomalies to which the dry climates are subject are controlled mainly by the atmospheric general circulation, and only to a minor extent by local circumstances. It follows that the root causes of drought or excessive rainfall are largely global, and that the problem of predicting them is also global. International effort on a large scale, already a tradition in meteorology, is needed as a basis for such predictions. But there is much that can be done, both regionally and nationally, to sharpen the focus of such enterprises as the Global Atmospheric Research Programme (GARP).

III. OBSERVED VARIABILITY OF THE DRY CLIMATES

42. To put recent climatic fluctuations in perspective, one must look at past variations on many time-scales. Essentially we are concerned in this review with the present-day climate, and with the kind of iterations or fluctuations that can occur within a few years, or at most with a decade or two. Much can be learned, however, from a study of longer periods – even out to the scale of tens of thousands of years. In this section we look at these long-term fluctuations first, and then approach the 1970s, with their large anomalies of climate. A more detailed chronology of variations in the main dry regions of the earth is given in Appendix II.

43. Weather varies rapidly from place to place, especially in the dry tropical countries, where thunderstorm rainfall is very patchy. But climate and its fluctuations are spatially coherent. Drought periods or excessively wet episodes typically affect large contiguous areas simultaneously. Moreover, there may be statistically significant spatial correlations between such regional anomalies (Pittock, 1975, 1976; Trenberth, 1976). This fact may be important for economic and political action against the consequences of drought or other climatic hazards; the geographical extent of extremes of similar sign is not a random process, but has a definable relationship to the atmospheric general circulation.

44. Variability in time and space is not without value to man, plants and animals. Obviously there is a large suite of organisms whose physiology and behaviour are adapted to such variability. Human nomadism and transhumance are useful responses to it. Complete climatic stability and predictability, if they existed, might prove harmful both to man and his support systems.

45. In the following account we shall adhere to a standardized nomenclature for the various time and space scales of climatic variation (Landsberg 1976). It is outlined in Table 1.

A. LONG-TERM VARIATIONS (GEOLOGICAL SCALE)

46. The antiquity of the world's deserts and semi-deserts is not in question, but they have shifted in latitude during recent geological history, and have also undergone considerable changes in aridity. The variation of most importance to the present study is that of the most recent glacial phase of world climate, called the Wisconsin in North America, and the Würm or Weichsel in Eurasia.

47. This phase lasted from about 75,000 B.P. until a little over 10,000 B.P. It was marked by several advances of extensive continental icesheets over Europe and North America, with lesser but still extensive montane glaciation in Asia and even tropical Africa and South America. There were similar expansions of montane glaciation in Australia and New Zealand, while the Antarctic icesheet was at least as extensive as today. The world as a whole was significantly cooler than at present, but not enough to imperil the tropical biota, which mostly survived the phase.

48. The general atmospheric circulation during this long period must have been more vigorous and rather different in conformation from that of today. Reconstruction of the vegetation and faunal zones of north and central Africa – on a fragmental basis – suggests that the Sahara was occasionally narrower, and at times about 6 degrees of

TABLE 1. *Terminology used in this paper to describe climatic variations of different origin and duration. Adapted from Landsberg (1976)*

Term	Duration	Known or potential origin
Climatic revolution	$> 10^6$ years	geotectonic activity (continental drift, orogeny, large-scale changes in land and water distribution). Possibly solar variations.
Climatic change	$10^4 - 10^6$ yrs	changes in solar emission – aperiodic or periodic (period $> 10^4$ years) changes in extraterrestrial insolation due to long-period changes in orbital elements (orbital eccentricity, ecliptical inclination, precession)
Climatic fluctuation	$10^1 - 10^4$ yrs	all other *natural* climatic variations with durations in excess of 10 years aperiodic : volcanic activity quasi-periodic : changes in solar emission (sun-spot rhythms), magnetic declination cycle slow deep ocean currents terrestrial feedback mechanisms
Climatic iteration	< 10 years	very short-term quasi-periodic *natural* variations quasi-biennial oscillation (2–3 years), terrestrial – possibly atmospheric/oceanic interaction
Climatic alteration	10–? years	anthropogenic causes – effects can be global, regional/subcontinental or local in scale global scale : increases in atmospheric concentrations of CO_2, NO_X, halocarbons and particulates regional scale : power production, industrialization, urbanization, clearing of vegetation local scale : urbanization, agriculture (clearing, tilling, drainage, irrigation), grazing, water storage, deforestation, afforestation

latitude further south, as were the Sehelian and Sudanian dry belts. Studies of lake levels in the Chad basin and the East African Rift valley imply aridity in late Wisconsin times (after 20,000 B.P.), terminated by a marked increase in rainfall and runoff beginning after 15,000 B.P. and continuing into early historic times. In the U.S.–Mexican dry belt, by contrast, parts of the late Wisconsin epoch appear to have been wet, and very high lake levels occurred in many of the inland basins that now have shrunken, saline lakes, or are dry. The Dead Sea was also high (and probably fresh) at this time (Broecker and Kaufman, 1965; Kaufman, 1971; Butzer *et al.*, 1972; Servant *et al.*, 1969). Australia appears in the main to have been arid in Wisconsin times, but there is evidence of occasional high water levels in certain interior lakes, most of which are now usually dry. It has been suggested that these high water levels may have been due to lessened evaporation due to low temperatures rather than increased rainfall (Bowler *et al.*, 1976). The very narrow extent of the South American dry belts make reconstruction more difficult, but it is known (Olivares, 1967; Heusser, 1974; Mercer, 1976) that the extent of Andean glaciation varied in a fashion similar to that of North America and Eurasia in the millennia before 10,000 B.P. For dynamical reasons it is likely that the Atacama desert is of great antiquity.

49. If the fragmentary records are correct, there was a major, natural desiccation of much of dry north and central Africa lasting thousands of years in the period ending about 15,000 B.P. This area recovered substantially thereafter, and the prosperity of early human settlement of the Sahara and its borders, and of parts of the Arabian peninsula, presumably depends on this (Wendorf *et al.*, 1976). It is still only possible, however, to make the most tentative correlations between the events of the late Pleistocene and the evolution of man and of human material culture (Butzer, 1971)

B. POST-GLACIAL VARIATIONS: THE PAST 11,000 YEARS

50. The modern era of world climate appears to have been ushered in rather suddenly a little over 10,000 B.P. There was a dramatically abrupt readvance of both North American and Scandinavian ice-sheets dated at about 10,800 B.P., but in a few centuries climate had recovered its warmth, and an exceedingly rapid melting of the ice set in that was to clear the Cordilleran area of North America by 10,000 B.P., Scandinavia a millennium and a half later, and central Canada by 6000 B.P. In South America the equivalent warming may have been a millennium earlier, but was equally dramatic. This is the only event in the past 11,000 years to which the term *change* should be applied. This period of warmth culminated, at least in the northern hemisphere, in the so-called *hypsithermal interval* from 7000 to 5000 B.P. Since then there has been a small and erratic decline in air temperatures, especially in high northern latitudes. Throughout the period the Greenland and Antarctic ice-sheets remained intact (though the west Antarctic ice may be slowly disintegrating) as did many high mountain glaciers. To this extent the present climates are still glacial.

51. The tropical dry climates underwent many fluctuations during these events. In North Africa the sequence involved substantial variations of rainfall, as they did in north-west India–Pakistan, though probably not of temperature and radiation income.

52. According to Grove (1973) the semi-arid and sub-humid areas of tropical Africa were significantly wetter than today 12,000 to 7000 B.P. So also were parts of the Arabian peninsula and the Iranian highlands. The central Saharan massifs had substantial rainfalls. Schove (1973) finds the second highest stage of Lake Chad – at 320 m above present sea-level – to have occurred roughly in the epoch 11,500–9300 B.P. It is believed that some of the non-saline groundwater of the Sahara was accumulated at this time. The Geyh-Jäkel (1974) radiocarbon chrononology for the Sahara confirms that the more humid phase lasted until 700 B.P., but finds evidence for a renewed rainy phase between 6000 and 4700 B.P. The earliest historical traditions from the Nile civilizations confirm that there were progressive losses of savanna grasslands in the centuries following this apparent rainfall failure. In many parts of tropical Africa, therefore, the phenomenon of desiccation, if not of desertification, is almost five thousand years old. This desiccation has been neither uniform in space, nor regular in time. It is quite certain that periods of severe drought have alternated with moister periods throughout the epoch. Moreover, the

present climate is not the most arid that has been experienced in some areas. Thus Grove (1971) finds that shifting sand dunes of the western Sahara extended at some epoch 5–6 degrees of latitude further south than they do today. Similar shifts in the margin of dune-stabilization are well-known in Australia and India.

53. It seems clear from the approximately simultaneous nature of these variations across all Africa south of the Sahara that their cause must have lain in variations in the penetration northwards of the summer monsoonal current, which in turn must have arisen from changes in the summer latitude of the sub-tropical high pressure belt, and possibly also by longitudinal cellular changes.

54. In north-west India–Pakistan very significant shifts of the moist-arid boundary have also been proven, as have their relation to the rise and fall of the Indus civilizations (Singh, 1971; Joshi and Singh, 1972; Singh, Joshi, Chopra and Singh, 1974; Bryson and Baerreis, 1967; Bryson, 1976a). The area appears to have been arid in late Pleistocene times, but after 10,300 B.P. there began a protracted moist phase, presumably derived from increased monsoon rainfall, that stabilized dunes and created freshwater lakes. Profound desiccation began about 3800 B.P., and except for short periods of amelioration the trend towards aridity has not significantly reversed since that time; rainfall has not exceeded about one-third of its early Holocene levels (Singh, 1976).

55. The very large scale of these Halocene changes, and their wide geographical extension, make it certain that they are of natural origin. They arose from aberrations of the climatic system in the wake of the abrupt ending of the last glacial age ten thousand years ago. Yet agriculture and pastoralism, at least in their old world guise, had their beginnings in the early post-glacial millennia. The early innovations took place not far from the desert margin in the fertile crescent of the Middle East, in the Indus Valley, and in well-watered valleys of the Anatolian–Iranian plateaus. The seeds of present-day desertification were sown early.

56. In recent years research has demonstrated that the major variations of climate of late Pleistocene and Holocene times have often taken place over a few centuries. It is quite likely that the next major variation, when it occurs – and it *may* have already done so, though this is disputed by most authorities – will go unrecognized until it is well established. We have no certain way of distinguishing between short-term fluctuations and the onset of persistent change, at least in the latter's early stages.

C. CONTEMPORARY SHORT-TERM VARIATIONS

57. The Sahelian–Ethiopian drought of the late 1960s and early 1970s focused world attention on the vulnerability of agriculture and pastoralism in the drier regions of the earth. The question has naturally been asked – what is the nature of this drought? Is it an aspect of the normal climate of the earth? Or does it foreshadow a hostile change that will lessen the carrying capacity of the dry regions for centuries, or perhaps forever?

58. High variability from year to year is characteristic of dry climates. A frequently-used measure of the variability is the coefficient of variation, defined as

$$V_\sigma = \frac{\sigma}{\bar{P}} \times 100 \qquad (2)$$

where σ = standard deviation of annual precipitation, and
$\bar{P}$ = mean annual precipitation.

The coefficient is over 25 per cent in most parts of the dry world, and exceeds 40 per cent along most desert margins. More than a third of all years will have precipitation outside the range of $\pm 1\sigma$, if rainfall is normally distributed. In most parts of the dry world the latter does not hold. Wallén (1967) and Wallén and Perrin de Brichambault (1962) made use instead of the interannual variability

$$V_I = 100 \frac{\Sigma(P_{n-1} - P_n)}{\bar{P}\,(N-1)} \tag{3}$$

(where n is an individual year in a series of N years)
which is the percentage ratio of the mean interannual precipitation-difference, (taken regardless of sign) to the mean annual precipitation. For the Middle East they found that this ratio exceeded 50 per cent along the desert margin, and was generally in the range 25–35 per cent in dry farming areas. Empirically they showed that at present, when P is in millimetres,

$$V_I = 0.07\,\bar{P} + 22 \tag{4}$$

along the dry margin of arable farming.

59. Discouraging though these high interannual variabilities may be, they are not necessarily crippling. Natural ecosystems in arid and semi-arid climates have evolved considerable resilience in the face of such variability. Nomadic pastoralism, too, can take big interannual changes in its stride. Matters become more difficult if more than one drought year occurs successively. In many parts of the dry world droughts may last several years. Jenkinson (1973) and Walker and Rowntree (1976) analyzed rainfall since 1911 in the Sahelian and Sudanian zones of West Africa. For a group of stations with mean annual rainfall 400–1000 mm and for the period 1911–74, they found eight runs of 2 years above or below median rainfall, one of 5 years, two of 7 years, and one of 10 years. The probability that such persistence would occur randomly is low (9 per cent) – i.e. there is a distinct tendency for abnormal wetness or drought to persist from one year to the next or succeeding years.

60. For five stations in the 130–320 mm belt – the true Sahel – they found since 1945 a persistence ratio[4] of 2.58, which indicates a probability of only 0.1 per cent that the runs occurred as a result of random processes. In West Africa, therefore, *it appears that there is a marked tendency – especially near the Sahara – for successive dry years to occur in a row*, sometimes out to a full decade.

61. The same is probably true of most other dry climates, but a full statistical analysis has not been done. In Fig. 9 we plot, for a selection of dry climate stations, the rainfall for each year 1945–74, to emphasize the severity of the recent drought in the Sahel, and the tendency for anomalies of the same sign to persist. At Agadez, Abéché and Alice Springs, moreover, there are periods of up to two decades in which a linear downward

[4]The inverse ratio of the number of runs with similar sign departures to that exhibited by a random series. The same is true of moist spells, as is made clear on Fig. 8, which shows the overall behaviour of rainfall over Sahelian West Africa (Manson, 1976).

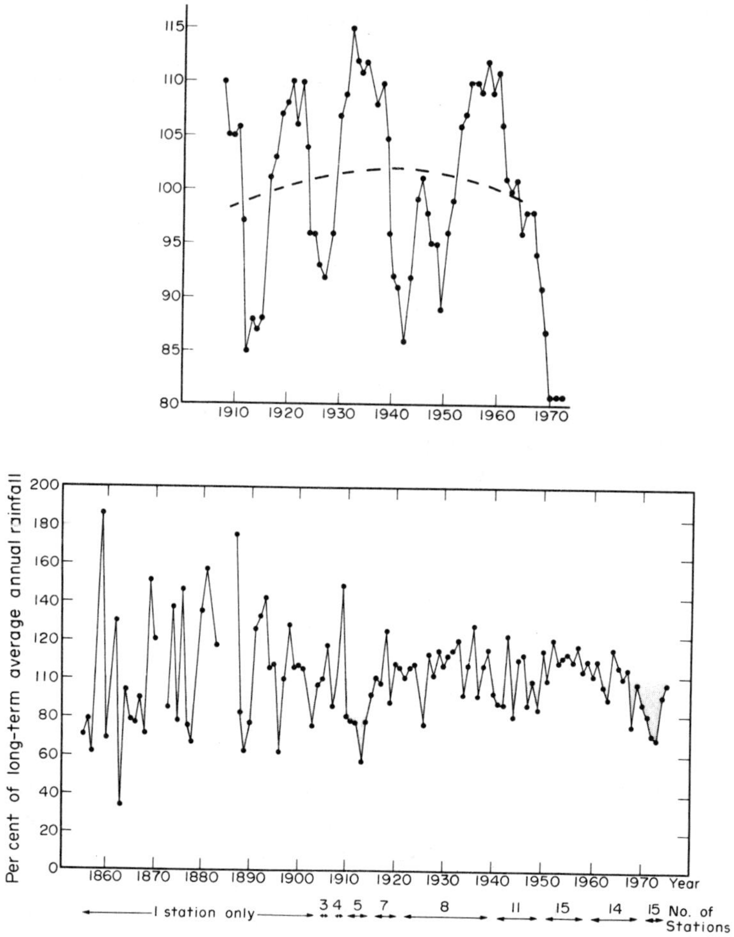

Fig. 8. Overall spatial average precipitation in Sahelian West Africa (after Mason, 1976). Upper curve shows 5-year values 1905–70, expressed as percentages of the average for the period, with long-term trend added. The lower curve gives annual averages for all available stations since 1857. The Sahelian drought of the early 1970s, though severe, is not unprecedented.

trend occurred, ending abruptly in each case. Similar linear trends have been analyzed in other parts of Australia (Tucker, 1975). It is easy to see why local inhabitants, confronted with such prolonged desiccation, should be persuaded that it is a permanent effect. Yet at Phoenix and Jodhpur, and at many other stations throughout the dry world, no trends are apparent.

62. Several other workers have attempted to answer the question: Is the recent Sahelian drought part of the normal climate, or does it signify a change of climate? Landsberg (1975), for example, concluded that the drought of the 1970s was not unprecedented, but had to be accepted as part of the normal climate. He used the long record (since 1887) at Dakar, Senegal, as evidence, and showed a tendency for drought or abundant rain to recur at 2–3- and 10-year intervals – presumably the quasi-biennial and sunspot oscillations. Bunting, Dennett, Elston and Milford (1976) did not find these periodicities in a group of stations (Niamey, Zinder, Sokoto, Kano and Maiduguri) in the

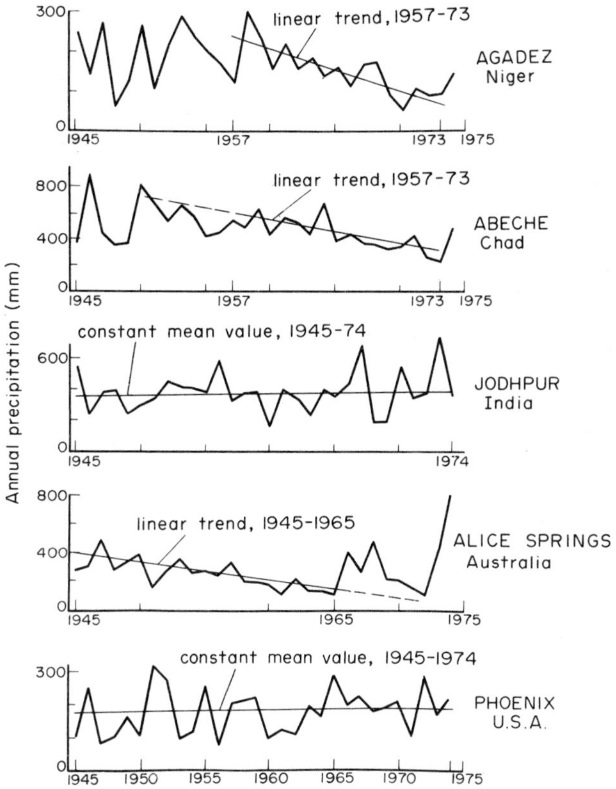

Fig. 9. Rainfall variations since 1945 at selected arid-zone stations. Note persistent linear downward trends lasting many years at Abéché and Agadez (Sahel) and Alice Springs (Australia) terminating in heavy rains, and lack of visible trend at Jodhpur and Phoenix.

540–875 mm range of annual rainfall, nor did they find much interannual persistence. But they, too, concluded that the recent drought fell within statistical expectation (see also Igeleke, 1975). Sircoulon (1976) arrived at a similar conclusion from a thorough analysis of streamflow in the great rivers of the area – the Niger, Chad and Senegal, most of whose flows are exotic. Winstanley (1973) pointed out, moreover, that the widely used 1931–60 normals for precipitation give an unrepresentatively high figure for annual totals throughout the region.

63. Severe and prolonged droughts have been reported from other dry regions within the past decade, but there is nothing quite comparable with the Sahelian tragedy. There were widespread monsoonal rainfall failures in India in 1972, the peak year of the Sahelian drought. Bryson (1974) has reported a sharp rise in the incidence of rainfall less than half normal from northern India since 1970 – back to levels typical of the first three decades of the century, and in striking contrast to the largely drought-free 1940s, 1950s and early 1960s.

64. Australia is a continent dominated by the southern sub-tropical high pressure belt, with an annual shift of latitude and marked cellular structure. Gibbs (1975) has analysed the major Australian droughts of this century, classifying a drought year as one whose annual precipitation would be exceeded in 9 out of 10 years (i.e. the lowest decile

range). Three major droughts were identified, each about three years long: 1900–2, 1912–15, and 1965–7. Each produced disastrous crop failures and stock losses. Very severe drought recurred in 1972, affecting much of the continent. But then began, early in 1973, a spectacular reversal. Heavy and frequent rainfall, derived from unusually strong, prolonged and deep penetration of moist tropical air from the north, soon expunged the drought, and serious flooding replaced parched dry streambeds. The rain continued through the 1973–4, 1974–5 and 1975–6 rainy seasons. Lake Eyre, which drains much of inland Queensland, reached its highest levels since European settlement began, and water has appeared from time to time in such dry "lakes" as Frome, in eastern South Australia. Higher levels of Lake Eyre were recorded about 10,000 B.P., however, so that the event is not unprecedented (Wopfner and Twidale, 1967; Twidale, 1976).

65. These extraordinary aberrations have raised the question whether we have entered a phase of markedly greater variability of climate, which is distinct from questions about the warming or cooling of global temperatures. Are the 1970s – and potentially the 1980s – more prone to such extremes than were the 1950s and 1960s? For the dryland farmer and nomadic pastoralist this is a vital question, and one that matters a great deal for the programmes of United Nations agencies. Unfortunately it cannot be answered firmly. There is some evidence that the tendency towards extremes is upward, and it is wise to assume that this greater variability will continue, i.e. that world affairs will unfold in the next decades in a climate at least as variable as that of the 1970s.

66. Interannual persistence of drought or excessive rainfall may have two possible causes:

(i) *Interannual persistence of general circulation anomalies,* which are well-authenticated realities. In particular air–sea interaction includes certain processes that work on a multi-year timescale (Namias, 1974). It has already been suggested (Gilchrist, 1975) that the recent Sahelian drought may have been connected with a persistent sea-temperature anomaly in the North Atlantic;

(ii) *feedback mechanism*, whereby drought feeds on drought, rainfall on rain. Several mechanisms have been suggested that should cause drought, once established, to amplify itself – and some of these involve man and his desertification capabilities.

67. Experience suggests that it has been fluctuations of climate on the several year to decadal scale, like the Sahelian drought, and the North American Dust Bowl of the 1930s, that have been most effective in accelerating the desertification process. Clearly both the above causes work on such scales.

IV. CAUSES OF THE OBSERVED VARIABILITIES

A. NATURAL CAUSES OF MACROCLIMATIC VARIATION

68. We have already suggested that the major controls over the dry climates are rooted in the general circulation, and are hence hemispheric or even global in character. The effects of global circulation iterations and fluctuations are strongly modulated, however, by local physiographic, ecological and hydrological influences. Even in the complete absence of human interference this interplay makes the response of regional climates to external changes very complex.

69. Two recent authoritative reports have been published on the causes of climatic variation. These are *Understanding Climatic Change* (U.S.A., National Academy of Sciences, 1975), and *The Physical Basis of Climate and Climate Modelling* (GARP, 1975). We have drawn extensively on both in what follows.

70. The atmosphere responds to the external forcing of the sun, whose beam has the average intensity at the top of the atmosphere (normal to the solar beam) of 1360 ± 20 W m^{-2}. Any change of this *solar constant* will produce variations of climate, as will significant changes in the composition of the solar beam (whose spectrum outside the atmosphere is still imperfectly known). The atmosphere interacts with the earth's surface, soil, rock, ice, water and biota, and climate also depends critically on the gaseous and particulate composition of the atmosphere. Changes in any of these properties will cause climatic variation. Even on the short time scales of importance here the last items may vary enough to produce such responses.

71. The oceans and enclosed seas also affect terrestrial climate. Ocean and atmosphere form a coupled system, though for most practical purposes – especially in mathematical modelling – we have to uncouple them. Similar links exist between glacial and sea ice and the atmosphere. The biota on the land surface, chiefly vegetation, interact chemically and physically with the boundary layer, troposphere and even the stratosphere. Climate, in fact, is a vast interactive response system, with many feedback loops of both signs.

72. There is hence no naive or intuitive way of guessing how the planetary system will respond to a change of external forcing, for example by a change in the solar constant, the addition of some new absorptive gas to the air, or the opening up of a new marine strait. It is first necessary to model the system mathematically, and then make some inspired guesses as to the questions to ask of the models. As we shall see later, this is already being attempted for the desertification process.

73. It is not even certain whether observed variations are externally forced. It may be that within the present external conditions – i.e. solar constant, present physiography, hydrology and ecology – there are more than one stable state for climate, and that the atmosphere may switch from one state to another rather abruptly. The suddenness of some past variations lends weight to this hypothesis, which, however, requires much further investigation.

74. On the short time-scale of importance in the study of present-day desertification, one can be more specific about causes. Figure 10 (U.S.A., National Academy of Sciences, 1975) indicates that snow-cover, the properties of the ocean surface layer, and

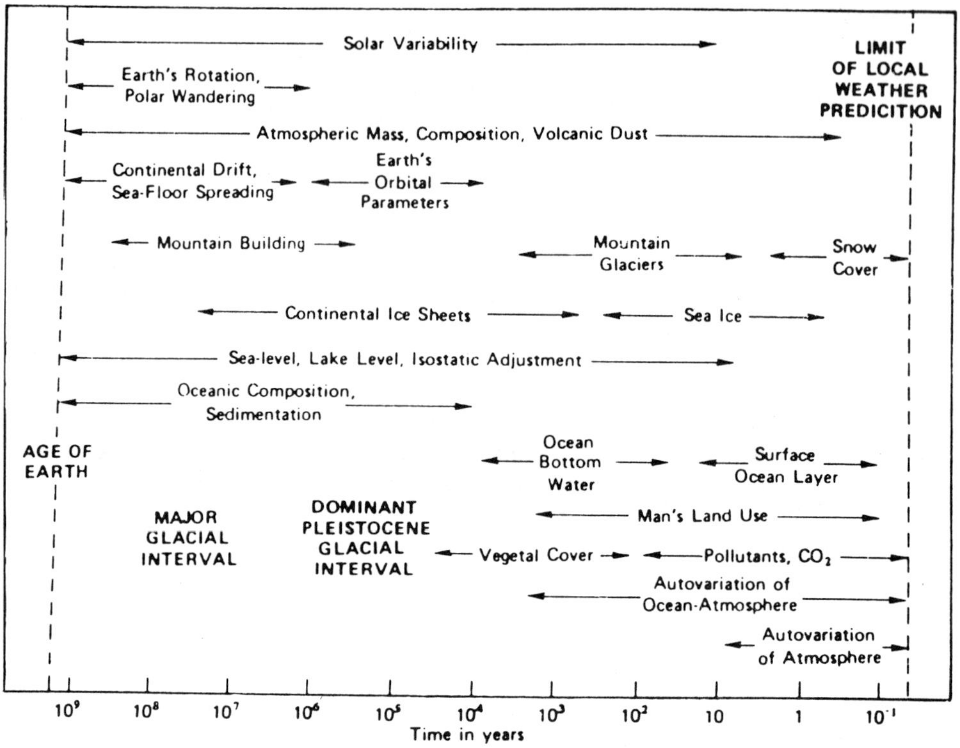

Fig. 10. Processes that may cause climatic variation, plotted against their characteristic time-scale (after United States National Academy of Sciences, 1975).

the internal behaviour (called auto-variation) of the ocean and atmosphere may produce climatic variations on the short and decadal scales. To these are added possible alterations due to human interference, especially to the effect of pollutants (including carbon dioxide and possibly aerosols), and changes in the use of land.

75. These specific processes will be examined in Section V, which deals with modelling attempts. Clearly they bear mainly on the large-scale behaviour of the atmosphere, i.e. on its general circulation, and hence on what was earlier in the text called the macroclimate, or dynamic climate. Desertification and the natural spread of desert surfaces must both be influenced by any changes that occur in this large-scale behaviour. But the reverse may also be true, since feedbacks link the large-scale processes with those operating in the physical climate at the surface, especially in the dry lands.

B. ALTERED SURFACE EXCHANGE PROCESSES IN THE DRY LANDS

76. The consensus among informed workers is that recent desertification has not originated as a series of pulses emanating from the desert cores. It results instead from the spatially uneven pressure exerted by man on soil and vegetation, especially at times of drought or excessive rainfall. The dust-bowl in the U.S. Great Plains, degeneration in the Sahel, in the Ethiopian plateaus, and in Mendoza Province of Argentina (Tanquilevich, 1976) are all manifestations of human misuse of the land at times of climatic stress. To quote Anders Rapp (1974) "a common mechanism of desertification is the following:

"1) Expansion and intensification of land use in marginal dry lands during wet years. These actions include increased grazing, ploughing and cultivation of new land, and wood collection around new camps or settlements.

"2) Wind erosion during the next dry year, or water erosion during the next maximum rainstorm."

77. The ecological changes that take place under these conditions are also microclimatic variations. Physical climate, bioclimate and soil climate all vary as vegetation is degraded or removed, and as the soil is blown or washed away. Moreover, as these essentially local, patchwork processes repeat themselves at innumerable sites across great areas, they may initiate feedback processes that do indeed affect the large-scale atmospheric motion, and hence the global climatic system.

78. The desertification process described by Rapp[5] has the following climatological implications:

(i) *Increased grazing during wet years.* This will tend to compact the soil near water holes and wells, and along the pathways invariably beaten out by grazing animals. As livestock numbers increase still further there may be pressure on perennial plants during the dry seasons, so that the plant cover late in that season will be far from complete, even in moist years. The result is to expose surface soil to erosion by wind – especially of the fine-grained materials. Nevertheless, if there are successive wet seasons the effect may not be large. Percolation (N_0) may, however, be decreased in favour of N'_0, the surface run-off, because of lowered infiltration capacity, and albedo (a) may increase (see Fig. 1 for identification of these processes).

(ii) *Increased cultivation during wet years.* The effect here is greatly to enhance the prospects for wind erosion of fine soil materials during the dry seasons, and possibly to increase evapotranspiration (in the case of water-demanding crops), unless the land is fallowed under mulches. In general there will be pressure on water stored in the soil, and a lowering of organic content – which in turn reduces water storage capacity. Albedo is in general increased.

(iii) *Wood collection.* There is already severe pressure on woody species in many parts of the dry world, for dry season grazing by domestic animals, and for firewood. The latter is in short supply everywhere. The removal of woody species increases direct solar heating of soil (surface temperature – T_0 – increase) and may considerably decrease evapotranspiration. The desert surfaces of parts of the Hindu Kush and north-west Himalayas are the end-product of such assaults upon a once prosperous forest (Flohn, personal communication).

(iv) *In the ensuing dry years* the acceleration of wind erosion by the above processes still further reduces water-storage capacity by removal of topsoil. The loss of some or all the perennial plant cover substantially lowers infiltration rates, and hence potential percolation (N_0). During subsequent rains surface run-off, N'_0, is increased, with loss of stored water for subsequent use by shrubs, herbage or crops.

79. The tendency for climatic drought to persist through two or more consecutive rainy seasons, which we have seen to be typical of such climates, immensely aggravates process (iv) above. But it can be argued that the parallel tendency for groups of wet years to occur is equally dangerous, because such a sequence permits unwise multiplication of flocks and herds, and extension of arable acreage. Vaccination of herds may allow

[5] This text should be read alongside the component review on ecological change.

numbers to climb even in the absence of favourable rainfall, so that the range carrying capacity is greatly exceeded. Prole (1967) speaks of one such case in Kenya, where cattle population rose to 700,000 on range with a carrying capacity of 400,000. A brief drought in 1960–1 led to the death of most of the stock. (See also Kowal and Adeoye, 1973; Kowal and Kassam, 1973; de Leeuw, 1974.)

80. Nor are such catastrophes confined to the developing countries. In Australia, for example, the 1900–2 drought reduced the 1903 wheat yield in New South Wales to 1.2 bushels per acre, and sheep dropped from 42 million to 27 million in 1902. The 1965–7 drought reduced the wheat harvest of Victoria, New South Wales, and Queensland by 40 per cent, and oats, barley and rice by 60, 30 and 15 per cent respectively. Sheep population fell by over 20 million (Gibbs, 1975). The tendency to overextend in sequence of favourable years is as much a reality in the wheat-exporting countries as it is in the subsistence economies – though the consequences are more serious for the latter. Perry (personal communication), arguing from research at Alice Springs, Australia, defines a drought as a period when there is a shortage of forage, and finds that about 4 weeks of soil moisture sufficient for good growth of forage is enough to carry over to the next summer. On this basis 25 out of a sequence of 92 years were droughts. Assuming random distributions of such years, he predicts that 2-year droughts can be expected four times in a century, 3-year droughts once in a century and 4-year droughts once in 360 years – under central Australia conditions. As we have seen, interannual persistence may exceed random expectations elsewhere, and hence the risk of a failure of forage for such long periods should be greater.

81. Vegetation and soil respond to climatic variation, and in regions of climatic stress such responses are the key to ecosystem survival. The ability of desert, savanna and scrub vegetation to outlive protracted drought is well known, and there is a large literature devoted to the adaptations of plants and animals to this kind of stress. Relatively little of the literature deals, however, with the explicit microclimatic aspects of the adaptations. There appear to be many obvious but only partially answered questions, some of which are discussed below.

82. Why, for example, is dry land vegetation characteristically discontinuous and patchy? On some scales there are obvious but suspect answers. In the Sonoran desert of Mexico and the United States, for example, the vegetation cover habitually appears to be below 50 per cent, and sometimes below 10 per cent, even in areas of 100–150 mm rainfall a year. Yet after rains ephemeral annuals carpet the entire landscape, only to die after seeding a few weeks later. It is the perennial herb-shrub layer that persists, rarely occupying as much as half the land surface. Similar open-structured ecosystems abound throughout the tropical dry lands. Competition among individuals for soil water is no doubt a reality – but is not a sufficient cause, since in some cases the horizontal root systems are themselves discontinuous, leaving areas of the terrain unoccupied by perennial plants even after rain.

83. From the climatic standpoint such open structure has certain implications. Each individual herb or clump is a kind of oasis, and is subject to advection of heat from the hot bare ground surrounding. To some extent the foliage of the plants shades the floor, and hence keeps the root zone cool; but dry belt vegetation is often dominated by species in which the leaf canopy is thin, discontinuous and even suppressed (as a means of restricting transpiration). Hence considerable sunlight penetrates to the floor, even directly beneath the plant. In many ways the functional utility of ecosystem structure, as

an adaptation to drought, is obscure. In some cases the structure of plant communities seems to maximize the climatic stresses. This is in striking contrast to the anatomy and physiology of individual desert species, most of which have some specific adaptations tending to protect individuals against excessive moisture losses.

84. Open structure leads to ready access to the soil surface by the wind. The velocity profile is, of course, dependent on the spacing and size of the clumps of the plant cover. Aerodynamic roughness may be quite high, which tends to reduce surface wind-speeds. But during periods of strong wind the bare soil patches will inevitably be subject to frequent impacting gusts. Hence drifting and blowing soil (deflation) is a characteristic of windy, dry-season weather even in undamaged areas.

85. Dust-laden airstreams emerging from the world's dry lands have been known from earliest antiquity. The khamsin, ghibli and scirocco winds of the Mediterranean shores are well known, as is the frequent phenomenon of "blood rain" over Europe, in which Saharan dust discolours the water. In this case the deflation is caused by cyclones of the westerly belt (Hare, 1942), and occurs mainly in winter and spring. Similar dust storms occur in the United States and Canada, and are sometimes reported in Australia. Geologically this process led at certain periods to the accumulation of substantial loessic deposits of soil in temperate latitudes.

86. Major deflation also occurs along the southern flanks of the sub-tropical highs, and in the monsoon areas. Rapp (1974) has analysed the transport of dust by persistent Saharan airstreams out of North Africa towards the Atlantic (where the volume increased sharply after the onset of the Sahelian drought). Though raised by the Harmattan north-easterlies, the dust is transported out of the continent mainly by mid-tropospheric easterlies (Fig. 11). In summer it is reddish, and is derived from the Sahara, but in winter the colour is darker, and the deflation comes mainly from the Sahel-Sudanian belts. Charles Darwin discussed this effect in *The Voyage of the Beagle.* There is also a

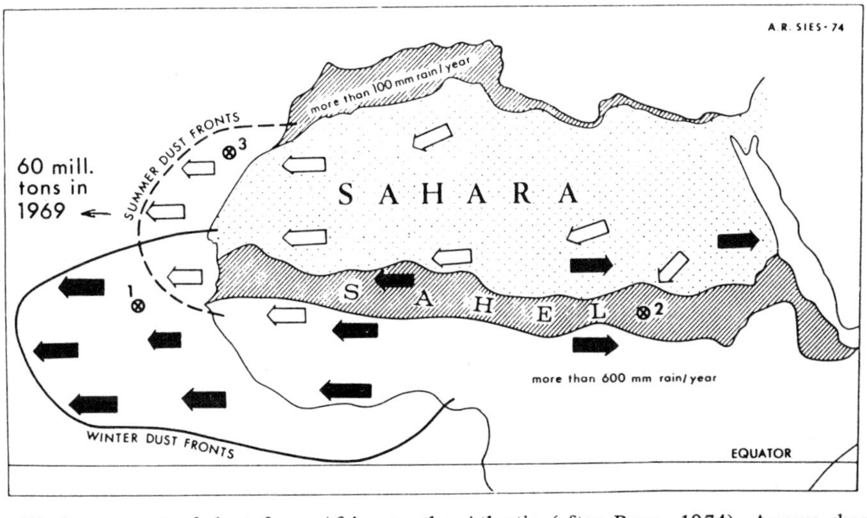

Fig. 11. Wind transport of dust from Africa to the Atlantic (after Rapp, 1974). Arrows show mean 3 km wind direction in July (white arrows) and January (black arrows). Summer and winter dust-fronts over Atlantic show limits of 10–15 per cent days with observed dust haze. Crosses 1, 2 and 3 are proposed monitoring sites for dust.

significant transport in summer of North African and Arabian dust across the Arabian sea towards north India–Pakistan. Rapp estimates that the west-bound deflation amounts to 60 million tons per annum, which is about one-third the annual sediment transport of the Nile at the inflow to Lake Nasser.

87. This deflation of fine materials has been in progress for a very long time. Diester-Haass (1976) has examined cores from the floor of the Atlantic within a few degrees of longitude west of the coasts of Senegal, Mauritania and Sahara (latitudes 15° to 27°N). She has used variations in the amount and grain size of terrigenous materials, the amount of calcium carbonate, the quartz-to-mica ratio, and the numbers of desert quartz grains, to infer climatic history for back into Pleistocene times.

88. Visibility statistics make it clear that similar deflation occurs over the Australian, South American and Kalahari deserts and adjacent dry regions. A world assessment of the process has been given by Griffin, Windom and Goldberg (1968). Bryson and Wendland (1970) and Bryson (1974) have raised the possibility that accelerated deflation due to the desertification process has so raised the dust content of the atmosphere as to alter significantly its optical properties, and to influence the general circulation.

89. These processes ought to bc incorporated into some system of modelling, presumably of the general families called *simulation* or *climatonomic.* A model of the desertification process should comprehend (i) the surface physical climate, including the hydrological regime; (ii) a reasonable representation of the surface vegetation cover and animal life; (iii) land use practices, pastoral, arable and others, including use of fire; (iv) a sketch of the potential impact of disease or parasites, such as locust invasions; and (v) the resulting ecological and geomorphic disturbance, which is the desertification itself. Such a model could be related rather easily to an atmospheric model of the general circulation type.

90. Unfortunately no comprehensive model of the process yet exists, though many attempts are being made (see for example ISAID, 1976). The Australian *Socio-Technical Model of a Region in which Grazing is a Major Industry* (Benyon, 1976), being carried out under SCOPE Project No. 5 (Simulation Modelling), is treated in the component review on ecological change.

V. MODELLING ATTEMPTS

91. In order to test hypotheses such as those discussed in Sections III and IV one must have effective mathematical models of the dry climates. These models must incorporate the essential dynamical and physical processes that are known to bear on the problem, and must be capable of predicting the consequences of specific changes of external forcing, including the inadvertent or deliberate work of man. An excellent account of the models useful for such exercises is given by GARP (1975, pp. 13–24, and Appendix 2, pp. 115–162).

92. Experiments conducted so far comprise:

(i) *General circulation models (GCMs) applied to the specific problems of the causes of climatic fluctuations in the dry climates.* These models simulate in three dimensions the general circulation of the hemisphere or globe, and can be subjected to chosen perturbations, such as sea surface temperature anomalies, to determine the effects.

(ii) *Investigations of specific feedback processes*, such as changed albedo, or variations of soil moisture storage. The models seek to predict the effect of the feedbacks on circulation and precipitation over the arid zone.

93. In practice, most attention has been given in these exercises to the Sahelian problem, but the methods used are quite general, and can be applied to any area where desertification is feared, or has occurred. A supplementary review is given in Appendix III.

A. GENERAL CIRCULATION MODELLING RELATED TO THE ARID AND SEMI-ARID ZONES

94. Since aridity arises largely from persistent subsidence in the general circulation of the atmosphere, one might expect that the various existing general circulation models (GCMs) would have much to say about the causes of droughts such as that which recently affected the Sahelian–Sudanian and Ethiopian zones of tropical Africa.

95. GCMs portray the three dimensional circulation of the global atmosphere (though there are related models that confine themselves to a hemisphere or a region of a hemisphere). They are designed in such a way that when they are allowed to run for long periods they produce a set of global statistics of temperature, pressure, wind, and precipitation that resemble those of the present-day climate. The external inputs to such a model can be varied deliberately, to see how the model responds. It is then possible cautiously (even sceptically) to infer how the real atmosphere will react to a real perturbation of the same kind. In view of the importance of air–ocean interaction, such models ought to incorporate the sea, but the very different response times of the two media, plus the differences of computational procedure required, makes this a difficult thing to do properly. It is also crucial that they deal effectively with the coupling between the extent of polar snow and ice surfaces and world temperatures.

96. One recent attempt at this kind of experiment has been reported by Gilchrist (1975), using the United Kingdom 5-level GCM. He investigated the speculation that rainfall variations over north and central Africa were related to sea surface temperature

anomalies over the tropical Atlantic. Correlations of this general sort have been long supported by Bjerknes (1966, 1969), Namias (1974, 1975) and others.

97. The U.K. GCM satisfactorily predicts the July surface wind régime over Africa and the adjacent oceans, though it drives the West African monsoon a little further into the Sahara than happens in reality. Gilchrist introduced an extensive sea-temperature anomaly over the tropical Atlantic as far north as 25.5°N, cooling temperatures at the centre of the anomaly by 2°C. This area is known (Brown, 1963) to be prone to prolonged anomalies of this sort. He then ran the model to see how precipitation would be affected over North Africa. He found that amounts were lower west of 5°W than in a control case with normal sea temperatures, but were higher east of 5°W. Namias (1974) and Jenkinson (1975) have reported that higher than normal pressure tends to occur over the north-east Atlantic with drought over Africa. The experiment produced such a pressure anomaly.

98. This experiment proves little, but points the way towards further proof, and, in the words of the author, changes "the speculation that there is a causal link between cold Atlantic sea-surface temperatures and drought in West Africa into a credible hypothesis", which he finds "more likely than not".

99. A similar effect has been found for the Indian monsoonal rainfall in relation to lowered sea-temperature over the Arabian Sea. Shukla (1975) applied the Geophysical Fluid Dynamics Laboratory (GFDL) GCM to a postulated cold anomaly in the western Arabian Sea, and found a sharp, statistically significant reduction of rainfall over India in the following month. Records show that over the entire period 1901–61 a cold Arabian Sea in July tended to be followed by low August rainfall in India – which should prove of value in rainfall prediction.

B. TELECONNECTIONS

100. An acid test of GCM experiments is their ability to explain observed teleconnections. Several instances have already been given of what appear to be meaningful links between climatic variations in two or more different localities – for example, the relation between Sahelian rainfall and sea surface temperature anomalies over the Atlantic, and extreme weather events over the Caribbean area. Such teleconnections, if real, are important for two reasons – they may be useful in prediction, and they may have significant economic consequences (if, for example, drought in one region tends to be accompanied by excessive rainfall in another).

101. Teleconnections are, in fact, manifestations of specific patterns of variation of the general circulation of the atmosphere, and good models of the latter will predict their occurrence and duration. Unfortunately the existing models do not yet function well enough for us to rely on them in this way. Instead we have to establish teleconnections empirically, from the examination of the climatological record.

102. An example of high importance to the desertification process is the *Southern Oscillation* effect first identified by Walker (1924) and subsequently reexamined by Troup (1965). Flohn (1971) showed from more recent upper air data that there were significant correlations between the behaviour of the general circulation over the eastern tropical Pacific and over South-east Asia and tropical Australia, including the arid and semi-arid interior.

103. Pittock (1975, 1976) has been able to demonstrate that this effect is real when applied in far more detail to surface rainfall in Australia. He used two crude but objective

indices, one expressing the southern oscillation effect, and the other the latitude of the sub-tropical high pressure belt over Australia. He was able to show that rainfall in eastern and northern parts of Australia – including the semi-arid parts of Queensland and Northern Australia – was correlated at above the 95 per cent confidence level with the crude index of the southern oscillation. Rainfall along the southern and eastern flanks of the arid heart of the continent was, by contrast, strongly correlated with the latitude of the sub-tropical high pressure belt.

104. Results such as this imply that the cellular effects in the sub-tropical high pressure belt referred to previously respond to very large-scale standing-wave-like longitudinal perturbations of the general circulation, emphasizing once again that drought or excessive rainfall in the dry countries are influenced by planetary processes that demand a world perspective if they are to be understood and ultimately predicted.

C. FEEDBACK PROCESSES

105. Most recent modelling attempts have been related to specific feedback processes that may augment or retard naturally-induced climatic variations along the desert margin. Most of these have been concerned with the effect of changed albedo (i.e. reflectively of the ground with respect to solar radiation), and other consequences of the degradation of vegetation cover.

106. The albedo feedback hypothesis was introduced by Otterman (1974, 1975) who reasoned that the destruction of vegetation and exposure of soil would increase albedo (as suggested above), and hence lower surface temperatures. This would lower the sensible and latent heat fluxes to the atmosphere, and suppress convective shower formation. By examining NOAA satellite scanning radiometer measurements of surface temperature over an area in the Middle East he was able to show that in summer daytime conditions the western Negev, with about a 35 per cent vegetation cover, was warmer than nearby northern Sinai, with 10 per cent cover – in one case by over 4°C. Otterman's interpretation was challenged by Jackson and Idso (1975), who claimed that denuded soils in the Sonoran Desert of Mexico and the United States were always warmer than vegetated soils. They suggested that bare soil may have an emissivity (see Fig. 1) as low as 0.88, and that this invalidated Otterman's result. They concluded that "the denuding of soil may have thermal and climatic effects just the opposite of those that he has postulated".

107. Regardless of the validity of Otterman's hypothesis, others have seized on albedo feedback as a mechanism of desertification. Charney (1975a, 1957b) noted that the central and northern Sahara, eastern Saudi Arabia and southern Iraq actually have a *negative* radiation balance *at the top of the atmosphere* on hot summer days, in spite of the intense input of solar radiation through the cloudless atmosphere. This deficit arises (i) because outgoing terrestrial radiation is enhanced by high surface temperatures (the Jackson–Idso position, rather than Otterman's), cloudlessness and low humidities, and (ii) because the high surface albedo (of order 0.35 over sandy deserts) increases the reflection back of solar radiation. Hence the North African–south-west Asian deserts are actually heat sinks relative to the rest of the hemisphere. Even the arctic has a positive radiation balance at this season. Only Greenland and Antarctica, with perpetual ice, compare with the desert belt as radiative heat sinks in the summer hemisphere. To quote Charney (1975b) verbatim:

"Since the ground stores little heat, it is the air that loses heat radiatively. In order to maintain thermal equilibrium, the air must descend and compress adiabatically. Since the relative humidity then decreases, the desert enhances its own dryness, i.e., it feeds back upon itself! . . . A bio-geophysical feedback mechanism of this kind could lead to instabilities or metastabilities in border regions, to advances or retreats of the borders themselves, which might conceivably be set off or maintained by anthropogenic influences."

108. Charney set up a limited sector radiative-dynamical model of North Africa, which suggested that an increase of albedo would enhance meridional overturning such that the rate of subsidence in tropical latitudes (a maximum at about 5 km) could be more than doubled in the summer season. He then used the Goddard Institute for Space Studies (GISS) GCM (which includes clouds, precipitation and evaporation) – with P. Stone and W. Quirk (1975) – to simulate the effect of a change of albedo from 15 to 30 per cent, and found sharp reductions of cloud and rain following the increase. In Figs. 12 and 13

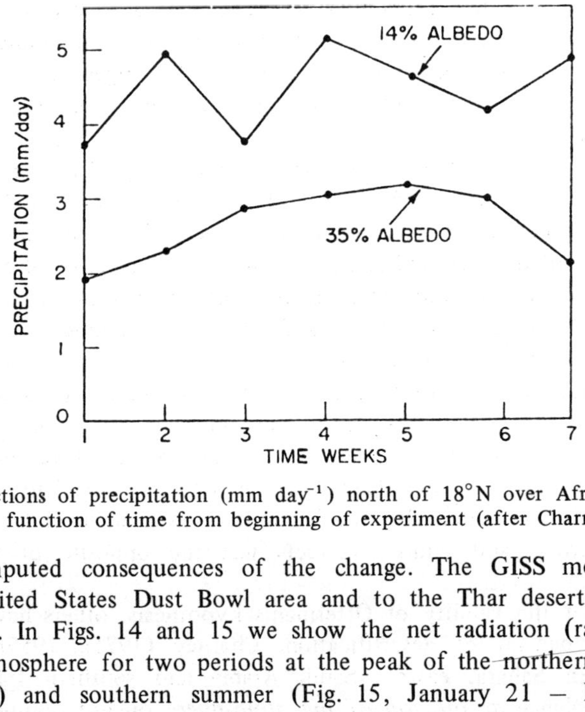

Fig. 12. Model predictions of precipitation (mm day^{-1}) north of 18°N over Africa for albedos of 14 and 35 per cent, as a function of time from beginning of experiment (after Charney, 1975a).

we show the computed consequences of the change. The GISS model has also been applied to the United States Dust Bowl area and to the Thar desert of Rajastjan, with very similar results. In Figs. 14 and 15 we show the net radiation (radiation balance) at the top of the atmosphere for two periods at the peak of the northern summer (Fig. 14, July 16–31, 1969) and southern summer (Fig. 15, January 21 – February 3, 1970), from NIMBUS-3 satellite observations (from files of V. Suomi, personal communication). On Fig. 14 it is clear that there is no net cooling over the Sonoran and Rajasthan deserts, where the values are typical of the latitude belt. Evidence from the Southern Hemisphere is sparse, but on Fig. 15 all the dry areas for which there are plotted values have positive net radiation. This specifically includes the Kalahari and the easternmost desert surfaces in Australia. It may be, therefore, that only the Saharan–Arabian–Iraqian desert tract regularly produces the radiative effect specified by Charney.[6]

[6] In some cases the other areas, while showing positive values, still appear as *relative* minima. See also Appendix III.

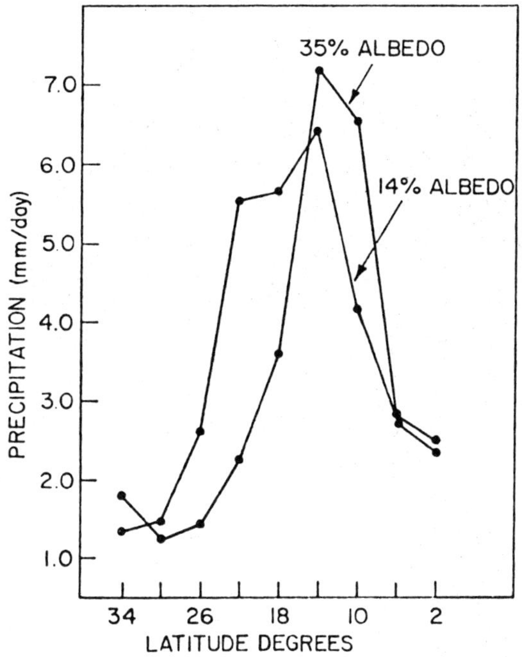

Fig. 13. Similar to Fig. 12, but showing the average precipitation as a function of latitude (after Charney, 1975a).

109. There have been other investigations of the albedo feedback process. Elsaesser, MacCracken, Potter and Luther (1976) used a two-dimensional model (longitudinally averaged) for an earth in which 30 per cent of the 15–25°N latitude belt was given an albedo raised from 14 to 35 per cent. In the latitude of the Sahara their model predicted a 22 per cent decrease of rainfall. The same model applied to the case where the entire tropical rainforest was removed (hence raising equatorial albedo from 7 per cent to 25 per cent) predicted a small *increase* in precipitation in the 5 to 25 degree latitude belt (Potter, Ellsaesser, MacCracken and Luther, 1975). Washington (1976, see Appendix III of this review) imposed an increase of surface albedo to 45 per cent over much of North Africa and the dry areas of western North America, in the GCM used by the National Center for Atmospheric Research at Boulder, Colorado, United States. He found, in agreement with Charney, a large decrease in precipitation over the Sahara, especially over the Sudanian and Sahelian zones. The model predicted statistically significant teleconnections in the form of rainfall *increases* over the Caribbean and Mexico.

110. Ripley (1976) criticized the Charney hypothesis, and by implication the other experiments just described, on the ground that the modelling dealt inadequately with the role of evapotranspiration in regulating surface temperature, and hence the Bowen ratio (i.e. the ratio of convective sensible heat to latent heat fluxes set off by the absorbed solar radiation). He cited experimental results from a savanna area in north-eastern Uganda, reproduced here as Table 2.

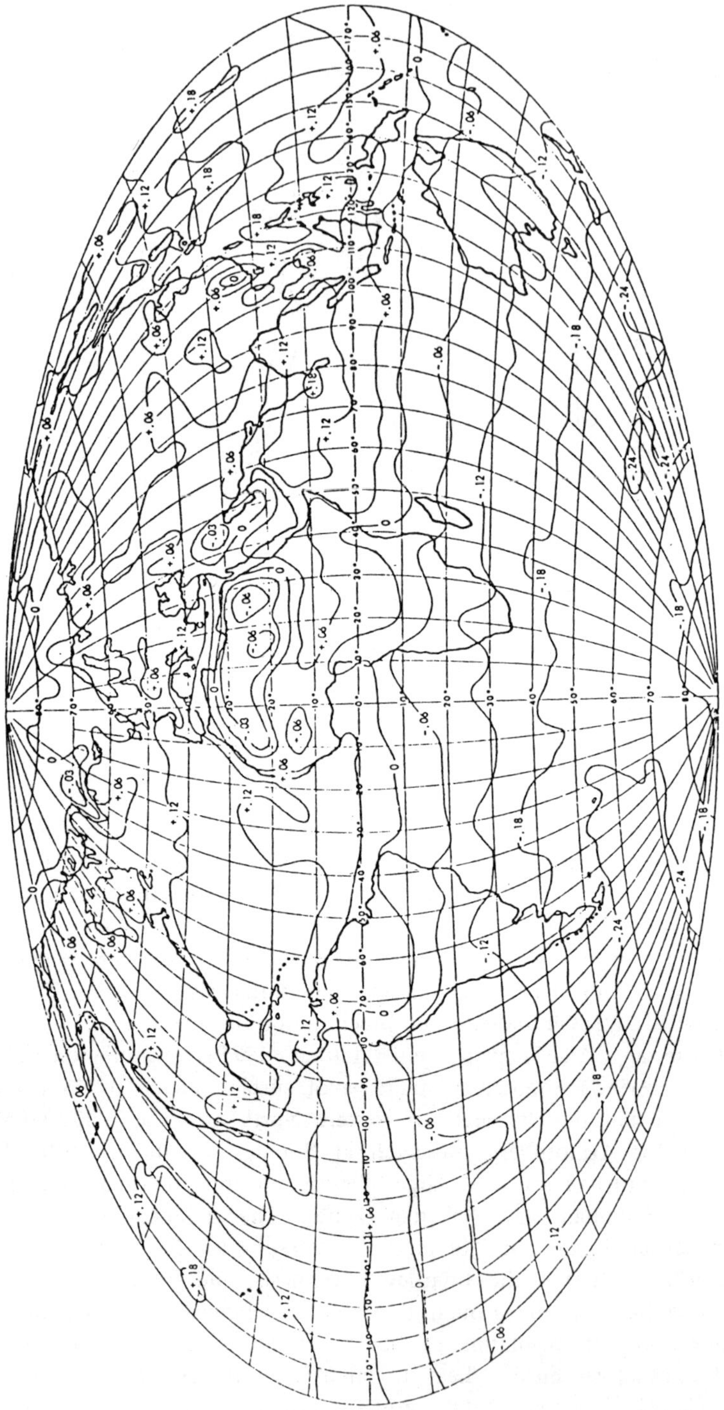

Fig. 14. Radiation balance (i.e. net radiation) at top of atmosphere (cal cm^{-2} min^{-1}) as measured by Nimbus-3, July 16–31, 1969. Note net cooling above Sahara and Iraq (1 cal cm^{-2} min^{-1} is equal to 697 W m^{-2}). (Courtesy V. Suomi).

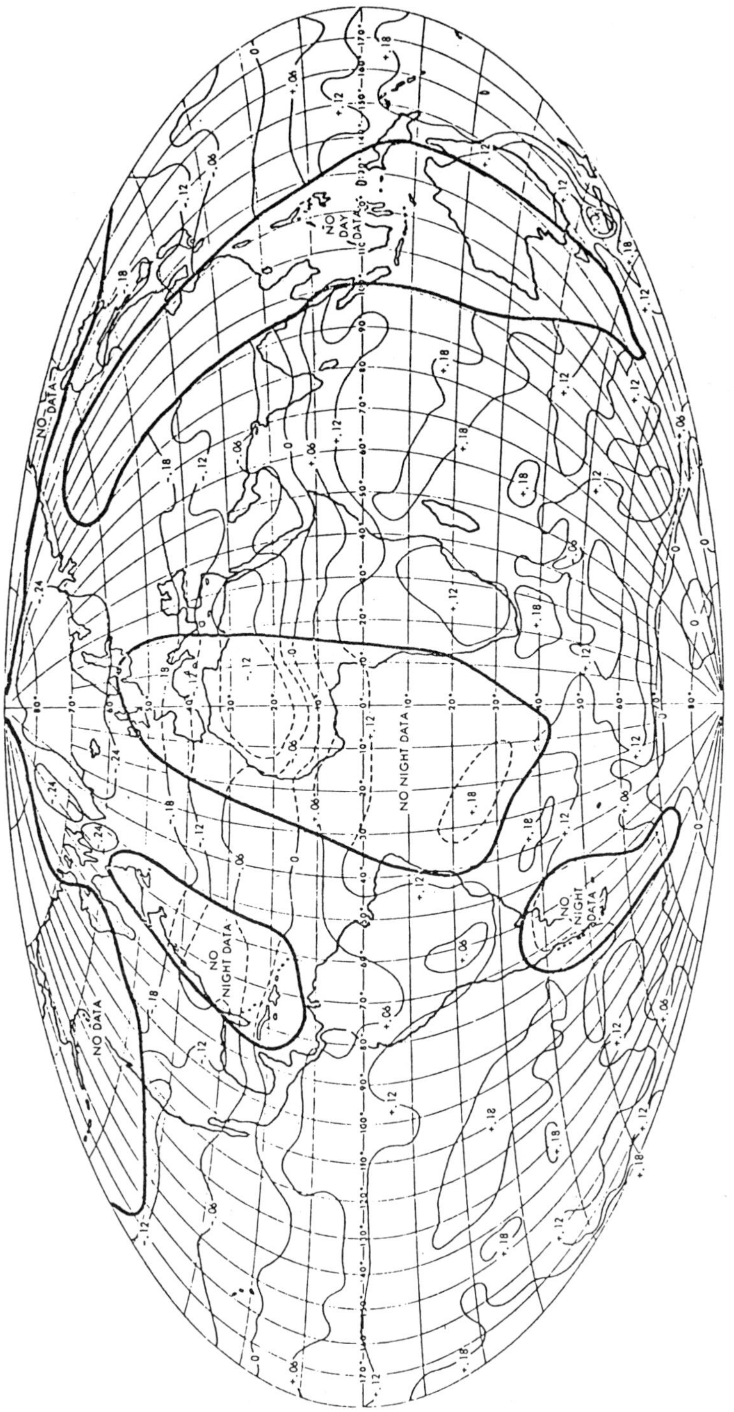

Fig. 15. Similar to Fig. 14, but is for January 21–February 3, 1970. Southern hemisphere dry areas are not radiatively cooled.

TABLE 2. *Mean daily energy balance for four days (0700–1900 local time) for a savanna area in north-eastern Uganda in January, 1964*

Surface condition	Global solar irradiance (J m^{-2})	Net all-wave irradiance (J m^{-2})	Albedo	T_0 * (deg C)
Grazed	2340	1256	0.20	34.7
Ungrazed	2340	1549	0.15	29.2

*Effective surface temperature assuming unit emissivity.

There is no doubt that surface temperature (and hence longwave cooling and convective heating) depend sensitively on the amount of available soil water, and on an effectively transpiring ground cover. The albedo changes are also much smaller than those assumed by Charney. This is a general limitation to all the experiments described. Excessively low and high albedos have to be assumed because of the limitations of the models.

111. The mechanism by which albedo is changed may, of course, be removal of vegetation by drought, by overstocking, by cultivation or all three; or it may be desiccation of the soil itself, soil albedo being linearly and inversely related to soil water content, according to Jackson and Idso (1975). In practice the three mechanisms are likely to occur simultaneously, so that soil moisture content may itself serve as a positive feedback for drought, wet soil favouring renewed rainfall derived from local evapo-transpiration.

112. Walker and Rowntree (1976) applied a limited-area eleven-level baroclinic model to West Africa to test the effect of variable soil moisture. As might be expected, the model predicts large variations of surface temperature between integrations for dry surface conditions, and those of wet surfaces. Figure 16 shows how the temperature diverged

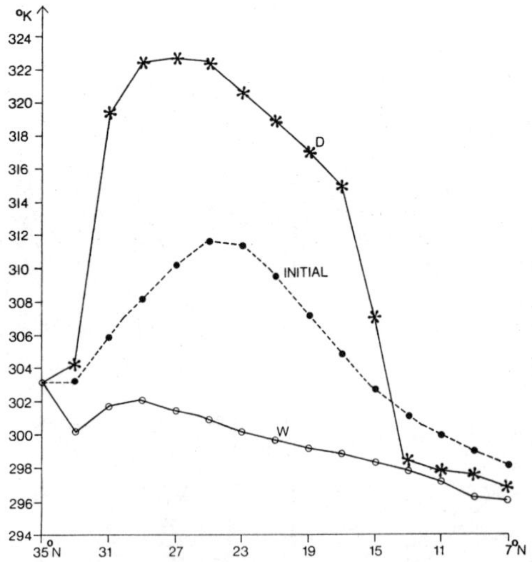

Fig. 16. Variation with latitude of longitudinally-averaged mean surface temperature over West Africa for days 5 through 10 of a numerical simulation by Walker and Rowntree (1976). "D" is for dry surface conditions, "W" for very wet. Third curve shows assumed initial temperature.

from assumed initial conditions 5–10 days after the beginning of the experiment. The dryland surface in Sahelian latitudes was predicted to be over 20°C warmer than a wet surface – which confirms the Jackson–Idso position rather than that of Otterman as regards the surface temperature. The model also predicted large differences in the behaviour of disturbances. With a wet surface, westward-moving cyclonic disturbances with warm cores (due to latent heat release in precipitating clouds) produced widespread rains in the Sahelian and southern Sahara latitudes, the entire system behaving like the oceanic trade-wind belt. The dry surface produced only shallow heat lows. A good case is hence made for saying that Sahelian rainfall feeds on Sahelian rainfall, not so much because of local evaporation, as of the behaviour of baroclinic disturbances[7] over the moist surface. A recent review of such disturbances by Pedgley and Krishnamurti (1976) considerably strengthens the Walker–Rowntree result.

113. All these hypotheses work by influencing the overall dynamics of the desert margin climates, essentially via their effect on rates of subsidence, and hence stability. The possibility remains, however, that cloud micro-physics may be affected by surface conditions. Schnell (1975) has suggested that the cumulus and cumulonimbus clouds of the Sahelian and Sudanian belts of West Africa are "seeded" by organic ice-nuclei raised from the vegetated surface below (Schnell and Vali, 1976; Vali *et al.* 1976). Removal of vegetation destroys the local sources of organic nuclei. Hence well-formed clouds remain unseeded, and remain rainless, thereby accelerating the decay of surface vegetation. This further positive feedback hypothesis is supported by laboratory results indicating that the organic nuclei are highly effective seeding agents at temperatures below about –8°C, which is warmer than for inorganic materials. They are hence more efficient than other nuclei. Nothing is known, however, of the actual microphysics of clouds in the area, and they may already be overseeded naturally – i.e. there may be a superabundance of suitable nuclei.

114. Ormerod (1976) finds little to support the hypothesis that denudation by locust invasions can lead to drought via the albedo mechanism. In fact he finds that major locust invasions of the Sahel–Sudanian belt have tended to *follow* drought periods.

115. In summary, modelling experiments suggest that there may well be a positive feedback process in operation along the desert margin, operating via the increase of surface albedo attendant upon destruction of the surface vegetation layer, and possibly also on decreases of soil moisture and of surface organic litter. The experiments conducted so far have dealt with exaggerated cases, in order to permit the present generation of models to operate with statistically significant results. But a *prima facie* case has been made out for supposing that widespread destruction of the vegetation cover of the dry world may tend to reduce rainfall over these areas still further.

[7] A baroclinic disturbance is a motion system containing horizontal density gradients in which large-scale rising and sinking motions convert potential to kinetic energy or the reverse.

VI. TOWARDS SOLUTIONS

116. Clearing the climatologist cannot hope to banish a phenomenon like desertification that arises largely from unwise human behavior and technology. Nevertheless, one can begin to discern potential climatological contributions in the following areas:

A. A REALISTIC STATISTICAL APPRAISAL OF DROUGHT INCIDENCE, OR OF PROLONGED ADVERSE OR UNUSUALLY FAVOURABLE WEATHER

117. It has been made obvious by recent events that governments, economists, chieftains, herdsmen and cultivators alike have dangerously short memories for adverse weather. It is hence essential that a statistical analysis of rainfall incidence be carried out for all the drier regions of the earth to offer objective advice on the probability of adverse conditions. In effect, this is a mobilization of the historic record, and is quite distinct from efforts to *forecast* change.

118. In many countries such analyses are carried out routinely. Yet even in the United States and Canada farmers, economists, marketing agencies and governments fail to make maximum use of the information derived. From 1955 until 1972 crop conditions in the interior of the North American continent were unusually good (McQuigg *et al.* 1973). It was widely assumed that the excellent crops of those years were due to better technology, and that the weather factor had been largely eliminated. Since then the return of drought and heat in several seasons has shown that crops are still vulnerable. In the winter of 1975–6 severe soil drifting took place in western Kansas and Nebraska, and at least a tenth of the winter wheat crop was damaged or destroyed by prolonged autumn and winter drought, with a complete lack of snow-cover. Fifteen consecutive years of good weather – the odds against which are of very high order – have rendered the land of the world's principal exporter of cereals vulnerable because most people had forgotten that drought recurs – though its return had been predicted on statistical grounds.

119. It is, of course, vital that the statistics be drawn from the right population, and this calls for an exceedingly discriminating approach. If there are discontinuities in the behaviour over time of the general circulation, for example, it is highly misleading to accept rainfall data on either side of the date of change as representative of the future. Hence the utmost skill and caution are required in using what seems deceptively to be an easy way of learning from the past.

120. **WMO should continue to take the lead in this matter, by recommending to all countries the most suitable format for such analyses, and by assisting developing countries to establish internal advisory services capable of reminding the competent authorities of the facts about drought incidence, or of prolonged excessive or favourable rainfall, since the latter tends to encourage unwise multiplication of stock, and extension of rain-fed cultivation into vulnerable areas.** This was, in fact, the first recommendation of the Regional Technical Conference of the FAO/UNESCO/WMO Interagency Project on Agroclimatology at Dakar in 1971 (WMO, 1971). Several excellent WMO Technical Notes (Perrin de Brichambault and Wallén, 1964; Cochemé and Franquin, 1967; Brown and Cochemé 1973) give a suitable account of rainfall variability throughout tropical Africa

and the Middle East, yet it is apparent that little official notice was taken of these documents; nor does their format lend itself to effective political use.

B. TECHNIQUES OF WEATHER AND CLIMATE MODIFICATIONS

121. These techniques have been developed largely in response to the needs of the drier lands. Cloud-seeding, the use of shelter-belts and techniques of dry farming have all been widely applied in these lands. It is doubtful, however, whether the overall effect has been beneficial, or even measurable in some cases. Professional opinion among meteorologists is pessimistic about weather modification except in highly special situations, *even when adequate experimental control can be achieved*. This is hard enough in cloud-seeding experiments. It is much more so when proposals for world climatic modification are aired (such as the removal of the polar pack-ice sheets by artificial means).

122. Cloud-seeding is the technique whereby a dry cloud – i.e., one whose droplets remain too small to fall as rain, the usual condition – is injected with some agent capable of causing precipitation. In mid-latitudes this agent is typically silver iodide particles released from ground generators, but a wide variety of techniques have been applied, many of them airborne (National Academy of Sciences, U.S.A., 1973). It should be emphasized that the major problem is to create the cloud in the first place. Seeding can only work when the atmosphere's dynamics and thermodynamics have created suitable clouds (that may well produce rain themselves even if left unseeded).

123. Countless experiments of this sort have been carried out throughout the world, many of them under uncontrolled conditions by private commercial interests. Relatively few controlled experiments have gone forward, and these have often produced inconclusive results. A careful study, however, has been carried out in Israel (Gabriel, 1967). Beginning in 1961 (mid-October to mid-April) rainy season clouds whose tops were colder than $-5^{\circ}C$ were seeded from below by aircraft generating silver iodide particles at the rate of 800 to 900 g ha^{-1}. In the first five and a half seasons, rainfall was 18 per cent greater during seeding than during non-seeding periods, and in the accessible central regions was 27 per cent greater. This is one of the most favourable results reported for cloud-seeding projects. On the other hand it demanded much flight time, good radar control and good ground observation. It was carried out, moreover, over a very small country.

124. **Wet season clouds in tropical savanna or desert margin lands differ somewhat from those of the Israeli experiment. Nevertheless there will be many occasions in which suitable dry cumulus cloud exists, and where there is some hope that seeding operations may work.** There is a long history of such exercises in Australia and southern Africa, as well as over the North American south-west. *Clearly such efforts should be pushed as hard as possible in drought-prone areas of the drier world.* But at best they only mitigate rainfall deficiencies; the overall control still depends on large-scale dynamics that seeding operations can barely touch.

125. **In 1974 a plan was drawn up (WMO, 1974) for a project for the artificial augmentation of precipitation in the Sahelian Zone, using such technologies. The abrupt end of the drought deferred action, but the WMO Executive Committee Panel on Weather Modification is now proposing an experiment in arid land precipitation enhancement that will probably focus on areas with rainfall above 250 mm. Niger actually undertook an independent exercise in 1973. This experiment should be strongly supported. It will gain in usefulness if it demonstrates economic feasibility for developing countries.**

126. Rainfall enhancement schemes seem most likely to be effective in areas of moderate rainfall, rather than along the desert margin. In the hot tropical countries this implies regions with precipitation marginally adequate for arable cultivation (400–500 mm; or a dryness index of less than 3). In the drier pastoral areas it is unlikely that enhancement programmes can be designed so as to "target" appropriate areas. Potentially seedable cloud systems are too scattered and too random in distribution in these areas.

127. The idea that forest plantations can halt the spread of desert is very old. There is no doubt that hardy stands of trees reduce surface wind speeds, and induce deposition of sand, dust and nutrients deflated from surrounding areas. The Australian flora, in particular, has evolved many species of drought-resistant trees that significantly reduce drifting and deflation during dry seasons and prolonged droughts. The *Casuarinas* (she-oaks), *Eucalypts* and *Acacias* of inland Australia play a major role in counteracting the effects of drought on soil. Mesquite and creosote bush play similar roles in comparable areas of Mexico and the United States.[8]

128. The effort to transform nature in steppe and desert conditions has been a preoccupation of Russian scientists since the late nineteenth century, when Dokuchayev and Voyeykov in particular laid the necessary foundations. In recent decades, under the Soviet régime, major emphasis has been placed upon controlled land-use practices in arid lands, including the deliberate planting of forest and shrub stands (Dzerdzeevskii, 1957; Kovda, 1961; Davitaya, 1962). The system of forest belts in the snow-covered steppes of the Soviet Union, though increasing evapotranspiration, derives the needed water mainly from the extra lodging or pitching of snow. Soviet writers are almost unanimous in advocating wind-breaking techniques in arid and semi-arid areas used for arable cultivation.

129. **Nevertheless the current proposals for green belts along the northern and southern margins of the Sahara are of dubious climatological value. Desertification does not spread outwards from the desert, and the green belt would serve neither as a cordon sanitaire nor as a Maginot Line (though it might well share the latter's fate).** Evapotranspiration would be increased, with consequent loss of valuable water for other purposes, but any increase in precipitation would probably at best be marginal. **This negative opinion urgently needs testing by suitable modelling experiments.**

130. Flooding desert basins is equally dubious. Precipitation depends primarily (though not exclusively) on far-travelled water vapour. Local sources do not moisten the atmosphere significantly. Mist from the Red Sea does provide certain cases along its hilly shores (M. Kassas, personal communication), but little or no change of climate has occurred round such artificial cases as the Salton Sea and Imperial Valley in California; and Lake Nasser and the Caspian do little to render their own shore-lines hospitable.

131. For 60 years or more there has been talk of flooding several of the Saharan enclosed basins with Mediterranean water, though in fact only a tiny area is below sea level. In Australia the Lake Eyre basin could also be flooded. Would such major inundations increase precipitation, and transform the climate? Occasional floodings of Lake Eyre (like that currently in progress) have not had this effect. After each flooding the lake obstinately vanishes. But the myth persists.

132. In 1974 the Rand Corporation of the United States (Rapp and Warshaw, 1974) carried out a simulation modelling of Lake Sahara, "flooding" a substantial part of the

[8] These effects are discussed in greater detail in the component review on ecological change.

inner desert. An integration was then carried out using the University of California at Los Angeles GCM, to test for rainfall changes. In fact the model predicted no significant change of rainfall around the lake, although rainfall was increased over an isolated mountain region 900 km from the shore.

133. Flohn, Henning and Korff (1974) have quantitatively examined a still more ambitious proposal for Africa – the diversion of 5000 m^3 s^{-1} of water from the River Congo into the Chad basin, where it would be used for irrigated agriculture. The water thus evaporated or transpired would mainly be transported westwards, after being carried upwards through the north-westerly monsoon current into the overlying equatorial easterlies. A rough calculation suggested that the most probable gain would be about 7 per cent of the mean annual rainfall. Given the huge cost and political complexity of such a diversion, this seems a small return.

134. A recent model calculation by Lettau (personal communication) along a trajectory of the monsoonal current over west and central Africa does, however, emphasise the role of reevaporated water in the water balance of the far inland parts of the Sahelian–Sudanian zones. Using a modified form of his own climatonomic model for evapotranspiration (Lettau, 1969), and incorporating the effects of advection and thermal structure, he found that 3000 km inland along a trajectory only about one-third of the incident precipitation was generated from water vapour imported from the tropical Atlantic. The rest fell from reevaporated rainfall.

135. Proposals for a sharp reduction in surface albedo by means of spreading carbon dust, or asphalting the surface, seem to involve such prodigious costs as to render them unfeasible.

C. CONTROL OF SURFACE COVER

136. Such control is unquestionably the key to the desertification process (Kovda, 1961). If desert pavements – i.e. the relatively secure wind-stable surfaces of some desert areas – or a reasonably complete vegetation cover (even if dead) can be maintained, soil drifting and deflation are minimized. It is precisely these protective covers that overstocking, unwise cultivation and the use of overland vehicles weaken and ultimately destroy (Rapp, 1974; MacLeod, 1976). The real value of the proposed green belts lies in the added protection they will give to the soil.

137. **The conservation of surface microclimate, on which the usefulness of the land depends, rests on transformations of desert technology, rather than on action within the climatological domain. The ability of the surface to respond quickly and generously to renewed abundance of rainfall depends on the soil's capacity to retain nutrients, organic substances, find materials, high infiltration capacity, and, of course, viable seeds as well as a surviving root system. A surface litter of organic debris may also be important (Schnell, 1975) for precipitation mechanisms, and has some bearing on the surface radiation balance.**

D. MAXIMUM POSSIBLE USE OF MODERN METEOROLOGICAL TECHNOLOGY

138. This is another avenue that must be travelled. The applications of such technology – especially satellite data – are largely in support of short-term forecasting and are hence outside the scope of the United Nations Conference on Desertification. In one respect,

however, such technology bears directly on measures that may be taken by affected countries to aid farmers and pastoralists. This is in the use of satellite imagery to track the major rainstorms of the rainy season, to study their habits and, if possible, to predict their displacement.

139. An excellent initiative of this kind was undertaken by the Agence pour la Securité de la Navigation Aérienne (ASECNA) at Dakar, which began in 1972 using ESSA and NOAA satellite imagery to track disturbance lines across West Africa. Aspliden, Tourre and Sabine (1976) have used similar data to compile a climatology of these rain-inducing disturbances during the GARP Atlantic Tropical Experiment in the summer of 1974. Their results make it clear that **further cooperative action of the west and central African countries to make wider use of such data (already received by several meteorological services in the area) could materially improve forecasting of the displacement of the major rainstorms.**

VII. RESEARCH AND THE FUTURE

140. It is unpopular today for a scientist to argue that more research is needed to help solve social and economic problems. Often such an argument is regarded as a self-serving utterance aimed at perpetuating the flow of research funds. Nevertheless, we must assert categorically that *more research is needed into the relationship between climate and the desertification process* and that the cost of such research is trivial when set against the costs of desertification. The end of continuous drought in the Sahel in 1974 seemed to vindicate those who had maintained that normality would return, and to make fools out of those who argued that a fundamental change of climate was in progress. In fact it did neither. But it *did* lead to a significant slackening of effort and political will, so short is the political timescale. Drought will return to the Sahel, to monsoon India, to inland Australia, to the dry lands of Brazil, Chile, Argentina and Mexico. Moreover, much of the potential for desertification is built up in spells of favourable weather, when herds and crops are increased unwisely. The world community must commit resources now to the effort to learn how to live prosperously in the drier parts of the earth. Doing this demands an increase in productivity of the natural and non-exploited ecosystems, and this implies that even "normality" is not enough; if one can achieve it, one should work towards *improved* surface microclimates.

141. In climatology the lead clearly rests with WMO, supported by the various unions of ICSU. These bodies are already the sponsors of the Global Atmospheric Research Programme (GARP) whose emphasis has shifted increasingly towards the problem of climate and climatic prediction. The climate dynamics sub-programme of GARP is the key to the understanding of the dry climates, considered globally. The first major project carried out under GARP, the GATE experiment, was aimed at a better understanding of the physical and dynamical processes at work in the tropics, and specifically over the tropical Atlantic. Major international attention has been focussed so far on the oceans rather than the deserts, and on atmospheric dynamics rather than surface physical and bioclimatic processes. Neither of these things is wrong. As we showed above, the deserts and the adjacent areas that can be desertified by human misuse, owe their origins to atmospheric subsidence that is an integral part of the general circulation. Equally we showed that ocean temperature anomalies may be related to drought phases on the continents – more generally, that ocean and atmosphere are a coupled system that we have yet to couple up successfully in our models. The same is true of the polar ice surfaces. Successful prediction of events in the drier tropical regions will in the end depend on these worldwide couplings.

142. **A further experiment is needed that specifically examines the problems of the general circulation over the dryland areas of the sub-tropics. The monsoon sub-programme of the forthcoming FGGE experiment of GARP includes specific experiments covering the circulation over south-east Asia (MONEX, see Fig. 17) and West Africa (WAMEX).** A recent numerical simulation of the entire Asian–African winter monsoonal structure by Washington (1976) emphasizes the need to incorporate Africa into the experiment, and ends with the words: "To progress further in our understanding and simulation capabilities, we require more observational material of the type proposed for the GARP Monsoon Experiment . . . and continued improvement of models such as the one used in this study."

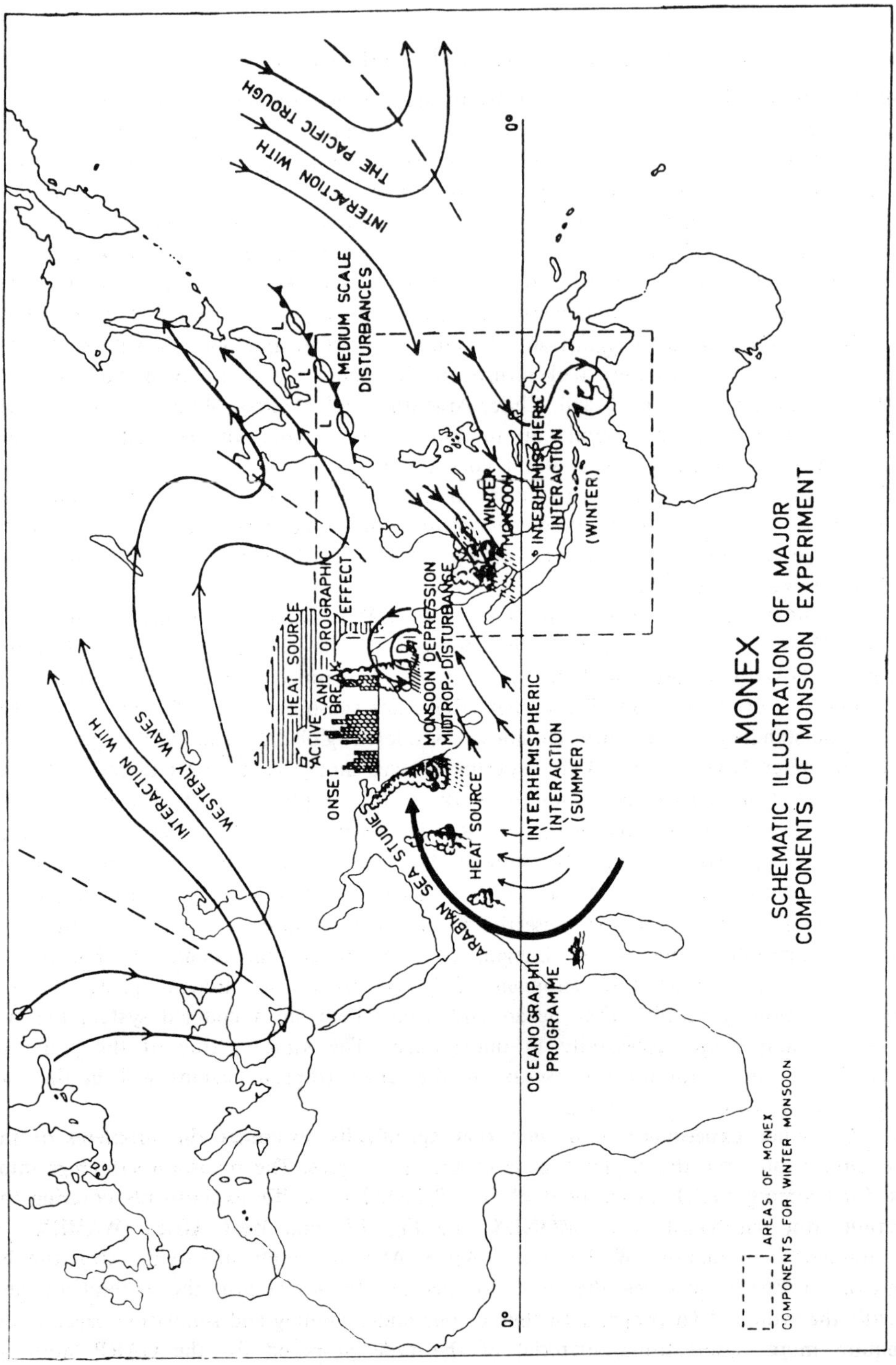

Fig. 17. Sketch of major winter components of MONEX component of the monsoon experiment in GARP.

143. The subjects investigated and the methods used in such studies will be significantly different from those of GATE. Major emphasis will need to be placed in three dimensions on the radiative regime, on aerosol distribution and on teleconnections. It is also necessary to adapt existing GCMs to specific use in these areas. In many ways the Australian continent is the perfect site for such an experiment. There is already a good radiosonde network, and a first-class dynamical research group. It is more difficult, however, to make southern hemisphere GCMs work effectively, in part because of poor data coverage over the oceans.

144. **There is a need, in addition, for a more concerted focus on the physical climatology and bioclimatology of the dryland surfaces, and in particular for a closer synthesis of climatology with dry belt geomorphology, soil science, hydrology and ecology.** This is a reasearch tradition strongly developed in the USSR, from whom major guidance can be expected. Within the WMO family it has been the Commission for Agricultural Meteorology that has had responsibility for this area. Much has already been done, but the results have acquired neither the desirable visibility nor sufficient sophistication insofar as the dry climates are concerned. **A closer working relationship between the climatologist and the various crop research institutes is also to be desired.**

145. **In the modelling area it is essential that multiple avenues be explored. The work discussed in this report shows that the dynamical modellers are already aware of the need to apply their skills to the dry belt. There is a need, however, for climatonomic models of the sort developed by Lettau that are capable of rapid sensitivity analyses, and which are capable of permitting judgment on proposed or apprehended changes in macroclimate, land use or population density. The hierarchy of models into which climatic input is needed extends to the ecosystem and environmental management arenas.**

146. **Finally it is desirable that atmospheric scientists become aware of, and hence sensitive to, the various economic, social and political issue that climatic instability aggravates.** Unless they do, it is quite certain that the world of action will take inadequate notice of climatic factors – as it has always done in the past. This will call for participation in policy and management directed multidisciplinary research, of the sort discussed by Glantz (1976b), whose closing words bear repetition here: "While a reliable long range forecast is not yet possible, it may not even be desirable until that time when some essential prerequisite adjustment to existing social, political and economic practices are being made." Not the least of these adjustments should be made by the scientific community itself.

VIII. CONCLUSIONS

147. We began by posing four questions:

(a) What is the present state of knowledge about secular or long-term shifts of climate?

(b) To what extent can climatic changes be attributed to man's influences?

(c) What are the prospects for longer-term (seasonal and longer) forecasts?

(d) What is the likelihood of significant human amelioration of present conditions?

Answers to these questions will now be given in the light of the previous sections of the report.

(a) *What is the present state of knowledge about secular or long-term shifts of climate?*

148. It is well established that throughout the arid, semi-arid and sub-humid parts of the earth – the areas prone to desertification – rainfall is highly erratic in spatial distribution, and comes during heavy but brief and unevenly spaced showers during the rainy seasons. There are large interannual differences, with some tendency towards quasi-biennial rhythms – i.e. for events to repeat themselves every 2 or 3 years. These rhythms alone, however, are not strong enough for useful local forecasts.

149. Overriding these weak quasi-biennial rhythms are longer periods of persistent drought or abundant rainfall that do not occur with any firmly established frequency. In some areas there is a suggestion of 10-, 20- or 30-year recurrences whose regularity is again too low for useful forecasting. It is unavoidable with present knowledge to regard these longer phases as occurring without rhythm, and to pay attention instead to their historic incidence:

(i) In most areas where analyses have been performed, there may be up to four consecutive drought or favourable years several times in each human lifetime, and once or twice these phases may be prolonged several years beyond. Thus the Sahelian drought lasted 6 years at most stations, as did the abnormally wet period following 1960 in the East African highlands. Much the same is true of monsoonal failures in North India and Pakistan, though the record suggests shorter periods of abnormality. The Australian experience reveals very intense but less prolonged phases of drought or excessive rain, as does the record in the winter rainfall areas of North Africa, the Middle East and Latin America. Regardless of details, all these aberrations trace back to the general circulation of the atmosphere, and its interaction with the oceans. They also affect humid climates, but the presence in these areas of several competing sources of rainfall mitigates their effects.

(ii) The historic and prehistoric record of climatic change reveals much more prolonged abnormalities of either sign. Within some millennia there is evidence of phases of drought or unusually abundant rain lasting one to two centuries. Certainly this has happened in the Great Plains of North America, and there is scattered evidence of similar phases along the margins and in the massifs of the Sahara. In the Rajasthan Desert there is evidence of a two-century failure of the monsoon. These events have a similar time scale to the Little Ice Age in northern latitudes, which lowered temperatures substantially for two and a half centuries. Once again the causes lie in the complex interaction between atmosphere, biota, ice, ocean and soil. Given their overwhelming potential impact, it is vital that we try to understand the mechanisms behind such prolonged fluctuations, and to predict their onset (Brysonn, 1976b).

(iii) On the longest time scales of importance to man – the life span of civilizations – there is evidence of natural desertification of areas once habitable, notably along the edges of the Sahara and in Rajasthan. The present arid phase of climate in both North Africa and Rajasthan is about four millennia old, though there have been briefer fluctuations of rainfall within that long period. In Australia it is much older, and the remarkable aboriginal cultures of the dry interior had perhaps the longest of all human experiences of an unchanging, harsh climate.

(b) *To what extent can climatic changes be attributed to man's influences?*

150. As just stated, the major shifts of climate are related to global changes in the general circulation, and its oceanic interactions. Most authorities agree that these are beyond the influence of man, but in several ways this opinion needs to be qualified:

(i) It is certain that overstocking and unwise ploughing of dryland surfaces expose the soil to deflation (during drought periods) and to sheet erosion and gullying (during heavy rains). Perennial vegetation, the natural protector of soil, is weakened or destroyed. Both processes introduce a new microclimate hostile to the re-establishment of vegetation and stabilization of soil. Desertification is thus a circular process that feeds itself, even on this local scale.

(ii) If such changes effect large areas, positive feedback processes are likely to be set in motion tending to accentuate anomalies in the macro-climate. Most of these depend on increase of surface albedo, which tends to increase the subsidence that is the prime cause of aridity. Albedo increases are due to decreases in vegetation cover, and to the drying out of soil. On the other hand, global general circulation processes can override these man-related feedbacks. The resumption of normal or excessive rainfall in most parts of the Sahel in 1974 is an illustration of the capacity of global events to reassert a more acceptable macroclimate. At this macroclimatic level man has not yet been proven a major actor (see (iv) below). But at the land-surface itself, where he is forced to live, his activities may well dominate the local microclimate.

(iii) The experience of North America during and after the Dust Bowl years of the 1930s should not be forgotten. Natural drought, combined with overstocking and unwise ploughing, produced a disastrous desertification phase from northern Mexico across the Great Plains of the United States into Alberta and Saskatchewan. Much soil was lost, dunes were set in motion, prairie vegetation was destroyed. Land was abandoned on a scale resembling the tragedy of the Sahel. But normal rainfall eventually returned to all areas. It took many years to rebuild soil and vegetation, and hence reestablish the favourable microclimate that had been lost. In many parts of the area there are once again levels of stocking that could not be sustained through a long drought, and much marginal land is sown annually to wheat. In the opinion of some authorities the western plains of North America are again at risk, though there are far more economic hedges to protect the farmer today. In many places in North Africa, southern Asia and Australia deliberate exclosure of animals by fencing has led to the reestablishment of a stable vegetation cover, even in dry periods; and in the fenced and intensively managed dry rangelands of North America fence-lines are visible (for a variety of reasons, not all adverse) from high altitude aircraft at immense distances, and in LANDSAT imagery.

(iv) At the macroclimatic level, i.e. the general circulation and its regional anomalies, the influence of man is not yet proven, though a variety of hypotheses have been put forward and defended (see GARP 1975 for a definitive review). Major attention is now being given to the accumulation in the atmosphere of carbon dioxide (from fossil fuel consumption) and various halocarbons (such as the chlorofluoromethanes used in spray cans and refrigeration). These gases are strong absorbers of terrestrial infrared radiation. They tend to raise global surface temperatures, and hence to affect the normal latitude of the sub-tropical high pressure belts. This is a global process dominated by industrialized societies. Man is also raising the aerosol load of the atmosphere, and in this respect tropical agriculture and pastoralism is believed to play a major role, notably through deliberate burning. Bryson (1974) calculated an annual addition of 60 million tonnes of carbonaceous matter to the suspended load. Meteorologists are now inclined to a majority view (still disputed by some) that this particle load will have only a secondary effect on atmospheric temperatures, and may even tend to raise them slightly. To repeat: the role of man at this global level is still controversial, and existing models of the general circulation are not capable of testing the effects in a conclusive way.

(c) *What are the prospects for longer-term (seasonal and longer) forecasts?*

151. There is as yet no method of forecasting climatic fluctuations on the seasonal and longer time scales. It would be wise to assume that this state of affairs will improve, but only slowly.

152. On the other hand determined attempts are being made to disprove this pessimistic conclusion. It *may* be that climatic fluctuations are as predictable as shorter term weather changes. Certainly they respond to rather different physical controls. The failure of numerical weather prediction to extend its skill beyond a few days is no proof that climate is unpredictable (Lorenz, in GARP, 1975).

153. At least three avenues of progress offer hope. One, already discussed, is the *development of general circulation models* that predict, not specific weather events, but the ensemble of events that we call climate. **It is imperative that such models be developed further as fast as resources will permit.** This work cannot be done in many places, because of data, computer and manpower requirements. All such development deserves the support of the entire world community. **Another possible avenue is the detection of meaningful periodicities. The Quasi-biennial cycles are the only ones that appear at this moment to have enough regularity to offer hope of useful application, though longer periods have been demonstrated in both west and east Africa. A third avenue is that of teleconnections, i.e., of spatial and temporal correlations of climatic events.** Sir Gilbert Walker's Southern Oscillation, involving such teleconnections, was for long thought to offer a means of foreshadowing Indian monsoon rainfall. As we have seen, this same oscillation affects northern Australian rainfall. **Highly significant longitudinal correlations of rainfall, streamflow, pressure and sea surface temperature in tropical latitudes have been demonstrated, and are now being tested in dynamical models.** These three avenues are not really separate. All converge. Fully competent dynamical models of the atmosphere should be capable of predicting anomalies, periodicities and teleconnections. Until such models are in current use we have to explore the other avenues.

(d) *What is the likelihood of significant human amelioration of present conditions?*

154. The most promising avenue for amelioration seems to be that leading towards improvement of surface microclimates by controlled land use. The green belt proposals

before this conference move in this direction, but are unlikely to have a significant impact on large-scale climate. The planned large-scale experiments in precipitation enhancement (to be conducted under adequate control, and with good observational back-up) offer some hope in the more humid areas of the dry regions, i.e. within the zone of arable cultivation rather than near the desert limit. Available satellite technology, already used in some areas, should be put to good use in support of such experiments, and should be developed internationally in support of pastoralists and cultivators. Proposals for climate modification by the creation of large lakes in dry regions will involve high expenditures for small returns (that are in any case hard to predict). Even when coupled with schemes for irrigated agriculture such proposals are unlikely to lead to appreciable amelioration of climate outside the irrigated area and its immediate surrounds. Finally it is important for the dry lands that decision-makers should have access at all times to realistic appraisals of drought incidence, and of prolonged adverse or favourable weather. Even quite crude estimates are better than none at all at the level of central decision making.

APPENDIX I. THE DRYNESS RATIO AND ITS WORLD DISTRIBUTION

155. The dryness ratio is derived (Budyko, 1958) from the heat and water balance equations for the earth's surface. In simplified form these are

$$R = H + LE + \text{heat storage} \tag{5a}$$

and

$$P = N + E + \text{water storage} \tag{5b}$$

where H = turbulent heat flux to air,
E = evapotranspiration,
LE = the turbulent latent heat flux, and
N = surplus water (i.e. run-off or percolation to ground water).

156. Over a period in which neither temperature nor stored water alters significantly (broadly true of annual values) the storage terms vanish. The equations then say that the radiative energy received is used to heat the air (H) or to evaporate water (LE in energy terms, E in terms of water), and that the precipitation runs off, percolates to ground water, or evaporates. These are, in fact, simply statements of the laws of mass and energy conservation.

157. We now introduce two other numerical ratios: (i) the Bowen ratio, B, which is the ratio of the turbulent fluxes of heat and latent heat, H/LE; and (ii) the run-off ratio, C, which is the water surplus divided by the precipitation, N/P. Then for periods over which storage does not change

$$R/LP = D \text{ (the dryness ratio)} = (1 + B)(1 - C) \tag{6}$$

an equation due to Lettau (1969) which relates the dryness ratio to the key energy and water exchange parameters at the surface.

158. Evaporation from the earth's surface requires the use of energy, derived mainly from the net radiation. Priestley and Taylor (1972) proposed the following evaporation equation:

$$LE = \alpha S (R - G) \tag{7}$$

where α = Priestley-Taylor parameter;
S = parameter related to the rate of variation of saturation vapour pressure with respect to temperature; and
G = flux of heat to the soil.

159. If mean annual values are used in (7), suitably adjusted for non-linearities and time covariances, it can readily be shown that

$$D = \frac{1 - C}{\alpha S} \tag{8}$$

Hence the Budyko–Lettau dryness ratio D is inversely related to the Priestley–Taylor parameter α. Under tropical desert or semi-arid conditions D is approximately $5/4\alpha$. In some ways α is a better aridity index than D. Under unrestricted soil moisture conditions (i.e. with potential evapotranspiration, and in the absence of strong advection i.e. warm winds) it approaches an upper limit of $\alpha_p = 1.26$ (conveniently approximated as 5/4, the suffix p indicating potential). Along the margins of the warm deserts it falls to about 0.35 or 0.40 (when applied to annual totals), but is zero in their rainless cores.

160. In his well-known UNESCO map of the arid zone Peveril Meigs III (Meigs, 1961) made use of another index, Thornthwaite's (1948) moisture index I_m, which is defined (Mather, 1974) as

$$I_m = 100\left(\frac{P}{E_p} - 1\right) \tag{9}$$

where E_p = potential evapotranspiration,
from which it follows that

$$I_m \times 10^{-2} \cong \frac{1}{\alpha_p SD} - 1 \tag{10}$$

In warm climates (with effective mean temperatures near 25–30°C) S approaches or exceeds 4/5. Hence (10) can be approximated as

$$I_m \times 10^{-2} \cong \frac{1}{D} - 1 = \frac{\alpha S}{1-C} - 1 \tag{11}$$

If water surplus is zero, as it usually is in the arid climates, (11) reduces to

$$I_m \times 10^{-2} \cong 0.8\alpha - 1. \tag{12}$$

Along the desert margin this gives $I_m = -68$, when $\alpha = 0.40$, almost precisely the value chosen by Carter and Mather (1966) in their application of Thornthwaite's index to the arid/semi-arid border.

161. The approximations given in (11) and (12) are valid only for tropical and sub-tropical temperature régimes, and for advection-free conditions. The latter is patently not true of the desert margins, nor of oases. The effect of warm advection (as when heated desert air passes across moist surfaces) is to increase the value of α_p applicable in equation (10), perhaps from 5/4 to as much as 7/4 in desert areas. The effect is to increase D and decrease actual α (and hence I_m), but not by large amounts. The effect of having run-off finite (as it may be after rainfall of great intensity) is also to increase D and decrease α.

162. In spite of these reservations relations (10), (11) and (12) show that all three of the aridity indices described are roughly interchangeable. There is no inconsistency in using, for example, the Meigs map alongside the Henning map of the dryness ratio (Fig. 2). Isolines of I_m are also approximately isolines of D and α in warm areas.

163. Too much precision should not be attributed to annually-integrated indices of this sort. The seasonal distribution of rainfall, for example, is not allowed for, and 400 mm of rain falling in low-radiation, winter conditions may support a higher biomass than the same amount falling in summer. Distribution of rainfall *within* the rainy season is also

important but is ignored by the indices. Hence precise correlation between the annual index and biotic distributions and irrigation requirements is not to be expected.

1. Calculation of Ratio

164. Figure 2 of this review is based on revised calculations of D carried out by Henning at the Meteorologisches Institut of the University of Bonn. Figure 2 is a drastic simplification of Henning's detailed original, which is being published separately, with a comprehensive explanatory text. Only a summary of the method of preparation is given here.

165. Net radiation (R) is the first element required, and is not available as a measure quantity. Hence it must be calculated. Henning (1970) made use of a semi-empirical relationship due to Albrecht (1962) in calculating world-wide values. The Albrecht relation first calculates clear-sky solar radiation as a function of surface vapour pressure, solar elevation, solar constant, solar distance, surface pressure and surface albedo, and then corrects for the presence of normal cloudiness. Net long-wave cooling of the surface is calculated from observed screen temperature, cloudiness and vapour pressure. Finally net radiation is estimated from

$$R = I(1-a) - R_* \tag{13}$$

where I = computed global solar radiation
a = albedo, and
R_* = net long-wave cooling.

This equation was solved for 3600 stations around the globe.

166. The second input, precipitation, was taken by Henning from standard climatological stations, and from a variety of detailed published regional analyses. Since rainfall varies rapidly with height in hilly areas, whereas net radiation does not (outside snow-covered areas), the spatial detail in the original analysis, and on Fig. 2, largely reflects rainfall gradients.

APPENDIX II. A CHRONOLOGY OF CLIMATIC VARIATIONS IN AND NEAR THE MAIN DRY REGIONS

L. KEITH HENDRIE

INTRODUCTION

167. The recent emergence of two important facts has led to the generation of considerable concern over climatic variation and its possible impact upon man's future life style and food supply. Firstly, there is now realization that the period from the 1920s to the early 1950s, upon which we based our climatic "normals" and expectations, was unusually benign, with relatively little short-term climatic variation. This became clear during the 1960s following the emergence in the northern hemisphere of a significant cooling trend, initiated in the late 1940s to early 1950s. Even though this trend can probably be considered to have leveled off at this stage (J. Murray Mitchell, personal communication), there is now an apparent tendency towards greater variability in seasonal and annual means of climatic parameters. Secondly, palaeoclimatic studies (e.g. Olausson, 1975) have revealed that rather than a major climatic pattern gradually changing to another régime, it is more likely to undergo a fairly rapid shift, or flip, to quite different conditions.

168. While deserts and their sensitive semi-arid fringes have always existed on the earth, their spatial extent and location have varied considerably with time. These changes have been in response to different mechanisms, dependent upon the time scale involved. In the very long-term, they have been in response to geotectonic activities, such as continental drift and the formation of mountain barriers, aspects which will be excluded from this study. Over periods of a few millennia or less, the changes have been natural, and mainly in response to major alterations in atmospheric and oceanic circulation patterns. As mentioned previously, it is now known that climatic change and fluctuation of this type can occur quite rapidly. Aridity changes over relatively short periods of time, of centuries, decades, or even individual years, are responses to climatic fluctuations and interactions, the actions of man, or their interaction. It is this type of event that is currently forcing us to focus upon the problem of desertification.

169. This paper investigates the present state of knowledge of secular changes of climate and climatic fluctuation, as they relate to desiccation, for the Pleistocene, and in particular the Holocene. The term Holocene (for the last 10,000 years) is preferred here to its synonym Recent. This permits the unambiguous usage of Recent to refer to the last few years when discussing climatic iterations and alterations. All major arid and semi-arid regions of the globe are considered, although data for some areas are meagre, inadequately dated, or of questionable reliability for generalized extrapolations. The arid and semi-arid areas of prime consideration are those delineated as already affected by, or vulnerable to, desertification processes on the World Desertification Map, prepared under the auspices of UNEP by FAO in conjunction with UNESCO and WMO (1976). These areas are similar to those shown as semi-arid to extremely arid on the 1960 revision of Peveril Meig's maps appearing in McGinnies, Goldman and Paylore (1968).

170. All the forms of relevant palaeo-environmental data have been utilized in the consideration of climatic variations in these regions. Detailed information concerning the many palaeoclimatic indicators and dating techniques available, and the problems associated with their use can be obtained in Nairn (1961), Butzer (1971), Tite (1972), Kutzbach (1975), and the United States Committee for the GARP (1975). Because of the difficulty in determining palaeoclimatic indications of many climatic parameters, extrapolations of temperature and effective precipitation have been used in the main to suggest likely past desiccation events.

ATMOSPHERIC CIRCULATION AND CLIMATIC VARIATION IN ARID AREAS

171. The important atmospheric dynamic considerations that must be borne in mind during climatic reconstructions of arid areas are discussed in detail elsewhere in this report. In summary, the sub-tropical deserts (cold deserts and marine deserts will not be considered here) lie under the belt of sub-tropical anticyclones and are dominated by large-scale atmospheric subsidence which may be supplemented by other dynamic mechanisms. Their margins are controlled by the location and movements of the Intertropical Convergence Zone on the equatorward side (wet summers, dry winters), and the westerlies and associated cyclonic disturbances on the poleward side (wet winters, dry summers). Any changes in these anticyclonic systems will be reflected in changes in the patterns of subsidence and hence precipitation and desert fringes. Flohn (1964) suggests that shifts of these anticyclonic belts are related to changes in the equator to pole temperature gradient, a poleward movement occurring with a decrease in the temperature gradient, and vice versa. Two hypothetical models of the influence of climatic change and fluctuation upon these north and south desert margins have resulted. One proposes a latitudinal displacement of the whole zone, with one margin becoming more arid and the other wetter (Tricart, 1963). The second suggests expansion or contraction of the whole subsidence zone resulting in similar changes at each margin. Although there is some field support for these proposals, they are probably too simplistic (Fairbridge, 1972).

172. The weight of evidence now indicates widespread sub-tropical aridity during the late Pleistocene. However, from 20–15,000 B.P. the climate of sub-tropical North Africa was cool and very arid (Butzer, Isaac, Richardson and Washbourne-Kamau 1972; Grove 1972); but simultaneously in South Africa it was cool and wet (van Zinderen Bakker and Butzer 1973). Also a recent study of the late Quaternary climates of Australia and New Guinea by Bowler *et al.* (1976) indicates major aridity dating from 18,000 B.P., and suggests complex temporal relationships between glacial events in the mountains of both New Guinea (equatorward side of anticyclonic belt) and Tasmania (poleward side).

173. Thus it is necessary when considering climatic reconstructions of areas under the influence of subsidence, and hence vulnerable to desiccation procedures, to look at latitudinal variations in addition to regional fluctuations. Chu (1973) and Barry (1976) also propose careful examination of any longitudinal change.

QUATERNARY CLIMATIC CHRONOLOGIES OF THE ARID REGIONS

174. The areas of prime consideration are those portrayed on the World Desertification Map, and which for the purposes of this section have been subdivided into nine convenient regions. These regions are designated as follows: Africa (North), Arabian Peninsula, Turkestan Desert, Iran–Thar Deserts, China (Takla Makan–Gobi), North America, Africa (South), Australia, and South America. This is felt to be justified since most of the research reviewed has been conducted with regional scopes of no greater extent. Additionally, desertification is frequently only a regional process at most, and so less confusion should result from this approach. It is, however, recognized that at least some of these separated regions are in fact contiguous with respect to climatic processes.

175. In order to provide a concise summary of known aridity phases that have affected each of the nine regions over the last half million years, the accompanying chart (Table 3) was developed. It displays the most widely accepted palaeoclimatic interpretations of a great number of studies drawn from a wide variety of disciplines. Each line represents a particular climatic phase affecting a particular region of sub-region, and is accompanied by a code indicating at least the type of phase and the nature of the palaeoclimatic indicators involved in its determination.

176. As indicated previously, the main climatic parameter considered was effective precipitation, and in the absence of any estimation of it, temperature. One difficulty experienced, and not fully overcome, was determining whether the mention of "precipitation" in reports and papers referred to effective or absolute precipitation. This is of importance since changes of temperature affect evaporation rates and hence water availability. Thus if temperature and precipitation change simultaneously, opposite changes in effective and absolute precipitation may result.

177. When viewing the chart it should be remembered that its apparent temporal precision is subject to at least the same margins of error as were the original chronologies from which it was developed, and that a number of contentious issues of interpretation are camouflaged. Also the apparent regional generality of an event could well be misleading, since often the lines represent the interpretation resulting from palaeoclimatic information afforded by one site or locality which could exhibit local climatic peculiarities out of phase with those affecting the overall region. However, where possible, the use of this type of data has been avoided. Further, the absence of marked aridity phases for individual regions, particularly at the beginnings and ends of chronologies, should not be construed to infer that none existed – it can be the result of a lack of available data or of a quantitative chronology.

178. No referencing of information sources has been provided upon the chart since the extensive bibliography that would have resulted from this section would have exceeded acceptable limits for the total paper. However, complete referencing, along with greater detail and attempted regional graphs of change of precipitation and temperature will be provided in a future monograph.

179. The following general points emerge from the chart:

(1) The only region with anything approaching a sufficient data coverage, with frequent cases of independently dated materials of different types, is North America. Even here there are situations of awkward gaps and deficiencies in information, inadequate dating, and conflicts of interpretation.

(2) The Australian and Africa (North) regions, and to a lesser extent the sub-region of

TABLE 3. *Table of periods of aridity influencing each of the nine regions*

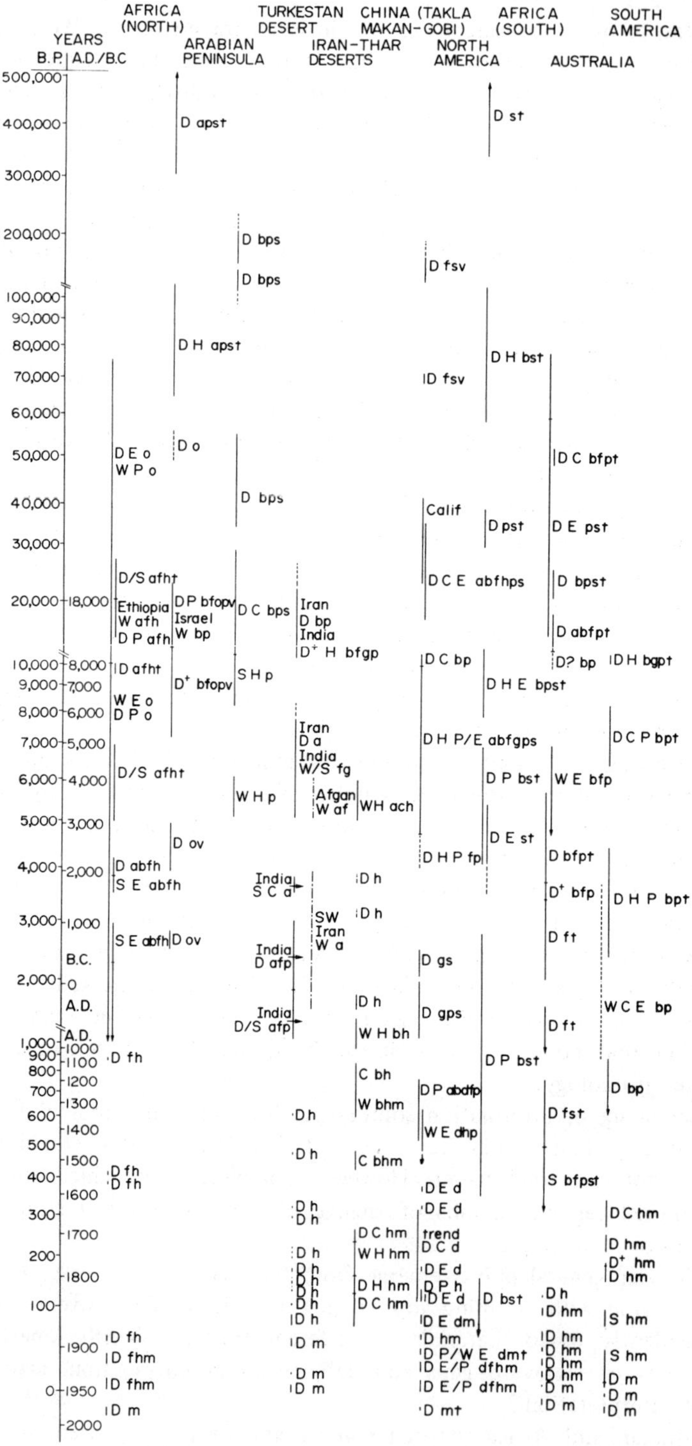

India, have moderate coverage, although the data are often related to limited localities and could well require considerable amendment in the light of future studies.

(3) Gross inadequacies related to lack of accessible data, limited types of palaeoclimatic information, inadequate dating and/or major conflicts of interpretation exist for each of the remaining regions. Much improvisation has been necessary, and the assessment of the climatic variations influencing them can at this stage be considered tentative at best.

(4) The fine detail of the effects of climatic iterations and alterations are evident only for the last 1–200 years, the coarser resolution of the impacts of climatic fluctuations are apparent for no more than the last 100,000 years, with only the very coarse detail related to climatic change evident prior to this. Hence the impacts of the different types of climatic variation are evident over a period of time no more than an order of magnitude greater than their maximum duration.

1. Africa (North)

180. This region encompasses the most extensive stretch of arid and semi-arid land on the globe. The recent devastating drought of the Sahel has given rise to most discussion and action with respect to desertification processes, although the overall area has been researched in only a spotty manner. The most comprehensive documentation of relevant climatic variations for this region are Butzer (1971, 1975) and Livingstone (1975).

181. In general the first well-dated major period of widespread aridity commenced about 7000 B.P., and extended to about 5000 B.P., and followed a warm, wet phase

---

KEY TO TABLE 3

Precipitation:
D Effective precipitation less than now (D^+ = much less)
S Effective precipitation approximately the same as now
W Effective precipitation greater than now

Temperature:
H Average temperature in excess of present value
N Average temperature approximately the same as at present
C Average temperature less than present value

Location:
E Equatorward side of the desert area
P Poleward side of desert area

Palaeoclimatic indicators/evidence from which the information was derived:
a archaeological
b botanical, faunal evidence
c fossil
d dendrochronological
f lake levels, floods
g glacial, ice sheets
h historical and other records
m meteorological, hydrological
o ocean cores (all types)
p palynological
s soils, pedological
t topographical, geomorphological, geological
v varves

which commenced with the onset of the Holocene. After a short break arid conditions were again prevalent from before 4000 B.P., and with minor fluctuations have persisted to the present. Pre-Holocene dating and information is insufficient to be really reliable. However, the general trend of southward migration of the belt of anticyclonic subsidence during the Würm, followed by a rapid northwards migration to its present position where it has remained during the Holocene, has been demonstrated by Diester-Haass (1976).

2. Arabian Peninsula

182. A relative deficiency of quantitatively dated palaeoclimatic indicators, and conflicting information, particularly between the Mediterranean coastal and other areas, have hampered the development of a reliable chronology for this region. The most comprehensive literature concerning relevant climatic variations for this region has come from Butzer (1971, 1975). There are indications of extensive aridity until about 300,000 B.P., and again from before 100,000 B.P. to about 65,000 B.P. The onset of the next major dry period occurred at roughly 24,000 B.P., although there remains evidence of this being a moist time in Israel. From about 12,000 B.P. to 7000 B.P. conditions became extremely arid, whereafter they moderated slightly, and with minor fluctuations have continued to the present.

3. Turkestan Desert

183. Analysis of the past climates of this region, as in the case of the Arabian Peninsula suffers from a shortage of data, and most of the arid periods must be considered tentative and requiring further confirmation. The main periods of aridity identified occurred around the latter portion of the Middle Pleistocene with a short moist interval near 150,000 B.P., from roughly 55,000 B.P. to 35,000 B.P., and from 28,000 B.P. almost to 8000 B.P. Assumedly more recent conditions have, like the adjacent regions, been similar to now with minor fluctuations.

4. Iran – Thar Deserts

184. Compared to Iran, the Indian sub-region has a relatively sound chronology developed from 20,000 B.P., and is well documented by Singh (1971, 1976) and Singh, Joshi, Chopra and Singh (1974). Additionally a greater variety of evidence is available to correlate with these events. The major periods of aridity coincide with extended failures of the monsoon, assumedly related to major shifts in the global atmospheric circulation patterns. Singh (1976) identifies two major desiccation episodes, the first extending from before 20,000 B.P. to the onset of the Holocene, and the second from between 4–3000 B.P., and lasting essentially to the present with only minor fluctuations. Historical records also provide an insight into recent arid iterations.

5. Takla Makan – Gobi

185. The long history and culture of the Chinese civilization has ensured an extensive documentation of extreme climatic events, phenological events, bird migrations, food storage procedures, odes, and religious festivals. These records, which date back almost

5000 years, have formed the major data source utilized in attempts at climatic reconstruction for the region, the most notable being that of Chu (1973). Some archaeological, palynological and glacial studies have also been undertaken. It has been difficult to assess the influence of climatic variations upon precipitation for the region, and recourse was made to Korean data to obtain some indication.

6. North America

186. This region has an almost voluminous literature related to climatic variation affecting it, a reasonable proportion of which is relevant to this study. Thus dating tends to be more precise, with more studies to provide correlations. Additionally there is a greater cross-disciplinary spread of relevant data. However, some conflicts of interpretation still occur, particularly related to the arid areas which have been less extensively researched and where data problems are greater.

187. The major periods of widespread aridity have been dated as from about 33,000 B.P. to 17,000 B.P.; from about 12,000 B.P. to 10,800 B.P. when it was cold and dry, followed by a warm, dry phase until about 4800 B.P., and in northern (polar side) areas until as late as roughly 4000 B.P.; from about 2600 to 1100 B.P.; from about 1100 A.D. to 1400 A.D.; and finally a dry and cold phase from about 1640 A.D. to 1850 A.D. through the Little Ice Age. Additionally, numerous dry iterations have been established accurately over the last 4–500 years by dendrochronological techniques.

7. Africa (South)

188. The South African region suffers from having insufficient data that have been soundly and quantitatively dated, and consequently the displayed trends are less definite. Much of the data have been gathered from geomorphological and pedological studies which have a relatively coarse temporal resolution, and also from cores that have been only stratigraphically dated. It appears certain that the results of future studies involving more exacting dating techniques will reveal more variation within the general trends displayed here. The most comprehensive relevant references are Butzer (1971, 1974) and van Zinderen Bakker (1973).

189. The main periods of aridity have been approximately from 40,000 B.P. to 30,000 B.P., from 14,000 B.P. to 8000 B.P., from 5500 B.P. to 4200 B.P., and from 2900 B.P. to 350 B.P.

8. Australia

190. Australian research into problems of climatic variations has begun to accelerate recently, with major contributions including those of Bowler (1975), Bowler *et al.* (1976) and Singh (1976). A greater variety of palaeoclimatic indicators are now being utilized and interesting results are beginning to emerge. However, for the sake of this topic the data remain meagre, being largely the product of a few isolated sites.

191. Major periods of aridity included from roughly 80,000 B.P. to 60,000 B.P., and then until 15,000 B.P. on the equatorward side (which provides an interesting correlation with the North African case as documented by Diester-Haass, 1976); from about

5800 B.P. to 2000 B.P.; and from about 1600 B.P. to the present with only minor fluctuations.

9. South America

192. This region is awkward because it is composed of deserts on either side of the Andes, each influenced by supplementary subsidence induced by local factors. It also suffers from a deficiency of suitable and adequately dated palaeoclimatic information, making the trends displayed tenuous at best. The most comprehensive papers are probably those of Heusser (1966, 1976) and Mercer (1976).

193. The major arid phases identified are roughly from 8500 B.P. to 6500 B.P., from 4500 B.P. to 2500 B.P., and from 1050 A.D. to the present with minor fluctuations.

CONCLUSION

194. While regional detail has been compiled, it is difficult to draw global implications and generalizations from the data at hand, probably because they are far more refined and of a greater resolution for some regions than others. The issue seems complex, as suggested by Fairbridge (1972), with latitudinal variations as shown for example by Diester-Haass (1976) and the North American case between roughly 650 B.P. and 500 B.P. This type of situation suggests latitudinal oscillations of the mid-latitude belt of anticyclonic subsidence, with it being displaced equatorward during periods of increased poleward temperature gradient, and vice versa, thus supporting the proposal of Flohn (1964). Longitudinal variations are also evident, as shown for example by the Indian–Iran differences, and the E-W variations across the Arabian Peninsula. A further point of interest is the simultaneous aridity on the equatorward sides of the North African and Australian desert regions during the later Würm. This suggests a simultaneous equatorward movement of the subsidence belts in the two hemispheres. However, this is not supported by the trends in South Africa (van Zinderen Bakker and Butzer, 1973).

195. Thus it seems that more research is necessary to refine the trends currently evident and to broaden the areal scope of data acquisition. Consequently support should be provided for research that may reveal these necessary details.

APPENDIX III. MODELLING THE DRY CLIMATES

TZVI GAL-CHEN

This technical Appendix is designed to supplement, and to be read alongside, Section V of the text.

INTRODUCTION

196. The theoretical study of natural systems is often facilitated by introducing various idealizations. Some of these consist of suppressing certain supposedly irrelevant details, so that the important processes may be more readily examined. In other instances certain features which may not be irrelevant at all are nevertheless omitted in order to render the theoretical treatment less awkward. One then distills from among the complex interactions which characterize the system those relatively few ingredients that permit the construction of simplified models that will have a strong relevance to natural phenomena. This approach in itself can not prove very much and the next step is some sort of experimental or observational verification, or both. In the process of verification one must demonstrate conclusively that the conclusion reached about the simplified system can be extrapolated and applied to the real system.

197. Rigorous techniques of experimental and observational verification are often impossible to follow, since many natural systems are unique and cannot be duplicated. Consequently, one must use partial analogies and accumulated evidence. It is for this reason that one cannot yet answer quite simple questions such as: "What is the percentage change in solar constant required to initiate an ice age or to melt the polar ice caps?"

198. In principle, if one could find a planet which differs from ours only by its solar distance, the answer would be relatively easy. In reality planet earth has a unique earth–atmosphere system, which interacts with the incoming (short waves) solar radiation and the outgoing long wave radiation. Thus the incoming solar radiation is only one ingredient of the heat budget (see e.g. Hess, 1959, pp. 155–160).

199. Limitations of this kind have in the past made climatic theory a weak and ineffective handmaiden of observation. The advent of high speed electronic computers made it possible to solve numerically the non-linear hydrodynamic equations of motion, and the equations of radiative transfer. Since the large-scale dynamics of world climate are governed mainly by the global distribution of radiative energy and the inequalities of land and sea, numerical simulation of the physical laws governing the large-scale climate should be an important tool in climatic theory.

200. If such numerical simulations are successful they will produce an artificial system that is sufficiently close to the natural system. Unlike the natural system, however, one can deliberately change the parameters that control the model's climate (e.g. solar radiation, chemical composition, topography) and thus get an insight into the physical controls of the real climate. The model has, therefore, the potential of becoming a vital link between theory and observation.

201. Unfortunately even the most detailed General Circulation Models (GCMs) are only crude simulations of the climate system. Consequently the results of GCM experiments, or

any other climate models, must be interpreted with extreme caution. It follows that when one is conducting a numerical experiment with a particular climate model, one must be aware of the model's inherent limitations. In the context of modeling the dry climate it must be realized that, though the fundamental causes are primarily dynamical, their details depend on the interaction between the atmosphere and the earth's surface. Extremely important are vegetative cover, which determines surface albedo and hence affects the radiation balance, and surface hydrology, which affects surface evaporation and hence precipitation. The space scale needed to model the surface climate correctly is typically several orders of magnitude smaller than large-scale atmospheric motions, and the associated time scale is much larger. These constraints dictate a rather crude representation of the surface conditions in a GCM. Notable examples are a simplified ground hydrology and specification rather than prediction of surface albedo.

202. Precipitation is quite obviously an important climatic element. In the arid and desert climates of the world, it comes mainly in the form of convective rains, squall lines and thundershowers. Again, this involves scales of motion (3–100 km) which cannot be resolved explicitly in a GCM.

203. Precipitation itself may be critically dependent on the spectrum and size of cloud condensation nuclei, as well as other microphysical considerations such as collision and coalescence. This involves scales of 10–1000 mm. The above cited examples demonstrate the kind of difficulties which are faced when one tries to develop a realistic model which incorporates the essential dynamical and physical processes of a dry climate.

204. In the subsequent parts of this Appendix, a hierarchy of models that may be used to explain deserts and desertification is outlined. Some of these models, mostly of the GCM type, have already been applied to this exercise and their results are critically reviewed. It is argued that progress toward an understanding of the dominant factors in desertification will require steady development of an almost continuous spectrum of models of increasing physical and mathematical complexity.

A HIERARCHY OF MODELS

205. Several classes of models may in principle be applied to study the dry climate. They may be classified as follows:

1. *General circulation models* (GCM) simulating the large-scale features of the global circulation.
2. *Mesoscale and regional models* simulating sub-synoptic phenomena, such as development and progress of a thunderstorm array, land-sea breezes, heat islands, hurricanes, fronts, and others.
3. *Convection models* simulating the fine structure of the mesoscale phenomena up to and including individual cloud, haboobs, tornadoes, waterspouts, and others.
4. *One-dimensional cloud models* with height (z) and time (t) as independent variables. These are essentially microphysical models with much simplified dynamics.
5. *Climatonomic models* of the surface and storage processes of the sort discussed by Lettau (1969).
6. *Mechanistic models* that investigate specific feedback mechanisms such as change of albedo, variation of soil moisture storage, increased aerosol load, and comparable mechanisms.

To date only models of the type 1, 5 and 6 have been applied to the desert problem.

206. Models of the type 1 through 5 may be broadly categorized as simulation models, where in the context of the phenomena of interest they include as many interacting physical processes as possible, the primary objective being to simulate observed phenomena.

207. These models are often too complicated to allow unambiguous results. It is thus difficult to trace cause and effect relationships to interpret which physical process or processes are responsible for causing a particular effect. Thus the mechanistic models are used to suggest the relative importance of various terms in the simulation models. Although the ultimate objective is to include jointly all the coupled feedback processes in a realistic fashion, it is unlikely that this can be done successfully without the understanding derived from simpler models of individual processes. The most celebrated and successful example of a combined use of a mechanistic and simulation model is the testing by numerical model of Charney's theory (Charney, 1975a and b), which relates changes in surface albedo to drought in the Sahel. Charney's theory neglects advection, precipitation and evaporation, all obviously very important, and to test his theory the specific feedback mechanism is tested in a detailed GCM (Charney, Quirk and Stone, 1975).

208. We now focus our attention on a more detailed description of the simulation models, so as to point out the potential applicability and limitations of the models to a variety of desertification problems.

General Circulation Models

209. It is generally accepted that the root causes of drought or excessive rainfall are global. It thus follows naturally that global circulation models are the natural vehicle for studying desiccation. Figure 18 displays schematically the components of the climatic system. These are broadly grouped as (i) thermal properties, which include the temperature of the earth's surface, oceans and atmosphere; (ii) kinetic properties, which include wind, ocean currents and motion of ice masses; (iii) aqueous properties, which include humidity, cloudiness, ground water, lake levels, and the water content of snow and sea ice; (iv) static properties which include pressure, density, salinity, topography and the composition of the air, in particular minor constituents like CO_2, O_3 and H_2O, which affect the thermal properties of the atmosphere; and (v) the living cover (biomass). For a complete discussion of the various processes and their role in climate modeling, see e.g., U.S.A. Committee for the GARP (1975), GARP (1975) and Schneider and Dickinson (1974).

210. A cursory examination of Figure 18 reveals few of the major problems associated with climate modeling. It is apparent that one has to deal with physical processes that are occurring on time scales ranging from a few seconds (small scale turbulence) to a few thousand years (the advance and retreat of the glaciers). Tentatively this problem may be solved by invoking the notion of external versus internal processes (e.g. Leith in GARP, 1975). The internal processes are the rapidly and relatively rapid fluctuations, which are embedded in the external system, including the oceans, which provide relatively slowly changing external influences on the internal system. As far as the numerical computations are concerned, the external forces are fixed. It is clear, however, that the only genuine external process is the solar radiation at the *top* of the atmosphere.

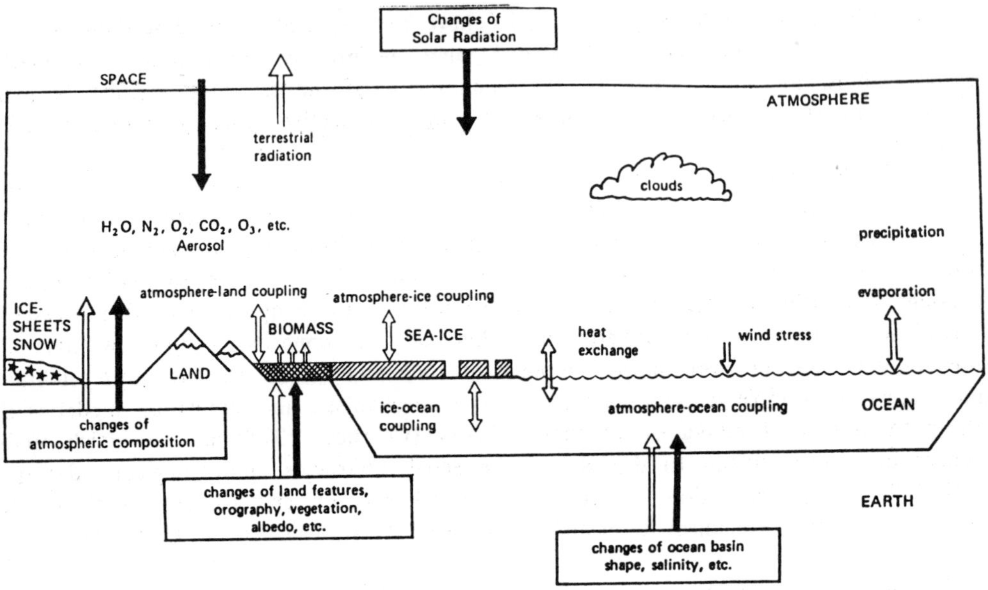

Fig. 18. Schematic illustration of the components of the coupled atmosphere–ocean–ice–earth climatic system. The full arrows (➞) are examples of external processes, and the open arrows (⇒) are examples of internal processes in climatic change. Source: GARP (1975).

211. More seriously, however, a further examination of Fig. 18 reveals that full understanding involves space scales ranging from few micrometres (cloud condensation nuclei, aerosols), up to the planetary scale (10,000 km). Figure 19 gives a schematic illustration of the various scales of atmospheric motions. It is clear that the simultaneous representation of all relevant scales is impossible with current and near future high-speed electronic computers.

212. The resolution[9] of current generation GCMs is typically a few degrees of latitude and longitude horizontally and a few kilometres or so vertically. Processes occurring on scales larger than this are explicitly treated, whereas phenomena occurring on smaller scales (e.g. cumulus convection, latent heating, turbulent transport) are parameterized. Following Somerville *et al.* (1974) we define parameterization as an algorithm[10] for uniquely determining the effect on the large-scale atmosphere of the small-scale phenomena, given the large-scale fields as calculated by the model.

213. Hydrological cycle components, radiative transfer, resolved boundary layer processes, large-scale dynamical processes and land surface temperature are all computed simultaneously, as is described in a review by Smagorinsky (1974). Atmospheric GCMs are currently able to simulate well many of the observed large-scale features of the atmosphere. A striking example is given in Fig. 20 which compares two climatic charts. The upper chart was generated by a GCM (Manabe, Hahn and Holloway, 1974) whereas the lower reflects actual observation. They depict the Köppen climatic classification of the earth. It can be seen that with the exceptions of southeast China and the southern United

[9] I.e. the "grid" interval, or distance between successive sampling points.
[10] I.e. a set of instructions specifying the details of the computational procedure.

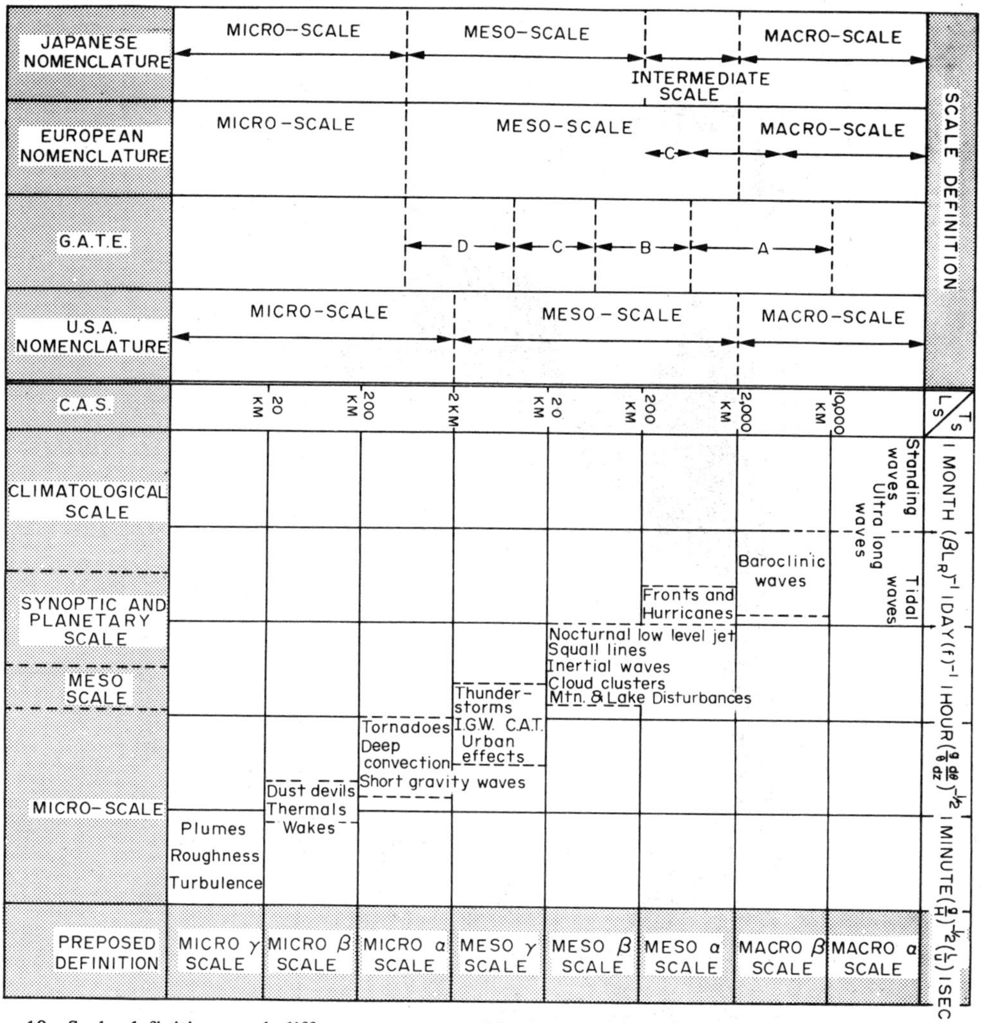

Fig. 19. Scale definitions and different processes with characteristic time and horizontal scales. Source: Orlanski (1975).

States, the model simulates very well the observed climatic regions. Considering the fact that precipitation, for instance, is highly intermittent in space and time, and its proper simulation involved microphysical consideration which, unlike moist convections, are not even parameterized in the model, the skill of this numerical simulation is impressive.

Mesoscale Models

214. Impressive as they are, GCMs under optimal conditions can simulate only the large-scale flow that they explicitly resolve. The use of ordinary second order difference equations for approximating the hydrodynamical equations requires 5 to 10 mesh points to adequately resolve a single Fourier (wave) mode (Kriess and Oliger, 1973). The grid size of a typical GCM is of the order 250 km, thus only the macroscale (Fig. 19) is

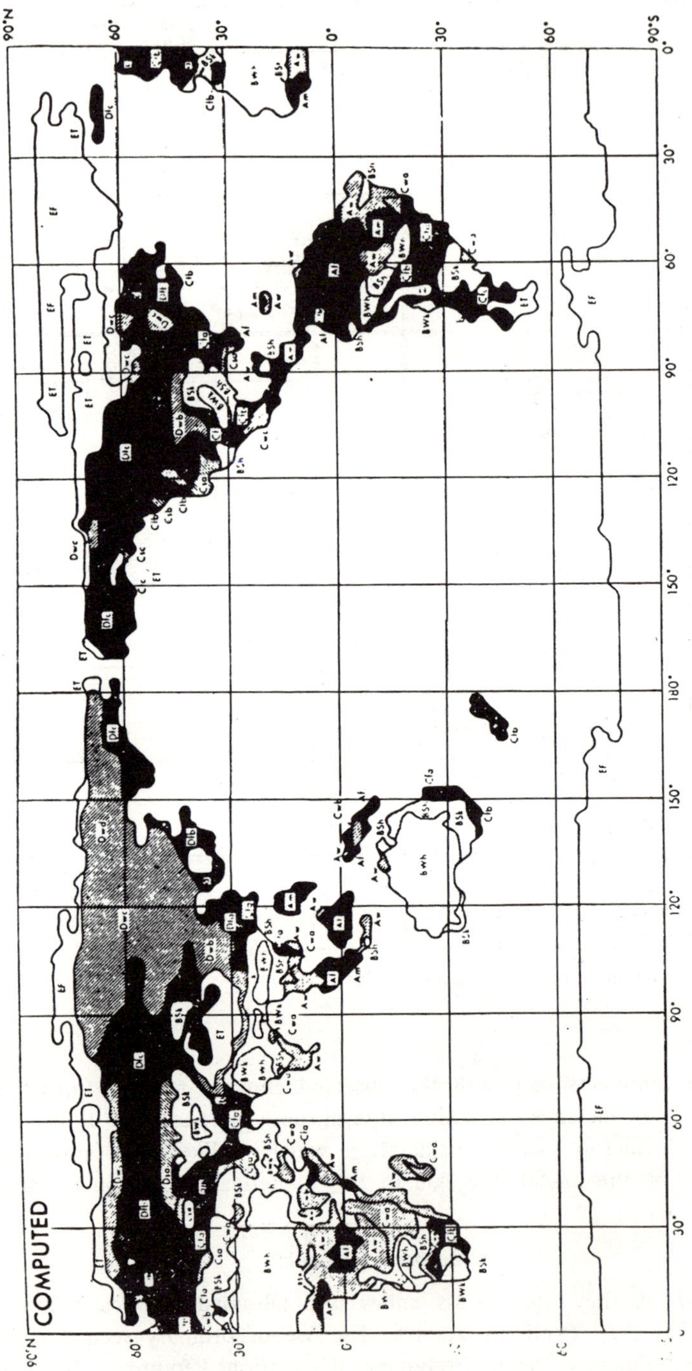

Fig. 20a. Global distribution of Köppen climate types based on model simulated data.

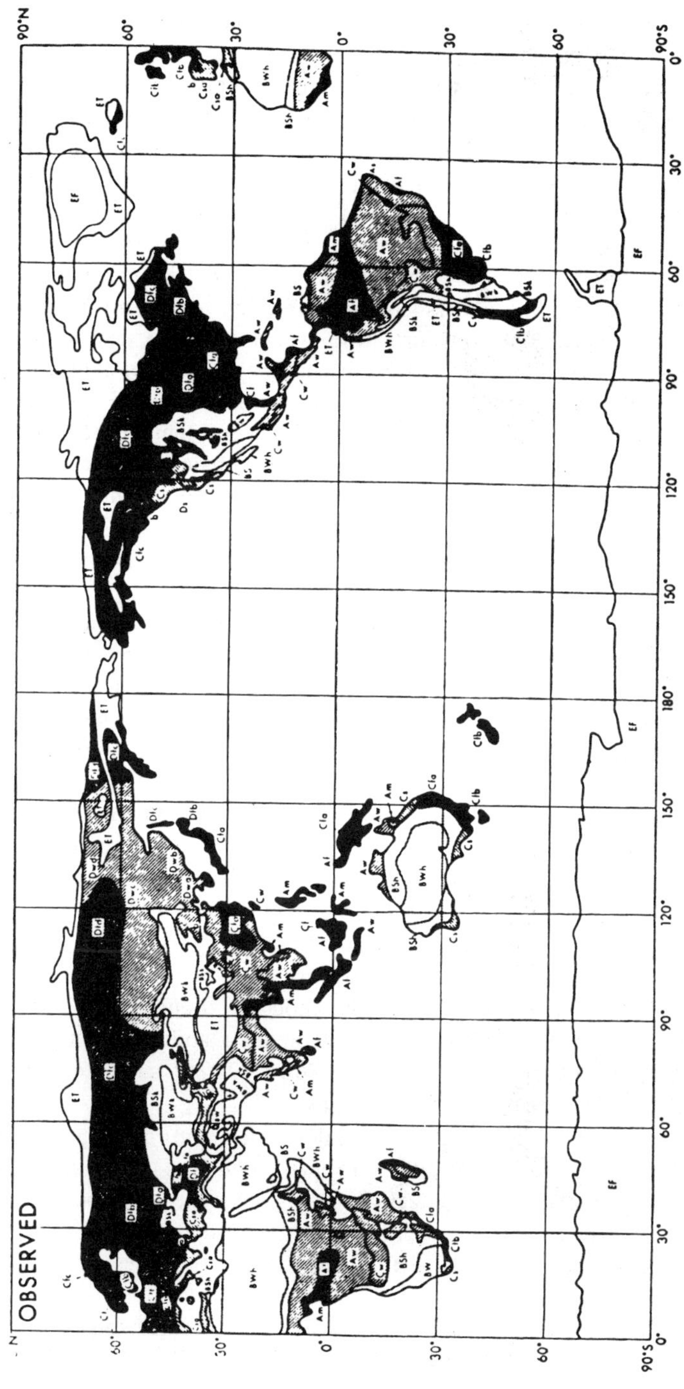

Fig. 20b. Global distribution of Köppen climate types based on observed data. Source: Manabe in GARP (1975).

properly resolved. Unfortunately, drought and excessive rainfalls occur many times as regional mesoscale anomalies. This is shown vividly in Fig. 21 which shows the crop moisture index of the continental United States for August 14, 1976. The negative values correspond to below normal moisture, while positive values correspond to above normal values. A drought condition occurs when the index is between –3.0 to –4.0, in which case potential yield is severely cut by drought. One can see that extreme drought prevails in South Dakota while in the neighbouring states of Montana, Wyoming, Nebraska and Iowa normal or above normal conditions prevail. Scattered showers in southern Minnesota and parts of Wisconsin helped some crops and the index is –1.0 to –2.0 which means some improvement, but still too dry.

215. This demonstrates that in areas which depend on convective precipitation, predictions from current GCMs, unless supplemented by other information, will be of little or no use to the individual cultivator who wants to make decisions about planting or other matters on his farm. Glantz (1976b) discusses the value of long-range forecast for the Sahel and concludes tentatively that due to social, political and economic constraints, the forecast will be of little use. Assuming, however, that socio-political economic constraints can be changed, the scientific problem still remains. How can we better model and understand regional and mesoscale features of the dry climates? It appears that limited area regional models are the natural answer.

216. Results obtained from the large-scale models provide the boundary conditions for such regional models.

217. Mesoscale models are being developed by a number of teams or workers, for example by Kreitzberg (1975), Anthes (1975), Pielke (1974, 1975) and Tapp and White (1976). There is no single pattern to such developments, because of the variety of mesoscale phenomena of interest or concern. The main thrust of development, however, is toward improvement of the very short range forecasts (5–10 hours), with a particular emphasis on severely damaging weather.

218. It is interesting to note, however, that GCMs evolved out of the early operational numerical weather prediction models (for description of these early models see e.g. Haltiner, 1971). In fact, one of the ostensibly difficult problems associated with very short range weather prediction, balance and initialization[11] (e.g. Haltiner, 1971; Richardson, 1922) does not exist for climate models. This is so since the ergodic presumptions[12] imply that the climate is independent of the initial conditions. It is perhaps for this last reason that mesoscale models, though developed for weather prediction, have been evaluated and tested based on their ability to simulate mesoclimate. Thus for instance, Pielke (1974) and Tapp and White (1976) simulated the sea breeze in Florida; Lavoie (1972) simulated lake-effect storms.

219. Probably the two most difficult problems for the development of mesoscale models are convective parameterization and the open boundary conditions. The importance of convection and its proper parameterization is obvious. The open boundary condition is a fundamental problem in computing science and applied mathematics. Whether one keeps the boundary conditions fixed at their climatological values, or computes them

[11] In a balanced field the complete state of the fluid can be deduced from the temperature field. Due to observational errors the initial conditions from which the forecast is made do not always reflect proper balance, hence the problem of initialization.

[12] In any ergodic system time averages over infinite period are equivalent to an ensemble (statistical) average of all the possible realizations of the system.

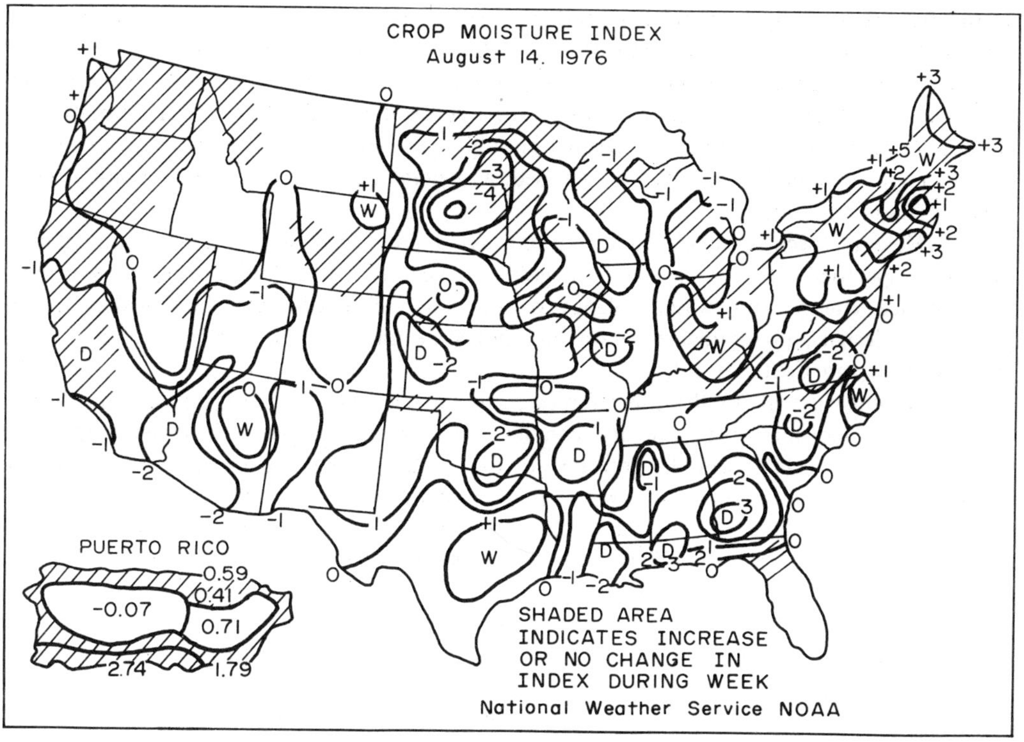

SHADED AREA: INDEX INCREASED OR DID NOT CHANGE
Above 3 Excessively wet, some fields flooded
2 to 3 Too wet, some standing water
1 to 2 Prospects above normal, some fields too wet
to 1 Moisture adequate for present needs
to −1 Prospects improved but rain still needed
−1 to −2 Some improvements, but still too dry
−2 to −3 Some relief but drought still serious
−3 to −4 Drought continuous, rain needed
Below −4 Not enough rain, still extremely dry.

UNSHADED AREAS: INDEX DECREASED
Above 3 Some drying but still excessively wet
2 to 3 More dry weather needed, work delayed
1 to 2 Favorable, except still too wet in spots
to 1 Favorable for normal growth and field work
to −1 Topsoil moisture short, germination slow
−1 to −2 Abnormally dry, prospects deteriorating
−2 to −3 Too dry, yields severely cut by drought
−3 to −4 Potential yields severely cut by drought
Below −4 Extremely dry, most crops ruined.

Fig. 21. Crop moisture index, August 14, 1976 (U.S.A., National Weather Service).

dynamically, disturbances propagate from the area of interest to the boundaries. There they must be allowed to propagate further. If they are not treated properly, at least part of their energy will be reflected back to the main area of interest as numerical noise. Orlanski (1976) discusses this problem in great detail.

220. Since computer models cannot run without boundary conditions, the problem is

in practice being solved in an engineering sense. The fact that every numerical modeler has his "own little trick" is an indication that the problem is not yet solved.

221. Since orographic effects are an important source of mesoscale disturbances, they ought to be included in mesoscale models. For hydrostatic models the σ transformation of Phillips (1957), or its various variants should be used. For non-hydrostatic models Gal-Chen and Somerville (1975a, b) devised a generalized non-orthogonal, non-Cartesian tensorial transformation.

222. Most mesoscale models to this point use a prescribed surface temperature that is arbitrarily specified. If these models are going to be useful for climate studies, they must have surface energy and water balance of the sort discussed later. For further details about the design and implementation of mesoscale models, the reader should consult Pielke (1975) and Cotton (1975).

223. Another useful application of mesoscale models is in testing proposals for climate and weather modification in the dry climates. The various proposals have been reviewed by Glantz (1976a). Of particular interest are those which are based explicitly or implicitly on some sort of a heat-island, natural or man-made (e.g. Malkus, 1955; Landsberg, 1974). The basic idea appears to be trivially simple. The differential heating will produce local circulation and convergence. When moisture conditions are adequate, clouds will form, often capped by a stable inversion layer. If the convergence could give a few of these clouds an impetus enabling them to penetrate the stable layer, they would often find themselves in an unstable layer and would grow rapidly into a mature cloud further downwind.

224. It must be realized, however, that the energy per unit mass which is needed to overcome the stable layer is proportional to $wg\Delta\theta$ where w is the vertical velocity, g – gravity and $\Delta\theta$ – the potential temperature discontinuity which separates the sometimes moist adiabatic layer below (say the monsoon air mass) from the sometimes almost dry adiabatic layer above (say the Saharan air mass). If the potential temperature discontinuity is too large or the convergence (which is proportional to w) is too small, the clouds will not be able to penetrate. Results from project METROMEX (Metropolitan Meso Scale Experiment) reported by Braham (1976) lend some support to heat-island-precipitation enhancement proposals. Measurements of the tallest echoes in the area every half hour gave a strongly bimodal top height distribution with one mode at 6.7 km and the other at 13 km (the approximate tropopause height for this area); however, the distribution of tops of the tallest echo from each of the days showed only the mode at 13 km. About half of the rain-producing days, and hours prior to the tropopause-reaching storms on other days, had echo-producing clouds that were stopped in their growth by mid-level stable layers. It is therefore natural to assume that the heat island gives a few of these clouds extra energy and enables them to penetrate the stable layer. This will be a logical explanation for both the rain excesses and the increased number of thunderstorms, observed over urban areas.

225. In the context of weather in West Africa, Fig. 22 may be illuminating. It depicts the West African summertime mean surface pressure, gradient-level wind (just above the planetary boundary layer 1.2 to 1.6 km above the surface) rainfall, daily surface temperature and meridional cross sections of zonal wind for August. There is a strong surface baroclinic zone, and a reversal of the meridional temperature gradient in the middle troposphere in response to which an east wind maximum developed near 600 mb. In addition (Burpee, 1972) there is a negative correlation between the meridional

component of the wind, v, and temperature and between temperature advection and moisture advection. Thus north winds are warm and dry, south winds cool and moist. Consequently, baroclinic instability obtains in the easterly flow and waves develop to the eastern end of the zone and intensify as they move across it (Burpee, 1972). In addition, barotropic instability (Charney and Stern, 1962) is also operative due to the lateral wind shear created by the existence of the wind maximum near 600 mb. Overall, there appear to be *enough disturbances* to produce lifting. The combination of monsoon air on the surface, with an approximately moist adiabatic lapse rate, followed by Saharan air aloft with an approximately dry adiabatic lapse rate indicate favourable conditions for thunderstorm activities, provided that the inversion layer can be penetrated.

226. The above cited situation makes a weather modification experiment very exciting, but clearly detailed observations about the dynamical and microphysical structure of the thunderstorms in this area are needed. An attractive possibility exists that one might simultaneously run a regional fine grid model, embedded in a global coarse model. It stems from the fact that there are large areas over the globe where the current coarse resolution is acceptable. Finite elements techniques (e.g. Strang and Fix, 1973) may be adequate since, at least for simple problems, they allow variable resolution without creating large truncation errors[13]. However, this kind of research is still in a very early stage.

Convection Models

227. If the GARP experiment is broadened to include the west African monsoonal climate, as is being proposed, some detailed observation of cloud dynamics and microphysics may be available. Understanding the details of the physics and dynamics of clouds may be of considerable importance in both operational and experimental weather modification. It will make the seeding experiment less dependent on statistics. If we have no idea about the internal physics, we need a large statistical sample to test hypotheses; the more we know the smaller the sample we need. It must be borne in mind, however, that three-dimensional dynamical cloud models with even a fairly crude representation of the microphysics are very expensive in terms of computer time. This is mainly due to the genuinely three dimensional nature of a cloud, as opposed to the quasi-two-dimensional structure of the large-scale disturbances. Three dimensional cloud models are described in considerable detail by Cotton (1975).

One-dimensional Cloud Models

228. The extreme complexity of the three dimensional convection models make it desirable to have a parameterized one-dimensional analogue. These models are capable of rapid sensitivity analyses. If a mini-computer is available in the field, they may be used to indicate the actual occurrence of favourable seeding conditions. Due to over-simplified dynamics they allow almost any degree of complexity in the microphysics. Therefore they may be particularly useful in testing theories like those of Schnell (1975) which attempt

[13]Truncation errors are defined as errors created aue to the fact that one approximates differential elements by finite increments (e.g. finite grid size as opposed to infinitesimal grid size).

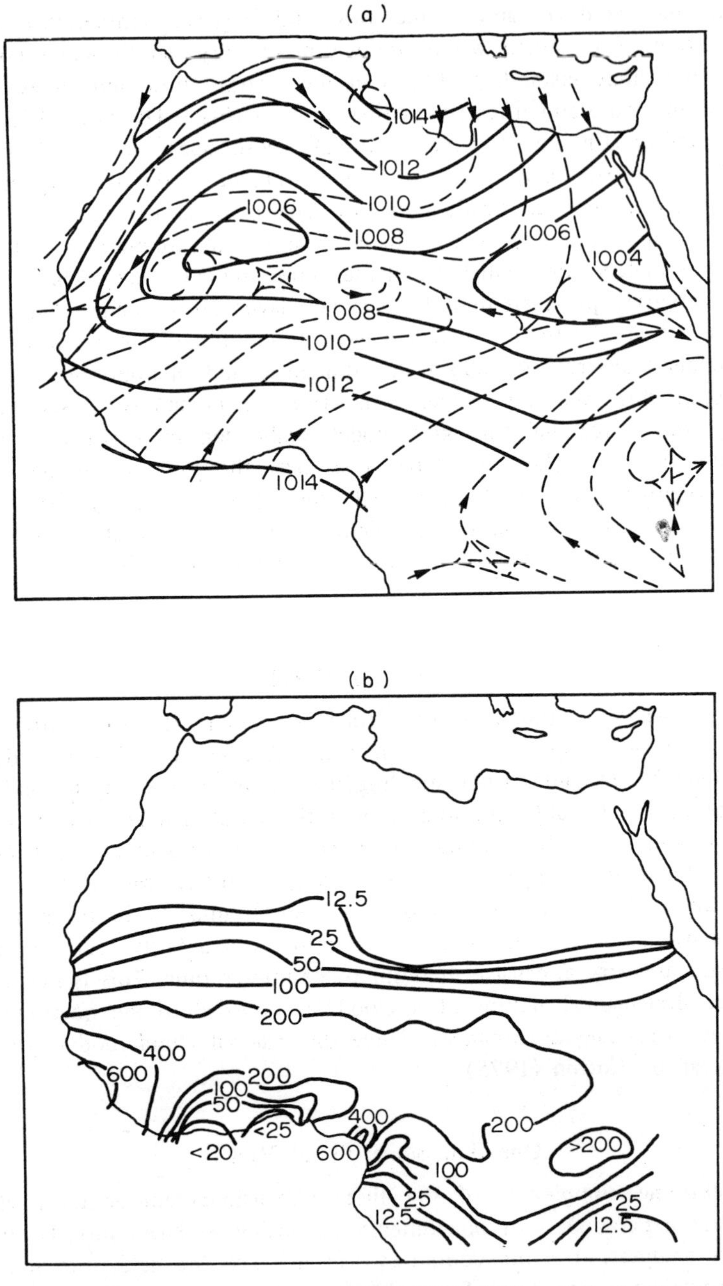

Fig. 22. Monthly mean maps for August: (a) surface pressure (mb) and gradient-level wind; (b) rainfall (mm).

to explain desertification as due to lack of one form or another of efficient rain producing nuclei.

229. The basic assumptions about the cloud dynamics of the one dimensional models are that the motion is axially symmetric, that horizontal pressure gradients can be

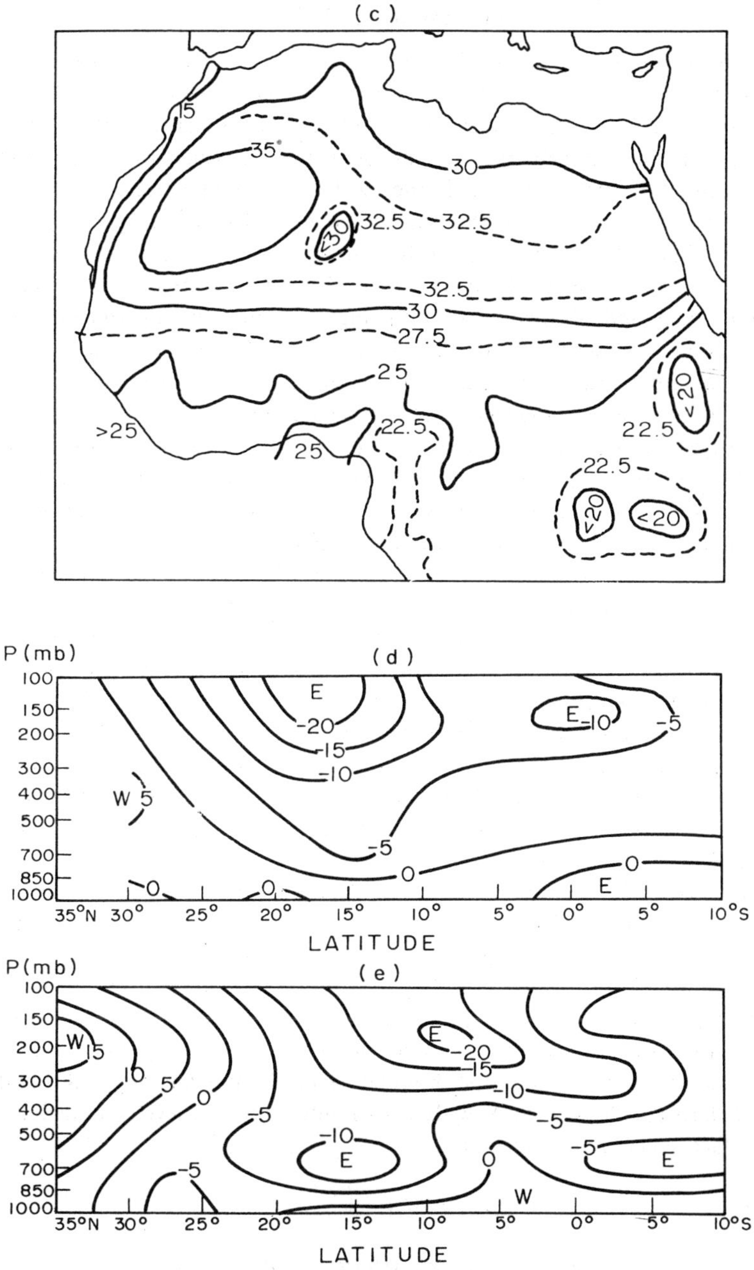

Fig. 22. Monthly mean maps for August: (c) daily average surface temperature (0°C); and (d) and (e) meridional cross-sections of zonal wind (m sec^{-1}) at 35°E and 5°E, respectively (Burpee, 1972).

ignored, and that entrainment through the side boundaries can be estimated from gross properties, such as mean vertical velocity. Cotton (1975) critically discusses the one-dimensional cloud models.

Climatonomic Models

230. A detailed understanding of the surface properties would involve a modeling just as complex as for the atmosphere itself. In this section, however, we discuss only those aspects of the surface properties that affect significantly the dynamics and thermodynamics of the atmosphere. This includes the exchange of momentum, heat and moisture between the atmosphere and the surface, the injection of various kinds of aerosols into the atmosphere, and the surface albedo. The exchange of momentum and heat and the surface albedo affect directly the momentum and heat budget. The water and aerosols, however, have an indirect effect. The water affects the heat balance via release of latent heat and absorption of outgoing infrared radiation (greenhouse effect). It also creates clouds, which affect the radiation balance. The aerosols affect the precipitation and cloudiness by acting as cloud condensation nuclei (CCN) or ice nuclei, and they also affect the coalescence process. Depending on their size and type aerosols can enhance, suppress or have no effect on the precipitation. Aerosols also can affect significantly the radiation balance since they reflect sunlight (cooling effect) but they also absorb (depending on their size) both short wave (incoming) and long wave (outgoing terrestrial) radiation. There is general awareness among global modelers of the central role played by moist convection, cloudiness and aerosols (Schneider and Dickinson, 1974), but the effect that aerosols may have on the global or regional climate via enhancement or suppression of precipitation is not considered. Apparently some dynamicists still believe that clouds are always naturally over-seeded, and that precipitation exclusively depends on favourable dynamical conditions.

231. Various procedures have been proposed for parameterizing the surface fluxes of momentum, heat and moisture for GCMs. Some of them are directly applicable to the other classes of models which have been discussed. They have been recently reviewed by Bhumralker (1976). The actual method depends mainly on whether or not the boundary layer is included explicitly in the numerical model. It also depends on the proximity of the first internal GCM level from the ground. If the lowest internal layer is close enough to the ground, so that it lies within the so-called constant flux layer (approximately 10–50 m above the ground), it is possible to use various variants of the Monin–Obukhov formulation (see e.g. Monin, 1970). Otherwise the structure of the planetary boundary layer must be regarded as determining surface fluxes in terms of the lowest internal level.

232. In the following discussions the subscript s applies to the top of the surface layer and the subscript h refers to the first internal level. In the Geophysical Fluid Dynamics Laboratory (GFDL) (in Princeton, New Jersey) model the two levels coincide.

Case 1: Surface layer bulk formulas

$$\begin{aligned} \tau_0 &= \rho_s C_D \mathbf{V}_s |\mathbf{V}_s| \\ H_0 &= -\rho_s C_p C_h |\mathbf{V}_s| (T_b - T_s) \\ E_0 &= -\rho_s (GW) C_E |\mathbf{V}_s| (q_b - q_s) \end{aligned} \tag{1}$$

Here τ_0 is surface stress, H_0 is sensible heat flux and E_0 is moisture flux; $\mathbf{V}_s$, T_s, q_s and ρ_s are velocity, temperature, moisture and density respectively at level s, T_b and q_b are temperature and moisture at the ground, C_p thermal heat capacity of dry air at constant pressure. GW is a ground wetness parameter which lies in the range [0,1], C_D drag coefficient, C_E surface layer moisture transfer coefficient and C_H surface layer heat flux (sensible) coefficient.

Case 2: Boundary layer bulk formulas. Following Clarke (1970), the surface values of turbulent fluxes can be obtained from

$$\begin{aligned} \tau_0 &= \rho U_*^2, \\ H_0 &= -\rho\ C_p\ C_T\ |\mathbf{V}_h|\ \Delta\theta, \\ E_0 &= -\rho\ C_Q\ (GW)\ |\mathbf{V}_h|\ \Delta q. \end{aligned} \tag{2}$$

Here $\Delta\theta$ and Δq are, respectively, the difference between potential temperature and mixing ratio at the lower boundary and the first internal level, and U_* is friction velocity. C_T and C_Q are, respectively, boundary layer heat flux and moisture transfer coefficients. One obvious advantage of parameterizations like (1) and (2) is that they avoid the necessity to specify the Bowen ratio B defined by $B = H_0/LE_0$.

233. Returning now to the surface energy balance, it is written as

$$G = R - H_0 - LE_0 - \Delta A + \phi \tag{3}$$

where G is the soil heat flux, R net radiation, H_0 sensible heat flux, LE_0 latent heat flux, ΔA advection (horizontal net heat transport), and ϕ secondary processes. While on yearly average basis G can be safely neglected, this is in general not true on a monthly or daily basis. In order to calculate G, one has to consider several layers of the soil and also must take into account the type of soil. In the GFDL GCM (Manabe, 1969) G is neglected.

234. Another important factor to consider is GW (ground wetness). For wet (saturated) surfaces, $(GW) = 1$ and E_0 is then the potential evaporation E_{0p}. When the soil moisture content SM is greater than some critical value SM_k (which depends on the type of soil) it is found that evapotranspiration proceeds at the potential rate E_{0p}. A parameterization for evaporation often assumed is

$$\begin{aligned} E_0 &= E_{0p},\ SM > SM_k \\ E_0 &= E_{0p}\ (SM/SM_k),\ SM < SM_k \end{aligned} \tag{4}$$

for which we conclude that

$$\begin{aligned} GW &= 1, \qquad SM > SM_k \\ GW &= SM/SM_k, \qquad SM < SM_k \end{aligned} \tag{4'}$$

According to Priestley and Taylor (1972)

$$SM_k = 5 \text{ cm}.$$

235. In the water balance calculation, it is important to take into account the capacity of soil for plant-available moisture, SM_F. This varies widely according to soil type (between 5 and 20 cm according to Priestley and Taylor). Manabe (1969) devised the following formulation for the water balance in a GCM calculation:

$$SM_F = 15 \text{ cm.}$$
$$SM_k = 0.75 \times (SM_F) = 11 \text{ cm.} \tag{5}$$

and in the absence of snow cover

$$\frac{\partial(SM)}{\partial t} = P - E_0, \; SM < SM_F$$
$$\frac{\partial(SM)}{\partial t} = 0, \; N = P - E_0, \; SM > SM_F \tag{6}$$

where P is the rate of rainfall, E_0 is evaporation, and N is the run-off rate.

236. Manabe found in his model a positive feedback between rainfall and soil-moisture, a desert-forming mechanism. Walker and Rowntree (1976) also found the same effect.

237. Lettau (1969) devised a more mechanistic model, whereby the forcing functions are the short wave radiation F which is absorbed at the ground and the precipitation P

$$F = I\,(1-a)$$

where I is the solar radiation (direct and diffuse) at the ground, and a albedo. From this, using conservation equation for water and energy, he is able to obtain an output which consists of monthly means and annual averages of evaporation run-off and sensible heat fluxes.

238. It is beyond the scope of this paper to repeat his derivation; nevertheless it will be instructive to outline the empirical closure assumptions that were used to derive a unique output.

239. The first assumption involved is

$$C^* = 1 - \tanh D^*$$
$$C^* \equiv \bar{N}/\bar{P} \text{ Run-off ratio}, \; D^* \equiv \bar{R}/L\bar{P} \equiv \text{Dryness ratio}, \tag{7}$$

where L is latent heat involved in phase changes of H_2O, R is the net radiation balance, P precipitation, and N run-off[14]. The bar indicates yearly average values. Next the evaporation E and run-off N are decomposed as

$$E = E' = E'' \; ; N = N' + N''.$$

The E', N' contribution represent a nearly immediate or direct response, while E'', N'' are delayed responses. Assuming that

[14]This is consistent with Fig. 1.

$$N'' = \overline{N''}\ (SM)/(\overline{SM})\ , \quad E'' = \overline{E''}\ (SM)/(\overline{SM}),$$

and adding, yields

$$N'' + E'' = (SM)/t_* \qquad t_* \equiv (\overline{SM})/(\overline{N''} + \overline{E''}) \tag{8}$$

Lettau interprets t_* as the characteristic residence time or turnover period and asserts that it is of the order of 2–3 months. The last simplifying assumption is

$$E' = e{*}\ PF/\overline{F}\ ; \quad \text{also } \overline{E'} = \overline{e{*}PF}/\overline{F}. \tag{9}$$

The nondimensional constant $e*$ is termed evaporivity and is the capacity of land surfaces to utilize a portion of solar energy for the *immediate* evaporation of the precipitation that has been received. Note that applying (9) for the delayed evaporation E'' is nonsensical since it would prohibit drying out of land surfaces. Relations (7), (8), (9) and the conservation equations for heat and water permit a unique evaluation of soil moisture, evaporation and run-off. Lettau applied his method to the central plains and eastern region of North America and obtained good to fair agreement with measurements of Rasmusson (1968).

240. Since the Lettau scheme is rather simple, it appears that a variant of this parameterization which also includes the effect of the wind on evaporation could be tested in a GCM, and compared to Manabe's parameterization.

DISCUSSION OF RESULTS

241. Having discussed the basic properties of the models which can be used to study desertification, we are now in a position to discuss some results which have been obtained from these models. The common pattern of the experiments to be described can be described schematically as follows:

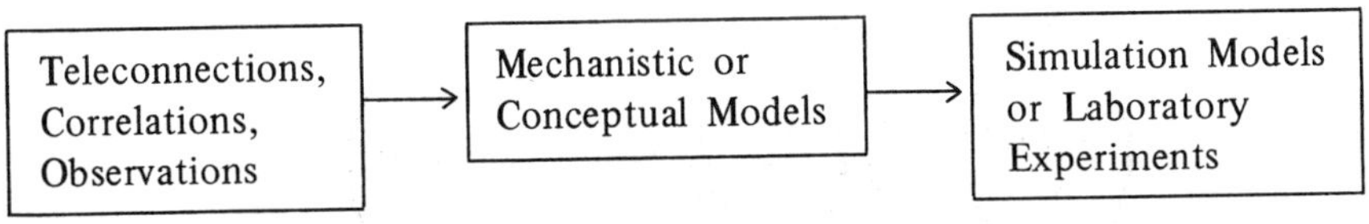

Statistical Significance Tests

242. Before discussing the details of the experiments, it is necessary to clarify an important point regarding the simulation models. It has to do with testing the statistical significance of climatic sensitivity studies (Leith, 1973). If we are interested in detecting the influence of external changes such as sea surface temperature or surface albedo, we have to be able to distinguish between signal, which is the change in the climatic means, and the "noise". In this case the "noise" is defined as the error in estimating the mean due to finite time averaging. The ergodic presumption (footnote 12) will ensure that averaging over infinite time interval of one time series is equivalent to *ensemble*

average of an infinite number of realizable states of the atmosphere, which correspond to the particular set of external forcing. Due to the fact that the averaging period is finite and the weather is unpredictable beyond 2 weeks (Lorenz, 1965), different initial conditions which have a finite probability to occur would produce time series with different statistics (i.e. different mean, standard deviation, etc.).

243. For a random time series $\Psi(t)$ with mean μ and standard deviation σ, the error $\phi(t)$ in estimating the mean due to averaging over a limited period T is given by

$$\overline{\Psi}(T = \mu + \overline{\phi}(T)$$

where

$$(\overline{\ }) = \frac{1}{T}\int_0^T (\ )\,dt,\ \mu = \lim_{T\to\infty}\frac{1}{T}\int_0^T (\ )\,dt;$$

$$\Psi(T) = \mu + \phi(t),\ \lim_{T\to\infty}\overline{\phi}\,(T) = 0.$$

Note that $\overline{\phi}(T)$ is the error (noise) in estimating the true mean μ. If $\overline{\phi}$ is normally distributed around the mean, then the standard deviation $\overline{\sigma}(T)$ of the time series $\overline{\sigma}(T)$ is an estimate of the noise due to finite time averaging, where again by definition $\lim_{T\to\infty}\overline{\sigma}(T) = 0$. Denoting $< >$ as an ensemble average we have then for every particular time t $<\phi(t)> = 0$ and,

$$<\phi(t)\phi(t+\tau)> = \sigma^2 R(\tau),\quad R(0) \equiv 1$$

It can be shown that

$$\overline{\sigma}(T)^2 = <\overline{\phi}(t)\overline{\phi}(t)> = 2\sigma^2/T\int_0^T (1 - \tau/T)\,R(\tau).$$

Using the value $R(\tau) = \exp(-\nu\tau)$ we get

$$\left(\frac{\overline{\sigma}(T)}{\sigma}^2\right) = \frac{2}{\nu T}\left(1 - \frac{1}{\nu T}\,[1 - \exp(-\nu T)]\right) \tag{10}$$

Leith estimates that $\nu = 0.3$ day^{-1}. At any rate if $\Delta\mu = \mu_1 - \mu_2$ is the difference between two model states due to two different external forcing, $\Delta\mu$ may be referred to as the signal and $\overline{\sigma}(T)$ is the noise. For the results of any climatic sensitivity study to have some significance the signal to noise ratio must be greater than 1.

244. It must be emphasized that a significant result does not necessarily imply causality. It merely puts an error bar on the "measurements". The possibility still remains that due to inadequate formulation of the physics the model response is different from the atmospheric response. It is also true that insignificant results do not necessarily indicate that the proposed climate change mechanism does not happen in nature. It only means that the "instrument" did not measure a detectable signal.

245. Due to the different assumptions employed in the various models, it is clear that if two or more models give similar response our confidence in the simulation should increase. Unfortunately we had great difficulties in trying to compare results from different models; this is due to the rather different way by which the modelers assess whether or not their results are statistically significant. Some modelers use stringent testing formalisms, while others use rather lax testing procedures. Some modelers apparently do not perform any significance test at all.

246. Shukla (1975) used equation (10) to estimate $\bar{\sigma}(T)$ from σ which is model's natural variability. In his calculation, σ is calculated from one control run. Chervin and Schneider (1976b) argue that this procedure is not strictly applicable for formal significance test and one has to perform more than one control experiment (with different initial conditions) in order to get the *probability* distribution of the mean.

247. Gilchrist (1975) applies the conventional "*t*-statistic" for formal statistical significance on one time series. In his notation,

$$t = [(\bar{x}_1 - \bar{x}_2)/S] \ (N/2)^{1/2}$$

where the $(\bar{\ })$ indicate time averages (54 days in his experiment) and S is the unbiased estimate of the standard deviation (the model's natural variability). N is the number of independent events. A t value of 2.68 is required for significance at the 1 per cent level. In this formalism, Gilchrist results are significant. As mentioned before, however, Chervin and Schneider (CS) object to applying *t*-statistics to one time series. In their formalism,

$$t = (\langle x_1\rangle - \langle x_2\rangle) / \left[S^2 \left(\frac{1}{N_1} + \frac{1}{N_2} \right) \right]^{1/2}$$

$$S^2 = \left[\sum_{l=1}^{\eta_1} (x_{1_l} - \langle x_1\rangle)^2 + \sum_{l=1}^{\eta_2} (x_{2_l} - \langle x_2\rangle)^2 \right] / (\eta_1 + \eta_2 - 2)$$

$$\langle x\rangle \frac{1}{\eta_l} \sum_{l=1}^{\eta_l} x_{i_l} = \text{Ensemble Average}$$

$$\eta_l = N_1, N_2, \eta_1, \eta_2 .$$

In the above, N_1 and N_2 are sample sizes, which is the number of experiments which are used to determine the prescribed changes and control population mean; and η_1 and η_2 are the sample sizes used to form the combined estimate of variance. The need to do more than two experiments is apparent in this kind of formalism.

248. It is quite possible therefore that an experiment which would be termed significant by Gilchrist formalism would be insignificant when tested by the Chervin–Schneider formalism.

249. Walker and Rowntree (1976) and Charney, Quirk and Stone (1975) apparently do not apply any statistical tests. However, the latter team presents zonally average values, thereby removing some of the noise. At any rate, we shall discuss the results of the various models, regardless of their statistical significance.

Sea Surface Anomalies and Response of the Dry Climate

250. It has long been speculated that there is a meaningful link between sea surface anomalies in the tropical oceans and climatic perturbations in continental regions, both tropical and extratropical (e.g. Bjerknes, 1966, 1969; Namias, 1974).

251. Schneider (personal communication) conducted a simple experiment with the NCAR GCM, aimed at studying the effect of sea surface anomalies on the structure of the zonally averaged Hadley cell in January. In the first experiment, he applied a uniform +2°C perturbation to the January sea surface temperature in all the oceans between latitude 10°N–20°N. The second experiment is similar to the first except that the perturbation applied is in latitude 0°–10°S. The experiments are compared with a control experiment in which the January sea surface climatology is specified. The land temperature is calculated explicitly. Figures 23 and 24 display the vertical velocity and precipitation for the three experiments. It is seen that the winter subtropical perturbation (10°N–20°N) has increased the vertical velocity (made it less negative) and consequently an increase in the subtropical precipitation occurs. In the ascending branch of the Hadley cell, the effect is opposite, decrease in vertical velocity and decrease in precipitation. To understand why this happens, one notes that the subtropical perturbation has created a local cell which is in opposite direction to the Hadley cell.

252. Examining the tropical perturbation (0°–10°S) one sees an increase in the vertical velocity in the immediate area of the perturbation. This is followed by a decrease in precipitation one strip further south. Since precipitation is critically dependent on vertical velocity, the precipitation is increased in latitudes (0°–10°S) and is decreased in (10°S–20°S). This is also explained in terms of the effect that the perturbation has on the structure of the Hadley cell. The tropical perturbation has increased the intensity of the Hadley cell and consequently there is more ascending motion in the tropics and more descending motion in the sub-tropics.

253. Gilchrist (1975), using the United Kingdom 5-level GCM, investigated the speculation that rainfall variation over north and central Africa were related to sea surface temperature anomalies over the tropical Atlantic. He found that amounts were lower west

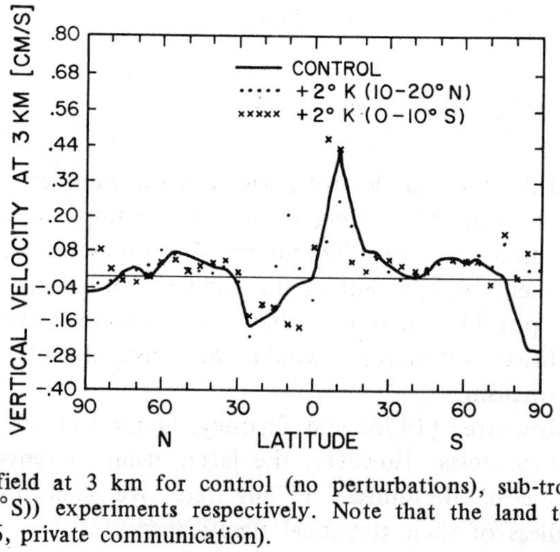

Fig. 23. Vertical velocity field at 3 km for control (no perturbations), sub-tropical (+2°K (10–20°N) and tropical (+2°K (0–10°S)) experiments respectively. Note that the land temperature is calculated explicitly. (Schneider, 1976, private communication).

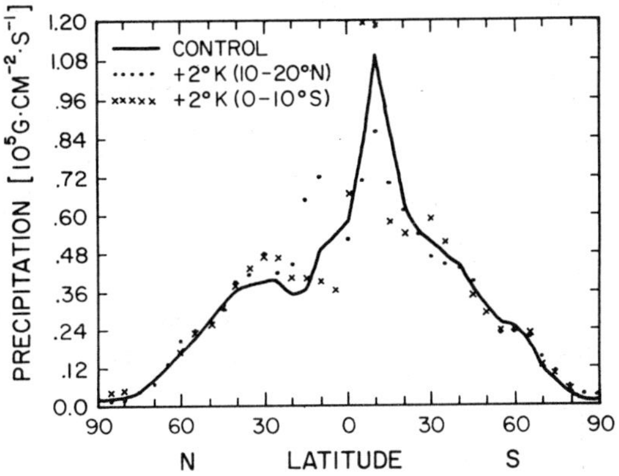

Fig. 24. Same as Fig. 23 but for precipitation rate.

of 5°W than in a control case with normal sea temperatures, but were higher east of 5°W. Jenkinson (1975) and Namias (1974) have reported that higher than normal pressure tends to occur over the north-east Atlantic with drought over Africa. The experiment produced such a pressure anomaly.

254. A similar effect has been found for the Indian monsoonal rainfall in relation to lowered sea-temperature over the Arabian Sea. Shukla (1975) applied the Geophysical Fluid Dynamics Laboratory (GFDL) GCM to a postulated cold anomaly in the western Arabian Sea, and found a reduction of rainfall over India in the following month. Records show that over the entire period 1901–61 a cold Arabian Sea in July tended to be followed by low August rainfall in India. In order to reduce the "noise", the meteorological parameters were averaged over the verification area (shown in Fig. 25). Figure 26 displays the atmospheric response of the verification area. It can be seen that indeed the response is pronounced at day 20–35, but after day 30 the two runs are

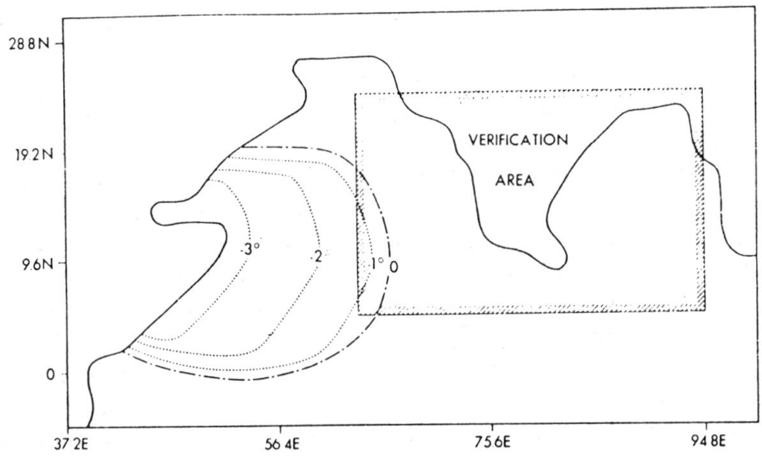

Fig. 25. Sea-surface temperature (°C) anomaly over western Arabian Sea and location of verification area (Shukla, 1975).

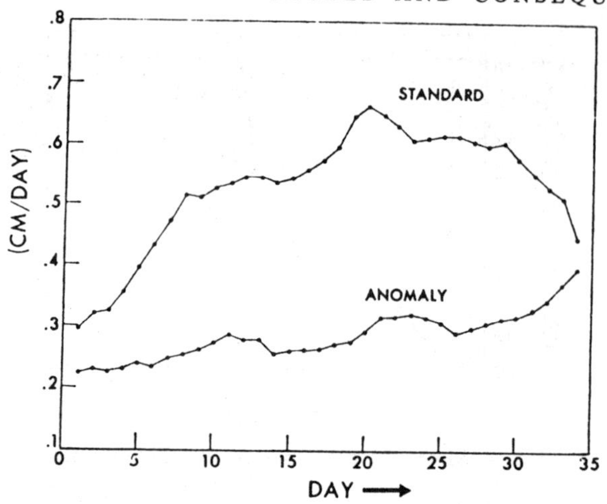

Fig. 26. Rate of precipitation (cm day^{-1}) averaged over the verification area for the Standard and Anomaly runs (Shukla, 1975).

apparently converging. Thus the response may have been a transient effect. This demonstrates the need to have an extended time integration, so that it is clear that the model is in a statistical steady state and is fully adjusted to the new external forcing. Chervin and Schneider (1976a) discuss this phenomenon in more detail.

Soil Moisture Content and Response of the Dry Climate

255. In the arid and semi-arid regions of the world, soil moisture content and consequently vegetation may play a non-negligible role in the water and heat balance of the atmosphere in that over dry surfaces evaporation is reduced and this in turn may reduce precipitation and release of latent heat.

256. Examining the resultant gradient level winds for July shows that in West Africa the convergence zone is in the Sahara desert (Fig. 27). The question arises: why does the maximum cloud cover (Fig. 28) appear so far south of the convergence zone? This leads to the speculation that if the desert is large enough to affect the global circulation, then aridity feeds on aridity. Manabe (1969) finds some support to this idea, but his conclusion is mainly based on a sensitivity study to check the effect of his ground hydrology parameterization.

257. Walker and Rowntree (1976) applied a limited-area eleven-level baroclinic model to West Africa to test the effect of variable soil moisture (see Section V above). They find large temperature differences between dry and wet soil conditions. Exciting as they are, these results should be viewed with a grain of salt. No statistical significance test was performed. The period of integration is too small and therefore they may have observed a transient response to a large shock (e.g. rapid and short flood). Their 20-day integration does not conclusively demonstrate that a statistical steady state is achieved. As far as their model's energetics is concerned, they find that the baroclinic energy conversion (eddy available potential energy → eddy kinetic energy) is dominant. This remains true even foı the wet case in which one expects decrease in the baroclinicity, due to the decrease in the desert-equator gradient.

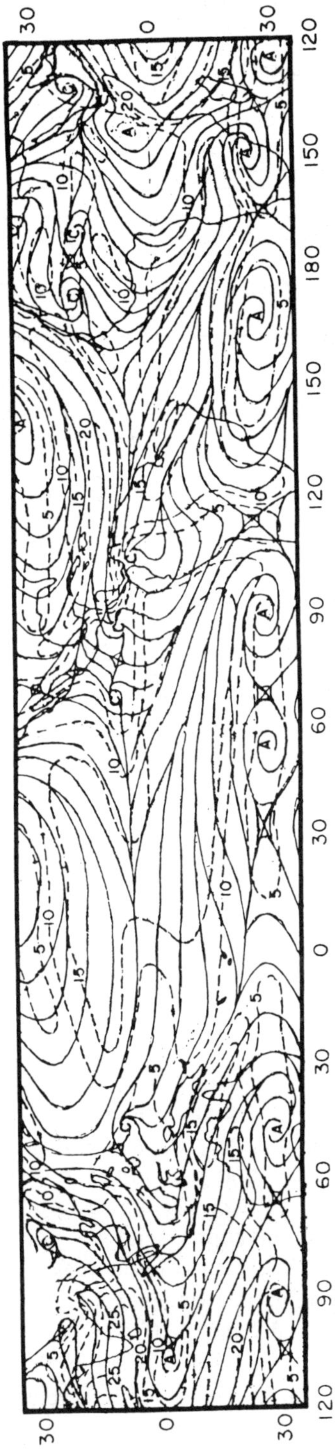

Fig. 27a. Resultant gradient-level wind for July (from AWSTR 215 |451).

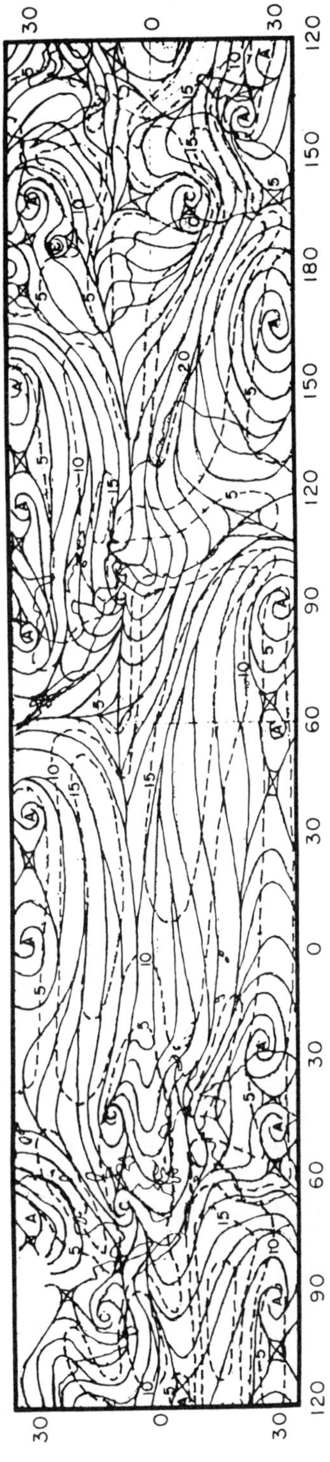

Fig. 27b. Resultant gradient-level for October (from AWSTR 215 |451).

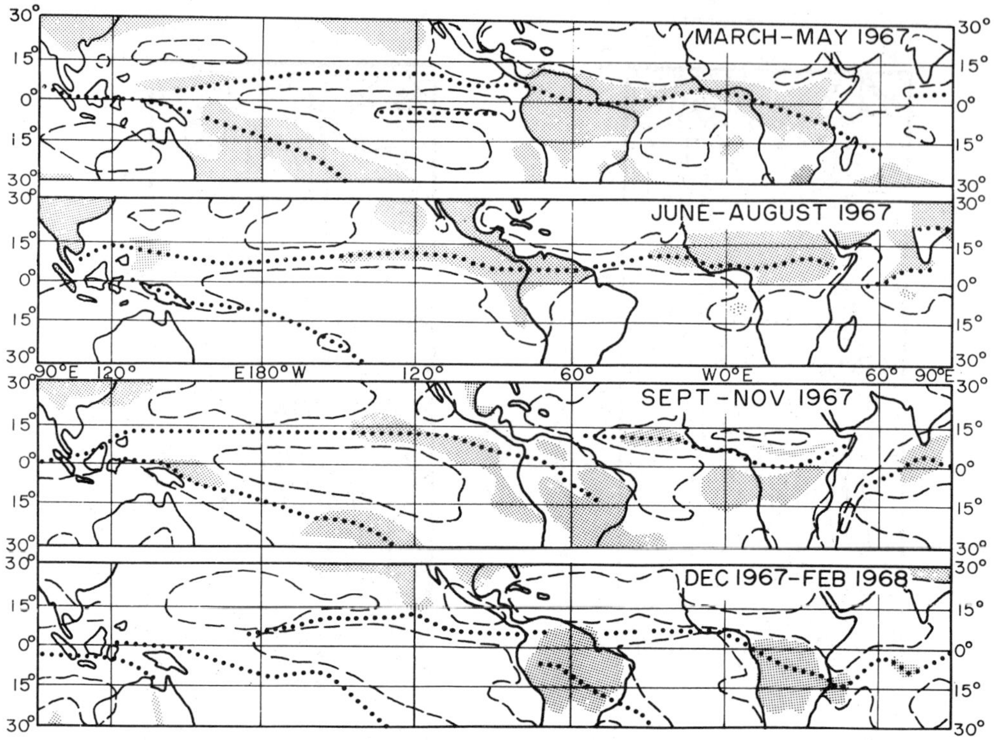

Fig. 28. Seasonal distribution of brightness from ESSA 3 and 5 digitized pictures over the tropics for the period March 1967–February 1968. The dashed lines enclose areas of minimum brightness, while stippling indicates regions of maximum brightness. The heavy dotted lines indicate the major zonally oriented axes of maximum brightness (or cloudiness).

258. Pedgley and Krishnamurti (1976), on the other hand, find from observation of the August 1972 monsoon cyclone that barotropic conversion (zonal kinetic energy → eddy kinetic energy) is also very important. As was mentioned before (mesoscale models), this latter conversion is due to barotropic instability of the easterly jet. It is true, however, that this particular cyclone produced a lot of disturbances and very little rainfall.

259. It is perhaps instructive to refer to Zipser's (1976, private communication) observations. Figure 29 shows the 850 mb July 1972 dew point anomalies over North Africa. The most striking feature is that *positive* anomalies occurred in the year in which the drought attained its full severity. It appears, therefore, that it is not lack of moisture which caused the drought (see also G. Macdonald's (1962) celebrated article, "The Evaporation-Precipitation Fallacy"). Instead, Zipser's observations tend to confirm the conventional meteorological view that lack of humid air streams are a major cause of deficiencies in rainfall. Figure 30 displays the longitudinal dependence of the 850 mb July meridional wind profile for band 2 (15°N–30°N) and band 3 (0°–15°N) in his area of study (20°W–40°E). The values are latitudinally averaged over the appropriate bands. Figure 31 displays the latitudinal wind profile (30°S–40°N) suitably averaged over longitude (25°W–15°E region 2, 15°E–55°E region 3). From these pictures one can deduce that in band 3 (Fig. 30) (0°–15°N) which include the Sahelian region, the

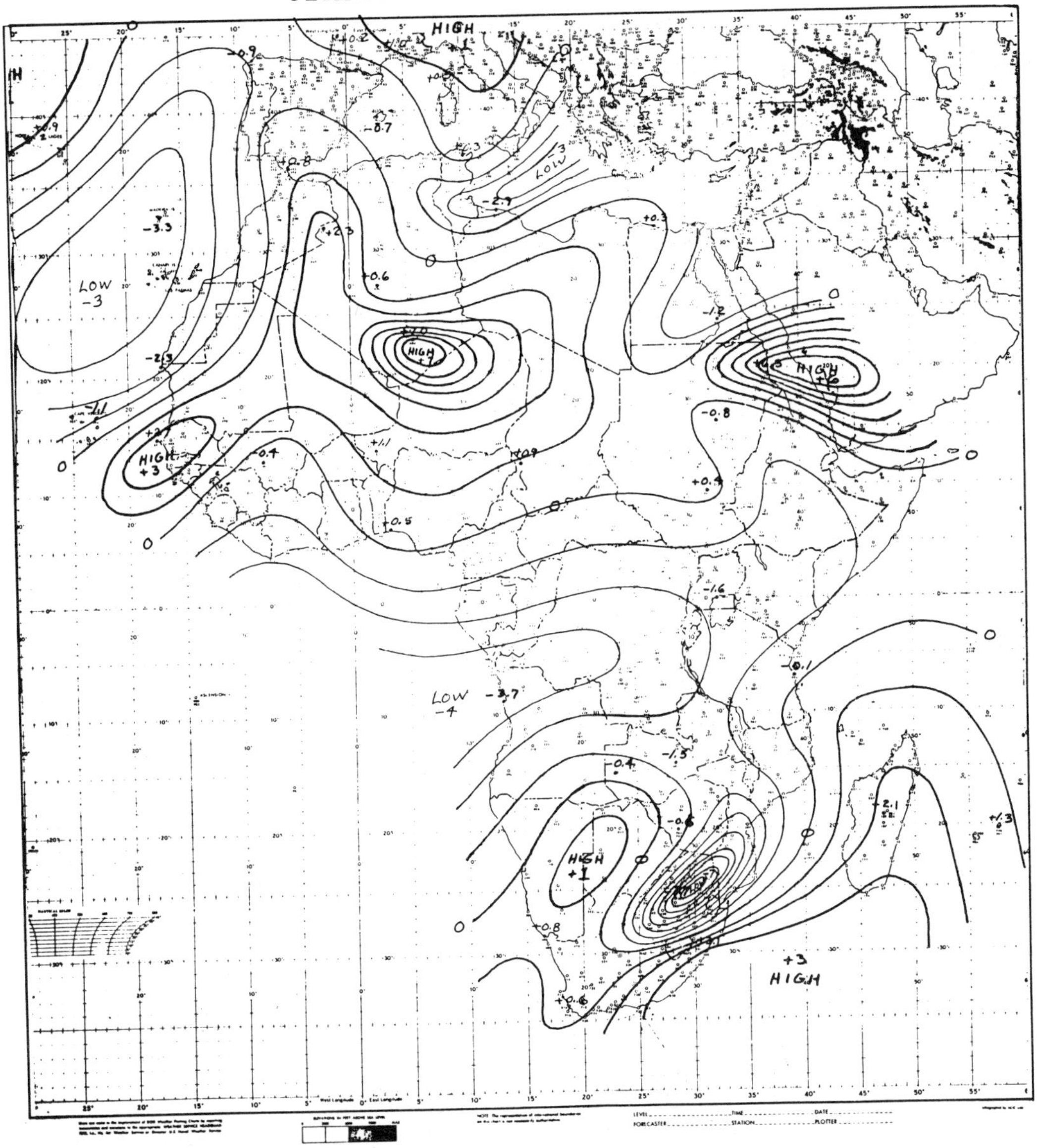

Fig. 29. 850 mb mean dew point anomalies for July 1972 (Zipser, 1976), private communication).

strength of the southerlies in longitude 20°W–10°E has decreased in the drought years 1970–2. In 1972 it actually changed sign. Inspection of Fig. 31 (region 2) reveals that in latitude 5°N–15°N the strength of the westerlies has decreased in the drought years 1970–2. Overall the strength of the moist, relatively cool, south-westerlies has decreased in the drought years. In 1972 the south-westerlies were replaced by the warm dry north-easterlies (Saharan air mass) and the drought was accelerated both in severity and in its extent.

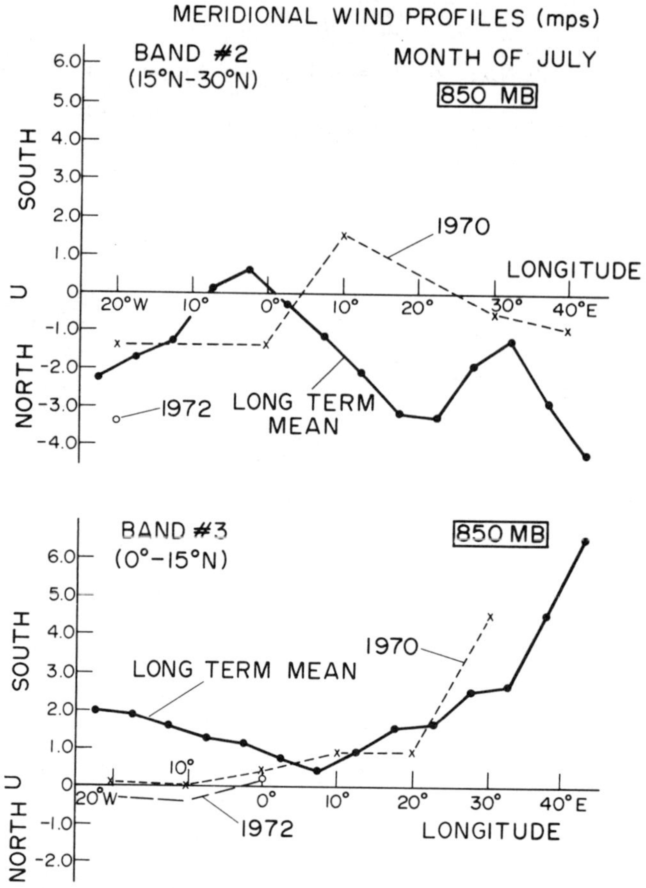

Fig. 30. 850 mb mean meridional wind profile, plotted as function of longitude. Results are shown differently for band 2 (upper part) and band 3 (lower part). The values are latitudinally averaged over the appropriate bands (Zipser, 1976, private communication).

Albedo Feedbacks on Vegetation and Rainfall

260. The albedo feedback hypothesis was discussed in great detail in Section V of the main report and the discussion will not be repeated here. Briefly stated, however, Charney's theory (1975a) postulates that overgrazing decreases the radiation balance (makes it less positive, or more negative). In order to maintain thermal equilibrium and in the absence of advection the air must descend and compress adiabatically (adiabatic warming) decreasing thus the relative humidity and precipitation. Even though Charney applied his theory to deserts with negative radiation balance at the top of the atmosphere, *the theory is also valid for deserts with positive radiation balance.* The main assumption is that adiabatic warming (descending motion), rather than advection, is the main compensation for the decrease in the radiation balance.

261. Similar results have been obtained using the NCAR GCM (Washington, 1976 private communication) and are shown in Figs. 32 and 33. Figure 32 displays the areas

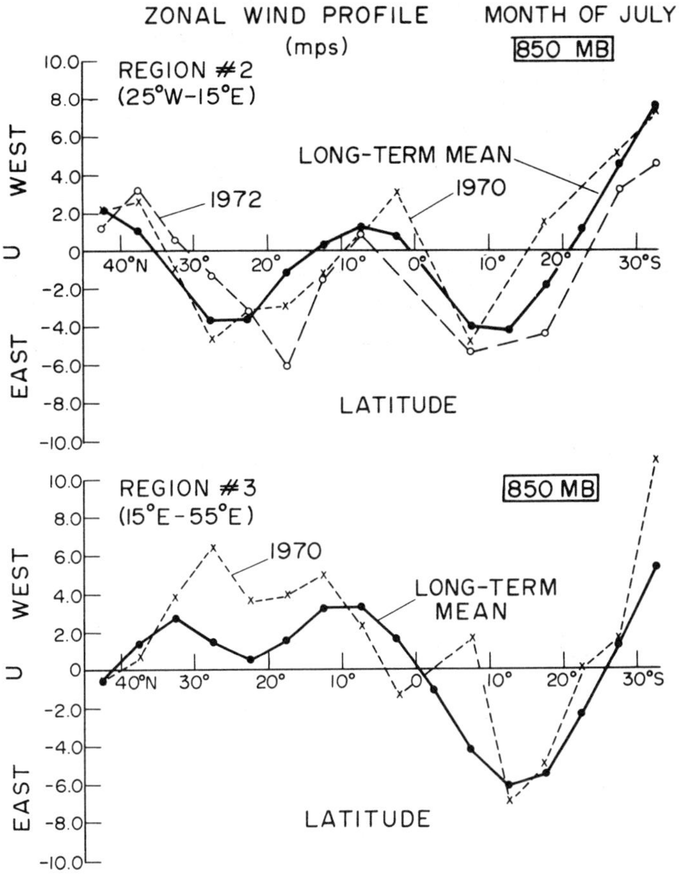

Fig. 31. Latitudinal wind profile (30°S–40°N) suitably averaged over longitude (25°W–15°E, region 2; 15°E–55°E, region 3). Note that the overall strength of the moist south-westerlies in latitudes (5°N–15°N) in West Africa is reduced (Zipser, 1976, private communication).

where the surface albedo has been increased to 45 per cent. It includes the African land mass north of 10°N and the North American dry belt. The unrealistically large perturbation both in terms of surface albedo and area of perturbation is necessary, if one wants to satisfy the stringent statistical significance test of Chervin and Schneider. Figure 33 displays the dramatic results. In agreement with Charney's theory there is a decrease in precipitation in the perturbed areas. But there are also teleconnections, with both positive and negative anomalies. Notably we see a decrease in rainfall in the Tibetan Plateau and an increase in precipitation further north in China (but these are not statistically significant according to Fig. 34). Other notable features are statistically significant increase in rainfall over the Caribbean and Mexico. Krishnamurti *et al.* (1973) discuss teleconnections between the 200 mb Tibetan High (reflecting the surface thermal low) and the African and Mexican (200 mb) highs.

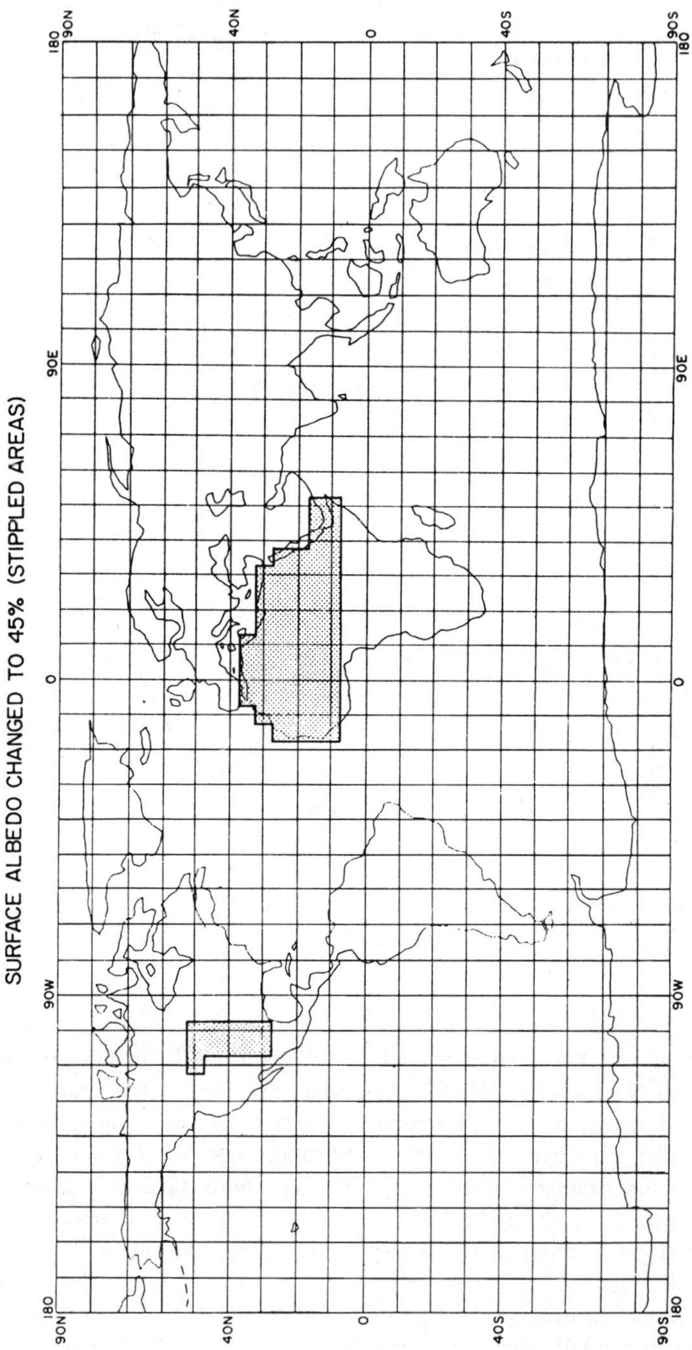

Fig. 32. Areas of changed surface albedo in an NCAR GCM experiment designed to test the effects of changing surface albedo on world precipitation (Washington, 1976, private communication).

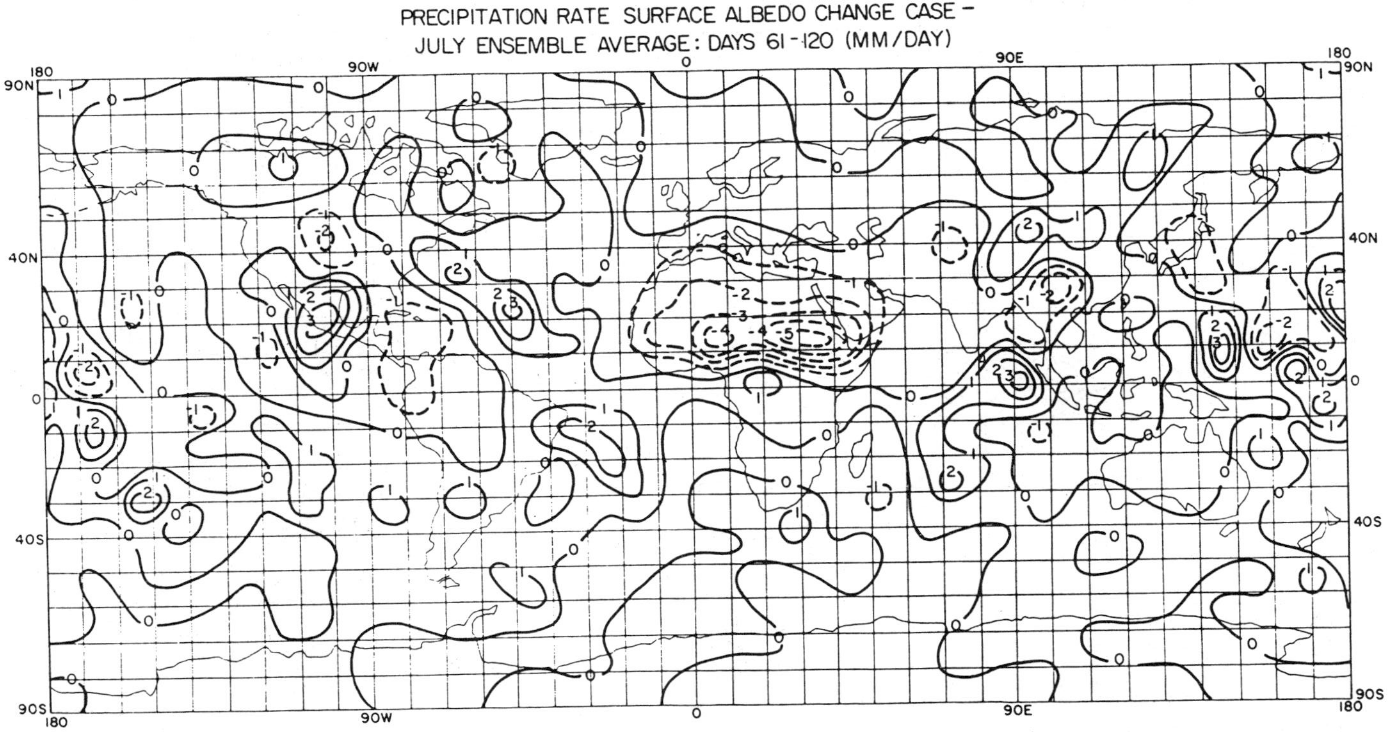

Fig. 33. Difference in precipitation rate between perturbed case (albedo 45 per cent) and the control case in which the observed July surface albedo is prescribed from observations. Note that in the NCAR GCM the soil moisture content is calculated explicitly (Washington, 1976, private communication).

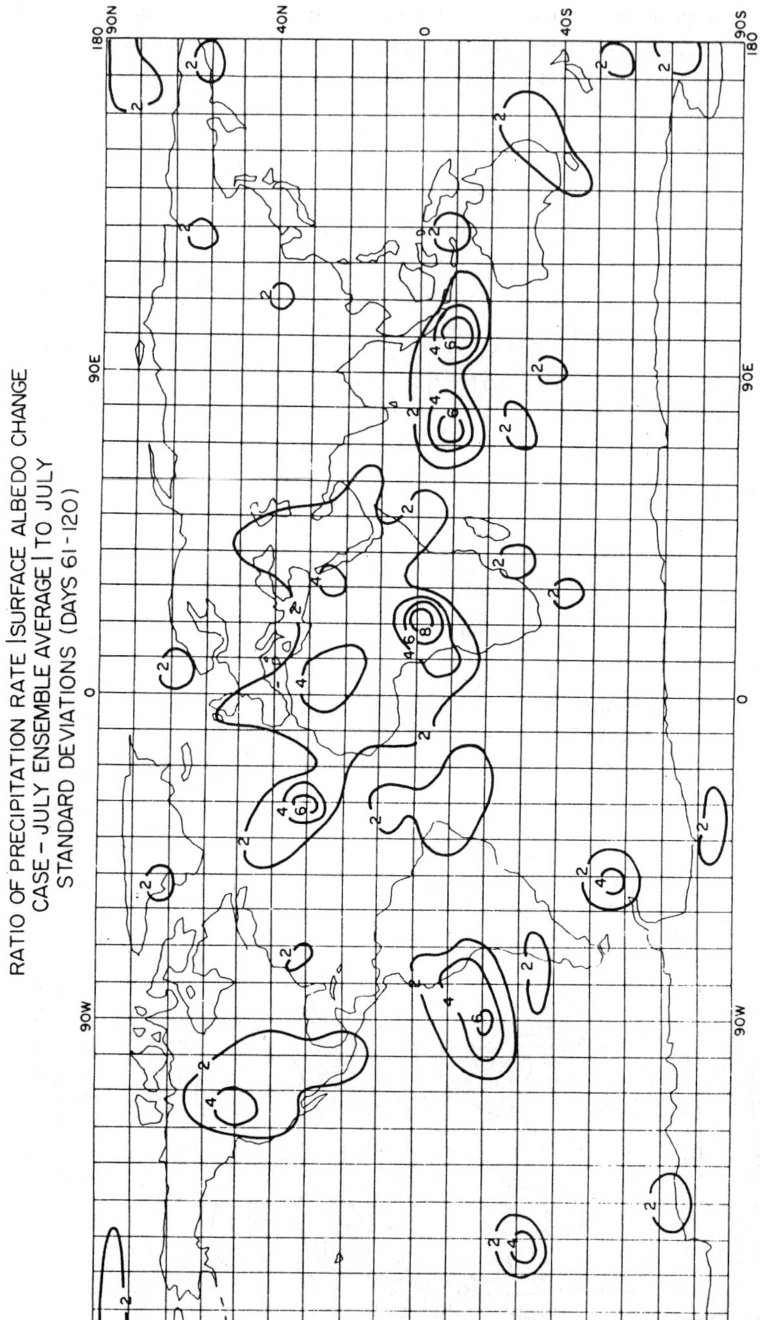

Fig. 34. Signal to noise ratio in the NCAR experiment. Note that values greater than one do not necessarily guarantee statistical significance. In this experiment, five Julys are used to estimate $\sigma 30$ (the noise). A value of 2 corresponds to 81.14 per cent significance level. See also section IIIA for further discussion.

Aerosols and Climate

262. All these hypotheses work by influencing the overall dynamics of the desert margin climates, essentially via their effect on rates of subsidence and hence stability. The possibility remains, however, that cloud microphysics may be affected by surface conditions. Schnell (1975) has suggested that the cumulus and cumulonimbus clouds of the Sahelian and Sudanian belts of West Africa are seeded by organic ice-nuclei raised from the vegetated surface below. Removal of vegetation destroys the local sources of organic nuclei. Hence well-formed clouds remain unseeded, and remain rainless, thereby

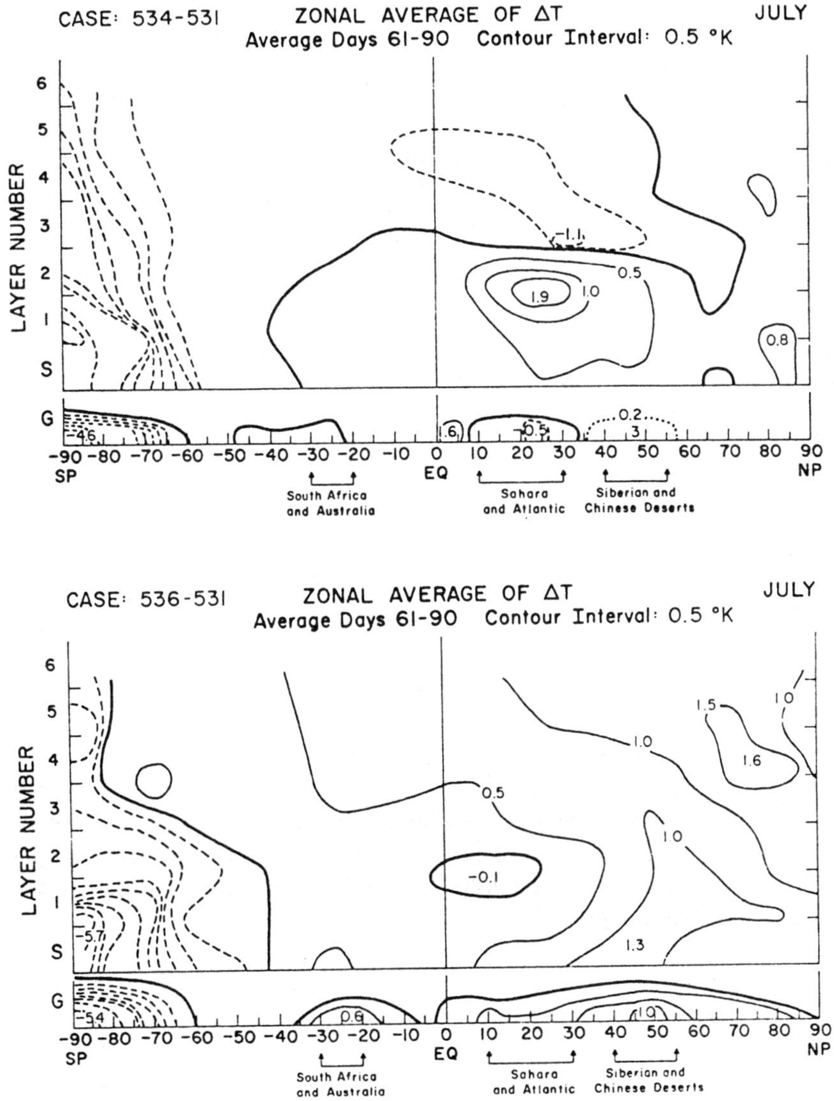

Fig. 35. Zonally averaged temperature deviations from a control case (531) which is July climatology as simulated by NCAR GCM. Case 534 corresponds to aerosols which are active in both the solar and terrestrial spectrum (Joseph, 1976).

accelerating the decay of surface vegetation. Schnell claims that this further positive feedback hypothesis is supported by laboratory results indicating that the organic nuclei are highly effective seeding agents at temperatures below about –8°C, which is warmer than for inorganic materials.

263. Aerosolas have long been thought to have significant effect on the global circulation. Recently Joseph (1976) performed experiments with NCAR's GCM to test this effect. Aerosols were placed over the major deserts, and over the Atlantic Ocean. It is well known that Saharan dust is transported to the Atlantic by the mid-tropospheric easterlies. In order to get high confidence level, ten times the observed amount of dust was assumed. Two experiments were carried out – one in which the aerosol is active in the solar spectrum only (Case 534) and one in which it is active in both the solar and terrestrial spectrum (Case 536).

264. Figure 35 shows the deviation of the zonally averaged temperature in Case 534 (visible only) and Case 536 (visible + infrared), from the control Case 531. As is expected in Case 534 the increased aerosol load has decreased the amount of incoming solar radiation causing ground cooling (level G in the figure). In the atmospheric layers, however, (S and above) the response is quite different. The aerosols placed aloft over the Atlantic Ocean have caused warming due to absorption of the short wave radiation.

265. In Case 536 (visible + infrared) the trend is reversed – the desert surfaces are warmer than the control case. This is due to the increased absorption in the infrared (greenhouse effect) which more than compensates for the decreased solar radiation. The previous large atmospheric heating over the Atlantic Ocean has disappeared and is replaced by slight cooling.

266. The different response of the two experiments demonstrates the need to know the exact size distribution and type of the aerosols so that one can determine the corresponding spectral properties. The spectral properties in return will determine whether the net effect of the aerosols is cooling or heating. These experiments also demonstrate that the vertical distribution of the aerosols is very important. Even though the Atlantic aerosols have the same optical depth as the land aerosols the former heat the atmosphere more effectively since they were placed higher (3–6 km) than the land aerosols.

ACKNOWLEDGEMENTS

267. Throughout the preparation of this Appendix discussions with the following people were extremely important: Robert Chervin, Edwin Danielson, Michael Glantz, Joachim Joseph, Cecil Leith, Stephen Schneider, Warren Washington and Edward Zipser. Many thanks are due to Joachim Joseph, Stephen Schneider, Warren Washington and Edward Zipser who provided pre-published research results which appear as Figs. 23, 24 and 29–35, in this manuscript.

BIBLIOGRAPHY

STANDARD REFERENCE WORKS

A. International Reports

Global Atmospheric Research Programme (1975) *The Physical Basis of Climate and Climate Modelling*, GARP Publication Series No. 16, World Meteorological Organization and International Council of Scientific Unions, Geneva, 265 pp.

Stamp, L. D. editor (1961) *A History of Land Use in Arid Regions*, UNESCO, Paris, 388 pp.

UNESCO, (1975) *The Sahel: Ecological Approaches to Land Use*, Unesco Press, Paris, 99 pp.

B. Books

Budyko, M. I. (1974) *Climate and Life* (edited by D. H. Miller), New York and London, 508 pp. and the following volumes of *World Survey of Climatology*, editor-in-chief H. E. Landsberg, Amsterdam, Elsevier:

Vol. 10, edited by J. F. Griffiths, 1971, *Climates of Africa*, 604 pp.

Vol. 11, edited by R. A. Bryson and F. K. Hare, 1974, *Climates of North America*, 420 pp.

Vol. 12, edited by W. Schwertfeger, 1976, *Climates of Central and South America*, 532 pp.

Vol. 13, edited by J. Gentilli, 1971, *Climates of Australia and New Zealand*, 405 pp.

WORKS CITED IN THE TEXT

Albrecht, F. (1962) Die Berechnung der Natülichen Verdunstung (Evapotranspiration) der Erdoberfläche aus Klimatologischen Daten, *Berichte des Deutschen Wetterdienstes*, **11**: No. 88, 19 pp.

Anthes, R. H. (1975) Some mesoscale modeling activities at Penn State, *Proceedings of SESAME Opening Meeting at Boulder Colorado Sept. 4–6, 1974*, D. K. Lilly, editor. National Oceanic and Atmospheric Administration Environmental Research Laboratories, Boulder, Co. 80302, pp. 302–317.

Aspliden, C. I., Tourre, U. and Sabine, J. B. (1976) Some climatological aspects of west African disturbance lines during GATE, *Monthly Weather Review*, **104**: 1029–1035.

Barry, R. G. (1976) Climate and climatic change in hot and cold deserts, Abstract for AMQUA Preliminary Conference Report, Tempe, Arizona, 4 pp.

Benyon, P. R. (1976) The use of semi-arid lands for grazing with the accompanying risks of deterioration, erosion and desertification, SCOPE Project 5, Australian contribution to International Workshop held in Indianapolis, MS, 10 pp.

Bhumralker, C. M. (1976) Parameterization of the planetary boundary layer in atmospheric general circulation models, *Rev. Geophy. and Space Phys.*, **14**: 215–226.

Bjerknes, J. (1966) A possible response of the atmospheric Hadley cell circulation to equatorial anomalies of ocean temperature, *Tellus*, **18**: 820–829.

Bjerknes, J. (1969) Atmospheric teleconnections from the equatorial Pacific. *Monthly Weather Review*, **97**: 163–172.

Bowler, J. M. (1975) Deglacial events in southern Australia: their age, nature and palaeoclimatic significance, in *Quaternary Studies*, R. P. Suggate and M. M. Cresswell eds., The Royal Society of New Zealand, pp. 75–82.

Bowler, J. M., Hope, G. S., Jennings, J. N., Singh, G. and Walker, D. (1976) Late Quaternary climates of Australia and New Guinea. To be published in *Quaternary Research*, **6**: (August 1976).

Braham, R. R. (1976) Modification of clouds and weather by a large metropolitan area, *Proceedings Second WMO Scientific Conference on Weather Modification*, Boulder, Colorado, August 1976, WMO Publication No. 773, pp. 435–442.

Broecker, W. S. and Kaufman, A. (1965) Radiocarbon chronology of Lake Lahontan and Lake Bonneville II, *Bulletin of the Geological Society of America*, **76**: 537–566.

Brown, L. H. and Cochemé, J. (1973) *A Study of the Agroclimatology of the Highlands of Eastern Africa,* WMO Technical Note 125 (WMO-No. 339), Geneva, 197 pp.

Brown, P. R. (1963) Climatic fluctuation over the oceans and in the tropical Atlantic, *Arid Zone Research* (UNESCO) **20**: 109–123.

Bryson, R. A. (1972) Climatic modification by air pollution, in *The Environmental Future*, N. Polunin, ed., London, Macmillan, pp. 133–154.

Bryson, R. A. (1974) A perspective on climatic change, *Science*, **184**: 753–760.

Bryson, R. A. (1975) The lessons of climatic history, *Environmental Conservation*, **2**: 163–170.

Bryson R. A (1976a) Some cultural and economic consequences of climatic change, *Proceedings Australian Conference on Climate and Climatic Change*, Royal Meteorological Society (Australian branch), Monash University, Dec. 1975, to be published.

Bryson, R. A. (1976b) Some lessons of climatic history, in *Atmospheric Quality and Climatic Change*, R. J. Kopec, ed., University of North Carolina at Chapel Hill, pp. 38–51.

Bryson, R. A. and Baerreis, D. (1967) Possibilities of major climatic modifications and their implications: north-west India, a case for study, *Bulletin of the American Meteorological Society*, **48**: 136–142.

Bryson, R. A. and Wendland, W. M. (1970) Climatic effects of atmospheric pollution, in *Global Effects of Environmental Pollution,* S. F. Singer, ed. Reidel, Dordrecht, pp. 130–138.

Budyko, M. I. (1958) *The Heat Balance of the Earth's Surface*, trs. Nina A. Stepanova, U.S. Dept. of Commerce, Washington, 259 pp.

Budyko, M. I. (1974) *Climate and Life*, D. H. Miller ed., Academic Press, New York and London, 508 pp.

Bunting, A. H., Dennett, M. D., Elston, J. and Milford, J. R. (1976) Rainfall trends in the west African Sahel, *Quarterly Journal of the Royal Meteorological Society*, **102**: 59–64.

Burpee, R. W. (1972) The origin and structure of easterly waves in the lower troposphere of North Africa, *J. Atmos. Sci.*, **29**: 77–90.

Butzer, K. W. (1971) *Environment and Archaeology: An Ecological Approach to Pre-history*, 2nd edition, Chicago, Aldine, 703 pp.

Butzer, K. W., Isaac, G. L., Richardson, J. L. and Washbourne-Kamau, C. (1972) Radiocarbon dating of East African lake levels, *Science*, **175**: 1069–1076.

Butzer, K. W. (1974a) Geology of the Cornelia Beds. *Memoirs van die Nasionale Museum, Bloemfontein, Republiek van Suid-Afrika*, No. 9, pp. 7–32.

Butzer, K. W. (1975) Patterns of environmental change in the Near East during late Pleistocene and early Holocene times, in *Problems in Pre-History: North Africa and the Levant*, F. Wendorf and A. E. Marks eds., SMU Press, Dallas, pp. 389–410.

Carter, D. B. and Mather, J. R. (1966) Climatic classification for environmental biology, *Publications in Climatology*, **19**: 305–395.

Charney, J. (1975a) Dynamics of deserts and drought in the Sahel, in *The Physical Basis of Climate and Climate Modelling*, WMO–ICSU, GARP Publication Series No. 16, pp. 171–176.

Charney, J. (1975b) Dynamics of deserts and drought in the Sahel, *Quarterly Journal of the Royal Meteorological Society*, **101**: 193–202.

Charney, J., Stone, P. H. and Quirk, W. J. (1975) Drought in the Sahara: a biogeophysical feedback mechanism, *Science*, **187**: 434–435.

Charney, J., Stone, P. H. and Quirk, W. J. (1976) Reply to E. A. Ripley, q.v., *Science*, **191**: 100–102.

Chervin, R. M. and Schneider, S. H. (1976a) A study of the response of NCAR GCM climatological statistics to random perturbations: Estimating noise level, *J. Atmos. Sci.*, **33**: 391–404.

Chervin, R. M. and Schneider, S. H. (1976b) On determining the statistical significance of climate experiments with General Circulation Models, *J. Atmos. Sci.*, **33**: 405–412.

Chu, Ko-Chen (1973) A preliminary study on the climatic fluctuations during the last 5,000 years in China, *Scientia Sinica*, **16** (2): 226–256.

Clarke, R. H. (1970) Recommended methods for the treatment of the boundary layer in numerical models, *Aust. Met. Mag.*, **18**: 51–73.

Cochemé, J. and Franquin, P. (1967) *An Agroclimatology Survey of a Semiarid Area in Africa south of the Sahara*, WMO, Technical Note 86 (WMO-No. 210. TP. 110), 136 pp.

Cotton, W. R. (1975) Theoretical cumulus dynamics, *Rev. Geophys. and Space Phys.*, **13**: 419–448.

Crutcher, H. L. and Davis, O. M. (1969) *Marine Climatic Atlas of the World, Vol. 8, The World*, Washington, Government Printer, 92 charts + text.

Crutcher, H. L., Jenne, R. L., Van Loon, H. and Taljaard, J. J. (1969) *Climate of the Upper Air, Part 1, South Hemisphere*, U.S. Naval Weather Service Command, NAVAIR 50–1C–55.

Davitaya, F. F. (1962) Transformation of nature in the steppes and deserts, in *Soviet Geography, Accomplishments and Tasks*, C. D. Harris, ed. New York, American Geographical Society, pp. 281–289.

Davy, E. G. (1974) Drought in West Africa, *WMO Bulletin*, **23**: 18–23.
Diester-Haass, L. (1976) Late Quaternary climatic variations in northwest Africa deduced from east Atlantic sediment cores, *Quaternary Research*, **6**: 299–314.
Dirección meteorológica de Chile (1976) *Regimen pluviométrico en la Cordillera de los Andes entre los 19° y 25° de latitud*, Santiago, 17 pp.
Dzerdzeevskii, B. L. ed. (1963) *Sukhovei ikh proiskhozhdenie i bor'ba s nimi*, Izdotel'stvo Akademii Nauk SSSR, Moskva, in translation *Sukhoveis and Drought Control*, Jerusalem, Israel Program for Scientific Translations, 366 pp.
Ellsaesser, H. W., MacCracken, M. C., Potter, G. L. and Luther, F. M. (1976) An additional model test of positive feedback from high desert albedo, *Quarterly Journal of the Royal Meterological Society*, **102**: 543–544.
Fairbridge, R. W. (1972) Climatology of a glacial cycle, *Quaternary Res.*, **2**: 283–302.
Flohn, H. (1964) Grundfragen der paläoklimatologie im lichte der atmosphärischen zirkulation, *Geol. Rdsch*, **54**: 504–515.
Flohn, H. (1971) *Tropical Circulation Pattern*, Bonner Meteorologische Abhandlungen, 15, 55 pp.
Flohn, H. and Ketala, M. (1971) *Investigations on the Climatic Conditions of the Advancement of the Tunisian Sahara*, WMO Technical Note 116 (WMO–279), 31 pp.
Flohn, H., Henning, D. and Korff, H. C. (1974) Possibilities and limitations of a large-scale water budget modification in the Sudan–Sahel belt of Africa, *Meteorologisches Rundschau*, **27**: 97–100.
Gabriel, K. R. (1967) Recent results of the Israeli artificial rainfall stimulation experiment, *Journal of Applied Meteorology*, **6**: 437–438.
Gal-Chen, T. and Somerville, R. C. J. (1975a) On the use of a coordinate transformation for the solution of the Navier-Stokes equations, *J. Computational Phys.*, **17**: 209–228.
Gal-Chen, T. and Somerville, R. C. J. (1975b) Numerical solution of the Navier-Stokes equations with topography, *J. Computational Phys.*, **17**: 276–310.
GARP (Global Atmospheric Research Programme) (1975) *The Physical Basis of Climate and Climate Modelling*, GARP Publication Series No. 16, ICSU–WMO, Geneva, 265 pp.
Geyh, M. A. and Jäkel, D. (1974) Spätpleistozäne und holozäne Klimageschichte der Sahara aufgrund zugänglicher C^{14}-daten, *Zeitschrift für Geomorphologie*, **18**: 82–98.
Gibbs, W. J. (1975) Drought, its definition, delineation and effects, in *Drought*, Special Environmental Report 5, WMO–403, Geneva, pp. 1–40.
Gilchrist, A. (1975) Two climate change experiments, Met.0. 20, Technical Note II/50, MS, 14 pp.
Glantz, M. H. (1976a) Climate and weather modification in and around arid lands in Africa, International Geographical Union Working Group on Desertification in and around Lands Symposium, Ashkahabad, USSR, 20 July 1976.
Glantz, M. H. (1976b) Value of a reliable long range climate forecast for the Sahel: a preliminary assessment, National Center for Atmospheric Research, MS, 44 pp.
Glantz, M. H. and Parton, W. (1975) *Weather and Climate Modification and the Future of the Sahara*, MS, (NCAR, Boulder), 21 pp.
Griffin, J. J., Windom, H. and Goldberg, E. D. (1968) The distribution of clay minerals in the World Ocean, *Deep-Sea Research*, **15**: 433–459.
Grove, A. T. (1971) *Africa South of the Sahara*, Oxford University Press, London, 2nd edition, 280 pp.
Grove, A. T. (1972) Climatic change in Africa in the last 20,000 years, in *Les Problèmes de Développement du Sahara Septentrional*, Vol. 2, Université d' Alger, Inst. de Géographie, Algiers.
Grove, A. T. (1973) Desertification in the African environment, in *Drought in Africa, Report of the 1973 Symposium*, D. Dalby and R. Harrison-Church eds., Centre for African Studies, U. of London, pp. 33–45.
Haltiner, G. J. (1971) *Numerical Weather Prediction*, John Wiley & Sons, Inc., New York.
Hare, F. K. (1942) Atlas lee-depressions and their significance for Scirocco, London, Air Ministry, *Synoptic Divisions Technical Memoranda No. 43*, 33 pp.
Hare, F. K. (1961) The causation of the arid zone, in *A History of Land Use in Arid Regions*, L. Dudley Stamp ed., UNESCO, Paris, pp. 25–30.
Hastenrath, S. (1976) Variations in low-latitude circulation and extreme climatic events in the Tropical Americas, *Journal of the Atmospheric Sciences*, **33**: 202–215.
Henning, D. (1970) Comparative heat balance calculations, International Association of Scientific Hydrology, *Proceedings of the Reading Symposium*, pp. 361–375 and 80–87.
Hess, S. L. (1959) *Introduction to Theoretical Meteorology*, Holt, Rinehart and Winston, New York, 362 pp.
Heusser, C. J. (1966) Polar hemispheric correlation: palynological evidence from Chile and the Pacific North-west of America, in *World Climate from 8000 to 0 B.C.*, J. Sawyer ed., pp. 124–141.

Heusser, C. J. (1974) Vegetation and climate of the southern Chilean lake district during and since the last interglaciation, *Quaternary Research*, **4**: 290–315.

Heusser, C. J. (1976) Palynology and depositional environment of the Rio Ignao nonglacial deposit, Province of Valdivia, Chile, *Quaternary Research*, **6**: 273–279.

Igeleke, C. A. (1975) Annual rainfall pattern and periodicities in the drought affected areas of the north, *Agrometeorological Bulletin*, January 1975, Nigerian Meteorological Service, pp. 16–28.

Jackson, R. D. and Idso, S. B. (1975) Surface albedo and desertification, *Science*, **189**: 1012–1013.

Jenkinson, A. F. (1973) A note on variations in May to September rainfall in West African marginal rainfall areas, *Proceedings Symposium on Drought in the Sahara*, Centre for African Studies, U. of London, pp. 31–32.

Jenkinson, A. F. (1975) Some quasi-periodic changes in rainfall in Africa and Europe, *Proceedings WMO/IAMAP Conference on Long-term Climatic Fluctuations*, Norwich, England, pp. 453–460.

Johnson, A. M. (1976) The climate of Peru, Bolivia and Ecuador, in W. Schwertfeger ed., *q.v.*, pp. 147–217.

Joseph, J. H. (1976) The effect of a desert aerosol on a model of the general circulation, *Proceedings Symposium on Radiation in the Atmosphere*, (IAMAP), Garmisch-Partenkirchen, F.R.G.

Kaufman, A. (1971) U-Series dating of Dead Sea carbonate, *Geochimica et Cosmochimica Acta*, **35**: 1269–1281.

Kovda, V. A. (1961) Land use development in the arid regions of the Russian plain, the Caucasus and Central Asia, in L. D. Stamp ed., *q.v.*, pp. 175–218.

Kowal, J. M. and Adeoye, K. B. (1973) An assessment of aridity and the severity of the 1972 drought in northern Nigeria and neighbouring countries, *Savanna*, **2**: 145–158.

Kowal, J. M. and Kassam, A. H. (1973) An appraisal of the drought in 1973 affecting groundnut production in the Guinea and Sudan savanna areas of Nigeria, *Savanna*, **2**: 159–172.

Kreiss, H. and Oliger, J. (1973) Methods for the approximate solution of time dependent problems, WMO–ICSU, GARP publication series No. 10.

Kreitzberg, C. (1975) Regional and mesoscale modeling at Drexel, *Proceedings SESAME Opening Meeting at Boulder Colorado Sept. 4–6, 1974*, D. K. Lilly ed., National Oceanic and Atmospheric Administration Environmental Research laboratories, Boulder, Colorado, pp. 289–301.

Krishnamurti, T. N., Daggupaty, S. M., Fein, J., Kanamitsu, M. and Lee, J. D. (1973) Tibetan high and upper tropospheric tropical circulations during northern summer, *Bulletin of the American Meteorological Society*, **54**: 1234–1249.

Kutzbach, J. E. (1975) Diagnostic studies of past climates, in *The Physical Basis of Climate and Climate Modeling*, GARP Publ. Series No. 16, ICSU/WMO, World Meteor. Organiz. Geneva, Switz., pp. 119–126.

Lamb, H. H. (1976) On the frequency and patterns of variation of climate: an opening statement on climatic change, *Climate Change, Food Production and Interstate Conflict*, Rockefeller Foundation Working Papers, pp. 40–58.

Landsberg, H. E. (1974) Man made climate changes, *Physical and Dynamic Climatology*, WMO Publication No. 347, pp. 262–303.

Landsberg, H. E. (1975) Sahel drought: change of climate or part of climate? *Archiv der Meteorologie, Geophysik und Bioklimatologie*, B, **23**: 193–200.

Landsberg, H. E. (1976) The definition and determination of climatic changes, fluctuations and outlooks, in *Atmospheric Quality and Climatic Change*, R. J. Kopec ed., University of North Carolina at Chapel Hill, Studies in Geography No. 9, pp. 52–64.

Lavoie, R. (1972) A mesoscale numerical model of lake-effect storms, *J. Atmos. Sci.*, **29**: 1025–1040.

de Leeuw, P. H. (1974) Livestock development and drought in the northern states of Nigeria, *Nigerian Journal of Animal Production*, **1**: 61–73.

Le Houérou (1974) North Africa: past, present and future, in *Arid Lands in Transition*, H. Dregne ed., Amer. Assoc. Adv. Sci., Washington, pp. 227–278.

Leith, C. E. (1973) The standard error of time-average estimates of climatic means, *J. Appl. Meteorol.* **12**: 1066–1069.

Lepple, F. K. and Brine, C. J. (1976) Organic constituents in Eolian dust and surface sediments from Northwest Africa, *Journal of Geophysical Research*, **81**: 1141–1147.

Lettau, H. (1969) Evapotranspiration climatonomy: 1. A new approach to numerical prediction of monthly evapotranspiration, run-off, and soil moisture storage, *Monthly Weather Review*, **97**: 691–699.

Livingstone, D. A. (1975) Late Quaternary climatic change in Africa, *Annual Review of Ecology and Systematics*, **6**: 249–280.

Lorenz, E. N. (1965) A study of the predictability of a 28 variable atmospheric model, *Tellus*, **17**: 321–333.

Louis, J. F. (1974) Mean meridional circulation, in U.S. Department of Transportation, Climatic Impact Assessment Program, Working Papers, *The Natural Stratosphere of 1974*, CIAP Monograph 1, pp. 6–21 to 6–29, especially fig. 6.8.

MacLeod, N. H. (1976) The processes and stages of desertification, in ISAID, *q.v.*, 7 pp.

Malkus, J. S. (1955) The effects of a large island upon the tradewind air stream, *Quart. J. Roy. Meteorol. Soc.*, **81**: 538–550.

Manabe, S., Hahn, D. G. and Holloway, J. L. (1974) The seasonal variation of the tropical circulation as simulated by a global model of the atmosphere, *J. Atmos. Sci.*, **31**: 43–83.

Mason, B. J. (1976) Towards the understanding and prediction of climatic variations, *Quarterly Journal of the Royal Meteorological Society*, **102**: 473–498.

Mather, J. R. (1974) *Climatology: Fundamentals and Applications*, McGraw-Hill, New York, 412 pp.

Meigs, P. (1961) Map of arid zone, in L. D. Stamp, *q.v.*, pp. 14–15.

Mercer, J. H. (1976) Glacial history of southernmost South America, *Quaternary Research*, **6**: 125–166.

Miller, A. (1976) The climate of Chile, in W. Schwertfeger ed., *q.v.*, pp. 113–145.

Monin, A. S. (1970) The atmospheric boundary layer, *Ann. Rev. Fluid. Mech.* **2**: 225–250.

McDonald, J. (1962) The evaporation-precipitation fallacy, *Weather*, **17**: 169–175.

McGinnies, W. G., Goldman, B. J. and Paylore, P. (eds.) (1968) *Deserts of the World: An Appraisal of Research into their Physical and Biological Environments*, U. of Arizona Press, Tucson.

McQuigg, J. D., Thompson, L., Le Duc, S., Lockard, M. and McKay, G. (1973) *The Influence of Weather and Climate on United States Grain Yields: Bumper Crops or Droughts*, Associate Administrator for Environmental Monitoring and Prediction, NOAA, Washington, 30 pp.

Naqvi, S. N. *et al.* (1959) *Integrated Survey of Isplingi,* Report 1, Arid Zone Research Section, Pakistan Meteorological Department, 38 pp.

Nairn, A. E. M. (ed.) (1961) *Descriptive Palaeoclimatology*, Interscience Pub., New York, 380 pp.

Namias, J. (1974) *Contribution to Report of International Tropical Meteorology Meeting*, Nairobi, American Meteorological Society, Boston, pp. 141–144.

Namias, J. (1977) The role of the oceans in generating continental drought, *Proceedings XVIth General Assembly, UGGI, Grenoble*, to be published.

National Academy of Sciences, U.S.A. (1973) *Weather and Climate Modification*, Washington, 258 pp.

National Academy of Sciences, U.S.A. (1975) *Understanding Climatic Change: A Program for Action*, Washington, 239 pp.

Oguntoyinbo, J. S. (1974) Land-use and reflection coefficient (albedo) map for southern parts of Nigeria, *Agricultural Meteorology*, **13**: 227–237.

Olausson, E. (1975) On the course of a glacial age, in WMO, *Proceedings WMO/IAMAP Symposium on Long-term Climatic Fluctuations*, Norwich, 18–23 August 1975, pp. 47–52.

Olivares, R. B. (1967) Las glaciaciones cuaternarias al Oeste de Lago Llanquihue en el sur de Chile, *Revista Geográfica*, **67**: 100–108.

Orlanski, I. A. (1975) Rational subdivision of scales for atmospheric processes, *Bulletin of the American Meteorological Society*, **56**: 527–530.

Orlanski, I. A. (1976) A simple boundary condition for unbounded hypergolic flows, *J. Computational Phys.*, **21**: 251–269.

Ormerod, W. E. (1976) Ecological effect of controls of African trypanosomiasis, *Science,* **191**: 815–821.

Otterman, J. (1974) Baring high-albedo soils by overgrazing: a hypothesized desertification mechanism, *Science*, **186**: 531–533.

Otterman, J. (1975) Reply to Jackson and Idso (1975), *q.v.*, *Science*, 1013–1015.

Pedgley, D. E. and Krishnamurti, T. N. (1976) Structure and behavior of a monsoon cyclone over West Africa, *Monthly Weather Review*, **104**: 149–167.

Perrin de Brichambault, G. and Wallén, C. C. (1964) *Une Etude d'Agroclimatologie dans les Zones Arides et Semi-arides du Proche-Orient*, WMO Technical Note 56 (OMM-No. 141. TP. 66), 69 pp.

Peterson, J. T. and Bryson, R. A. (1968) Influence of atmospheric particulates on the infra-red radiation balance of north-west India, *Proceedings First National Conference on Weather Modification*, Albany, New York, American Meteorological Society, 532 pp.

Philips, N. A. (1957) A coordinate system having some special advantages for numerical forecasting, *J. Meteor.*, **14**: 184–185.

Pielke, R. A. (1974) A three-dimensional numerical model of the sea breezes over south Florida, *Monthly Weather Review*, **102**: 115–139.

Pielke, R. A. (1975) A survey of fine-mesh modeling techniques. *Proceedings SESAME Opening Meeting at Boulder, Colorado, Sept. 4–6, 1974*, D. K. Lilly ed., National Oceanic and Atmospheric Administration Environmental Research Laboratories, Boulder, Colorado 80302, June 1975, pp. 219–232.

Pittock, A. B. (1975) Climatic change and the patterns of variation in Australian rainfall, *Search*, **6**: 498–504.

Pittock, A. B. (1977) Interactions between the general circulation and climate on smaller space scales, *Proceedings Australian Conference on Climate and Climatic Change*, Dec. 1975, Monash University, Royal Meteorological Society (Australian branch), to be published.

Potter, G. L., Ellsaesser, H. W., MacCracken, M. C. and Luther, F. M. (1975) Possible climatic impact of tropical deforestation, *Nature*, **258**: 697–698.

Priestley, C. H. B. and Taylor, R. J. (1972) On the assessment of surface heat flux and evaporation using large-scale parameters, *Monthly Weather Review*, **100**: 81–92.

Prohaska, F. (1976) The climate of Argentina, Paraguay and Uruguay, in W. Schwertfeger ed., *q.v.*, pp. 13–111.

Prole, J. H. B. (1967) Pastoral land use, in *Nairobi: City and Region*, W. T. W. Morgan ed., Oxford University Press, pp. 90–97.

Rapp, A. (1974) *A Review of Desertization in Africa – Water, Vegetation, and Man*, Secretariat for International Ecology, Stockholm, SIES Report No. 1, 77 pp.

Rapp, R. R. and Warshaw, M. (1974) *Some Predicted Climatic Effects of a Simulated Saharan Lake*, Rand Corporation, Santa Monica, 40 pp.

Rasmusson, E. M. (1968) Atmospheric water vapour transport and the water balance of North America: II. Large-scale water balance investigations, *Monthly Weather Review*, **96**: 720–731.

Ratisbona, L. R. (1976) The climate of Brazil, in W. Schwertfeger ed., *q.v.*, pp. 219–293.

Ripley, E. A. (1976) Drought in the Sahara: Insufficient biogeophysical feedback? *Science*, **191**: 100.

Richardson, L. F. (1922) *Weather Prediction by Numerical Process*, Cambridge University Press.

Schneider, S. H. and Dickinson, R. E. (1974) Climate modeling, *Rev. Geophy. Space Phys.*, **12**: 447–493.

Schnell, R. C. (1975) *Biogenic Ice Nucleus Removal by Overgrazing: A Factor in the Sahelian Drought?* Final Report to Directors, Rockefeller Foundation, MS, NCAR, Boulder, Colorado, 12 pp.

Schnell, R. C. and Vali, G. (1976) Biogenic ice nuclei: Part I. Terrestrial and marine sources, *Journal of the Atmospheric Sciences*, **33**: 1554–1564.

Schove, D. J. (1973) African droughts and weather history, in *Drought in Africa, Report of the 1973 Symposium*, D. Dalby and R. Harrison-Church eds., Centre for African Studies, U. of London, pp. 29–30.

Schwertfeger, W. editor (1976) *Climate of Central and South America, World Survey of Climatology*, Vol. 12, Amsterdam, Elsevier, 532 pp.

Servant, M. *et al.* (1969) Chronologie du Quaternaire récent des basses régions du Tchad, *Comptes Rendus, Académie des Sciences de Paris*, **269**: 1603–1606.

Shukla, J. (1975) The effect of Arabian Sea-Surface Temperature Anomaly on Indian Summer Monsoon: a numerical experiment with GFDL model, *Journal of the Atmospheric Sciences*, **32**: 503–511.

Singh, G. (1971) The Indus valley culture seen in the context of post-glacial climatic and ecological studies in north-east India, *Archaeology and Physical Anthropology in Oceania*, **6**: 177–189.

Singh, G., Joshi, R. D. and Singh, A. B. (1972) Stratigraphic and radiocarbon evidence for the age and development of three salt lake deposits in Rajasthan, India, *Quaternary Research*, **2**: 496–505.

Singh, G., Joshi, R. D., Chopra, S. K. and Singh, A. B. (1974) Late Quaternary history of vegetation and climate of the Rajasthan Desert, India, *Philosophical Transactions of the Royal Society, London*, Series B, **267** (889): 467–501.

Singh, G. (1975) Late Quaternary vegetation and climatic oscillations at Lake George, New South Wales, Paper presented to the Australasian Conf. on Climate and Climate Change (Roy. Meteor. Soc.), Monash University, Melbourne, Australia, 11 pp.

Singh, G. (1976) Stratigraphical and palynological evidence for desertification in the Great Indian Desert, Special Number of the *Annals of the Arid Zone* on Desertification, Jodhpur, India, 16 pp.

Sircoulon, J. (1976) Les sécheresses et les étiages: la récente sécheresse des régions sahéliennes, presented to the Société hydrotechnique de France, 30 pp.

Smagorinsky, J. (1974) Global atmospheric modeling and the numerical simulation of climate, in *Weather Modification*, W. N. Hess ed., John Wiley, New York.

Somerville, R. C. J., Stone, P. H., Halem, M., Hansen, J. E. Hogan, J. S., Druyan, L. M., Russell, G., Lacis, A. A., Quirk, W. J. and Tenenbaum, J. (1974) The GISS model of the global atmosphere, *J. Atmos. Sci.*, **31**: 84–117.

Stamp, L. D., editor (1961) *A History of Land Use in Arid Regions*, UNESCO Arid Zone Research, Publication XVII, Paris, 388 pp.

Strang, W. G. and Fix, G. J. (1977) *An Analysis of the Finite Element Method*, Englewood Cliffs, N. J., Prentice-Hall.

Tanquilevich, R. F. (1976) La desertización antrópica, Paper presented at first meeting of the panel on the Transnational Project to Monitor Desertification . . . in Critical Areas in South America, Instituto Argentino de Investigaciones de las Zonas Aridas, Argentina, MS, 6 pp.

Tapp, M. C. and White, P. W. (1976) A non-hydrostatic mesoscale model, *Quarterly Journal of the Royal Meteorological Society*, **102**: 277–295.

Tetzlaff, G. (1974) Der Wärmehaushalt in der Zentralen Sahara, *Berichte des Instituts für Meteorologie und Klimatologie der Technischen Universität Hannover*, No. 13, 113 pp.

Thornthwaite, C. W. (1948) An approach toward a rational classification of climate, *Geographical Review*, **38**: 55–94.

Tite, M. S. (1972) *Methods of Physical Examination in Archaeology*, Seminar Press, London, 389 pp.

Trenberth, K. E. (1976) Spatial and temporal variations of the Southern Oscillation, *Quarterly Journal of the Royal Meteorological Society*, **102**: 499–513.

Tricart, J. (1963) Oscillations et modifications de caractère de la zone aride en Afrique et en Amérique Latine lors des périodes glaciers des hautes latitudes, *Arid Zone Res.*, **30**: UNESCO, Paris, pp. 415–419.

Troup, A. J. (1965) The Southern Oscillations, *Quarterly Journal of the Royal Meteorological Society*, **91**: 490–506.

Tucker, G. B. (1975) Climate: is Australia's changing?, *Search*, **6**: 323–328.

Twidale, C. R. (1976) *Analysis of Landforms,* Sydney, John Wiley & Sons, pp. 484–489.

United States Committee for the Global Atmospheric Research Program (1975) *Understanding Climatic Change: a Program for Action*, Nat. Acad. Sciences, Washington, D.C., 239 pp.

Vali, G., Christensen, M., Fresh, R. W., Galyan, E. L., Maki, L. R. and Schnell, R. C. (1976) Biogenic ice nuclei. Part II: Bacterial sources, *Journal of the Atmospheric Sciences*, **33**: 1565–1570.

Van Zinderen Bakker, E. M. and Butzer, K. W. (1973) Quaternary environmental changes in Southern Africa, *Soil Science*, **116**: 236–248.

Walker, G. T. (1974) Correlations in seasonal variations of weather, IX, *Memoirs of the Indian Meteorological Department*, **24**: 333–345.

Walker, J. and Rowntree, P. R. (1976) The effect of soil moisture on circulation and rainfall in a tropical model, MS, U.K. Meteorological Office, 23 pp.

Wallén, C. C. and Perrin de Brichambault, G. (1962) *A Study of Agroclimatology in Semi-Arid and Arid Zones of the Near East*, FAO/UNESCO/WMO Interagency Project on Agroclimatology, 185 pp. plus annexes.

Wallén, C. C. (1967) Aridity definitions and their applicability, *Geografiska Annaler*, **49A**: 367–384.

Washington, W. M. (1976) Numerical simulation of the Asian-African winter monsoon, *Monthly Weather Review*, **104**: 1023–1028.

Wendorf, F., Schild, R., Said, R., Haynes, C. V., Gautier, A. and Kobusiewicz, M. (1976) The prehistory of the Egyptian Sahara, *Science*, **193**: 103–114.

Winstanley, D. (1973) Rainfall patterns and general atmospheric circulation, *Nature*, **245**: 190–194.

Winstanley, D. (1974) Seasonal rainfall forecasting in West Africa, *Nature*, **248**: 464–465.

WMO (World Meteorological Organization) (1971) *Agroclimatology in the Semi-Arid Areas South of the Sahara.* Proceedings of the Regional Technical Conference, FAO/UNESCO/WMO Interagency Project on Agroclimatology, Geneva, 253 pp.

WMO (1974) Report of the Third Session of the EC Panel on Weather Modification/CAS Working Group on Cloud Physics and Weather Modification, Toronto, Canada, Appendix F.

Wopfner, H. and Twidale, C. R. (1967) Geomorphological history of the Lake Eyre basin, in *Landform Studies from Australia and New Guinea*, Canberra, Australian National University, pp. 119–143.

ECOLOGICAL CHANGE AND DESERTIFICATION

ANDREW WARREN and JUDITH K. MAIZELS

University College London

CONTENTS

ACKNOWLEDGEMENTS

We should like to thank the following for reports contributed to the present study:

Mr. J. R. Blackie and Mr. R. T. Clarke and Mrs. Celia Kirby, Institute of Hydrology, Wallingford, U.K.
Dr. P. Bradley, University of Newcastle-upon-Tyne, U.K.
Dr. D. W. Goodall, C.S.I.R.O., Wembley, Western Australia.
Dr. T. Ingold, University of Manchester, U.K.
Mr. Paul Richards, International African Institute, London, U.K.
Mr. J. Swift, University of Sussex, U.K.
Mr. S. Wilson, University College London, U.K.

We should also like to thank the following whose comments and suggestions have greatly helped us: Professor J. A. Mabbutt, University of N.S.W., Kensington, N.S.W.; Professor R. Slatyer, A.N.U. Canberra, Australia; Mr. O. B. Williams, Dr. R. Winkworth, Dr. K. Myers, Dr. A. Newsome and Dr. J. Leigh, also of C.S.I.R.O., Canberra; Messrs. Newman, Condon and Cunningham, Soil Conservation Service, N.S.W.; Mr. Lendon and his team, C.S.I.R.O., Alice Springs; Mr. G. Webber, Department of Agriculture, South Australia; Mr. K. Fitzgerald, Department of Agriculture, Western Australia; Mr. K. Hyde, Department of Agriculture, Northern Territory, Australia; Dr. R. Perry and his team, C.S.I.R.O., Wembley, Western Australia; Mr. A. Blair Rains and Mr. R. Lawton, Land Resources Division, Ministry of Overseas Development, Tolworth, U.K.; Mr. D. Hall and his colleagues of the M.O.D., London, U.K.; Mr. S. Sandford, Overseas Development Institute, U.K.; M. Floret, FAO, Tunisia; Mr. F. Mouttapa, Mr. Poloicarpou, M. Perot, Mr. Arnoldus, Mr. R. Griffiths and M. P. Auriol, FAO, Rome; Dr. M. Hadley, M. M. Batisse and M. Sasson, UNESCO, Paris; Professor M. Evenari and Dr. E. Noy-Meir, Hebrew University of Jerusalem; Professor F. Bourlière, Faculté de Médecine, Paris ouest; M. Lacroûts, M. G. Boudet and M. Chadelas, IEMVT, Maisons Alfort, France.

I. INTRODUCTION

1. A simple and graphic meaning of the word "desertification" is the development of desert-like landscapes in areas which were once green. Its practical meaning, used in this report, is a sustained decline in the yield of useful crops from a dry area accompanying certain kinds of environmental change, both natural and induced.

2. In dry ecosystems, where a shortage of water is the main constraint on biological productivity, desertification is commonly the result of the reduced availability of water to crops, but it is important to distinguish between the effects of drought alone and those of desertification. If damage is due merely to drought, then short-term palliatives will suffice, even if they are periodically necessary over the years, but if there is a sustained fall in yield, then a much longer-term strategy is called for.

3. Dry ecosystems that are only slightly tampered with have the capacity to recover from natural disturbances such as a run of dry years, but are not able to recover readily from the effects of others such as the erosion of a water-absorbent topsoil that may follow misuse. Such a loss, which might induce a progressive decline in biological productivity, would be an example of desertification. In some cases it would not be feasible to reverse this decline and desertification would be permanent. In other cases recovery may be possible.

4. Desertification, though often aggravated by drought, can occur in good as well as in bad years, since the loss of a soil-moisture reservoir by erosion, which reduces the water supply available to crops, could happen even in years with above average rainfall. The loss of yield due to such processes, however, is usually difficult to detect because of the fluctuations in production due to varying annual rainfalls. Only over periods greater than a decade can desertification be clearly distinguished from the less-lasting effects of drought.

5. The idea of the yield of useful crops incorporates a socio-economic value and their production involves processes in which cultural factors are as important as ecological ones. Although the yield from a dry ecosystem can change if less water or nutrients are available, it can also change if a new technique or organizational framework is introduced, if a new demand is made of it, or if a new perception of risk is adopted. Indeed ecosystems, and the economic superstructures that depend on and manipulate them, are so complex that a decrease in yield could come about in many different ways. In some pastoral economies, for example, men depend upon milk produced by domestic animals from ingested herbage and water. The forage plants of the cattle live in communities which have evolved in response to particular climates and soils and to grazing by wild animals. The soils, their surface processes and hydrology are themselves in balance with the overlying vegetation, wild fauna and climate. Men have selected the stock and their numbers to achieve some optimum in productivity, manageability, palatability, marketability, mystery or beauty. Diseases, parasites and predators have also selected them and, ideally, have achieved some kind of symbiosis with them. The milk yield might decrease in response to change in any one of these vulnerable components, the change radiating through the web to and from the teats of the cows, yaks, goats, sheep or camels.

6. The spread of the effects of a drought through a pastoral system is illustrated in

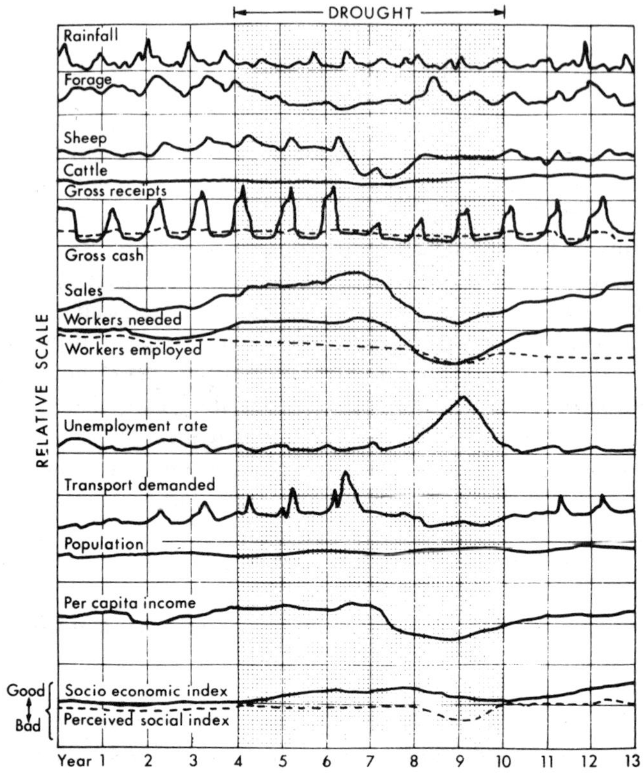

Fig. 1. A socio-technical model of a region in which grazing is a major industry: Charleville Region, Australia (Benyon, 1976).

Fig. 1 by the results of a numerical model. A similar, though sustained, decline in yield would follow desertification.

7. Desertification is a process involving environmental and not merely social change, but it is seldom easy to discover the causes of a decline in yield in complex ecological and economic systems. Increasing expectations of living standards or population pressure might induce an expansion in cropland; pastoralists accustomed to using this land would then be forced onto hitherto ungrazed pastures; to maintain or increase their living standards they might demand more watering points in order to exploit the new areas; these provided, cattle numbers could increase and disease spread more readily; veterinary authorities would be called upon to control the diseases; the ensuing increased numbers of cattle could hammer the range and destroy it; some of the nomads might then decide to settle and plough new fields. This process would have been the result of a progressive and destructive interplay of culture and environment. Despite the difficulty of the task, it is important to discover the causes of change, since only if aimed at the real culprits will action prove effective.

8. Desertification could be the result of a secular change of climate, or of the inexorable loss of nutrients from ancient weathered soils. Since there seems to be no good evidence that there has been a long-term change in climate (Hare, 1976) and since the

natural loss of nutrients from ancient soils is very slow, this report concentrates on desertification induced by cultural practices.

9. Figure 2 shows the ways in which yields might fall in several land-use systems. Figure 2A is a general statement of the idea drawn from Held and Clawson (1965) among others. Figure 2B is an approximation of the process on pastoral land. It incorporates the concepts used by Benyon (1976) and Swift (1974). A variable input of rain would be matched, under relatively light usage, by the supply of forage, but heavy stocking might so damage the grazing that there would be an enforced decrease in cattle numbers when the rains failed. If the damage were severe, neither forage nor cattle numbers could recover and there would be a sustained fall in yield. Figure 2C generalizes the history of agricultural land misuse in many parts of the temperate desert margins: persistent cropping at first overtaxes the nitrogen supply and drains the organic matter that binds the soil; eventually the unprotected soil is blown or washed away and all cropping must cease. Figure 2D shows the pattern that has occurred in some shifting agricultural systems on tropical desert margins: growing population or enforced immobility increases the demand for production from a smaller total area; to achieve this fallows are curtailed and peak production is therefore reduced. A vicious downward spiral of yields and acceptable standards ensues. Figure 2E is a generalization of what often happens in irrigation projects

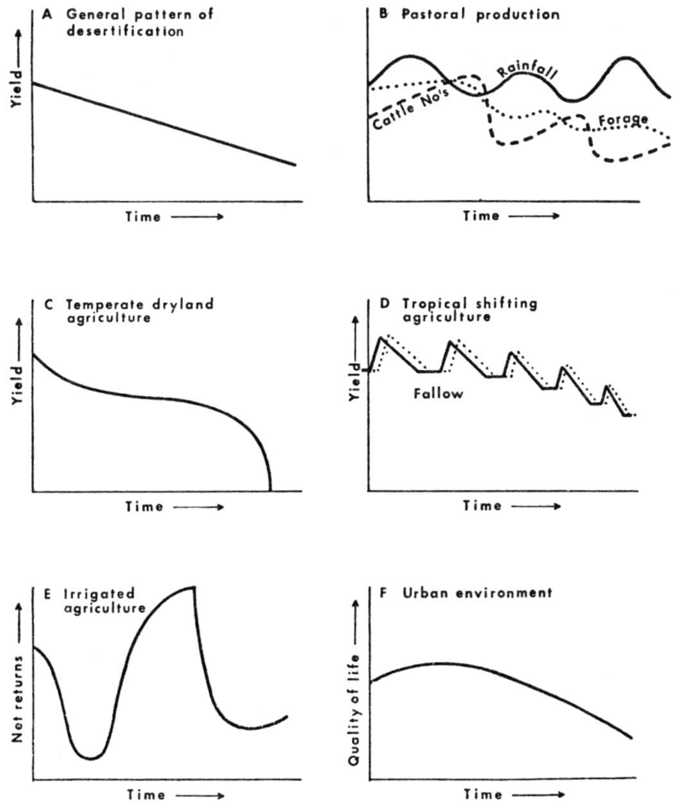

Fig. 2. Changes in yields, returns or quality of life during desertification in five major land-use systems.

in which sufficient drainage is not installed; the first flush of increased production is followed by a sharp fall as the water-table nears the surface and salinization and waterlogging begin to affect crops. Figure 2F shows the kinds of pattern that might accompany urban and industrial development in dry areas: after an initial improvement, life becomes more difficult and unpleasant.

10. Although this report looks at desertification processes like these only from the ecological viewpoint, it cannot ignore cultural factors, and, indeed, must analyse the ecological strategies and impacts of different economic systems. Cultural change, which is really the main initiator of desertification, is examined in another review. The ecosystem concept which views the whole complex of interacting processes allows the multiple repercussions of single events to be seen clearly. The concept can profitably illuminate problems and point to solutions for pastoralism where there is little technological input; but in agriculture, problems and solutions are more technological, whether simple, intermediate or advanced, and can only partly be clarified with an ecosystem approach.

11. The areas of concern are the warm deserts and their semi-arid margins. Because there are more people in the wetter margins, they are the more vulnerable, and will be the more closely scrutinized but serious problems in densely populated irrigated areas and even in sparsely populated ranges of the true deserts will also be examined.

12. The report will first look at the evidence for decreasing yield (Chapter II). If the evidence is found, then there is a mandate for what follows. Patterns of production and strategies of resilience in dry ecosystems will then be discussed (Chapter III). These need to be safeguarded and, indeed, emulated by management. The changes induced by exploitation must then be explained and evaluated by this yardstick (Chapters IV and V) and, finally, a prescription for wise use based on the lessons pointed by the preceding description must be formulated (Chapter VI).

II. EVIDENCE

A. INTRODUCTION

13. This chapter briefly reviews the evidence for desertification. It will soon become apparent that this is not an easy task. Statistics are seldom in the right form, are hard to come by, and even harder to believe, let alone to interpret. Evidence from surveys on the ground is no less reliable, and historical and archaeological evidence raise their own hosts of conflicting interpretations.

14. The only incontestable proof of desertification would come from experimental plots on which all but the environmental factors in production were kept constant. The experiment would have to continue for at least a decade to encompass very wet and very dry years, and would have to be so arranged that the effects of rotational grazing or cropping and the complementarity of different vegetation types in a grazing strategy were allowed for. Needless to say, such data do not exist.

15. Aggregated statistics for production by nation, province or region hide the causes of any trends that may be apparent: declining output can be assigned as much to changes in the market, the application of different techniques, or the deployment of labour, as to desertification. Because of these difficulties, alternative sources of evidence must be tapped and cross-checked. Because they are the easiest to collect, the most accessible of these are direct observations (at a number of levels of detail) of the state of soil, vegetation and other environmental characteristics. Common as it is, this kind of evidence needs to be treated with great caution. In a complex ecosystem one symptom, such as the absence of grass or the occurrence of a gully, could have many causes.

16. An example of a glib conclusion can be taken from Arizona where many observers have blamed the over-use of the range for the appearance of gullies at the turn of the century. A close examination of the evidence led Cooke and Reeves (1976) to a much more complicated explanation. They found that there had been little gullying during the period of heaviest stocking levels, and the gullies had only begun to develop well after the numbers of cattle had fallen. In search of another explanation, they first looked to the rainfall and found, with earlier workers, that although there had been little sustained change in annual amounts during the period of gullying, there had been a consistent increase in intensity at some seasons. Nevertheless, this alone had probably not been enough to form the gullies, for the history of each particular case showed that there had almost always been some local disturbance of the surface such as the cutting of a roadside ditch or the insertion of a culvert.

17. Driven by enthusiasm and a sense of impending crisis, observers of dryland condition seldom have the time for these niceties, and although they are constantly adjusting their perceptions to accommodate the findings of detailed research, any use of their results must take account of the fact that they have to be based on a considerable degree of personal judgement. One has simply to invoke the classic survey of soil erosion undertaken by the early Soil Conservation Service in the United States after the dust bowl years of the 1930s, which showed large areas of the country to be damaged, to realize that gross distortions can be made by skilled observers in very good faith, and

with the best of intentions (Held and Clawson, 1965). If that survey was to be believed most of the High Plains would still be a desert.

B. PASTORAL PRODUCTION

1. Declining Yield

18. The longest runs of pastoral statistics for dry areas and therefore those that might show the effects of desertification rather than of drought come from the dry parts of the United States and Australia. These areas have undoubtedly experienced great reductions in *numbers* of domesticated animals since they reached a maximum in the late nineteenth century. Perry's figures for sheep numbers in the Western Division of New South Wales are shown in Fig. 3. The figures are corroborated for the cattle areas of North Australia (Payne *et al.*, 1974). In the United States, Hastings (1959) showed that in 1882 the south-western range supported 500,000 head of cattle but only 253,000 in 1890, and for the saltbush desert of Utah, Holmgren (1973) noted that, although the maximum number of sheep had been reached by 1905 and were maintained until 1930, a continuous decline has occurred since then.

19. The use of these data as evidence for desertification needs to be qualified in three important ways:

(1) The numbers of beasts on the range is as much a function of economics as of forage. The present huge populations of cattle in central and northern Australia are due in part to a run of wet years, but more to a fall in the price of beef to below the cost of transporting it to market. A similar relation of supply to demand is said to have maintained the huge numbers of sheep apparent on Perry's graph in the period of the 1890s. By the same token, low numbers of stock can occur either with a rapid through-put of beasts to an avid market, or when a long-term decline in demand removes the incentive to produce. Such a process seems to be happening in the Gascoyne catchment case study area in Western Australia.

(2) In the developed dry lands there has been a major effort to introduce better techniques to pastoralists, bringing a major improvement in production per beast

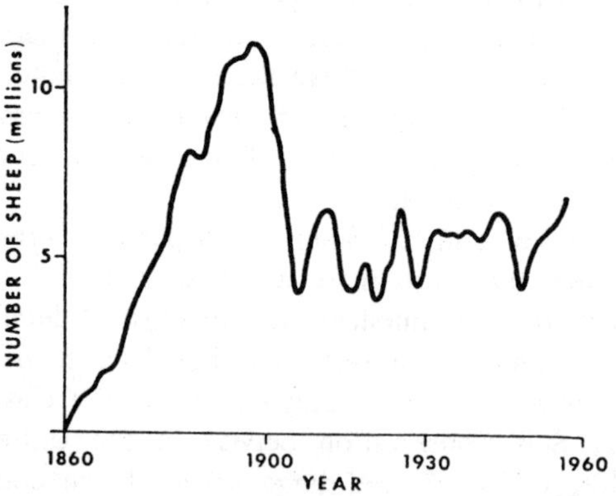

Fig. 3. Sheep numbers in the Western Division, New South Wales, 1860 to 1955 (Perry, 1968).

(Holmgren, 1973) which may well offset the decline in numbers. Its effect on the range is another question. In the second place there has been a concerted, though admittedly only partly successful, effort to impose safe stocking levels by law. Although this may have averted disaster, its effects cannot at the same time be taken as evidence for desertification. The prophylaxis cannot be taken as a symptom of the disease.

(3) The statistics are far from perfect. In western Nebraska, for example, the yearly count of cattle is taken in mid-winter before the cows have calved, and before cattle are brought in from the west to prepare for later fattening in the corn-yards of the east. Statistics taken for taxation purposes are always to be suspected, and very often the station owners themselves do not know how many stocks they have on the open range.

20. These cautions, and especially the last, apply *a fortiori* to less developed countries. Methods of estimating cattle numbers are improving, but the improvements throw very grave doubts on earlier figures. In particular the very large and more or less illegal movements of cattle into and out of census districts probably invalidate much of the picture shown in Fig. 4. In the recent drought the Fulani are said simply to have moved south across international frontiers and then to have drifted north again in the good years. Despite the imperfection of the statistics it can hardly be doubted that the Sahelian

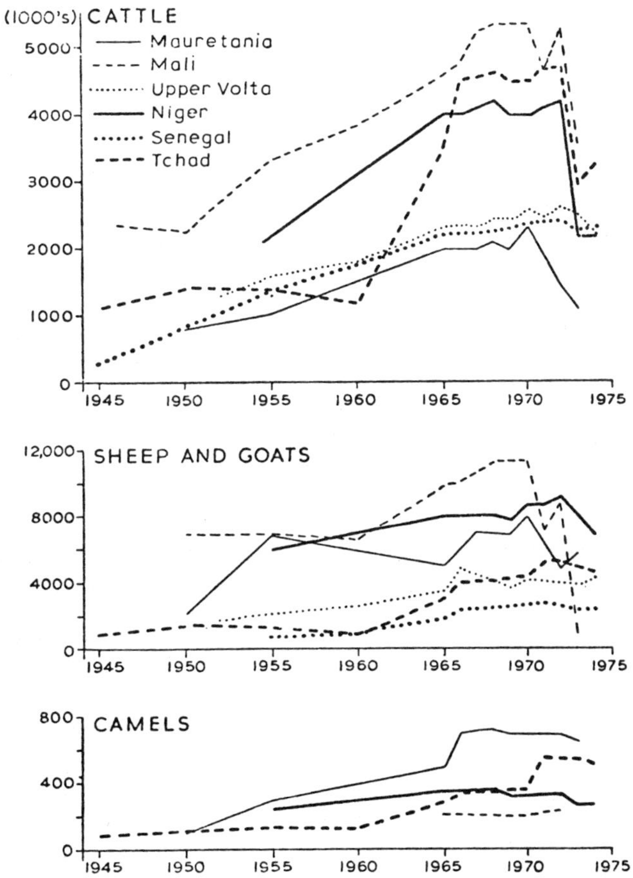

Fig. 4. Livestock numbers in the six Sahelian countries, 1945 to 1974 (SEDES, 1975).

flocks suffered very badly in the early part of this decade, and the general pattern in which slow increases are followed by drastic declines is corroborated by accounts from many other areas (Heady, 1969; Alpay, 1974; Pearse, 1970; Le Houérou, 1970).

21. Although the recent statistics cover the period of only one drought, the even less accurate estimates for losses in earlier droughts, for example in 1913, suggest smaller losses and so underline the gravity and perhaps greater permanence of the decline in the early 1970s. Nevertheless, the question as to whether there has or has not been desertification can clearly not be answered with these data alone.

2. Herd and Range Surveys

22. Surveys of individual herds show that, at the very least, their recovery from the recent drought in the Sahel will be slow. Both Garcia (1974) and Temple and Thomas (1973) found that very few young animals had survived the drought so that the herds were going to take at least 10 years to recover their former population structure, let alone their numbers. Such observations, though important, tell nothing of the capacity of the range to support the slowly recovering herds.

23. Evidence of desertification must therefore be supported with observations from the pastures themselves, and this must come from surveys of ground cover, age structure of the vegetation, and the occurrence of palatable and unpalatable species. The reasoning behind these surveys and their methods will be discussed in Chapters IV and V. Here we need only note the general conclusions about degradation that they have reached.

24. With results of exclosure experiments as reference points, and with a close acquaintance of the local flora and environment, ecologists and range managers have made wider surveys to estimate the condition of the range and its carrying capacity. It is these surveys which usually paint a gloomy picture of the state of the grazing resources of the dry world. Examples of gloom come from the results of Aubréville (1973) in the central Sahel, of Boudet and Duverger (1961) in the Hodh region of Mauritania, of Boudet *et al.* (1969) in the Dallol Mauri in Niger and of Boudet's (1972) wide-ranging surveys of the Sahel. In North Africa the surveys of Le Houérou (1959, 1974), Floret (1971) and Floret and Le Floc'h (1972, 1973) reveal widespread degradation as do the surveys of Long and Perry (1975) in Iran and of Lamprey (1976) in the Sudan. The results of the survey by Newman and Condon (1969) in central Australia, given in Table 1, show that

TABLE 1. *Percentage of rangeland types in particular degeneration classes in Australia (Newman and Condon, 1969)*

	Present pastoral value in relation to expected pristine condition			
	Little or no degeneration	Minor degeneration	Moderate degeneration	Severe degeneration
Grasslands	30	55	15	negligible
Shrublands	10	25	40	25
Low woodlands	40	30	20	10
Flood-plains and alluvial pans	40	30	20	10
Spinifex grassland	70	20	5	5
Mountains and hills	50	20	20	10

large areas of Australia are in need of rehabilitation. Acocks' (1975) description of the South African Range gives us yet another depressing picture.

25. Surveys in the United States, particularly those from the first half of this century, produce a similar diagnosis of the condition of the western rangelands (Tueller, 1973). It seems to be agreed that the effects of overgrazing in the late nineteenth century and in the first three decades of this century finally had their inevitable effect after the droughts of the 1930s, and that the range has never recovered. Hastings and Turner (1965) give a particularly vivid account of the change that occurred in Arizona. Ogden (1973) and McKell and Goodin (1973) noted a decline in productive potential between the 1930s and 1970s. One of the best pieces of evidence for sustained damage comes from Utah where a careful examination showed that the drought had had little effect on an ungrazed area but that a heavily grazed plot had suffered serious damage (Stewart *et al.*, 1940).

26. Rangeland survey information is often corroborated by other observations of the environment, but because these aspects are even more remote in the ecological network from the producing animal, they need to be treated with much greater care as evidence for desertification. Some authorities have noticed that there has been a distinct change in the hydrology of areas that are apparently over-grazed. Hastings (1959) noted an increase in flooding in Arizona between 1882 and 1890, and both in the Karamoja district of Uganda and in Australia it has been noted that stream flows have become much more irregular (e.g. Pereira, 1961). Delwaulle (1973a) has reliable data to show that flooding became much more erratic on the Niger near Niamey after the mid-1960s. Falling well-levels, presumably due to increased drawdown for cattle, but perhaps due to less infiltration, were noted in the central Sahel by Aubréville (1973) and were Stebbing's (e.g. 1953) strongest support in his quest for the creeping desert in Africa. Rapp (1974) has noted that there seem to have been increasingly dense clouds of dust coming from West Africa and moving out over the Atlantic in the period since the start of the recent drought there, and suggests that this may be a way to monitor desertification.

27. Although the surveys and some of the other observations must undoubtedly be believed in their assessment of great damage, it is a more complex task to determine just how permanent this is. Where, as in the hilly Ethiopian Province of Tigre, there has been extensive and deep gullying, it cannot be doubted that recovery will be very slow indeed, but the sandy lands of the Sahel may recolonize and recover within a few years, and other areas on harder soils can recover quickly if helped by some simple soil conservation practices. Ultimately, assessments of recoverability must take account of economic and social factors.

C. RAINFED AGRICULTURE

28. The problem of disaggregating the effects of market forces, changing techniques, drought and desertification is multiplied several times in the interpretation of trends in dryland agricultural production.

29. The recent drought in the Sahel was certainly accompanied by a fall in output of most rainfed crops (Fig. 5) and, although there could be many reasons for this, it seems likely that the drought was to blame. Fournier (1963, in De Vos, 1975) claims that there were falling yields of millet in Upper Volta well before the drought but he is not borne out by the aggregated national production statistics published by FAO which show a

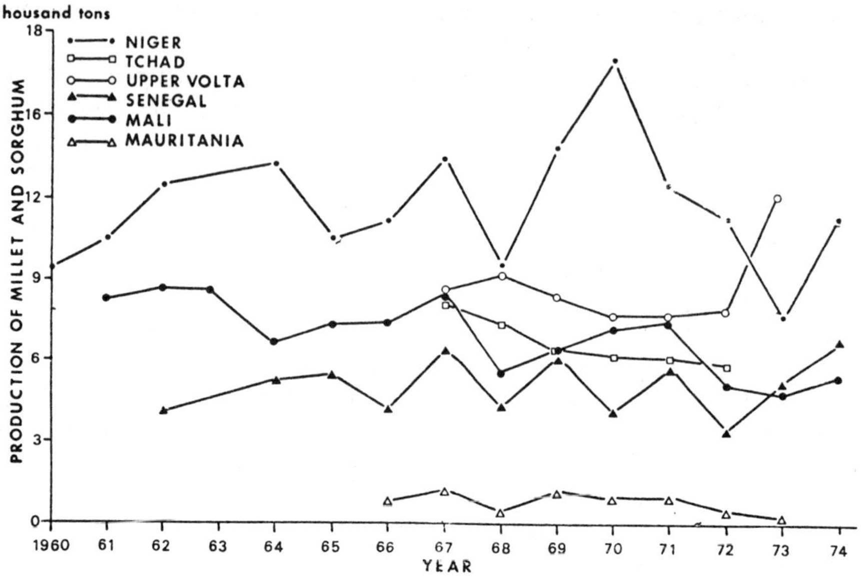

Fig. 5. Millet and sorghum production in the six Sahelian countries, 1960 to 1974 (SEDES, 1976).

consistent rise in yields per hectare for millet and sorghum in Senegal, Mali, Niger, Upper Volta and the Sudan between 1948 and 1964. Lericollais (1970) shows that in the Serer region of Senegal groundnuts and many other crops were failing even before the present drought. In Kordoan in the Sudan, too, groundnut and sesame yields fell in the 1960s (Le Houérou, 1975).

30. Several authorities have reported the extension of cultivation into unsuitably dry lands in parts of the Middle East and North Africa, and a concomitant decline in yield (e.g. Floret and Le Floc'h, 1972 Le Houérou, 1970, 1975). Their observations are corroborated by the FAO statistics for the period between 1948 and 1964 which show consistent declines in overall average yield of wheat in all the countries of this area where dry farming is prominent (except in Israel, Tunisia, and Turkey), although this too may merely reflect the inclusion of poorer land in the farmed area. The situation in India is less clear, for despite persistent reports of "desert spread" in northern Rajasthan, Mann (1975) reports that there is little supportive evidence.

31. The data for dryland agriculture in the development world are even more difficult to interpret, because of the higher capitalization and technical awareness of farmers. Figure 6 shows declining returns from wheat farms in South Australia up until 1900; this is attributed to a loss of soil fertility under the dryland farming techniques practised until then (Williams, 1976), but the introduction of phosphatic fertilizers and of a rotation with legumes has reversed the trend and production has continued to increase despite droughts. A similar picture can be painted for the High Plains in the United States where severe soil loss during the drought of the 1930s was followed by a drop in production, but where fertilizer applications, soil conservation techniques, nitrogenous crops in rotations and improved varieties of crop reversed the downward trend (Held and Clawson, 1965). There may again be declining returns after the current drought in these

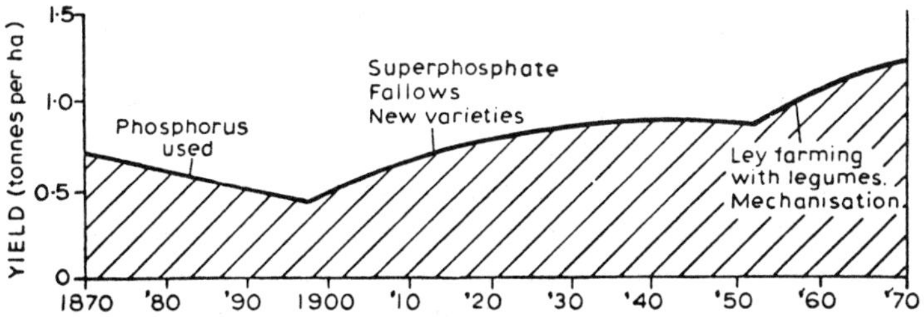

Fig. 6. Trends of wheat yield in South Australia from 1870 to the present day (South Australia Dept. Agric.).

areas, but agricultural technology may well make good the losses. The figures both in the United States and in Australia are now, however, considerably influenced by quota schemes and subsidies to suppress production.

32. Stories of technological success should not be accepted complaisantly. If greater care had been taken in earlier times in South Australia or in the High Plains, then production potential would undoubtedly have been higher today. Local surveys reveal that some fields were irreparably ruined in this period.

D. IRRIGATED AGRICULTURE

33. Technology has had even greater success with irrigated agriculture than with dryland farming, at least matching desertification in some areas with feats of agronomy or engineering in others. Nonetheless, desertification, in the very real sense of almost irreversible damage to the productive potential of the land, has affected thousands of hectares that were once irrigated. It is a poor commentary on technology that it must add new acres rather than protect those already reclaimed.

34. While the many accounts of irrigation that paint a gloomy picture of desertification are undoubtedly correct (e.g. Kovda, 1961; Allison, 1964; Michel, 1969; COWAR, 1976), the problem is by its nature hard to quantify. Growth may be inhibited by salt or by a high water-table for only part of the year and the pattern of damage is almost always extremely irregular, with highly productive crops standing next to saline slicks on which plant growth is impossible. Salinity begins to restrict production at a low level, and becomes an increasing constraint at higher levels, but can be countered up to a point by replacing sensitive by salt-tolerant crops. Above all, salinity is only one constraint among many in the agricultural production process, and its effects are hard to disaggregate from those of the others. Because of these difficulties, surveys of salinization and waterlogging must rely on rather sweeping estimates of affected acreages, rather than on actual figures for reduced production.

35. The most often quoted examples of the growth of salinization and waterlogging come, not surprisingly, from the world's greatest irrigation schemes along the Indus River in Pakistan (e.g. Dorfman *et al.*, 1965; Karpov, 1964; Michel, 1969). The estimates are subject to the limitations noted above, but are nonetheless alarming: out of a total irrigated area of 15 million hectares (9 million irrigated and sown each year), 10 million are affected by waterlogging or salinity in some way, and 2 million are severely affected. The annual rate of destruction is somewhere between 20,000 and 45,000 hectares (El Gabaly, 1976; Michel, 1969).

36. El Gabaly (1976) gives a good summary of the damage to irrigated areas elsewhere in the Middle East. Egyptian agriculture did not begin to suffer until the ancient basin irrigation schemes of the Nile Valley were replaced by perennial irrigation from dams. The estimate is that out of a total irrigated area of 2.7 million hectares some 1.2 million are affected in some way. In Iraq the problem is many times more serious, for half the irrigated area is affected (Fig. 7). Jacobsen and Adams (1958) give a graphic account of the affect that salinity and silting had on the fortunes of the ancient Mesopotamian civilizations. Kovda (1961) and Gerasimov (1968) give estimates of the large areas affected by salinization in central Asia and China, where again many of the problems are ancient. Despois and Raynal (1967) describe the frightening spread of salinity in the Macta Plain irrigation scheme near Oran in Algeria.

37. The problem of salinization is by no means confined to the Old World. Allison (1964) estimated that 27 per cent of irrigated land in the United States was affected by

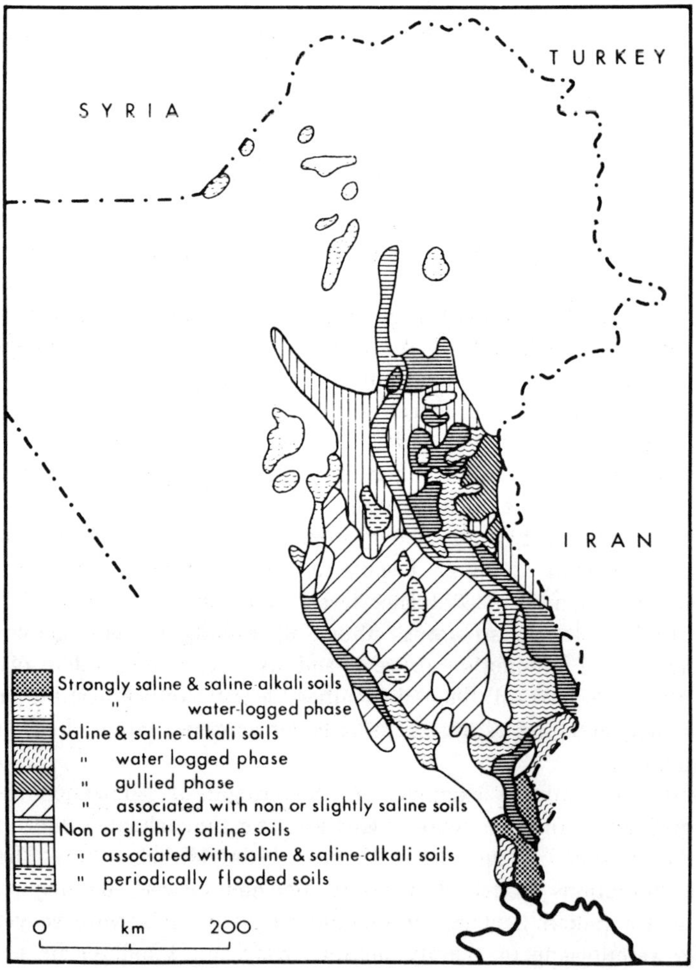

Fig. 7. Provisional map of saline and waterlogged soils in Iraq (El Dujaili and Ishmail, 1971, in Vink, 1975, p. 105).

salt in some way, and that between 1929 and 1939 alone about 400,000 hectares were rendered useless. Only 18 years after the opening of the huge Imperial Valley irrigation scheme in southern California, 80,900 hectares had been abandoned. Pels and Stannard (1976) note similar problems in irrigation schemes in Australia.

38. A further kind of desertification, in this case deferred but unavoidable, exists in irrigation schemes fed by deep wells which extract water accumulated over millennia; this "fossil" water is a non-renewable resource, so that the wells can have only a finite life. This is the case at Kufra in southern Libya.

E. HISTORICAL EVIDENCE

39. Because of changing perceptions of landscape and the very incomplete survey techniques employed by early explorers, historical accounts are usually very unreliable evidence for desertification. Greater and greater circumspection is needed in the interpretation of increasingly ancient records.

40. It is nevertheless hard to discredit some accounts. Depierre and Gillet (1971) and Grove (1973) have assembled a body of evidence that clearly points to desiccation in the Sahel of West Africa since the mid-nineteenth century. The Batha river in Chad, for example, was then perennial and supported resident fisherfolk. The river filled a lake, which was itself perennial until 1913. Although this period of history was very turbulent in Chad (Hugot, 1965), the desiccation certainly predates popular cultural culprits such as the restriction of nomadic movement, the digging of new wells, and the introduction of better health care. The implication is that there may have been a recent climatic shift in that area.

41. In East Africa, although the historical record is one of periodic severe drought, there is little clear sign of progressive desiccation and evidence only of local desertification (Baker, 1973). In North Africa, where the history of land use is long and better documented than elsewhere on the continent, there is considerable disagreement about whom to blame for the undoubted environmental degradation. Le Houérou (1970), summarizing the work of many others, and Mensching and Ibrahim (1973) think that the environment was damaged by cultivators in Roman and even earlier times, before the arrival of Arab pastoralists, although Johnson (1973) disagrees. Kassa is inclined to attribute desiccation to the decline of central authority after the withdrawal of the Romans from northern Egypt. Vita-Finzi (1969) believes that climatic changes, possibly in Mediaeval times, were partly responsible for the loss of soil, while Butzer (1974) is inclined to blame the pastoralist. In the Middle East there appears to be greater consensus that pastoralists have done great damage (e.g. Whyte, 1961; Butzer, 1974).

42. The early explorers of the new lands in Australia and the Americas left accounts of verdant landscapes where now there appears to be a semi-desert, but these tales need to be read with caution. The traveller's perception is often suspect and many of them were journeying through areas that had just experienced a series of very wet years (Heathcote, 1975). Archaeological and palaeobotanical evidence is a better guide; Gill's (1973) account of environmental degradation since European colonization in south-western New South Wales is a convincing story.

43. On balance the historical record since the neolithic cannot be called upon for any firm conclusions about desertification. There may well have been climatic desertification in

some parts of the world, and doubtless cultural desertification in many others, but the detailed indictment is hard to document.

F. CONCLUSION

44. The evidence for desertification is diffuse and almost impossible to quantify. There can be little reasonable doubt that many environments have suffered serious damage, and this mostly by cultural practices, but the persistence of the effects is much more debatable. In particular, the losses that followed the recent droughts in the Sahel, and elsewhere, cannot be classed as desertification until clear evidence emerges that the yield of useful crops in the ensuing good years has been depressed beneath that of the preceding wet period, and that the losses are due to environmental causes.

45. Despite the paucity of hard evidence, a common pattern of desertification emerges. Misuse that may go undetected or unremarked over a good period is accelerated and made evident by drought. The signs and extent of degradation must be observed and corrected in good, as well as in bad, years, if only to avoid accelerated distress in the droughts that will inevitably recur.

III. DRY ECOSYSTEMS

A. INTRODUCTION

46. Low returns per hectare on invested capital and labour mean that there is a reluctance to invest in dry ecosystems and management is thinly spread. In this situation yield can be sustained by utilizing the strategies that natural systems themselves have evolved to survive in harsh environments. To do this, dryland managers need to adopt an ecosystem frame-work.

47. In an ecosystem, energy, carbon dioxide, water and nutrients are absorbed and apportioned through a ramified network in which slopes, soils, plants, herbivores, carnivores, decomposers, and man exist in self-stabilizing, interdependent relationships (Fig. 8). Only those ecosystems that have evolved strategies to recover after unpredictable but recurrent disasters have reached states of long-term equilibrium in which each particular combination of species and external controls achieves its own balance. The character of the ecosystem may deviate from the balanced position in response to accustomed variations in the external controls and, if the deviation is not great, will always tend to return to, or oscillate regularly around, its original state once the pressure is released, the mode in which this occurs being described as its *resilience* (Holling, 1973). Unaccustomed and strong variations may, however, jerk the system so far from the original or may induce such violent fluctuations that it may lift itself beyond recall into the zone of a new equilibrium which may have more, but usually has fewer, possibilities for economic yield. In more familiar ecological language the balance achieved in an ecosystem by long-term evolution is its *climax*, systems in process of returning to the climax are in various *successional* stages, and the equilibrium points reached after disturbance are *disclimaxes*.

48. Although few ecosystems today are allowed to regain their climax condition and many have been irreparably altered, these concepts have many uses:

(1) Resilience and long-term equilibrium have distinct economic utility, for whereas the highest biological productivity and economic yield almost always occur in a successional stage, systems nearer the climax need fewer economic inputs to maintain productivity over the long-term. The energy available for productivity in a successional stage would be used for maintenance near the climax, and this is especially so under the stress of aridity. It may in future be possible to create artificial self-stabilizing, productive and even low-input communities of plants and animals, but for the moment the optimal resilience and long-term stability of the climax must be balanced against the optimal yield of a successional stage.

(2) Resilience includes the idea of low vulnerability. Heavy rainstorms have minimal effects on the soils and plants of systems near the climax, but may be very damaging in a successional stage. If nutrient- and water-holding topsoils are lost by erosion, return to the previous climax may be slow or impossible.

(3) Resilience and long-term stability are often (though not necessarily) associated with the greatest diversity of species and so the largest number of management options. If, for example, seed sources of a plant species preferred by a particular domestic animal are lost

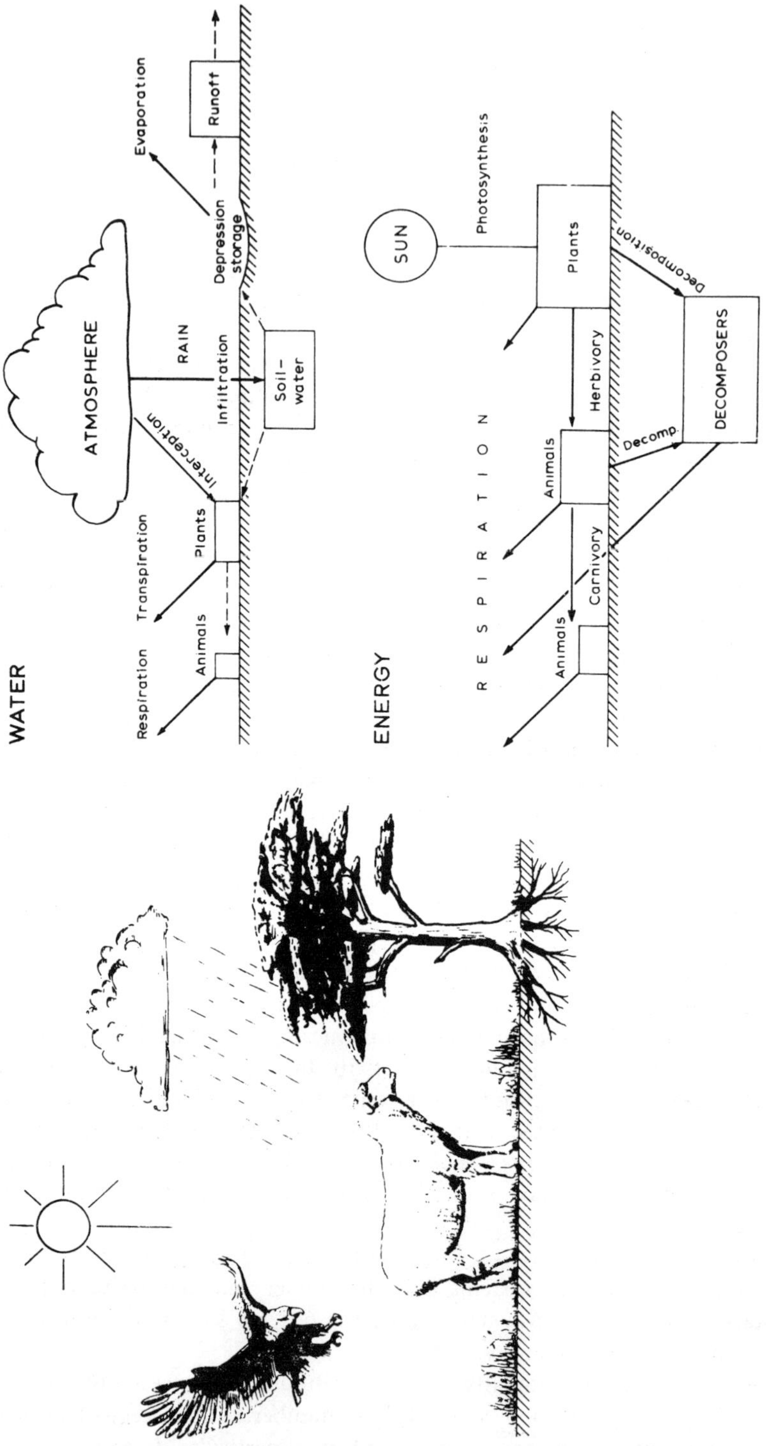

Fig. 8. A diagrammatic representation of water and energy flows in the ecosystem.

in the adjustment to a disclimax, it may not be possible to introduce the animal even if the market were to demand it.

(4) The idea of the succession helps in understanding the dynamics of vegetation (Long, 1974). If a change in management were contemplated, for example, by reducing grazing, an understanding of the succession would enable the manager to tell whether the vegetation would become a more productive sward or a less productive scrub.

49. An understanding of resilient mechanisms and of the climax is the start of any discussion of ecosystem management: the climax is a reference point against which to measure change or the potential outcome of proposed management strategies.

B. THE REQUIREMENTS FOR AND CONSTRAINTS ON BIOLOGICAL PRODUCTION

1. Water

50. Since carbon dioxide, radiation and even temperature are rarely limiting to biological production in warm deserts, the primary limitation is water. This is so because of the stomatal control mechanism in plants, for no production takes place if the plant is not transpiring, and it is also important to the nitrogen cycle which further encourages plant growth.

51. In rainfed agriculture, water supply is closely related to crop yield (Fig. 9), although yield depends on other variables as well such as the seasonality of rainfall, the volume of evaporative loss, crop variety and cultivation techniques. In irrigated agriculture water can actually be superabundant, but there are very complex relationships here between the amount and timing of water supply and the gross yield of various parts of the plant and its quality in terms of nutrition (e.g. Tadmor *et al.*, 1972).

52. Animal production is also limited by water supply since the amount of forage eaten by an animal is controlled by the amount of water it takes in. In wet seasons stock may gain all their moisture from forage, but in dry seasons they must take up to 90 per cent of it from wells and pools.

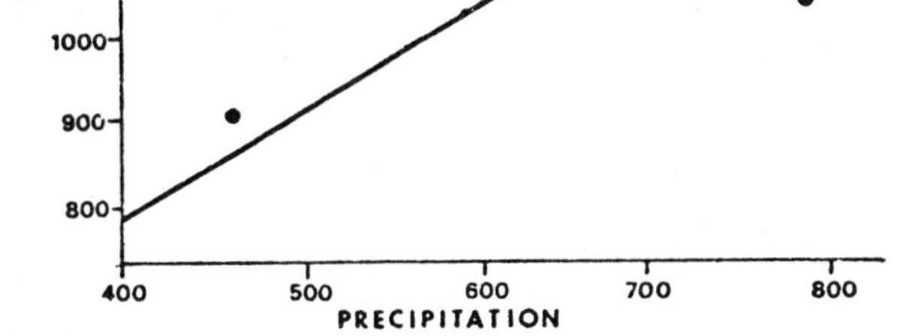

Fig. 9. Millet yield and rainfall in the Sahel (Cochemé and Franquin, 1968).

2. Nutrients

53. Some soils in Africa and Australia are very poor in mineral nutrients such as phosphorus and their administration to both plant and animals can promote increases in yield (Boudet, 1975a), but other dry lands are more fertile.

3. Salinity

54. Many dryland soils are imperfectly flushed of any excess salt which may be leached from geological deposits or be brought in by the wind from the ocean. Some soils and most low sites inevitably accumulate salt to the point where only a limited number of wild or cultivated species can survive or even where no growth is possible at all.

55. Saline forage or drinking water is a limitation to animal production. Many animals living in dry lands are genetically adapted to a large salt intake and others can acclimatize, but they both do so at the expense of a higher water consumption to flush the salt through their bodies.

C. THE GEOMORPHIC FRAMEWORK

1. Introduction

56. The massive configuration of the land (mountains, plains, major valleys) changes so slowly that it can be regarded as a permanent framework within which ecosystems function.

57. Land managers and planners, particularly of range land, have found it useful to divide the landscape systematically into areas with similar natural characteristics and responses to management (Mabbutt, 1968; Perrin and Mitchell, 1969). The smallest of these is an "eco-unit" or "land-element", defined as an area with a uniform vegetation type. These units fit into an ascending hierarchy in which the next size of area is known as a "facet", this being characterized by a particular combination of soil, hydrology, microclimate and vegetation types. Facets fit at a higher level into repeating patterns within "land systems" and these themselves fit into "land regions" (Fig. 10). A similar approach, often in the guise of a soil map, has been applied to many parts of the dry world and is now being actively used for pastoral land-use planning in many countries, notably in Australia (Condon *et al.*, 1969).

58. Land for agricultural use needs a more careful treatment and here the information from a soil survey can be abstracted to produce land capability units for more detailed management planning (Klingebiel and Montgomery, 1962; Moorman, 1974).

2. A Simple Classification of Dry Landscapes

59. The principal "land regions" defined by three authorities are listed in Table 2. A synthetic and very simplified classification together with a description of the accompanying ecological patterns is listed in Table 3. This system will be used in Chapter IV to describe the responses of dry landscapes to exploitation.

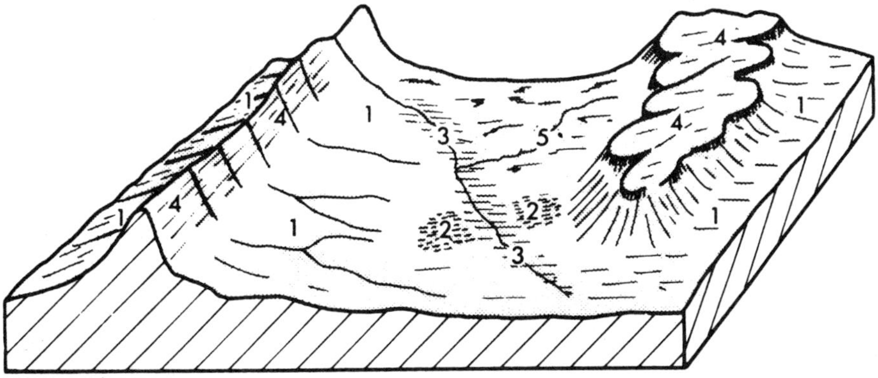

Unit	Land Form	Soil
1	Slopes 1 to 7%, stony surfaces locally dissected.	Texture-contrast.
2	Alluvial fans, extensively scalded surfaces, gullying and clay pans.	
3	Valley floors, shallow drainage depressions with local gullying.	Variable, coarse-textured.
4	Cuestas and mesas with silcrete duricrust, concave hill slopes with closely-spaced gullies.	Outcrop and shallow stony soils.
5	Vigorously dissecting channels.	

Fig. 10. Examples of a land system description from the report on the survey of the Alice Springs area (Perry, 1967).

60. Table 3 is, of course, extremely generalized. Each Land Region contains many small facets in which the hydrology, soil and vegetation are very different from the overall pattern. The table is useful merely as a very general guide to dryland ecology and as a basis for understanding the context of degradation, as explained in Chapter IV.

D. FUNCTIONING ECOSYSTEMS

61. An ecosystem functions as a whole, but can be divided for convenience into a number of subsystems each with its unique value to overall resilience and long-term stability.

TABLE 2. *Classifications of dry land regions*

Mabbutt (1969, 1974, 1976) (Australia)	Perrin and Mitchell (1969) (Arid lands in general)	Land Resources Division (U.K.) (e.g. N.E. Nigeria, 1972)
Uplands and piedmont	Mountains	Mountains and cones
Lowlands and footslopes	Hills	Hills and dissected areas
Shield and tableland (stony)	Piedmonts	Piedmonts
Clay plains	Plateaux	Upland plateaux
Sand country	Plains	Plains
Riverine plains	Aeolian areas	Dunes and sand plains
Lakes (pans)	Valleys	Alluvial land
	Enclosed depressions	
	Intertidal land	

TABLE 3. *A simplified classification of dry landscapes*

Land regions	Hydrology	Soils	Vegetation
Mountains	rapid run-off; springs and pools in main valleys	very thin	very sparse on slopes; very abundant in valleys
Hills	slower run-off; fewer springs and pools	thin	sparse on slopes; abundant in valleys
Piedmonts	slow run-off; groundwater abundant	stony or loamy	sparse but locally abundant
Plateaux	slow run-off; deep groundwater	stony, thin	sparse
Plains	slow run-off; groundwater variable	deep, fine	> 50% cover depending on rainfall and land use
Alluvial valleys	flooding occasionally; groundwater abundant	deep, fine or sandy	abundant, woody
Enclosed depressions	flooding periodically; saline	very saline	absent
Aeolian sands	no run-off; occasional deep groundwater	deep, sandy	> 50% cover depending on rainfall and land use

1. Slopes

62. At the scale of the individual hillside the land is very much a part of the functioning of the ecosystem.

63. Because fine soils have pores which are too small to absorb intense rain showers, they shed water after many rainstorms and the surface form of the slopes they cover is a response to the action of running water. A cover of plants considerably modifies this activity, for where they are close together, the surface is so rough and disturbed by the activities of organisms which live off their litter that most of the rainfall has time to percolate (Emmett, 1970); the channels to which the little run-off drains are widely spaced and slopes long and gently convex. But where plants are widely spaced they barely protect the surface from pounding and sealing by raindrops, fast run-off occurs after most showers, and the channels being more closely spaced and steep-sided become gullies. Slopes are short and concave or straight and soils thin and very mobile. Such landscapes are quite natural in steep hilly terrain in dry lands, but in gentler piedmont or plateau country they are found only when there has been interference with the water balance or after very rare intense storms.

64. Rates of soil removal by water, even under climax conditions, are greater in semi-arid areas than anywhere else on earth (Langbein and Schumm, 1958; Fournier, 1962). This is because in extreme deserts rain is too infrequent to sustain high erosion rates, even if the sparsity of vegetation does mean that rare showers are very destructive, while in humid areas dense vegetation prevents fast run-off; semi-arid areas, being protected neither by the infrequency of showers nor by plants, lose their soil more quickly.

65. Landscapes in which the wind is the principal agent of erosion are less extensive than those formed by water, but are still important. In the sub-Saharan semi-arid zone of Africa, 70 per cent of the area between the White Nile and the Atlantic is covered by

aeolian sands. Vegetation cover is as vital to the stability of such sandy areas as it is to the stability of water-worn slopes: it reduces wind speeds close to the ground, its roots bind the soil, it adds organic matter to the surface and it traps fine dust brought in by the wind.

2. Soils

66. In dry areas most biological activity in the soil is very close to the surface and is very intense in short wet periods (Charley and Cowling, 1968). The richest soils are patchily distributed beneath a few large trees or shrubs. Here shade slows the decomposition of litter and conserves moisture, and nutrient-rich leaf-fall fed by the deep ramified roots of the tree or shrub itself feeds a small community of grasses and forbs. The shrub canopy curtails erosion (Garcia-Moya and McKell, 1970), and the protected zone may actually accumulate soil splashed in from outside (Rapp *et al.*, 1972). Tree and shrub stems funnel water to the soil beneath (Lyford and Qashu, 1969) and nutrients are imported in the droppings of animals that shelter from the heat of the day.

3. Hydrology

67. Figure 8 gives a simplified picture of the hydrological "cycle" in arid areas. Only a small proportion of rainfall is intercepted and held on sparse dryland vegetation, the rest reaching the ground and there either percolating, running off or being evaporated. Although much of the intercepted rain will be lost directly by evaporation, at least some may be collected by grazing animals. Water storage in surface depressions may be very significant to dry ecosystems because of the high proportion of the small amounts of incoming rain that they can hold.

68. The soil is the most vital and complex store of water. It has already been explained how the inward flux to this store is controlled by the character of the surface. Soil texture is also vital in this respect: i.e. sandy soils may absorb almost every shower, while clays, even with considerable storage capacity, cannot imbibe water quickly enough and much will flow away. The deeper the soil the more water is stored: shallow soils lose a high proportion to evaporation and in intense showers can overflow and lose water by run-off.

69. When the surface is bare more solar energy is lost by reflection, and less is available for evaporation than when the surface is dark and covered by plants. This, and the fact that the flow of water through plants is more efficient than through the soil because the plants tap deep sub-soil sources, means that bare surfaces lose less water. This is held in check, to some extent by the effect of shade and the protection of a dense, high stand of plants which cuts down wind speeds and so the power of the atmosphere to transport away absorbed moisture. The overall effect of these various balancing processes is that, while a dense vegetation cover loses more water than a bare surface, more of it is channelled productively through the plants.

70. Water that escapes transpiration, evaporation or loss to deep soil layers runs off the surface. The number of times this happens in a year clearly depends on the character of the stores and the intensity of the rain showers. In a relatively vegetated catchment near Alice Springs in central Australia, run-off happens on average only once a year.

Intense tropical showers will yield run-off frequently, while gentler cyclonic rain or snow-melt will yield it more occasionally.

71. Run-off is redistributed across the surface in many arid ecosystems from bare areas to collecting zones. Bare areas absorb little moisture, so shedding it to zones where, because almost all that is received is absorbed, vegetation can thrive. There are several ways in which this comes about, few of them well understood, but the effect creates an unmistakable pattern in the vegetation when seen from the air: "groves" or "stripes" in which trees, grass or shrubs are quite dense are separated by bare zones with hard water-shedding surfaces. These vegetation patterns (under various names) have been described from many areas (see Cooke and Warren, 1973 pp. 147–149 and Slatyer 1962).

72. Water that manages to leave slopes reaches streams. The sediment from fast erosion in dry ecosystems clogs their channels with deep, coarse alluvium. Absorption of run-off in this channel alluvium and the very spotty nature of rainfall mean that few flows reach beyond the valleys of small streams.

73. Although some of the ground water reaches deep aquifers, much of it is held in the alluvium from where it is readily accessible to the simple technology of pastoralists and, of course, to the vegetation. Low spots in the landscape are therefore small oases in which primary production may be several times greater than in the nearby interfluves (Kassas and Girgis, 1965; Gillet, 1967; Fig. 11).

74. In general, dry ecosystems considerably smooth out the pulses of rainfall delivered to them: deep soils, mature wind-slackening vegetation, roughened soil surfaces, and deep valley alluvium all help to store water. It is said that as ecosystems approach "climax" state they control more and more of their own micro-environment (Whittaker, 1970).

4. Plants and Animals

75. The physical environment feeds and interacts with inter-dependent communities of plants and animals. The relationships between these ensure resilience and further suppress irregularity of the environment.

76. Plants and animals in dry ecosystems have to cope either physiologically, behaviourally, or communally with three outstanding environmental characteristics:

Low over-all possibilities of production.

Temporarily very favourable conditions but with low predictability.

Highly irregular spatial distribution of productive possibilities.

77. Physiological adaptations to irregular drought are numerous and well documented (e.g. Schmidt-Nielson, 1964 Brown, 1968; Noy-Meir, 1973a,b). Plant and animal species also have a number of behavioural adaptations to counter the unpredictability in the timing and size of showers. Most annual and ephemeral plants and some invertebrates produce large amounts of seed, in the hope that some at least will survive (MacArthur and Connell, 1966). The larger animals and perennial plants produce a few seeds every year, hoping that, in their long lifetimes, there will be enough good years to ensure survivors to replace them. Others may produce seed only after good years. Among the larger animals there are many other adaptations. The red kangaroo (*Megaleia rufa*), for example, does not conceive in bad years, but may do so with a young in the pouch in good years (Newsome, 1964). Other species aestivate, holing out during the summer

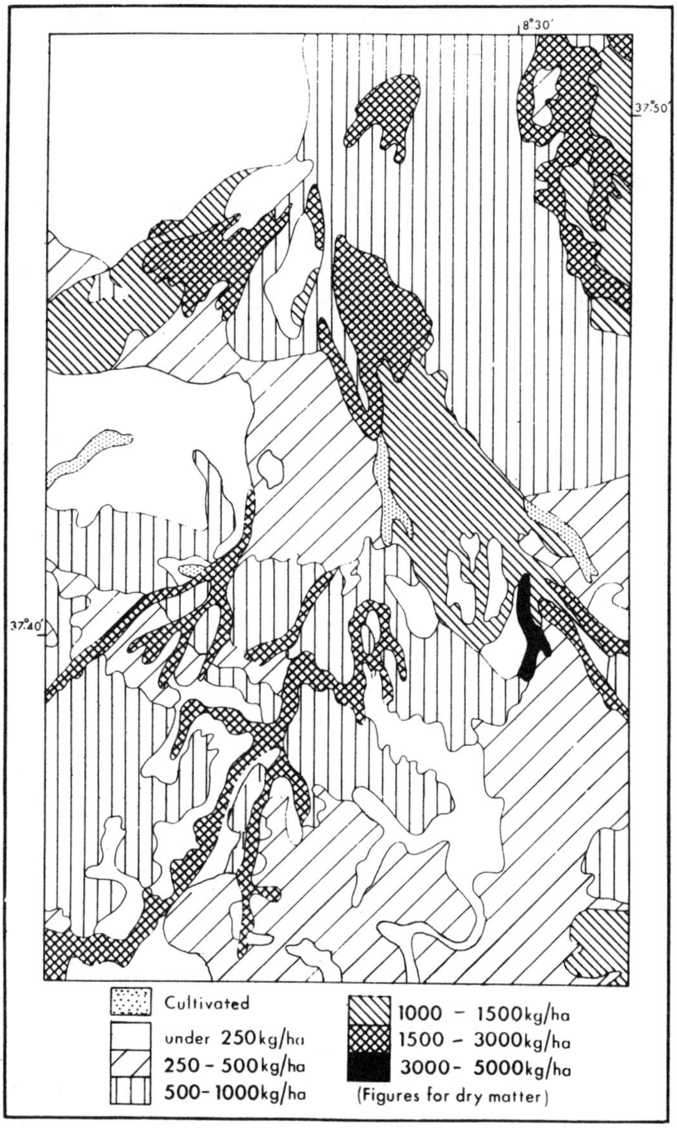

Fig. 11. Example of the variation in standing crop on different facets of the landscape in southern Tunisia (Floret and Le Floc'h, 1973).

drought in an inactive state, and many, if not most, species are active only at night. Many large animals adopt a policy of omnivory, so that food may be available to them even at the worst of times.

78. More important than individual survival is the reaction of the community as a whole to the challenge of drought. The importance of the small "oases" beneath some key shrubs and trees that has already been described has more than day-to-day importance, for in a drought they are refuges in which plants and animals hole out and from which recolonization can take place when the good years return (Muller, 1953;

Bille, 1974). The community also acts together to utilize the available resources of water and nutrients, with lichens, ephemerals and perennial grasses, shrubs and trees each utilizing a different set of nutrients and water and either passing on or utilizing the detritus of others.

79. The remarkably similar structure of vegetation in widely separate dry areas underlies an argument for the conservation of all the main plant groups: a structure of widely spaced, low, multiple-stemmed trees, low shrubs and sparse grasses seems to have been independently evolved as an efficient mode for utilizing sparse water and for surviving erratic drought. Its loss probably renders the system more vulnerable to stress.

80. In theory the numbers of herbivores that have no major predator should be limited only by their food supply (in most cases), but this is, in fact, rare in dry ecosystems (Slatyer, 1973a), perhaps because drinking water is more important to them than the scarcity of forage (Noy-Meir, 1973b). If this is so, it has considerable significance, because it implies that many plants have evolved under light grazing, and that the provision of more watering points to allow fuller exploitation of forage may be one of the chief ways in which dry ecosystems have been thrown off balance. Stability and resilience may depend on a very low conversion rate of forage to meat, the plant material left after a good season helping to tide the system over the bad years.

81. Herbivores and carnivores in dry ecosystems, like plants, evolve to fill certain niches. Among the herbivores, for example, some are grazers, some browsers, and many are even more specific in their tastes. In this way they, as a community, can achieve a dynamic equilibrium with the vegetation pattern.

82. One very characteristic adaptation of animals in dry ecosystems is mobility (Bourlière and Hadley, 1970). Among the grazing ungulates and marsupials, migration can be of hundreds of miles, making use of seasonal or more erratic patterns of rainfall and the complementary opportunities offered by different habitats. Noy-Meir (1973b) notes that in models of dry ecosystems, the introduction of the mobility factor often brings stability or at least resilience to an apparently hopelessly unstable system. Migration also allows the periodic resting of some parts of the range.

5. Conclusion

83. Five principal characteristics of dry ecosystems need to be accommodated if their utilization is not to cause long-lasting damage:

(a) Biological productivity, matching moisture supply, is highly variable over time. Only with irrigation can this characteristic be mitigated.

(b) Primary productivity (by plants) is very patchily distributed over the landscape, in response both to the spottiness of rainfall and to the run-off characteristics of the land. Because of this, a very small proportion of the land may produce a very large proportion of the available food.

(c) Indigenous plants and animals have evolved efficient mechanisms for resilience, both as individuals and as populations. These must not be damaged if the community is to survive.

(d) Recovery is much slower in dry ecosystems than in humid ones because of the generally lower species diversity of dry communities and the rarity of seasons with sufficient moisture for recovery.

(e) As in humid areas the catchment of a major river functions as a unit. Damage, as by over-grazing, in an upper catchment will lead to siltation lower down. Concentration of salts by evaporation in waters higher up a valley will cause salinization lower down.

E. MAN IN DRY ECOSYSTEMS

1. Introduction

84. *Homo sapiens* has greater genetic and behavioural adaptability to his environment than any other animal. With tools, weapons, language and the control of fire, hunting and gathering bands live so well off the land that they have large amounts of leisure time. The 'Kung bushmen of the Kalahari spend less time in producing as many, if not more, calories to consume than most shifting cultivators, or even many peasants cultivating permanent fields, and unlike them seldom suffer famine (Lee and DeVore, 1968). A deep knowledge of the local ecology underpins strategies that are more intelligent and cunning than those of other species, but are based on essentially the same ecological principles. Mobility and a knowledge of the terrain allow hunters and gatherers to exploit sites favoured by occasional rains. They are able to disperse to exploit widely spread resources, being untrammelled by property. Their varied diet is sufficiently nutritious to enable them to survive even in the worst of times. They use fire to manage the succession to attract game to good grazings and to round it up, although in other respects they still maintain a normal predator–prey relationship. And some features of their social organization help to keep their numbers within the bounds of their resources. With the possible exception of their use of fire, these people seem to have done little damage to the resilience of the ecosystems on which they have depended for millennia.

85. The origins of agriculture and herding are obscure and controversial, but it is not controversial to say that the start of serious environmental change by men is due to these techniques (e.g. Whyte, 1961). We can speculate that a chance, such as an inadvertent growth in human population, or the long continued settlement of a particularly favourable site, may have led men so to disturb the long-term equilibrium that the local ecosystem was moved to a low successional stage in which there was greater useful productivity. To maintain this productivity, new techniques, such as cultivation, would have to be developed. This could have been the start of a vicious circle in which increasing returns from the land were bought only with decreasing returns on labour. As the division of labour and structures of dominance became more elaborate, survival depended more on socio-economic than on simple ecological factors, so that the system became both more dynamic and erratic.

2. Pastoralism

86. People dependent mainly on the produce of animals live mostly in the earth's less productive and risky ecosystems, but it is a serious mistake to extend this rough determinism and to assume that pastoral people satisfy all their needs from poor

lands or that pastoralism is a simple and perfect ecological adaptation to them. In the often quoted words of Theodore Monod: "nomads live in the desert, but not from the desert".

87. Pastoralists obtain a higher yield from dry ecosystems than do hunters and gatherers. They manipulate their herds in a much more intense way than any other predator does its prey, and also considerably interfere in the relationship between the cattle and their forage. The herds can be managed to yield as near a continuous supply of milk as possible, and can also be manipulated either to maximize productivity by producing young stock or to maximize survival by concentrating on older beasts (Dyson-Hudson, in press). Herdsmen can bring about much greater rates of increase in stock numbers (10 to 20 per cent per annum) than are possible in natural populations of large herbivores (Allan, 1965). They can also manipulate vegetation with fire and other methods to yield more forage for their own species of stock.

88. The survival of this way of life for several thousand years is a measure of the success of many of its ecological strategies. Herdsmen make use of their great knowledge of their high risk environment to achieve optimal yield, and to try to conserve their resources. They often keep a number of different species of stock in order to crop different ecological niches, to ensure that they have some yield in bad times (Swift, 1973; Johnson, 1975), and when times get really bad they can resort to gathering a great range of plant foods and to the hunt (Bernus, 1975; Western, 1974). Most herdsmen use their stock to maximum energetic efficiency by milking (or bleeding) them rather than killing them for meat and they have selected stock, such as zebu cattle, that are able to maintain a yield while surviving in harsh environments (Seifert, 1975). Above all, pastoralists are highly mobile. At a small scale this means the use of rich bottom lands when other pastures are in poor condition, and upper slopes and waterless areas when rain brings on the grass there; and they can graze their stock on salt-rich or nutrient-rich fodder when it is needed in the breeding cycle (e.g. Boudet, 1975a). At a larger scale it means following the rains either seasonally or adventitiously, seeking the best pasture. Some of these movements can cover immense distances. In the Sudan the Kabbabish migrate hundreds of kilometres from the summer "Gizu" grazing to the dry season pastures in central Kordofan. In Mali, Gallais (1967) has described the wet season use of pastures on fixed dunes and then, as the Niger floods recede, the grazing of the highly productive *bourgoutières* among the channels of the inland delta. These, and many other ecological strategies, have allowed herdsmen to survive with what is, for most of the time a very healthy diet.

89. Beyond these facts, discussions about the ecology of pastoralism enter a field of controversy. Most of the arguments, being about the society and demography of pastoralism, are discussed in another report (Kates *et al.*, 1976), but arguments about the ecological damage said to have been perpetrated by pastoralists and the relationship between their culture and ecology must be briefly discussed here.

90. The history of pastoral depredation discussed in Chapter II raises one set of controversies. Nomadism has often been a one-way street: the Fúlani have been moving into new areas for the last two centuries, as have many East African groups, notably the Masai. The more ancient movements of the Mongols and of the Beni Hilal are well documented. The question that arises, but can probably never be satisfactorily answered, is: were these movements the result of degradation in the pastures these folk left behind?

91. More recent movements of pastoral people in many parts of Africa and Asia are

less controversial because better documented. They have been initiated by agricultural colonization and other restrictions placed on former grazing lands (e.g. Leeuw, 1966; Le Houérou, 1970; Bernus, 1975; Warren, 1975). This has undoubtedly removed the best grazings and increased pressure on the pastures that remain.

92. What is known of the history of pastoralism (e.g. Flannery, 1965) supports the hypothesis that it did not spring directly from a hunting way of life, but appeared either as an adjunct to agriculture or as a result of regular trade with agricultural peoples (Lattimore, 1951; Zeuner, 1963), and most herdsmen have retained a foot in the agricultural camp ever since (e.g. Barth, 1964; Le Houérou, 1970; Johnson, 1975). Many bedouin in the Middle East and the Turkana and Karamojong in East Africa sow grain themselves; the Twareg have strong client relations with oasis cultivators and most of the mountain nomads of the Middle East, the Fulani and many bedouin exchange pastoral products with farmers and others. They frequently have agreements with farmers that allow them to exchange animal products for grazing the stubbles, and make use of this facility particularly in bad years. And in addition they have always been able to exchange protection, transport, labour or advice with industrialists, anthropologists, masochists, generals or prophets.

93. These outlets add enormously to the over-all strategy of the pastoralist: he is partially protected from the effects of environmental irregularity, since the losses of a bad year can be made good by selling goods or services or by making use of somebody else's surplus production. The question arises: do these alternatives alter the relationship between the herdsman and the ecology of dry lands to the extent of damaging resilience or productivity?

94. While the availability of standby grazing in wetter climates may allow the more efficient cropping of the biological production that occurs in the good seasons in the dry lands, the resulting intense grazing in good years may remove reserves that would have been used to tide the ecosystem over the next bad patch. These reserves could also be endangered by the ability of the herdsman to "track" the good years, in other words to increase his flock at a faster than "natural" rate in order to eat up the forage when it is there. It is a common temptation not to suppress rates of herd increase as a drought intensifies, in the gamble that the drought will not be sustained, and this, being unlike the closer tuning of the natural system, may introduce severe instability and eventually an almost irreparable change in the ecosystem. Chapter II has given evidence that such changes do in fact seem to have taken place.

95. Another controversy concerns the problem of the personal property of stock and the common ownership of pastures. Personal herd ownership allows a more flexible response to the changing environment than if cumbersome group decisions were necessary, for it maximizes the use of skill, observation and knowledge; on the other hand it encourages the growth of large herds as an individual insurance against drought, and so does great environmental damage for little long-term gain to the community. The individual need not suffer, for the better kept his herd, the more likely it is to survive, and his gambler's instinct will be to keep as many as he can so that more can survive to build up the herd when the good years return. Whether or not increased expectations of living standards add to these problems is another debatable point. Nevertheless, it must not be supposed that the pastoralists' personal ethic is untempered with charity; most pastoral folk have an extensive network of relationships which help to keep the less fortunate alive.

96. The obverse of personal ownership of herds is the common ownership of pastures, and this is often said to be a conflict that leads to the destruction of the latter. But common ownership of the land is very often controlled by elaborate arrangements for restricted or rotational use. The Sudanese Nuer, the Iranian Bakhtiari and the Libyan bedouin are examples of groups who have very precise territorial rights (Bonté *et al.*, 1976). It is true, however, that among African cattle pastoralists territorial rights are less well defined, perhaps because of their history of recent migrations, and the very complex system evolved by Chikhou Ahmadu for the Fulani of the inland Niger Delta seems to be a prominent exception (Gallais, 1967).

3. Tropical Rainfed Agriculture

97. Although small plots are sown even in very arid parts of the Sahara, cultivation in West Africa, though still a gamble, can be carried out on higher slopes and can become the main form of subsistence where the rainfall is greater than about 250 mm (Cochemé and Franquin, 1968). In East Africa, where the rain comes in two seasons, cultivation becomes too much of a gamble where there is less than a 30 per cent probability of at least 500 mm rainfall in any one year (see Warren, 1967).

98. In these slightly moist environments indigenous cultivators have overcome the problem of low returns from dry and infertile fields by adopting, in shifting agriculture, another method that makes intelligent use of the ecosystem. Shifting agriculture is the dominant form of land use in the Sahel, for in Niger alone 8 million out of a total cultivated area of 12 million hectares is farmed in this way (Braun, 1973). In dry lands where few crops can survive, there are some widely adopted strategies among shifting agriculturalists. These include clearing small plots of above-ground vegetation only, recovering much of the mineral nutrients (especially phosphorus) bound in this material by burning it or allowing it to rot in place, mulching the soil surface with vegetation and only lightly tilling it with hoes. These methods maintain a return on labour that is at least as high as in many more intensive peasant systems, and they avoid serious erosion, and feed nutrients to the crops for a few years. After two or three (but sometimes even ten) seasons, the yields so tail off and weeds become so troublesome that returns decline to the point where the plot is abandoned for a new one and is left to recover its fertility in the ecological succession of the next 5 to 10 years. During this time, trees and shrubs coppice again from the stumps, and bring up nutrients from depth, the recovering vegetation adds litter to the soil to build up its organic matter and nitrogen content, and grass roots bind the soil, so helping it to recover its erosion-resistant structure, and pests and diseases are once more controlled by their natural predators (Nye and Greenland, 1960; Morel and Quantin, 1964; Allan, 1965; Mouttapa, 1973; Jones and Wild, 1975). Crop types and varieties including nitrogen-fixing legumes, such as cow peas and fail-safe drought relief crops, have been evolved to deal with each local condition (Cochemé and Franquin, 1968; Mortimore and Ologe, 1975) and the effects of excessive drought or rains are minimized by sowing several different crops and by doing so down the slope so that wet, intermediate and dry sites are all planted. Shifting cultivators, like hunters and gatherers, use a wide knowledge of their local ecology to play a complex game in which they pit crop varieties, planting times and labour and market requirements against the amounts and timing of the rains and against pests and diseases (Gould, 1965).

99. Beyond this direct involvement in the environmental game, these farmers also use

storage of varying effectiveness, and a supportive cultural superstructure of land rights and of family, clan and tribal obligations to help them weather their environment. In general they seem to have been able to produce a nutritious diet, resorting in the bad years to hunting and gathering often from the regenerating fallow (Harris, 1969). They can use this, too, for some commercial produce such as gum arabic (Depierre *et al.*, 1975). The system is a brave attempt to cope, but, like pastoralists, shifting agriculturalists seldom relied on one form of land use alone and had been able, until recently, to send out colonists to new empty areas (Mortimore and Ologe, 1975). In any event, the system can only support 25 persons per km^2 in dry lands, and comes under severe pressure when rotations are shortened and the nutrients and organic matter in the soil used up at an excessive rate.

100. On the wetter margins of these dry lands in both East and West Africa, shifting agriculture has been replaced by sedentary peasant farming which supports much higher population densities (cf. Boserup, 1965). Here human and animal wastes are used to maintain soil fertility, but the system is much more labour intensive and achieves little improvement in terms of returns per man day.

4. Temperate Rainfed Agriculture

101. On the cooler margins of the deserts dry farming of wheat and barley is the mainstay of another group of agriculturalists. Wallén (1967) has shown that in the Middle East, the risks appear to be acceptable where the precipitation/evaporation ratio reaches 0.80, but that an approach that uses mean annual figures alone such as the Meigs method overestimates the amounts of moisture available. Rainfall variability is a vital consideration, for where it is low, as in parts of Syria, dry farming is possible at mean annual rainfalls of 180 mm, but where it is high, as in the Zagros, the risks are too great even at 250 mm. Run-off farming exists as a very high-risk enterprise in areas much drier than these and in places such as the littoral of southern Tunisia, a viable olive culture exists with a mean annual rainfall of only 170 mm (Le Houérou, 1958).

102. Dry farming is the technique used in these semi-arid areas. By fallowing and mulching, the land is kept clear of weeds and the surface loose so that it is absorptive of rain and less liable to lose water by evaporation, and the rainfall of 1 or 2 years can be stored and used for 1 year's crop. In some systems stubble is left standing for grazing by the tenant's cattle or is sold to bedouin; the stock return crop-residues to the land and help the nitrogen cycle. In the older systems ploughing is wisely very shallow, leaving the richer upper soil layers undiluted with subsoil.

103. Risk is built in to the strategies of many of these farmers. In Tunisia less and less seed is sown and therefore hazarded as the rainfall probabilities decline to the south (Despois, 1961), and the fallow is not sown after a very dry season. An element of pastoralism gives added insurance, since the stock can be moved to more favoured environments in the bad times.

104. Dry farming has been extended to the New World, the Virgin Lands of Russia, and Australia with some modifications, and varying success. The use of deep ploughing in the new areas was probably unwise and undoubtedly added to the problems of the dust-bowl years in the Great Plains, and to the decline in production from South Australian wheat lands in the 1890s (Fig. 6), but the wise use of medicago and clover in

a fallow rotation to fix nitrogen and the addition of phosphatic fertilizers has reversed the downward trend in yield, and these systems, while still suffering in the bad years, seem to have achieved some kind of sustained yield with high technological inputs.

5. Run-off Farming

105. Agriculture is attempted, sometimes against heavy odds, in some areas with extremely low rainfall probability by encouraging the natural concentration of run-off in low places. There are several variants of the practice, depending on the climatic, soil and technical background of the farmers (National Academy of Sciences, 1974). At its simplest, the method depends merely on low dams in valley bottoms to trap silt and moisture (Monod and Toupet, 1961). More evolved methods include the feeding of run-off from sloping plots into small ponds around fruit or nut trees, or full-scale management of watershed run-off as described by Evenari, Shanan and Tadmor (1971). As the technical complexity increases, risks decrease, but investments of labour increase. It is unlikely that any of these schemes have led to environmental degradation in an important way.

6. Small-scale Irrigation

106. Some low spots even in the most extreme deserts are irrigated by springs or by seasonal or even perennial rivers flowing from wetter climates (Quezel, 1965). These natural systems need little manipulation to make them produce useful crops, and the simple methods developed for this have supported small populations in oases and along some river valleys for millennia with few signs of serious degradation (Capot-Rey, 1953). Many examples exist from all over the dry world, for example in Mexico and the American South West, in Peru, and in China. The success of these systems lay in such practices as the application of plenty of water and thorough natural drainage.

107. Even some large-scale valley irrigation schemes, as along the lower Indus and most notably along the Nile, survived millennia with little serious degradation. Others, as in Iraq, were less successful.

IV. THE IMPACT OF EXPLOITATION

A. INTRODUCTION

108. Most ecosystems can be cropped in some way without endangering their powers of recovery. Because a surplus is provided by some species as a means of ensuring their resilience after severe stress, the crop is often large in good times but is very small in bad. If a natural predator is removed, maintenance of the ecosystem may even depend upon cropping: if hyenas were to be eliminated, buffaloes would overgraze their range, if themselves not culled; and if buffaloes in turn were exterminated, cattle grazing might be necessary to prevent a decline of forage productivity. In more modified ecosystems, as in agriculture, much greater productivity of useful crops is bought at the price of labour, technology and imported materials.

109. At some level, exploitation ceases to benefit and begins to degrade productive potential. Degradation can be either chronic or acute. In pastoral production desirable species may be slowly eliminated as their competitive ability is sapped by grazing, and less valuable plant species may thrive; and in agriculture there may be a gradual loss of nutrients as they are lost from the farm in crops sold elsewhere. Pastoralism, rainfed agriculture and land cleared for industry or housing can all suffer the same kind of acute losses when rare heavy showers or strong dry winds erode in hours soil that has taken centuries to accumulate. Irrigated land suffers its own types of desertification. Each of these processes lends its peculiar imprint to the pattern of desertification.

B. PASTORALISM

1. Introduction

110. Sheep are selective grazers, tearing and biting grasses and forbs close to the surface, or even out by the roots. Their selectivity varies from ecosystem to ecosystem, because of different ranges of species available to them. Only when their preferred species have been eaten out or have dried up do sheep start to eat low browse. Sheep are the most gregarious of domestic stock, heavily grazing or trampling small areas. In Australia sheep need 3.5 litres of water a day (Graetz, 1973) in the dry season, restricting them to 2.5 to 3.0 km from watering points (Lange, 1969), though some Old World breeds may be able to travel twice this distance. In the dry season Australian sheep may move up to 8.8 km away from a well (Lynch, 1974).

111. Cattle must use their tongues to hold herbage while they bite and cannot remove herbage below 12 mm above the ground. They too prefer grasses and forbs (Ellison, 1960; Hansen and Reid, 1975), although in good seasons they are not in competition with sheep in the Sahel. They seem to move more readily to browse than sheep and can reach higher. Although the trampling damage by a single cattle beast is greater than that of a single sheep, cattle are more independent and only really bustle

together near watering points. In dry seasons cattle may need 23 to 32 litres every second day (Heady, 1960b), and can therefore move no further than 5 to 10 km from the watering point (De Vos, 1975).

112. Goats and camels are the best adapted of domestic stock to dry ecosystems, since they have very catholic tastes, but prefer browse and are well able to get it, the goat by climbing and the camel by using its weight to bring down quite hefty branches. Goats move quickly and in dispersed formation over the range. Camels can go for up to 10 to 15 days without water and exist on dry herbage alone on journeys of 25 to 30 km.

113. Stock movements over the range, when not dictated by waterpoints, are governed by a search for starch, sugar, nitrogen, phosphorus and fibre. Fibre is seldom scarce, and it is usually starchy and nitrogenous foods that are critical to cattle growth and movement. Green grass leaves can supply a balanced diet, and grass seeds are also nutritious even when dry, but in the dry season cattle and sheep join goats and camels in using the pods, seeds and leaves of trees and shrubs. By using these evergreen sources the stock can tide themselves over the bad seasons to times when the grass is green again.

114. It is almost a truism to say that grazing and simple management practices by graziers increase the productivity of wild pastures (e.g. Fig. 12). Light nibbling increases the vigour and growth of both the above- and below-ground parts of the plant (e.g. Pearson, 1965), and in some species, such as the valuable African grass, *Themeda triandra*,

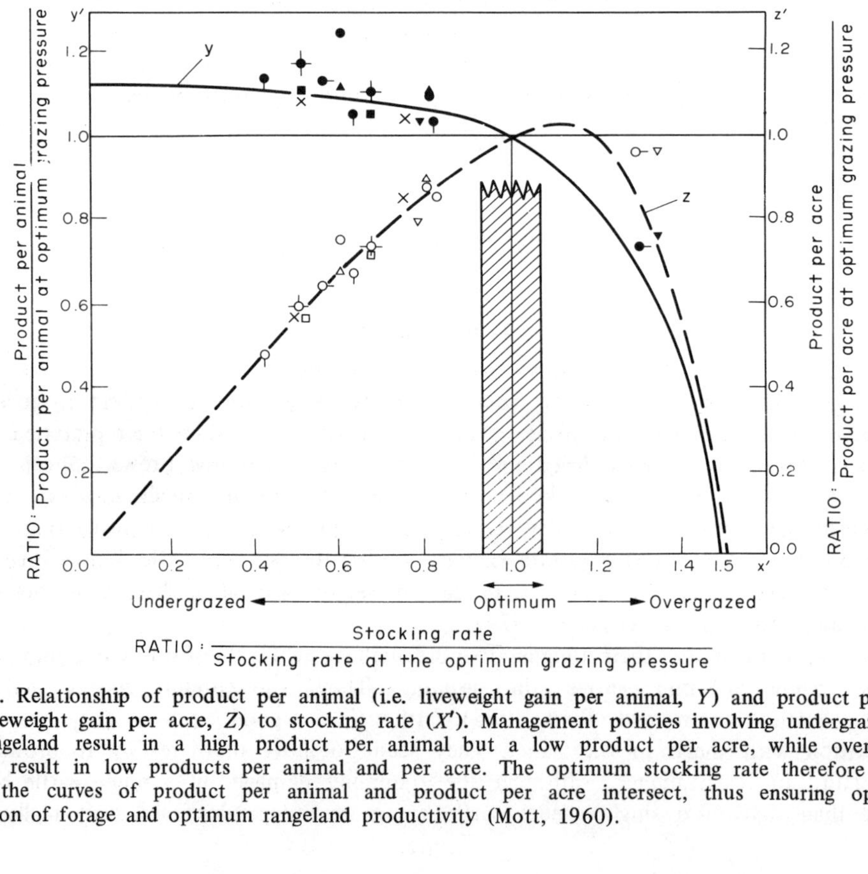

Fig. 12. Relationship of product per animal (i.e. liveweight gain per animal, *Y*) and product per acre (i.e. liveweight gain per acre, *Z*) to stocking rate (*X′*). Management policies involving undergrazing of the rangeland result in a high product per animal but a low product per acre, while overgrazing policies result in low products per animal and per acre. The optimum stocking rate therefore occurs where the curves of product per animal and product per acre intersect, thus ensuring optimum utilization of forage and optimum rangeland productivity (Mott, 1960).

the removal of coarse, dead stems allows the sprouting of new succulent shoots. The passage of herbage through the gut and out as faeces speeds the nitrogen cycle, and grazed pastures are richer in nitrogen than ungrazed ones (Leigh, 1974; Boudet and Coulibaly, 1975). The seeds of some valuable species are spread efficiently by being carried in the guts of cattle and are then placed in favourable seedbeds in dung or where they are trampled into the soil surface (Moss, 1969; Boudet, 1975a).

115. Every herdboy knows the value of burning as a management tool for increasing forage production (so does Roland, 1967, in Bourlière and Hadley, 1970). Like light nibbling, a quick, low-temperature fire can help to bring up new green shoots, remove old, tough ones, and help to speed nutrient cycling. More important, it keeps woody unpalatable species in check and unable to compete for water with palatable grasses.

116. Both grazing and burning can increase species diversity by opening out the community and creating more niches (Barker and Lange, 1969a; Walker, 1974). Increased production is characteristic of the early successional stage that is encouraged by these practices. It may seem to damage resilience little, and may in fact increase it, as more species spread in and as rates of erosion by water decrease under grass (Ingebo and Hibbert, 1974). But the loss of a framework of perennials may render the system much more vulnerable in bad years when annuals fail to germinate, when the drought refuges in the shade of the large trees have gone, and when the dry unprotected soil is liable to be blown away.

2. The Effects of Heavy Grazing, and Associated Management Practices

117. At some level of stocking pastures cease to improve and begin to degrade, sometimes drastically. This critical level varies very considerably according to the biological productivity of forage, its palatability, the way in which it is utilized, and the recent sequence of climatic events.

(a) Direct effects

118. (i) *Trampling*. Hooves are far more damaging than mouths, especially when stock are concentrated around watering points. The devastation around Sahelian wells, often attributed to over-grazing, is the result much more of trampling than of eating.

119. The first casualty in trampling is the vegetation. Some species are more sensitive than others and quickly disappear to leave a sparse cover of hardier, and usually less palatable, plants. The proportion of bare ground increases rapidly. The hooves then pound, powder or puddle the surface. In dry conditions soil aggregates and plant litter are reduced by pounding to a size where they are whipped up and away by the wind (Quinn and Hervey, 1970). When the rain comes, the remaining dust may help to seal the surface as a crust, but more serious is the puddling of wet sites and the further reduction of their infiltration capacity. The combined effect of dry season pounding and wet season puddling is an increase in bulk density of the surface. In South Australia and southern Africa, heavy trampling breaks up a lichen crust which otherwise protects dry soil from blowing and when wettened swells and roughens the surface and prevents violent run-off (Walker, 1974; Crisp, 1975).

120. Hooves trample, pound and disaggregate organic litter and soil humus so that it can be blown away or redistributed by the animals themselves (Packer, 1953; Barucha and

Shankanarayan, 1958; Ellison, 1960; Crowcroft, 1969; Whitman, 1971). The close herding of stock concentrates the nitrogen from their droppings and urine so that it denitrifies uselessly and is not returned to the system (Noy-Meir and Harpaz, 1976). These concentrations of nitrate can also seriously contaminate the water in some water-yards (Peyre de Fabrègues, 1971; Hunting Technical Services, 1974).

121. (ii) *Grazing*. Heavy grazing removes especially the leaves of grasses and forbs to such an extent that photosynthesis can be severely curtailed (Tueller, 1973). Browsing and lopping of shrubs and trees can distort their growth-form and may kill them altogether. The most serious threat to the resilience of the ecosystem comes from the inhibition of seeding observed in grasses after drought and heavy grazing by Poupon and Bille (1974) in Senegal, and in the valuable *Acacia albida* tree in Niger and Chad by Gillet (1975). If, even after this depredation, some seedlings do manage to establish themselves, they are often pounced upon by stock, and the reproductive capacity of the system is further placed in jeopardy. Many plant populations in heavily grazed areas have been observed to be ageing without being able to replace themselves (Crowcroft, 1969; Payne *et al.*, 1974; Depierre *et al.*, 1975) although in some cases this may be quite natural since regenerative possibilities like the occasional fire or a very good year may be very episodic (Slatyer, 1973b). If this is the case, then the decreased life expectancy of saltbushes in heavily grazed pastures observed by Crisp (1975) may be equally serious, since they may not be able to survive long enough to experience the rare years in which they can regenerate. If the fire that is so necessary for regeneration is curtailed by management, whole populations may age and die waiting for the event they need which never comes.

122. Grazing tends to damage the most palatable first and then progressively less palatable species, so reducing their frequency in the community. These are the "decreasers" of Dyksterhuis (1949). Not all palatable species, however, need be decreasers: some may actually thrive on grazing. The loss of palatable species means a loss in nutritional value of the forage (Granier, 1975; Boudet, 1975b) and therefore a decline in forage intake (Boudet, 1975a). In different communities decreasers are annuals, perennials, grasses, forbs, shrubs or trees, but because of their vulnerability in dry seasons and droughts when they are the only grazing, and because they do not always have the same strategy of reproduction and germination as the annuals, perennials are common decreasers. Because their loss is likely to herald erosion at the times when they are the only protection to the soil, and because of their value as a drought reserve, a great deal of attention has been lavished on perennial grasses (Halwagy, 1962a and b; Pereira *et al.*, 1962; Humphrey and Mehrhoff, 1958; Whitman, 1971; Walker, 1974; Boudet, 1972). The loss of perennial shrubs and trees is serious for similar reasons. They hold down wind velocities and so decrease both evaporation and wind erosion, and shelter and feed small oases in which animals and weaker plants can hole out in the drought and recolonize the deserts between the shrubs when the good years return or after heavy grazing is relieved.

(b) Indirect effects

123. (i) *Plants and animals*. When relieved of competition from palatable plants or plants liable to trampling damage, resistant and usually unpalatable species expand their cover. These are the "increasers" of Dyksterhuis (1949). Annuals are prominent among these, especially those with very short life cycles (Bentley and Talbot, 1948; Pratt, 1969; West, 1969; Rapp, 1974; Boudet, 1975b). In many ranges poisonous forbs (such as the notorious burroweed of the western United States) become dangerously common. A

characteristic change is an invasion by plants from more arid neighbouring regions (Williams, 1961; Boudet and Coulibaly, 1975).

124. One of the common consequences of intense pasture use is the invasion of woody shrubs. The reasons are many. Part of the story is that grasses compete with shrubs for water, but the competition is not simple: shrubs can tap deeper layers and use water at different times of the year (Rowe and Reimann, 1961). A more important factor is the suppression of burning that has sometimes been imposed without much thought on many ranges and has elsewhere followed the removal of grass fuel by grazing. This seems to be a prominent reason for the rapid spread of mesquite in the south-western United States and for similar invasions in many parts of the world (Heady, 1960a and b; Wilson, 1962; Buffington and Herbel, 1965; West, 1969; Acocks, 1975; Boudet, 1975a). But there are many other contributory factors, such as the spread of seed by cattle and rodents, or a reduction in browsing by wild or domesticated species. Whatever the reason for it, the spread undoubtedly reduces grass production. In Rhodesia, Walker (1974) quotes four-fold increases in grass growth after a scrub clearance. Opening up of pastures by over-grazing does seem to render them more liable to invasion (Williams, 1961). These are the "invaders" of Dyksterhuis (1949).

125. The repercussions of heavy grazing on wild animals is often disastrous, but seldom really serious to the domestic grazers. One potentially important impact is the effect on termites of mulga scrub clearance in southern Queensland. The grass-eating termites thrive in the new grass-rich habitat, but begin seriously to compete with sheep in droughts, and the bare ground round the termitaria both increases the erosion hazard and inhibits grass regeneration when the good years return (Watson and Gay, 1970). Clos-Arceduc (1956) also believes that termite activity follows over-grazing in the Sahel and forms the growth of *brousse tigrée* patterns. A similar effect has been noted in Uganda by Wilson (1962).

126. The marked and very considerable decline in large mammal populations in many parts of Africa is only partly attributable to intense grazing. An increase in the use of firearms, killing to control trypanosomiasis, and exclusion from watering points (De Vos, 1975; Sayer, 1976) have all contributed. In Australia, on the other hand, the introduction of a pastoral industry seems to have been accompanied by increases in some marsupial populations. The increase in red kangaroo (*Megaleia rufa*) is said to be due to the multiplication of watering points (Main, Shield and Waring, 1959) and the increase in euro (*Macropus robustus*) numbers in north-western West Australia is said to be due to the spread of spinifex after heavy grazing. However, there have been very marked declines in the numbers of most dryland marsupials in Australia (Marlowe, 1958).

127. In the western United States there have been declines in the numbers of some small mammals, such as the gopher, whose habitats have been damaged, but increases in others, such as the Merriam kangaroo rat (*Dipodomys merriami*) which lives beneath mesquite. A recent increase in the rat population of Senegal may be due to heavy grazing and the removal of a predator, but little is yet known about this phenomenon (De Vos, 1975).

128. A change in the character of vegetation habitats undoubtedly has a marked effect on invertebrate populations, though the data are scant (e.g. Ellison, 1960; De Vos, 1975). Hutchinson and King (1970) report declines of over 60 per cent in arthropod populations in heavily grazed pastures in Australia, with a radical alteration in species composition. Although little is known, these changes may herald a decline in the pollination and so germination of some plant species. Crawford (pers. comm.) suggests that a decline in the

diversity of species that accompanies heavy grazing and drought may have a serious effect on bees in the Sahel, and so on the potential for honey production. Bees need a continuous supply of flowers over the wet season. The desert locust, which needs bare ground for breeding, may expand in numbers after heavy grazing or trampling (Haskell, 1969; De Vos, 1975) and tsetse flies are another species that may increase, this time in areas of woody scrub invasion (Uvarov, 1962).

129. (ii) *Vulnerability of different communities to heavy grazing*. The dry parts of Australia are a lesson to anyone who tries to make sweeping generalizations about the effects of heavy grazing. There are some plant associations, notably those dominated by spinifex on both steep rocky slopes and on ancient dunes which, with no prompting from grazers, contain such unpalatable species that the communities are virtually indestructible by domestic stock (although they are in some danger from recreation). There are others, such as the mitchell grass plains of Queensland that, although not highly productive, have a framework of perennial grasses of moderate palatability, which would only be destroyed by extremely heavy use. And there are saltbush plains in South Australia, and Danthonia grasslands of New South Wales, each with its own characteristic response to grazing. A similar range of communities and reactions exists in the American West, and undoubtedly in different parts of the Middle East. Central Africa is more uniform, the species that are found on fixed dunes on the banks of the White Nile in the east occurring right across the Sahel to the Senegal River in the west, but here too, subtly different communities have subtly different problems.

130. One general principle is that the most valuable communities are often the first to suffer. The Gascoyne case study reveals that the suffering inflicted there occurs not on the poor upper slope sites but on the richer pastures of the valley bottoms.

131. (iii) *Hydrology*. Severe trampling and heavy grazing, by damaging soil surfaces, have a drastic impact on the water balance. Higher bulk density means less infiltration and this in turn depletes the soil moisture store. Where water is so vital to biological production, the restriction of water availability to a short period following each brief shower has far-reaching implications.

132. Because of its small volume, the loss of interception storage in plants is probably not serious, although it may mean thirstier beasts in the wet season. The critical change occurs to the infiltration capacity of the surface. On sandy soils, where trampling loosens and roughens the surface and grazing helps to impede root growth, infiltration is probably increased several fold, but the bared surface of a more silty or clayey soil is hardened and encrusted by the impact of raindrops as well as of hooves, and there is a very marked decrease in infiltration capacity. This process of surface sealing is seen by many to be the most destructive effect of the over-use of pastures, and the one most likely to be long-lasting (McGinnies *et al.*, 1963; Fitzgerald, 1968; Rauzi *et al.*, 1968; Trombie *et al.*, 1974).

133. Although the removal of plants helps to increase evaporative loss by allowing greater wind speeds to reach close to the ground, and by enabling solar radiation to heat the soil surface directly, it also replaces the relatively efficient flow of water through plant roots with the much less effective and interrupted flow through soil capillaries, and so sometimes reduces the amount of evapotranspiration (Buckhouse and Coltharp, 1968; Hay, 1969; Peck, 1975). In some cases, especially where deeper-rooting trees and shrubs are lost, water-tables may actually rise in pastoral and on agricultural land (Grove, 1973), and may induce dryland salting. On the other hand, falling water-tables, thought to be

due to decreased infiltration, are also widely reported especially from drier areas (Baker, 1973; Delwaulle, 1973a; Depierre *et al.*, 1975).

134. The absence of evaporating surfaces of all kinds allows air temperatures to rise, and because the soil is drier and heats and cools more readily, its biological activity is restricted (Ellison, 1960; Moore and Biddiscombe, 1964).

135. One potentially serious effect which may lead to the actual expansion of denuded areas is desiccation of groves of plants by winds heated by contact with areas already laid bare. Although this has not been observed in natural vegetation (but see Jenik and Hall, 1966), it is well known in irrigation agriculture where water consumption on the upwind side of a scheme next to a desert may be twice the consumption in fields farther downwind (Stanhill, 1965; Rijks, 1971).

136. Less infiltration, less evapotranspiration and lower surface roughness inevitably mean increase in run-off. These observations are widely corroborated. Lusby (1965) noted a 20 per cent greater run-off from a grazed than from an ungrazed basin in western Colorado, although he attributed most of the effect to trampling rather than to the actual grazing, Delwaulle's (1973a) data for the Niger have already been noted (section 2.2.2). In parts of the semi-arid world where there is a high demand for irrigation or domestic water, experiments in which plant cover is reduced in order to increase run-off have often been successful. Whether they are economic in the long-term is another matter (e.g. Brown *et al.*, 1974; Hibbert *et al.*, 1974).

137. More run-off means more erosion, and this in turn reduces the depth of soil and the amount of water it can hold. In some situations, erosion uncovers a more clayey or more cemented subsoil layer on which infiltration is lower than on the original soil surface (Condon, 1961; Marshall, 1972), and this also helps to increase run-off – there is a progressive and accelerating loss of water-holding capacity and so of biological production.

138. Increasing herds demand more water in the dry season and increased drawing of water depresses well levels (Stebbing, 1953; Aubréville, 1973; Boeck, 1974). The effect on deeper aquifers is difficult to determine. The dating of deep waters with carbon 14 is only just beginning, but it is already suspected that recharge is very much slower than present rates of draw-down. Certainly in the Great Artesian Basin of Australia, wells that were fully artesian when they were drilled some 50 years ago have become only sub-artesian since. In areas near to the sea (as in parts of southern Israel) the draw-down of a fresh aquifer can allow the infiltration of saline water.

139. (iv) *Fire*. There is no fire without fuel, and little fire with little fuel: only on the better-vegetated margins of the dry lands can burning be a significant factor in the ecosystem.

140. Where they can, pastoralists deliberately, and often effectively, use fire to achieve several objectives. A gentle burn every year or two at the start of the dry season removes old stifling growth and allows new green shoots to come up in the following wet season. It releases nutrients held in the old litter that would otherwise only be slowly recycled, and it does not seriously damage the soil humus (Reynolds and Bohning, 1956; Daubenmire, 1968). Fire discourages scrub and therefore both encourages grass and keeps away pests and diseases such as tsetse and trypanosomiasis (Heady, 1960b); indeed where there has been a deliberate prevention of burning, there has been such serious encroachment that burning has been re-introduced (Langdale-Brown *et al.*, 1964; Buffington and Herbel, 1965; Moore, 1973a and b). Another consequence of deliberate

fire prevention is that fuel accumulates, and given a chance spark or lightning strike, a very serious fire may result (Wright, 1974; Hopkins, MS). Agriculturalists may also use fire to clear bush before cultivation and to release the nutrients held in the wood (Clawson, 1963). Sometimes these fires get out of control and burn more than the intended area.

141. Despite its uses, fire presents some dangers. A hot fire, burning the accumulated fuel of a whole season, can remove all the litter and over 20 per cent of the soil humus, and may sterilize the soil's nitrogen-fixing bacteria (Reynolds and Bohning, 1956; Hopkins, MS; Wright, 1974). Although the nutrients may be released, they may blow away as ash in the wind or float away in run-off. The fire lays bare the soil for wind and water erosion, and, in gentle fires, distillates of organic matter may coat the particles of the soil surface and render them "unwettable", so reducing the soil's infiltration capacity and increasing its vulnerability to run-off erosion (Sampson, 1944; Bartlett, 1956). If, say, in a very hot fire, the wettability is not altered, then nutrients held in ash are more readily reached through the soil by infiltrating water than they would have been had they been held in the plants (Jacques-Félix, 1956). Burning changes the species composition of the range: it usually eliminates the deep-rooting woody plants which reach for nutrients deep into the soil and replaces them with grasses. Some of the fire-tolerant grasses may be unpalatable (Emberger and Lemée, 1962; Young and Evans, 1974), but most burning replaces unpalatable with palatable species (Hopkins, MS). Large mammals flee fires, and may then concentrate on and damage unburnt habitats (Curry-Lindhal, 1972).

142. It can be concluded that gentle restricted fires in the right season, repeated every one or two years, do not significantly damage the range, and indeed probably improve it, in the short-term. They may, however, attack its resilience by removing the key woody species which form shelters for annual plants and small animals, but, on the whole, it is far less damaging than heavy grazing or trampling (Hopkins, MS).

C. RAINFED AGRICULTURE

1. Introduction

143. Clearing and ploughing are such drastic disturbances to the ecosystem that virtually none of its original flora and fauna survives. Even nutrifying bacteria are destroyed in some farming systems. The new ecosystem that replaces the old is much simplified and possesses few of the complex checks and balances of the long-evolved climax. Stability and productivity must be maintained by frightening or killing pests, manipulating slopes to retain moisture and regulate erosion, weeding out crop competitors and turning the soil to create seedbeds and prevent evaporative losses. These are problems for technology, not ecology (see Anaya, 1976).

2. Nutrient Loss

144. If plant litter or animal faeces are recycled, agriculture need not lead to great losses of nutrients, but if crops are sold off the farm, nutrient loss may become too fast to be replaced naturally from dust or the weathering of minerals. Jones and Wild (1975),

summarizing the evidence for nutrient loss in West African agriculture, note that badly over-used soils will never recover with fallowing alone and that commerical sales of groundnuts from northern Nigeria lead to a loss of phosphorus which could only be replaced by 25,000 tonnes of single superphosphate fertilizer.

3. Dryland Salting

145. An instructive, but localized, hazard to temperature dryland agriculture has been experienced in Western Australia and in parts of the High Plains of the United States and Canada. The replacement of deep-rooting plants by shallow-rooting crops has reduced evapotranspiration losses from the soil to such an extent that water tables have risen, bringing with them salt accumulated over many millennia in deeply weathered mantles or in sedimentary rocks. Salty efflorescences and springs have appeared, which, although occupying a small proportion of the cropped land, have led locally to significant losses of good land (Peck, 1975).

D. EROSION

1. Introduction

146. Both pastoral and agricultural land use can, in the extreme, accelerate the loss of soil beyond the rate of replacement, and so initiate the most serious form of desertification. Unlike the slow losses of valuable range species or of nutrients from agricultural land, induced erosion usually occurs only in a few, intense storms.

147. Soil removal by wind or water accelerates when the surface is disturbed or laid bare by trampling or ploughing, overgrazing or clearance for sowing. Erosion is further aggravated if cropping saps the nutrients and organic matter which bind the soil into stable aggregates.

2. Water Erosion

148. Damage by water is the result of three principal processes. First, the impact of the raindrops crumbles weak soil aggregates which when wetted slake more readily; these processes together reduce the aggregates to smaller and more easily transported particles, which become plugged into pores in the surface to create an impermeable crust. Second, the splash itself projects the separated particles downhill, and third, they are carried off in any run-off that occurs. The effectiveness of these processes depends largely on the intensity with which the rain comes down: more intense showers have more energy to both pound the soil and splash it farther, and in intense showers the surface cannot absorb water fast enough to prevent run-off.

149. Slope angle and length and the inherent "erodibility" of the soil also affect the amount that may be lost from a field. Sandy soils are permeable and therefore rarely yield run-off, but when their infiltration capacity is exceeded, their poor aggregation

renders them very vulnerable. In the silty soils of much of the temperate desert margins the infiltration capacity is much more commonly exceeded by rainfall intensity and these soils are very liable to erosion. Clayey soils, because they have the toughest aggregation and generally occur on gentle slopes and in low spots in the landscape, are not usually badly eroded. They may even benefit from sediments and nutrients washed from higher sites.

150. The lost topsoil is the most fertile, the subsoil being poor in nitrogen and phosphorus or heavily cemented, is hard and impermeable.

151. Run-off sorts the soil it picks up, moving small stones only a short distance, depositing sand usually at the first break of slope, and taking the more valuable silt, clay and organic matter to distant lakes, reservoirs or the sea. Reservoir siltation is a serious problem in semi-arid lands: rapid natural and even more rapid induced erosion transfer débris into the traps provided by storage reservoirs (e.g. Chakela, 1974, 1975; Depierre *et al.*, 1975). In the western United States there have been alarming reductions in the storage capacity of reservoirs: New Lake Austin in western Texas lost 96 per cent of its capacity in only 13 years, and siltation is not confined to the reservoirs alone: disruption of the river régime above the huge Elephant Butte Reservoir on the Rio Grande River meant that over 1 m of sediment was deposited in the river channel at Albuquerque some 160 km upstream and below the dams there was extensive downcutting in the river channels (Maddock, 1960). In Algeria the Habra reservoir lost 58 per cent of its capacity in 22 years and in Pakistan some estimates of the life of the huge and essential Mangla Reservoir are as low as 100 years. In a dry part of Tanzania (Fig. 13) Rapp *et al.* (1972) calculated that the life of a 3,807,000 m^3 reservoir was only 30 years.

152. Increased run-off needs a more extensive network of streams to remove it. In an intense shower, the stream network may quite suddenly be expanded when steep-sided gullies cut back into fields. These gullies are usually discontinuous, cutting only into locally steeper slopes along river bluffs or terrace edges. Gullying removes less soil than surface wash by run-off, but the dissection of fields that results may make cultivation impossible (Task Committee, 1970). A spectacular and unpredictable form of gullying occurs on soils which have a high propensity for swelling and shrinking: underground "pipes" can develop when water penetrates to deep cracks; and a gully may suddenly appear as the roof falls on an enlarged cavern (Cooke and Warren, 1973, pp. 141–142).

153. It is not always easy to estimate by just how much these induced processes increase rates of erosion above the "normal", especially as few of the measurements given in the literature contain "control" plots (e.g. Jones and Wild, 1975, pp. 61–66). Some results from Upper Volta and Senegal do however show increases in the ratio of 1 : (6 to 80) : (100 to 200) from "forest" to "crop" to bare soil (Charreau, 1972), and further results from Tanzania, shown in Fig. 14, demonstrate an enormous increase in the erosion rate as land is brought into cultivation. Estimates of erosion damage arrived at by other methods also indicate that cultivation can do severe damage: Chalk (1963) showed that 20–50 per cent of an upland area in Sokoto Province in northern Nigeria was gullied. In the Mediterranean area great damage is indicated in the reports by Beaumont and Atkinson (1969) from northern Jordan, Mensching and Ibrahim (1973) from central Tunisia and Nir and Klein (1974) from central Israel. It should always be remembered, nevertheless, that the estimates of land damaged by erosion, made by the Soil Conservation Service in the United States in the 1930s, have been shown to have been more than somewhat exaggerated (Held and Clawson, 1965).

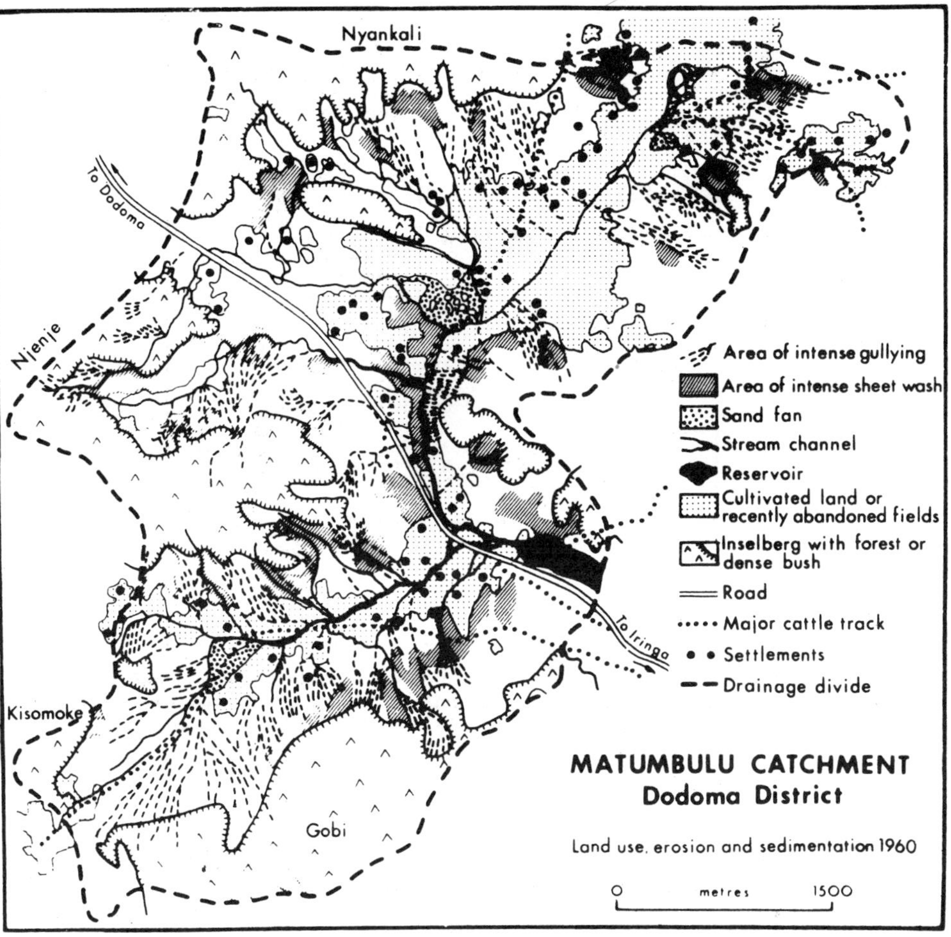

Fig. 13. Map of land use, erosion and sedimentation, Matumbulu catchment, Tanzania. Based on air photographs of 1960 and field checking. Note the zones of erosion and deposition: gullied upper pediments with intense sheet wash; cultivated lower pediments; stream channels with three sand fans; reservoir with heavy sedimentation (Rapp *et al.*, 1972).

3. Wind Erosion

154. Like run-off, the wind picks up, transports and sorts the soil, removing nutrients and clogging nearby areas with unwanted sediment. Clay, silt and organic matter may be removed far from the dry lands altogether: sands are moved close to the ground, blasting the remaining crop and accumulating around local bushes and hedges as small dunes with a much lower fertility and "tillability" than their parent soil. A form of veritable "desert creep" can occur on a small scale as the eroded sands blow onto, blast, desiccate and bury crops in neighbouring fields (Chepil, 1959). Soil loss usually continues until it is checked by a hard or cemented subsoil. The wind erosion hazard is greatest in temperate dryland farming where the soil may be left fallow for one or two seasons, but it is also a hazard in tropical dry areas where the soil is sandy, poorly aggregated and quickly drying.

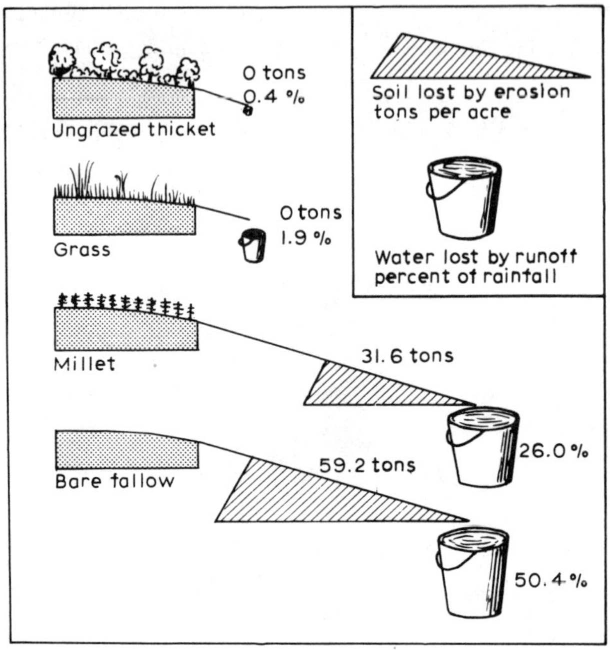

Fig. 14. Results of soil erosion tests on ground with different vegetation covers at Mpwapwa, Tanzania. Annual average of two years' records in erosion plots of 50 m² area of red sandy loam soil on a pediment slope of 3.5° gradient. Data from Staples, 1938. Design of diagram after Gilluly *et al.*, 1960. The grass cover effectively prevents loss of soil and water (Rapp *et al.*, 1972).

155. The formation of small mounds of sand and dust around bushes is, like gullying, an inconvenience to cultivation (e.g. Mensching and Ibrahim, 1973). These mounds are partly the result of the erosion of the less protected areas between the bushes and partly due to the accumulation of moving sand and dust in the calm area round the obstruction.

156. Wind and water erosion combine in some areas to form further kinds of impoverished landscape. Where the soil is stony, the removal of finer material by water or wind leaves stones littered on the surface as a "desert pavement" (see Cooke and Warren, 1973, pp. 120–129). On gentle slopes, notably in the Australian pastoral areas, shallow "scalds" develop in which the topsoil is stripped off by both wind and water. These areas, often a few hectares in extent, have such hard surfaces that plants find it hard to recolonize without the help of ploughing and seeding (Warren, 1975). Irretrievable scalding is said to cover 10 per cent of central Australia.

4. The Soil Balance

157. Losses of soil by both wind and water are balanced to an uncertain extent by incoming material. Considerable inputs of dust have been measured in Nigeria by Bromfield (1974) and in Israel by Yaalo and Ganor (1975). The dust in West Africa is said to be quite rich in sodium, calcium and magnesium but to be deficient in phosphorus (Jones and Wild, 1975, p. 145). On the wetter margins of the dry zone, increased erosion from agricultural land may, by bringing bedrock nearer to the surface, actually increase

the rate at which it is converted to soil; a new and more dynamic balance may be reached between weathering and erosion (Stamey and Smith, 1964).

5. Erosion in Different Climatic and Geomorphological Contexts

158. The danger of rainfall erosion on bare ground is greatest on the sub-humid margins of the dry lands, while wind erosion is the greater hazard in the extreme deserts (Fig. 15). This general pattern results from the greater frequency of erosive showers in the wet areas, and of loose, dry surfaces in the dry zones.

159. Little is known of more specific regional variations in the effectiveness of the two agents. The most thorough research on rainfall "erosivity" has been undertaken in the United States (Smith and Wischmeier, 1962). Much less is known about the rest of the dry world, although Fournier (1967) has summarized some valuable work from West African experiments which show that the Sahel, particularly in the west, is liable to suffer very intense showers. Rainfall intensities exceeding the threshold for erosion at one Sahelian site, for example, occurred for a total of 20 hours in 2 years. Rapp *et al.* (1972) describe further high intensities in Tanzania. Even less is known about the general picture of the magnitude and frequency of soil-carrying winds although some research has been published for the central United States and for Israel (see the summary in Cooke and Warren, 1973, pp. 236–245).

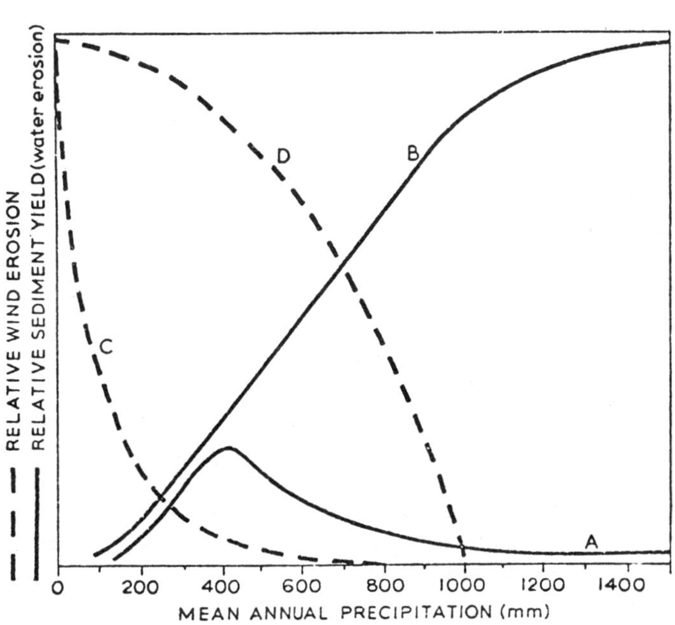

Fig. 15. Relationships of water erosion (continuous lines) and wind erosion (broken lines) with increasing mean annual precipitation. The curves for water erosion indicate the relationships with mean annual precipitation for (A) areas of natural vegetation cover and (B) bare ground (after Schumm, 1969). The curves for wind erosion indicate the relationships with mean annual precipitation for (C) areas of natural vegetation cover and (D) bare ground. These curves are based on what would be expected from the relationship of wind erosion to vegetation cover and to moist soil (Marshall, 1970, 1971 and Chepil, 1956) (Marshall, 1973).

160. The geomorphological context of erosion can be stated somewhat more clearly. The various erosional processes combine in different ways in different landscapes (Table 4).

161. Where inherent vulnerability coincides with intense land use, destructive erosion can be expected. Gentle slopes and favourable water supplies in hills and piedmonts attract agriculture, so rendering these areas liable to erosion. Under pastoral use, the perimeters of watering points, which are commonly located on piedmonts, alluvial valleys and areas of ancient aeolian sands, are the most commonly desertified. The gullying and sheetwash of piedmonts may be difficult, if not impossible to reclaim, but, if allowed to rest, many areas of aeolian sand could recover relatively quickly.

TABLE 4. *Processes of erosion in dryland regions*

Land region	Water erosion	Wind erosion
Mountains	landslides, coarse washout deposits	little, except on summits
Hills	gullying, sheetwash	little
Piedmonts	severe gullying, sheetwash, coarse washout deposits	soil loss, pavement formation, wind mounds
Plateaux	gullying and landslides on scarps, some sheetwash	soil loss, pavement formation
Plains	piping, some gullying	soil loss, pavement formation, wind mounds, scalding
Alluvial valleys	some gullying along terrace edges, sheetwash, silt accumulation	pavement formation, wind mounds, scalding
Enclosed depressions	silt accumulation	soil loss, wind mounding, scalding
Aeolian sands	rarely eroded – only on sub-humid margins	dune formation, wind mounds

E. IRRIGATED AGRICULTURE

162. The problems of irrigated agriculture are a perennial cause for concern. Threats to very high yielding farming systems engender a huge investment in international and national co-operative research and there are at least two major international groups currently constituted to consider irrigation (the Scientific Committee on Water Research, COWAR, and the International Commission on Irrigation and Drainage, ICID).

163. Successful irrigation depends on maintaining a flow of water through crops and soils. Enough water must be transpired by plants to produce a crop whose sale will allow a profitable return on large investments of capital and labour, and an additional 10–15 per cent of the applied water is needed to flush the salts left in the soil by transpiration and evaporation since the soil solution may be anywhere from 2 to 10 times

more saline than the applied irrigation water. If salts are not flushed away the osmotic pressure in the soil solution rises and roots find it progressively more difficult to abstract the water they need, and may even, at high osmotic pressures, lose water to the soil. Most of the salts involved are not themselves toxic to plants, although in some areas boron and sodium toxicity may occur.

1. Waterlogging and Salinity

164. Because of the difficulties of calculating optimum water inputs, and, more frequently, because of inefficiencies in the supply system, it is common for only a half of the water entering a canal system to reach the crops it is intended for. In Pakistan, for example, canals may lose, by seepage, one-third of the water that enters them. If soils were deep hȧd a high hydraulic conductivity, and were well ḋrained, the over-use of water would be a waste rather than a danger, but few irrigated soils are like this: most are in low positions in the landscape, where they can be "commanded" by canals, and many of the soils on these low sites are heavy clays with low hydraulic conductivity. When irrigation starts, the water table is usually very low, but seepage soon brings the "capillary fringe" to within striking depth of the surface (Bharadwaj, 1961). When this happens water can be drawn up from the ground water through minute tubelets in the soil to evaporate and deposit its salt load at the surface. Heavy soils have a greater number of these capillary pathways and so suffer more in this process. When net annual downward movement of water is replaced by net annual upward movement,, serious salinization problems can result. A further rise in the water table brings water into the root zone of plants and further restricts growth; and when the water reaches the surface it completely stops plant production.

165. Another danger is the accumulation of sodium ions. If irrigation or rain waters are applied after the soil has been salinized with a sodium salt, they may remove the anions with which the sodium is held, and leave the sodium attached to clays in the soil. When this happens the soil disperses when wet, and is rendered virtually impermeable. Some irrigation waters (with a high bicarbonate content, for example) are very dangerous to apply in this situation, precipitating calcium and magnesium and freeing sodium (see Allison, 1964). Another, though rarer, hazard is encountered with boron-rich waters.

166. Given the variability in total salinity, types of anions and cations, soil texture and clay mineralogy, and the drainage characteristics of irrigated sites, there are a large number of possible irrigation strategies. Water of extraordinarily high salinity can be tolerated if it is used in large quantities and on very permeable soils; but where the soil is clayey and the site low, even the sweetest water would be dangerous. Crops vary in their tolerance of salinity, and as salinity creeps up on a site, wheat may have to be replaced by rice and ultimately only palms will survive.

167. The most alarming, because the most extensive, problems occur in large canal irrigation schemes, but many smaller irrigation schemes using well-water have also suffered. In North Africa, for example, access to artesian supplies of water, the introduction of motorized pumps, and the development of a good market for horticultural crops and dates, have encouraged a great expansion in oasis cultivation, which has brought with it salinity and waterlogging problems (Cointepas, 1965; Achi 1969), elsewhere a falling water table (Flohn and Ketata, 1971).

2. Long-term Problems

168. Salinization and waterlogging can be alarmingly sudden afflictions, but there are also worrying longer-term problems in irrigated agriculture. One of these is the disposal of the salt that inevitably accumulates as water is evaporated and transpired. In Pakistan it is proposed to lower water tables by pumping from deep tubewells; if this water is saline, as it may well be, it will have to be led into special canals to take it to the sea. A similar proposal has been made to lead away saline drainage water from the irrigation schemes on the lower Murray River system in Australia (Leigh and Noble, 1972). More commonly saline return flow is simply remixed with the water in the trunk stream and the salinity problem is transferred downstream. The build-up of salinity in this way has become apparent on both the Rio Grande and Colorado Rivers in the United States and treaty obligations with Mexico are now necessitating millions of dollars of investment in desalinization plants and other programmes to avert the problem (Hay, 1969; Hotes and Pearson, 1976).

F. OTHER LAND USES

1. Wood Collection for Fuel and Construction

169. Caravans of camels, single donkeys, trucks and even railway trains converge on any sizeable settlement in the dry zone bringing with them firewood or charcoal from woods often 50 km away, and, in the case of Dakkar, from as far away as 400 km. The result is a vast "phryganasphere" of denudation that can create the most severe form of desertification. The fuel is used mainly for domestic cooking, but also on occasion for fish-smoking, pottery kilns, industrial steam engines or, until recently, for steam boats as on the Niger and the White Nile (Delwaulle, 1973a). Rapp (1974) quotes estimates of the resulting devastation in Tunisia: a conservative assumption that 1 kg of dry fuel is used per person per annum means that 2 million ha of woody steppes are devegetated each year, and although some of this, particularly in the wetter parts of the country, recovers by regeneration, it implies great damage to dry areas.

170. Wood collection, usually in the hands of small entrepreneurs, is the cause of the most obvious desertification around the larger towns in the dry world. The commercial pressure behind the destruction can be intense: a Niamey family may spend a quarter of its income on cooking-fuel alone (Delwaulle, 1973a). In Australia the environs of the early mining towns are surrounded by a zone, up to 50 km in radius, from which wood was systematically collected, and within which regeneration has not occurred. Even today train loads of mallee wood arrive in Adelaide each day to satisfy a demand for a cosy hearth.

171. The removal of a small crop of wood by resident shifting cultivators or by nomads has little or no long-term effect, and even the controlled collection for sale need not be destructive, for many species coppice or can regenerate successfully given seed sources and protection for a year or two. If trees and shrubs are removed there may be an increase in grass growth and wood collection could therefore in some cases be a complement to pastoralism. But if collecting is intensive, the small oases beneath each perennial shrub are removed, and with them one of the principal mechanisms for dry ecosystem resilience. If wind and water erosion follow, the capacity for recovery may be totally removed.

2. Urbanization

172. Towns and cities are taking an increasing share of the population in dry lands (Kates *et al.*, 1976), and although urbanization occupies a very small proportion of the area, it poses its own ecological problems. Fuel collection, mentioned above, probably has the furthest-reaching effect, but there are many more intense local effects. Building and road-construction involve a very considerable disturbance to the surface, and consequently induce an increase in erosion. Dagg and Pratt (1962) discovered a 1140-fold increase in sediment loss from a building site in northern Uganda, and many similar increases have been noticed in wetter parts of the world. Both wind and water erosion remove sediment from the bared sites, clogging reservoirs, streams and roadways, and increasing flooding and dustiness.

173. Water supply schemes for towns in dry areas often draw on only slowly recharging ground water supplies and so may be a potential desertification hazard. The disposal of waste water may also cause ecological problems. If drainage is not provided, sewage may collect in stagnant pools and by evaporation become a concentrated solution of salts and nutrients that can sterilize the neighbouring land (Beaumont, 1974). In some areas the effluent can and is being used for irrigation (National Academy of Sciences, 1974), but in hard-water areas the addition of soap to the water may make it dangerously saline, and high concentrations of nitrates or phosphates are an additional danger in some of these waters.

3. Mining and Industry

174. As with the clearing of land for building, its clearance for industry and especially for mining increases the area of bare ground and so leads to faster erosion. The stream channels round mines are usually choked with sediment, and the air surrounding mine dumps is full of dust. Problems such as these have been experienced round the towns of Kalgoorlie and Coolgardie in West Australia, the Rand in South Africa, and the iron-ore mines in Mauritania. Some industries and mines can, in addition, add toxic wastes to the environment and, at a stroke, create a desert where virtually no plants will grow. Traffic to and from mines and the much more widely ranging journeys in the exploration phase are an additional disturbance to soil surfaces.

4. Tourism

175. The income from tourism already far outstrips other sources of income in many dry lands. The American South-West is a vast playground and retirement home for the people of North America; Alice Springs thrives on camel rides and day trips; the Algerian Sahara and the dry southern Tunisian coasts are sunny alternatives to drab winter Europe; the East African Safari Parks bring in as much income as most of the other local exports. Some (e.g. Talbot, 1969) see tourism as one of the best hopes for dry lands.

176. Tourism, of course, has its environmental and social problems. The chief environmental problem is the damage inflicted by vehicles on fragile soils (Jewell, 1974). It is almost a cliché to mention that the tracks of Rommel's tanks can still be seen in the North African deserts, and that his campaigns produced a greater cloud of dust than ever before, detectable even in the West Indies. The present invasion of the deserts is

often as intense as that of the European armies of World War II, and serious concern is being expressed about the ability of some ecosystems to withstand it. The stability of the fixed dunes in the hitherto virtually unvisited Simpson Desert in Australia is in danger, and Ayer's Rock in Central Australia may not be able to cope with any more tourists (Hooper *et al.*, 1973; Lacey and Sallaway, 1975). With the high capital investment available to those who manage tourists, these should not be serious long-term problems.

5. Plant and Animal Introductions and Invasions

177. It is always a temptation to compare the performance of species in different continents, and then to try to introduce the evidently more successful ones into areas from which they were absent. This is often very successful; it is sometimes, especially in arid areas, a dismal failure; just occasionally it is a disaster.

178. The most often quoted examples of a disastrous introduction is that of the rabbit to Australia. The progeny of a few escapees had, within decades, populated almost half of the continent. Being unchecked by a major predator or disease they did enormous damage to the vegetation, and in drought years seriously competed with sheep for scarce forage (Myers, 1962, 1970; Myers and Poole, 1963; Myers and Parker, 1975). They have now been somewhat checked by the deliberate introduction of the disease myxomatosis, but their warrens can still be seen disfiguring pastures even in very dry areas.

G. PATTERNS OF DESERTIFICATION: CREEP OR RASH?

179. Stebbing (e.g. 1953) used the very emotive term "creep" to describe the desertification process that he thought he detected in parts of Africa, but the review of the literature presented here, while supporting the evidence that there has been degradation in some areas, has revealed a rather different pattern to the process. It is important to visualize this pattern before prescribing remedies for desertification.

180. Rangeland is at its most vulnerable in the dry season when the soil is dry and dusty, when only perennials protect it, and when they alone provide forage. It is just at this season too, that stock must be at their most concentrated around watering points. In consequence, the immediate surroundings of pools or wells, where the stock wait their turn to drink or to be led out again, are very severely damaged. The area around each perennial source of drinking water becomes a *piosphere* in which damage diminishes away from the focus (Lange, 1969; Fig. 16). The radius of the piosphere rarely exceeds 5–8 km (Peyre de Fabrègues, 1971) but varies with seasonal forage conditions, the cattle population and their condition, and the maximum distance travelled by the local race or species of stock (Osborn *et al.*, 1932; Heady, 1960b). FAO/SIDA (1974) reported some piospheres of 10–15 km diameter in the Sahel following the recent drought. Beyond the diffuse margin of each piosphere lies a relatively undamaged area, used only when rains fill local pools or are held briefly on leaves and stems; only along the tracks leading between perennial watering points is there serious damage.

181. Damage to pastures can also be concentrated, even in the wet season, on those areas that provide particularly delectable forage, even though they cover only a very small proportion of any range. In the Gascoyne catchment in West Australia, for example, it is the better pastures in the valley bottoms rather than the poor range on the upper slopes that have suffered most.

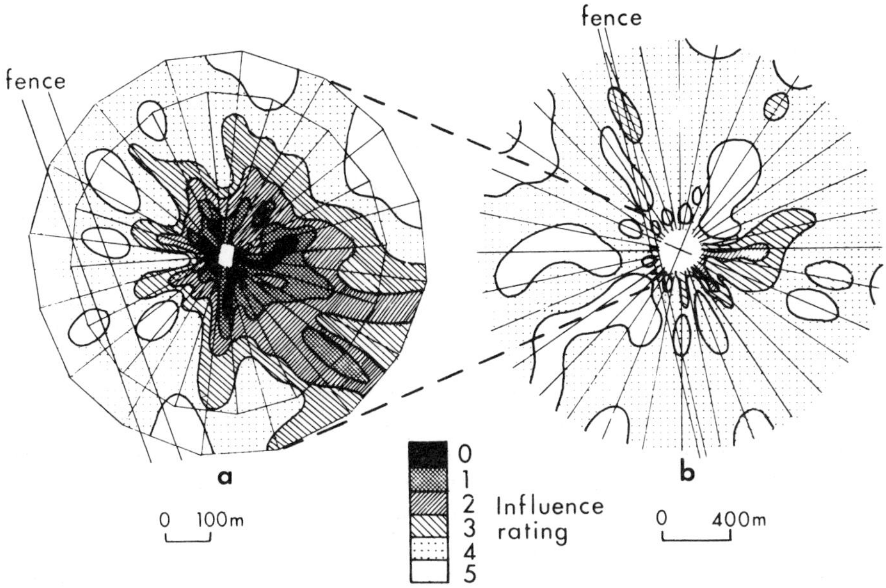

Fig. 16. A piosphere on an Australian rangeland. (a) Isotels (lines enclosing areas where effects are the same) for interactions with *Atriplex vesicaria* up to 250 yd from the trough; (b) the same up to 1300 yd.

182. Extreme damage to dryland agricultural fields is also localized, especially if some form of bush fallowing is maintained. Only where the fallows have been shortened and nutrients and organic matter not allowed to accumulate, and where cropping has extended to the poor land, does devastation become widespread. Even here the pattern tends to be rather like the piosphere in that the zone of heaviest damage closely surrounds a village and the devastation declines in intensity away from this centre. In irrigated agriculture it is the low spots in the landscape that are the first attacked; the bad practices of one farmer or canal sector engineer endanger a basin below a particular contour with salinity of waterlogging. The collecting of firewood from a *phryganasphere* around every town and the devastation brought about by urbanization and other land uses also tend to be localized.

183. In most of these cases desertification is worst at a centre of intense land use, the effects diminishing away from this point. The most common mode by which the damage is spread is by the opening up of new centres: new boreholes or tanks for water (Rapp, 1974), new agricultural settlements (Mortimore and Ologe, 1975; Bernus, 1975), new towns, tourist centres or mining settlements. In each case, it will be the favoured places that will be attacked first, not the most desert-like.

184. The processes of desertification are usually linked directly to land use; if the use diminishes in intensity, so will the rate of degradation, indeed in many places one could hope for recovery. Only in the following rather special conditions can desert spreading, for a short time, and over a small area, be a self-generating, cumulative process:

(a) Where sand dunes are freed of vegetation near to a watering point, and slowly migrate to engulf less damaged sites (Hurault, 1966);

(b) Where sand blowing from an over-cropped field moves on to desiccate, bury and sandblast crops in a neighbouring field;.
(c) Where sediment removed by run-off buries fields or pastures downstream;
(d) Where a hot, dry, desiccating wind blows off a bare area into a vegetated area; except in irrigated areas the effects of this are uncertain;
(e) Where water tables rise in irrigated basins, and salinize and waterlog the fields to such an extent that they are abandoned, leaving canals still pouring water onto the land.
(f) Where the surface albedo is changed over large areas, there may be a regional effect on climate. This effect is discussed at length by Hare (1976).

185. In none of these instances is desertification an advance of the desert along a broad front: desertification develops from within formerly greener areas.

186. Desertification, as explained in Chapter II, is usually made apparent by a severe drought, obviously an advance of the desert, albeit for a limited period. If one envisages a number of zones of potential desertification centred on watering points, villages and towns, then a drought, bringing dry conditions in from the desert, would tip the scales only at these points: a rash of degraded areas would appear. When the good years return the relatively undamaged areas would use their resilient strategies to recover, the degraded sites being left more permanently damaged.

187. Afforestation may well be a very useful tool in protecting areas from the spread of intense land use, or the spread of small centres of self-generating desertification, as in moving dunes. But a green belt along the desert edge (wherever that is) would not halt the attacks behind this Maginot Line.

V. DIAGNOSING DESERTIFICATION

A. INTRODUCTION

188. A policy for rehabilitation must be based on as adequate an assessment of the degree of degradation as is possible given the constraints of time, money, trained manpower and technology. Since all production depends on soils and vegetation, and since these, fortunately, are easier to observe than wild and domestic animals or climate and hydrology, the most ready way of estimating both damage and capacity for rehabilitation is to survey the range or the fields themselves. Such surveys nonetheless present many problems of definition, sampling, and interpretation. The least costly, and sometimes the only possible, method relies simply on the experience of a few trained observers, but the accuracies of this approach soon become apparent to all concerned and successively more systematic methods need to be evolved. Nevertheless, it is always necessary to balance speed against accuracy, time involved in survey against the urgency of problems, or the "gut" knowledge of men on the ground against the technical knowledge of experts from elsewhere.

189. This chapter suggests a quick and simple method of evaluation. The preliminary review shows that more elaborate and accurate methods exist but the evolution of a universally acceptable system will need much more research and discussion. Such a development should be an important priority in any plan of action.

B. PASTORAL LAND

1. Existing Methods

(a) Introduction

190. The assessment of "condition" and "trend" of pastures is probably as old as pastoralism. It is clear from accounts of pastoral communities that they have a very close knowledge of the ecosystems on which they depend, and know, for example, when to move on to better land, or when to allow a paddock to recover. Because seldom systemized, this knowledge carries the danger that it may be based on ill-formulated or dated assumptions, but the ecological know-how it embodies is probably much greater than that available to more highly trained range scientists, and should not be disregarded.

191. Running parallel with the methods of herdsmen, though not always in close touch with them, a series of more systematic approaches has developed and become progressively better adapted to meet real needs.

(b) Review of existing systematic methods of range assessment

192. (i) *The United States*. The longest tradition of systematic assessment of range degradation is American, for as early as 1930 semi-quantitative methods had already been proposed and tested. Three principal methods have been evolved (Lendon, MS):

(a) The U.S. Bureau of Land Management has developed the "Deming Two-Phase Method" (Deming, 1957 in Lendon, MS) in which Phase 1 is a survey of forage quality, quantity, vigour and reproduction, and Phase 2 adds other site variables such as cover, vulnerability to erosion, surface run-off and soil stability. Each variable is rated and given a value between 0 and 25 according to the most appropriate of five reference descriptions rated at 5, 10, 15, 20 and 25.

(b) The U.S. Forest Service employs the "Parker Three-Step Method" (Parker, 1951): Step 1 consists of range trend estimations along repeated 100-ft line transects; in Step 2 soil and vegetation characteristics are measured within loops at 1-ft intervals along the transect; Step 3 is a photographic record.

(c) The Soil Conservation Service of the U.S. Department of Agriculture estimates differences in the proportions of 30–40 range species compared with their proportions in a "climax" area as defined by the "Dyksterhuis Quantitative Climax Method" (Dyksterhuis, 1949).

193. (ii) *Australia.* Range assessment has only developed in Australia since the late 1950s. In place of the rather close attention to site conditions and to year-to-year or even seasonal carrying capacities that characterize the American and Sahelian systems, the Australians have been interested in optimizing long-term returns from pastoralism by minimizing pasture and surface degradation.

194. The Australians have adopted the fundamental concepts of climax and succession from the Americans. The Soil Conservation Service of New South Wales have devised a 7-point rating for each of six factors, using the land units provided by the CSIRO Land System surveys as a framework (Chapter III; Condon, 1968; Condon *et al*., 1969). Each factor is rated by comparison to a standard, "pristine" site on the same land system. A variant of this method was used by Wilcox and McKinnon (1974) in their assessment of the desertification in the Gascoyne Catchment in West Australia (a Case Study area for the present Conference).

195. The Australians are in the course of evolving a new and, hopefully, widely applicable method, termed "STARC" (Standards for Testing and Assessing Range Condition, Lendon, MS; Lendon and Lamacraft, MS). Twelve key variables are assessed for each eco-unit, facet or land system by comparing them with an undisturbed control, which is visited and observed at the beginning of each season. The method is a development of the Deming Two-Phase method with four more factors for topfeed and browse to assess range condition, and four further factors to estimate range trend. Lendon and Lamacraft (MS) and Perry (MS) found that botanical composition was the best single index, and in fact that only three species could act as indicators, but when a scale of degradation was needed vegetation and soil factors were also useful.

196. Quantitative analysis associated with the Australian and American methods has been described by Tueller (1973), Tueller and Blackburn (1974) and summarized by Shiflet (1973), Perry (MS) and Lendon (MS).

197. (iii) *Eastern and southern Africa.* The management strategies adopted in eastern and southern Africa start with the assumption that vegetation must first be improved and conserved; only then can production be optimized, by improving livestock management, introducing new land uses such as tourism, or even by the cropping of wildlife. The problems to be tackled include the elimination of tsetse, scrub encroachment and erosion (see Heady, 1960a and b; Abercrombie, 1974).

198. (iv) *The Sahel.* The primary objectives of range assessment in the Sahel have been to improve herd productivity. Assessment is almost wholly based on the amount of herbage production by palatable species during the rainy or dry seasons (Boudet, 1975b, 1976), and as such is geared to a much finer adjustment of stock numbers to pastures than some of the other schemes, but could be criticized for not bringing into account long-term safe carrying capacities. It does have the advantage of being able to make judgements about particular parts of the range that are especially valuable, although each assessment is time-consuming and needs a large amount of technical manpower.

199. (v) *North Africa.* Floret and Le Floc'h (1973) have devised a quantitative method of assessing condition and trend which aims specifically at combating desertification (see also Long, 1974, and Gischler, 1975). Observations are made over a 5-year period of the effects of grazing on changes within phyto-ecological (PE) units (similar to the "eco-units" of the Australians) on a 20,000 ha test area in Tunisia. The stages are: (1) mapping PE units and assessing their state of degradation; (2) sampling the above-ground biomass of perennial plants and calculating totals for the whole unit and assessing the state of erosion and the sensitivity of the vegetation cover on each unit; (3) calculating the rates of change and trends of the vegetation in each unit; and finally (4) estimating the probabilities of different kinds of change and, with "transition matrices", developing a predictive model of change over the next 25 years under each of several possible management strategies including: protection grazing, limited grazing, present levels of exploitation, increased grazing pressure, and encroachment by cultivation. The predictive nature of this method, especially when linked to alternative strategies, is of particular value.

(c) The translation of range assessment into "carrying capacity" for stock

200. (i) *Introduction.* The stocking rate and levels of management at which grazing ceases to benefit the range and begins to degrade it (section IV B2) are vitally important to a conservation policy for dry lands, but the underlying concepts are not easy to define, let alone to quantify. The central idea is that of *carrying capacity*.

201. The "carrying capacity" of a range can be defined as the level of stocking that it can support without degradation. This definition itself depends on an understanding of each of its terms: "level of stocking", "range" and "degradation"; the last term has been discussed in Chapter IV and earlier in the present chapter; the discussion below will try to clarify the other two terms.

202. The simplest understanding of the term "level of stocking" is the number of beasts on the range, but this immediately begs questions about the type of stock, and their condition. Since some beasts are larger than others, the easiest way out of this problem is to define them as multiples or fractions of a standard unit. The IEMVT (Institut d'Elévage et de Médecine Vétérinaire des Pays Tropicaux) in France define the basic unit for the Sahel as a Tropical Bovine Unit (TBU): one TBU is equivalent to one beast weighing 250 kg; for 60 per cent of the bovine herd and for horses and camels, one head is equal to one TBU, and for the remaining, younger portion each head is equal to one-half a TBU; a sheep or a goat is equal to one-tenth a TBU, and a donkey to one-half a TBU. As a rule of thumb for estimating the impact of a beast on the range this is obviously better than nothing, but, as it stands, it does not account for the

complementary use of the range by different species, the condition or breed of the stock, whether the stock is being held as a "store" awaiting better times or as a marketable product, or the age structure of the herds, all of which are very variable. It is probably true, in addition, that a small number of beasts in good condition can yield as much as a larger number of poor ones, and the small number may have less impact on the range. While one group of pastoralists or administrators may want to maximise returns in cash and so presumably concentrate on quality, another group may feel that mere numbers are more important. Clearly this aspect of the concept of carrying capacity can become very involved.

203. The idea of the "range" is even more difficult to define. A Fulani herdsman would be very puzzled indeed by a statement that he could graze only one beast on so many hectares; his first question would be "which hectares?" for he knows very well that his stock graze different ranges at different seasons and that they graze not alone but as a group. They will take forage according to the speed at which they are moved over the ground; in the dry season they will not stray far from the well; if they find a particularly attractive spot, they will ignore the sparse but adequate forage in nearby poorer pastures; they will not search out grasses if there is too much scrub; they like different kinds of forage at different times of the year, or at different times in the breeding cycle; when they are in poor condition they will be less efficient at searching for forage; unless they have enough of one kind of forage, they may not touch another; in fact forage intake is a highly complex process. The problems of measuring some of these variables can be, and are being, overcome by range scientists (see Boudet, 1975a and b), but the evolution of a fool-proof method will be a very complex and time-consuming task, and poses many difficult practical and theoretical problems. Some of the theoretical problems are examined by Noy-Meir (1975). Above all the problems is the definition of the whole range of a herd and the assessment of its carrying capacity, rather than the measurement of only a small section.

204. The way in which graziers themselves arrive at stocking density optima are very imperfectly understood. It is known that the ratio between animal and human populations varies very widely (Sandford, MS, 1976), so that mere subsistence is probably only one consideration among many, the others being cultural or economic. It has been asserted that the only constraint on the cattle numbers held by some African pastoralists is the labour required to herd them (e.g. Swift, 1975) and that, given a large labour supply, the only remaining constraint is the amount of forage at any one time. Because herdsmen can control predators and diseases, and because they can make maximum use of all the productive niches that the environment offers, this can mean that animal populations rise with frightening rapidity in good years and crash equally startlingly in droughts. But the indigenous pastoralist may be adjusted to this, often because he is accustomed to playing an involved game with the environment or because the crashes are rare enough to allow him to forget them or to believe that they can be balanced by the years of milk and honey between. There is, at any rate, a counter-argument that states that African cattle pastoralists only maintain enough beasts for a sustained and adequate diet of milk, blood or meat (Dyson-Hudson, MS). In Australia, and probably in other "developed" dry lands, actual stocking rates seem to be controlled more by the market than by any other factor.

205. Even where the pastoralist appreciates that his future depends on conserving his forage resources by limiting his stock, he may be constrained from such a policy by several factors: his imperfect knowledge of the environment, particularly of the weather;

his life expectancy; his control over the range, for if he owns it in common with many others or rents it from the government, his conservation measures may be taken advantage of or destroyed by other users; his ability to make forecasts of the demand for his products; his alternative sources of income, should his pastoral enterprise fail; and finally his personality and culture. The uncertainties involved in all these alternatives need experience and nerve to manage.

206. The official view is usually very different. The government usually has an interest in stability: it does not want periodically to have to bale out the indigent grazier, or to deal with gluts or shortages of pastoral produce. It can also afford a longer view: it wants to see production maintained into the future. Against this it is influenced by many other considerations: the supply of meat to hungry and politically active urban populations (Dyson-Hudson, MS); considerations of defence strategy (keeping the empty areas at least sparsely populated); and long-term questions of national economics. These considerations are interpreted in the field by a skeletal staff of pastoral inspectors whose financial and legal constraints vaguely reflect the politics of central government. These local officials are hampered as much as the pastoralist by an imperfect knowledge of the physical and economic environment, and their views of the gamble between short-term gains and long-term yield must still be governed by a host of personal judgements.

207. (ii) *Review of methods*. (a) Empirical estimates. This method, which is very widely used, is based on experience of the behaviour of the range and the stock on it, over a period of years.

208. Optimal stocking rates for United States rangelands have mostly been estimated in this way. The method can be made more precise by making comparisons of stock and range performance on one site with that on experimental ranges, enclosures or similar ranges elsewhere. The most frequently used procedure is to use records of stocking levels which are known either to have caused no obvious damage, or to have led to pasture degeneration and erosion (e.g. Stoddart and Smith, 1943; Ogden, 1973). Such methods can be continually readjusted with new experience, but, being often built on a knowledge of only a limited number of years, may not succeed in stopping slow changes that could be serious in the long-term. Estimates based on long-term experience, on the other hand, may, by anticipating the worst that could happen, dampen fluctuations in pastoral production, and mean that in really good years there are not enough cattle to make use of production that may then go to waste. The desirability of a glut of cattle that might follow such a good season might indeed be undesirable in Utah, but might be both a boon and of little long-term consequence in Karamoja.

(b) Semi-quantitative rating. This method has been developed to produce carrying capacity figures for an extensive part of central Australia (Condon, 1968; Condon *et al.*, 1969). The method starts with a Land Systems map and climatic statistics. Seven critical factors are abstracted: soil, topography, tree density, drought forage, pasture composition and condition, extent of barren areas and annual rainfall. Several characteristics of each variable are estimated from field and aerial photograph surveys, and their contribution to productive capacity is compared to a standard rangeland in which each characteristic is assigned a "par" value of 1. If a variable acts to increase the grazing capacity it is allotted some empirical rating value greater than 1, and if it appears to lower the capacity, when compared to the standard, it is given a value of less than 1. The product of all the values is then multiplied by the grazing capacity (safe stocking rate) of the standard area to produce a "standard grazing value" (SGV). By 1968 SGVs had been calculated for 88 Land Systems in central Australia. The safe grazing capacity was then estimated for these areas in optimal, present and drought years.

209. This complex method has the advantage that it forces consideration of and attempts to quantify a whole range of variables, and although the conversion of the data to the SGV involves some subjective judgements, the conversion ratio can be changed as knowledge increases. The method also has in its favour the fact that it has evolved to allow some "tracking" of good and bad years. Although acknowledged to be merely a first step in the evolution of a system, it deserves to be applied and adapted to other areas.

(c) Quantitative measurements of forage. Somê quantitative methods have been evolved for use in the Sahel and elsewhere. The IEMVT method is based on the amounts of palatable forage production on a pasture during the rainy and dry seasons, the forage requirements per kilogramme of unit livestock weight, the forage preferences of the stock and a 30 per cent reserve of material to allow regeneration (Boudet and Coulibaly, 1975; Boudet, 1975b, 1976). The number of forage days available in a particular pasture and active-season and dry-season grazing capacities are then calculated from these data. The method has many advantages for short-term management, and can be improved as better figures and observations of preference become available. The method does not at present seem to include a provision for pasture improvement, and it requires quite a large amount of technical manpower.

210. In Western Australia, Payne and his colleagues (1974) have evolved a similar method as have the Land Resources Division of the British Overseas Development Ministry in southern Africa (Abercrombie, 1974 Hyde, MS, 1975). There is also a long tradition of this kind of more quantitative approach in the United States (Humphrey and Lister 1941; Pearson, 1975).

(d) Summary. The suitability of any of these methods will depend on the scale of the problem, the resources to hand, and the type of area involved. The methods fall into a kind of sequence: the empirical method is perhaps the least that could be expected of any administration; the Condon method of semi-quantitative assessment is suitable for broad-scale planning of resources; the quantitative measurements are useful for closer control of grazing.

(d) Fundamental research

211. (i) *Introduction*. The systematic methods reviewed above depend on hosts of assumptions that cannot at present be supported by either theory or evidence. While such approaches are necessary in order to apply as much rationality as possible to ecosystem management, safe manipulation can ultimately depend only on sound research. Such work is time-consuming, but already promises new insights into dryland exploration.

212. Since the ecosystem functions as a whole, the primary requirement of research is that it too should be holistic. Many of the problems that are encountered in the application of research and technology to the exploitation of resources arise from piecemeal treatment of problems: the application of disease eradication programmes to cattle without a programme for limiting the explosion of cattle numbers that is sure to follow; the prevention of burning without the knowledge that this might encourage scrub encroachment; the eradication of game to control trypanosomiasis without research into utilizing the game itself as a source of meat; and the restriction of nomadism without an understanding of its ecological strategies.

213. The first requirement of a holistic approach to research is that there should be some model of how each component fits into the ecosystem. Not only does such a model give an indication of those components that are vital and those that are only secondary to the maintenance of productivity, resilience or long-term stability, but it also highlights what is not known and assigns some priority to the field or laboratory research that is required to solve these questions.

214. Any model of a dry ecosystem must be dynamic, for only then can it be used to predict the effects of proposed land use strategies. A dynamic model must be built from hypothesized relationships, not from simple regression equations between measured ecosystem characteristics. A model built from a large number of hypothesized relationships can be validated only as a whole against the reality of a test site. Only then can it be adjusted to approximate more closely to reality by changing the relationships within the model itself. Although computer modelling of dry ecosystems may give imperfect predictions of the effects of alternative land-use strategies, it is likely to be better than those arrived at in any other way, since the modelling process enforces clear and precise formulation of concepts that are then amenable to testing.

215. (ii) *Review of modelling in dry ecosystems*. One of the first computer models concerned specifically with dry ecosystems was developed by Goodall (e.g. 1970) for a grazing system in an area with 250 mm mean annual rainfall in Australia, but the modelling process was not able to begin in earnest until it was given the backing of the International Biological Programme (IBP).

216. The IBP Desert Biome study in the United States began by constructing small models to answer specific theoretical and practical problems such as grasshopper control, the adaptive strategies of annual plants, or the effects of a standard level of stocking on one type of range (Bridges, 1971; Wilcott, 1973; Wilkin and Norton, 1974). Later it proceeded to more generalized and widely applicable models consisting of a package of smaller models for different components within the whole ecosystem (e.g. Goodall, 1974; cf. Anway *et al*., 1972). Sub-routines or sub-models dealt with plant processes, animal processes, soil processes, hydrology and erosion. Each of these was tested on one or more of five test sites.

217. Modelling has been initiated in the arid zones of Israel (Tadmor *et al*., 1974; Van Keulen, 1975; Noy-Meir, 1975a and b); on the north-western coast of Egypt (Ayyad, 1975, 1976) and in Iran (Panahi, 1976). The Australian work has been continued: it has concentrated on hydrology (e.g. Winkworth, 1970), and now there are also models for the grazing of arid shrublands (Chudleigh, 1971; Fisher, 1974; Noble, 1975). At present the CSIRO are concentrating modelling efforts for dry ecosystems on a large experimental site north-west of Alice Springs (e.g. Ross, 1973). The Australians have also attempted a "whole system" model of a grazing system; this starts with the rainfall input, and then moves through forage and sheep production to employment and social indices of the quality of life (Benyon, 1976). The print-out of an early unvalidated attempt at this model is given in Fig. 1. Other programmes are discussed in the MAB (Man and the Biosphere) Reports No. 6 (1972, Annexe 5) and 25 (1974, Annexe 4) and by Long (1974).

218. (iii) *Field research*. Modelling depends heavily on research in the field. Such research in future should be concentrated on a few experimental sites in which the interaction of a wide range of variables can be studied by scientists from many disciplinary backgrounds. Research of this kind is already in progress in Tunisia (Flohn

and Ketaka, 1971; Floret and Le Floc'h, 1973; Novikoff *et al.*, 1973; Novikoff, 1974; Long, 1974), in Israel (Noy-Meir, 1975a), in Australia (e.g. Low *et al.*, 1973; Ross, 1973) and in the United States (Heady, 1973; Tueller, 1973). Such programmes are to be encouraged.

219. Research must also be tied firmly to practical problems. Goodall (pers. comm.) believes that the only successful modelling efforts are those that have a distinct site and problem in mind. Research efforts must constantly be checked against problems discovered by extension officers and through them by pastoralists in the field.

C. RAINFED AGRICULTURE

220. Cultivated fields can lose productive potential in two principal ways. The first, which is the slow loss of nutrients, can be estimated either by precise measurements of the availability of important nutrients, or by the potentially more accurate and much older "bioassay" method in which the appearance or disappearance of certain weeds is used to assess whether fertility is declining. A review of these methods is beyond the scope of this report: the first is a matter of detailed laboratory technique; the second is much less well documented and depends in any case on the use of different species in different areas.

221. The second mode of desertification in cultivated fields is the loss of topsoil by erosion. Because this danger is shared with pastoral land and because it forms the basis of the simplified method for evaluating desertification proposed in this report it will be discussed separately in the next section.

D. EROSION

222. Gullies, rills, ripples, dunes, stony surfaces and thin soils are both the most serious signs of desertification and the most obvious. Numerous authorities have used these indicators in estimating the extent of damage to both pastoral and agricultural land.

1. Review of Methods

223. Two formalized methods are reviewed here. There are many others, but their basic assumptions and categories do not differ significantly from those mentioned here at the scale of concern of this report.

(a) The United States Department of Agriculture

224. The Soil Survey Manual of the USDA (1951, pp. 261–268) contains a well-tried method for erosion assessment. Like many other methods for estimating desertification it depends on some reference site in a "pristine" condition, or some idealized model of an undisturbed soil on a similar site to the one being surveyed:

225. *Water erosion*

Class 1 (slightly eroded): The surface is slightly rilled, and the topsoil (where organic matter and nutrients are held) has been thinned in places, but has been totally removed only in a very small proportion of the area. Management is not usually restricted by erosion.

Class 2 (moderately eroded): Ploughs reach below the topsoil and bring up the less fertile material from beneath. There may be shallow gullies (these being features that, unlike rills, last from year to year). The topsoil has been lost over 25–75 per cent of the area.

Class 3 (severely eroded): More than 75 per cent of the topsoil has been removed. Gullies are deep and have removed some of the subsoil.

Class 4 (excessively eroded): The land is covered by an intricate pattern of deep and shallow gullies. Reclamation will be very difficult except in special circumstances.

226. *Wind erosion*

Class 1 (wind eroded): The wind has removed enough of the topsoil for ploughing to bring up subsoil material in places. The topsoil has been lost on 25–75 per cent of the area.

Class 1a (overblown): Recent loose deposits of wind-laid material overlie the soil so that its upper layers are significantly altered.

Class 2 (severely wind eroded): All of the topsoil has been blown away with some of the subsoil as well.

Class 2a (wind-hummocked): Wind deposits form a fine pattern of small hummocks and dunes. The drifting is fairly local.

Class 3 (blown-out land): The wind has removed most of the soil profile down to virtually unaltered parent material. The surface is very irregular, and cultivation impossible.

(b) Australia

227. The following categories were used by Condon *et al.* (1969) for areas of relatively low topography in central Australia:

228. *Forms of erosion*

Class 1. Negligible erosion on coarse structured clays and alluvial sands.
Cl ss 2. Scalding (on texture contrast and some silty alluvials).
Class 3. Scalding with drift.
Class 4. Wind sheeting.
Class 5. Wind sheeting with drift.
Class 6. Drift and dune activation.
Class 7. Water sheeting with rilling and gullying.

229. *Degree of erosion*

(a) Minor = rarely more than small areas affected.
(b) Moderate = frequent small areas or occasional large areas indicating moderate susceptibility to erosion.
(c) Severe = frequent large areas (often with rilling and gullying in Class 7) indicating high susceptibility to erosion.

E. IRRIGATED LAND

230. The classification of saline and alkali (sodium affected) land is well developed. Table 5 is a composite classification culled from three major authorities (Richards, 1954; Hunting Technical Services Ltd., 1961; Allison, 1964).

TABLE 5. *Classification of saline soils and associated crop production*

Electrical conductivity of the upper 50 cm of soil	0 — non saline — 2	2 — non saline — 4	4 — slightly saline — 8	8 — saline — 16	16 — very saline — 40	40 — ultra-saline
Effects on crops	Negligible	Yields of only very sensitive crops affected	Yields of many crops restricted	Only tolerant crops yield satisfactorily	Only very tolerant crops yield satisfactorily	Plant growth virtually impossible
Some examples of crops possible within each range	Peaches	Celery Beans	Wheat Rice Potatoes	Barley Spinach Date-palms		

F. SUGGESTED SIMPLE METHOD FOR DIAGNOSING SEVERE DESERTIFICATION

231. The evolution of a comprehensive method for evaluating the extent of desertification on pastoral, rainfed agricultural and irrigated land is beyond the scope of this review. Such a system is nonetheless urgently required and its development should have priority among requirements for research in the plan of action. The problems of development are not negligible: conditions vary widely; various national authorities have evolved their own methodologies on which their records may depend; and there are different cultural perceptions of what constitute desertification.

232. The damage that may be inflicted on arid ecosystems covers a very large spectrum from a minor dampening of the vigour of desirable pasture grasses to the formation of gullies cutting deeply into the infertile subsoil. Even if the more subtle vegetational components in this range are ignored, there remains a huge difference between the slightly accelerated loss of nutrient-rich topsoil and the gullied, duned or white salty surface of a completely unusable area. In the earlier stages of desertification, however, the signs are both more subtle and more easy to mitigate, and a simplified method of rapid assessment should thefore concentrate on signs which are both more obvious and more serious. The scheme proposed here is consequently confined to such indicators.

TABLE 6. *Increasing desertification*

	Slight	Moderate	Severe	Excessive
Water erosion	Rills, shallow runnels	Soil hummocks Silt accumulations	Piping coarse washout deposits Gullying	Rapid reservoir siltation Landslides Extensive gullying
Wind erosion	Fluting and small-scale erosion Rippled surfaces	Wind mounds Wind sheeting	Pavements	Extensive active dunes
Water and wind erosion			Scalding	Extensive scalding
Irrigated land		Minor white saline patches	Extensive white saline patches	Land unusable through excessive salinization

233. The scheme is set forth in Table 6. Similar methods have been used by Rapp *et al.* (1972) in their work in Tanzania (Fig. 13) and are being used by Dregne (pers. comm.) in the development of a system for mapping desertification on a world scale.

VI. ECOLOGICAL PRINCIPLES IN COMBATING DESERTIFICATION

A. INTRODUCTION. CONSERVATION IN DRY ECOSYSTEMS

234. Conservation, being a policy of sustained yield, should be the basis for any action in combating desertification. Five main principles are involved: the integrity of the ecosystem; collaboration, rather than confrontation, with the ecosystem; careful timing of use; careful distribution of use; and the matching of uses and resources.

1. The Integrity of the Ecosystem

235. The plants, animals, soils, slopes and hydrology in an ecosystem are dependent on each other in their self-stabilizing strategies. An alteration to one part at one time may have wide-ranging repercussions. Many human cultures have adjusted their activities to the constraints of their local ecosystems, and these in turn have been modified. The altered, relatively stable communities rely on continued use.

236. If new land-use practices are to be introduced into dry lands, their secondary as well as their primary effects must be anticipated. Well-drilling will be disastrous if, as a result of the water provided, the well perimeters are irreparably damaged by excessive trampling; mechanization of agriculture could accelerate erosion well beyond an acceptable threshold if new crops or ploughing patterns are not also introduced; irrigation is often hopeless in the long-term without adequate drainage; excessive grazing in the upper part of a catchment may threaten distant reservoirs or irrigation schemes.

237. **Comprehensive planning is necessary to maintain the all important integrity of the ecosystems. Traditional land-use strategies are often, but not always, integrated with the ecosystem, and their change may endanger even quite remote facets of the system.**

2. Collaborating with the Ecosystem

238. Communities of plants and animals adopt patterns of distribution, movement and reproduction that enable them to survive their harsh and unpredictable environment. Successful modes of land use adopt similar strategies and barely interfere in the ways in which the ecosystem plays the game of survival. Nomadism is a strategy used both by wild ungulate populations and by cattle pastoralists in East Africa. Shifting agriculture allows the indigenous ecosystem to refurbish the fertility of the soil during the period of bush fallow. In basin irrigation, drainage from the soil can take place in the season of low river flow. Higher yields from the same land are bought only with greater investment, and must be guarded with much greater care against the effects of erratic climate and long-term degradation.

239. **Where investment is meagre it is better to work with the ecosystem than to fight it. Where investment is available, higher yields can be achieved by making modifications to the ecosystem, but new strategies must then be evolved to counter degradation.**

3. The Timing of Use

240. The achievement of a long-term sustained yield is illustrated, very generally, in Fig. 17. Actual patterns would be complicated by fluctuating climatic factors, and would be difficult to illustrate. Three paths to sustained yield are shown.

241. Retrenchment, which is the first, is the only alternative in many areas where the range has been irretrievably damaged by stock numbers well in excess of the carrying capacity, where cultivation has been allowed in climates that are periodically too dry for any yield or where salinization and waterlogging are severe. Retrenchment may initiate a slow recovery, but seldom to past levels of cropping.

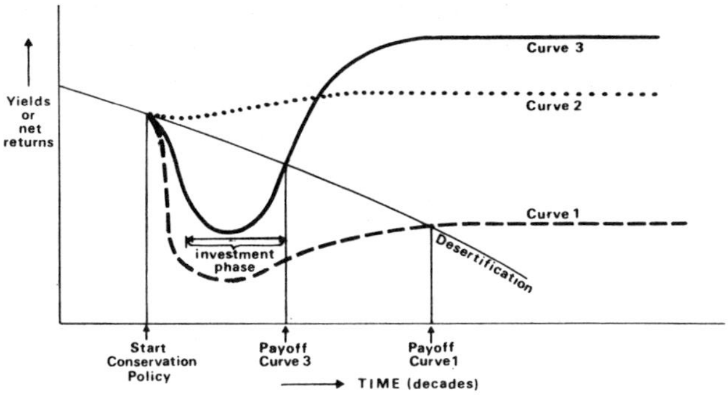

Fig. 17. A simple model of conservation policies applied to desertification in pastoral rangelands (see text for explanation).

242. The second option is possible in less damaged or less vulnerable ecosystems. Rotational grazing, culling of unproductive beasts, redistribution of ownership rights, careful ploughing or water application could, with little investment, halt desertification.

243. The third option applies only to more favoured environments. Here investment might be attracted by the promise of returns from wells, fences, terraces, irrigation, water-spreading, or drains. One critical factor in deciding on such an option will be the length of the initial period in which the loans must be repaid. If the breakeven point does not come within the accounting period of the investors, they will be reluctant to invest.

244. Timing of use is also important at the scale of a year or so. Many pastures or fields need to be rested periodically to allow the sward or soil nutrient levels to recover.

245. **Long-term sustained yield is the only sensible aim of management in dry ecosystems even if considerable fluctuations from the mean are inevitable. Resting and fallowing are important tactics where output is low and unpredictable.**

4. The Distribution of Use

246. Successful exploitation in dry ecosystems integrates production from different environments at both regional and local scales.

247. The produce of dry lands has commonly been traded for that of more humid areas, because the poor environments are not able to provide some of the basic needs of their inhabitants, such as grain. Although the effects of this trade may at times have been

damaging to the dry environments, it has often been the only way in which any production has been maintained.

248. Within the dry ecosystems themselves the complementarity of different environments is vital to the tactics of land use. Richer bottom lands can produce crops or dry season forage to sustain herds and herdsmen who use the sparser pastures for most of the year; some pastures provide salt, others roughage, protein or minerals; one area can be used while another is rested from either grazing or cultivation. Many traditional cultures have accommodated themselves to these patterns, and while many innovations may prove initially effective, they often fail because of the change they require in a spatial strategy. A less flexible system may not only involve less efficient cropping, but it may restrict pressure to a few localities which are then severely desertified.

249. **The economies of dry lands cannot exist on their own but, given fair processes of exchange, can contribute useful produce to wider economic systems. The most effective use of dry lands depends on the flexible use of many complementary facets.**

5. Land Capability

250. Because desertification is most commonly the outcome of an incompatibility between a use and its resource, it can be best countered by matching the two.

251. Capability is usually much less easy to define than incompatibility, for most land is capable of many uses. Hunters, gatherers, pastoralists (nomadic or highly commercial), shifting agriculturalists, plantation managers, co-operative farmers, nature conservationists, and tour-operators will each wish to use the same piece of land in a different way, depending on their social background, skills, technology, markets, capital and politics. Because of this it is impossible to be rigid in prescribing the use of an area. On the other hand, it is usually much easier to demonstrate that a particular use is incompatible with the land it uses. For example, it is not difficult to show that some parts of the Sahel are unsuitable for agriculture because of the erratic nature of their rainfall.

252. **Land capability (or incapability) assessment is a basic tenet of any conservation policy, but such a policy should be both dynamic and flexible.**

B. PASTORALISM

253. The following brief section is another in a series of recommendations about the ecological principles in pastoral land use. The recommendations of the EMASAR group (FAO, 1974) in particular should be seen as complementary to this report.

1. The Integrity of Pastoral Ecosystems

254. The integrity of the ecosystem is a particularly vital consideration in pastoral land use, as can be seen from the descriptions of the impact of grazing and trampling in Chapter IV. The integrity of ecosystems can be maintained in many ways. Here we mention only one salient strategy.

(a) The exploitation of many kinds of primary production

255. Most herdsmen in the dry zone recognize that a single species of stock utilizes

only a small part of available plant resources, and also that, where one species may be damaged in an unpredictable environment, it becomes progressively less likely that two, three or four species will suffer at the same time. This is the thinking behind the risk-spreading use of cattle, sheep, goats and camels by one family. It underlies the theory behind experiments with goats in Australia, which have the additional purpose of testing the ability of goats to keep down scrub and so allow a reasonable growth of grass for sheep: this becomes a strategy for managing the whole ecosystem instead of cropping only a part of it (e.g. Wilson *et al.*, 1975). In some cases new domestications from indigenous wild stock or the co-existence of game-cropping and domesticated herds may be the most effective mode of exploitation (see Anaya, 1976).

256. **Single-species herds are often both less effective at keeping the range in good condition and less productive than a variety of grazers on the same land.**

2. Collaborating with the Ecosystem

257. Although many pastoral societies in recent years either have been given the means of fighting the ecosystem or have been forced to do so, much of their traditional pattern of land use had evolved methods of collaboration by many trials and errors over thousands of years. Two particularly important principles underlie such co-operation.

(a) The maintenance of healthy herds

258. In wild ungulate populations, disease, predators or intra-specific competition ensure that most surviving animals are healthy and productive. Similar controls operate among the herds of traditional pastoralists, although cultural controls are also important. When diseases and pests are stringently controlled by more effective modern methods, however, numbers are controlled neither by the ecosystem nor by the culture of the herders, and as herd sizes increase, available forage per beast declines and the average productivity of the herd deteriorates.

259. The eradication of disease, like the multiplication of watering points, is often given the blame for desertification, but to conclude that disease should be left to endanger herds is a counsel of despair. One healthy individual will yield as much milk or meat as several unhealthy ones and will probably do much less damage to the range. While it is true that it is difficult to prevent pastoralists from taking advantage of a disease eradication programme to increase the size of their herds, some system in which the pastoralist pays the real cost of the programme might be a partial check (FAO/SIDA, 1974). These arguments apply to any programme that seeks to increase herd productivity (such as improved breeding, feed supplements, etc.).

260. **Maximum productivity comes from healthy herds, but healthy herds can, if uncontrolled, severely damage the range. Disease eradication and other herd improvement programmes should never be introduced without due regard to the control of their repercussions.**

(b) Catering for uncertainty

261. Uncertainty is the way of life for everything that lives in a dry environment. The ways in which uncertainty is accommodated by plants, animals and men have been

discussed in Chapter III. In particular, the following principles apply to the management of the different parts of the system:

262. (i) *The maintenance of ecosystem resilience.* The vegetation in any area comprises a community in which each plant species may play an important role. It is important to consider the consequences for the ecosystem as a whole before changing a management practice in any area. In many situations it may be acceptable to destroy a scrub or tree cover to encourage grasses, but the risks inherent in this must be recognized and tested before wide-spread programmes are carried out.

263. (ii) *Cultural alternatives.* The review in Chapter III showed that indigenous human cultures in the dry zone have evolved numerous non-ecological strategies to avoid drought. These include migration, either temporary or permanent, and the selling of other services, such as tourist facilities, transportation or manpower. These strategies allow more efficient exploitation of the dry zone, and as such should not be hampered, unless there is good evidence that they are leading to desertification.

264. **Strategies of catering for risk in plant, animal and human populations should be well understood before any change in land-use is planned.**

3. The Timing of Pastoral Use

265. The general approaches to the achievement of sustained yield that were outlined in section VI.A.3 apply as much to pastoral as to any other form of land use.

(a) A generalized policy

266. Figure 18a shows how a policy of retrenchment might apply. The patterns of sustained yield would be similar, but at different levels, for policies that involve stabilizing or increasing yield. The curve for stock numbers in Fig. 18a shows a simple approach to retrenchment in which destocking would be followed by a programme that would faithfully "track" the carrying capacity of the range. Such a policy might, however, present some problems:

267. (i) Accurate "tracking" would depend heavily on adequate monitoring of range condition, and the conversion of this information into figures for carrying capacity. Methods for doing this are undoubtedly being improved, but it is difficult to conceive of a system that would be totally reliable at predicting the carrying capacity of the highly unreliable and spotty distribution of forage production in dry lands. Flexibility of

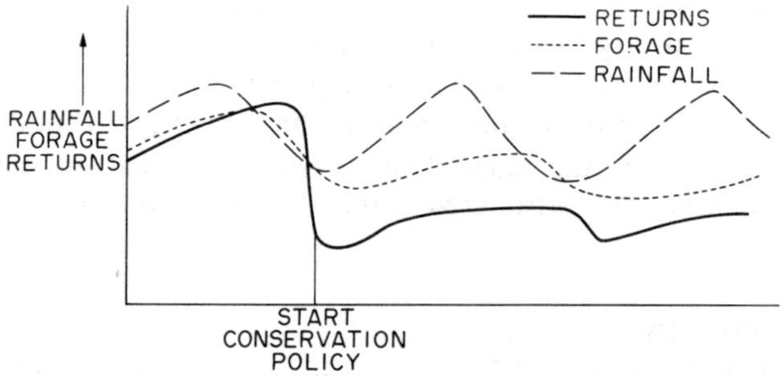

Fig. 18a. A model of the effects of a simple destocking policy.

response and mobility will always, therefore, be necessary exploitive strategies. If there is to be widespread "tracking" without serious "overshooting" of capacity, simple and rapid methods for estimating carrying capacity, usable by the pastoralists themselves, will have to be evolved.

268. (ii) A highly fluctuating yield of stock, especially on a year-to-year basis, would present economic difficulties. Beef "mountains" in the good years might be impossible to store or to transport, let alone to market. This is the situation in central Australia today.

269. (iii) There is still some uncertainty about the long-term effects of such a tracking policy on the ecosystem. Slatyer (1973a) suggests that the low conversion rate of primary to secondary production in ecosystems under the stress of drought is an adaptation to allow the storage of nutrients against bad years, and helps to add stability or at least resilience to the system. If the reserves are grazed out by domestic stock, the system might degrade. At least some of the reserves need to be retained, if only as drought relief.

270. Figure 18b illustrates a policy that might be capable of overcoming these difficulties. Superimposed on a base-line curve representing a minimum "safe" stocking rate are positive and negative deviations: the positive deviations represent the possibility of increased stocking rates, if extra forage is available, if ecological research shows them to be safe, and if the market allows; the negative deviations represent periods in which

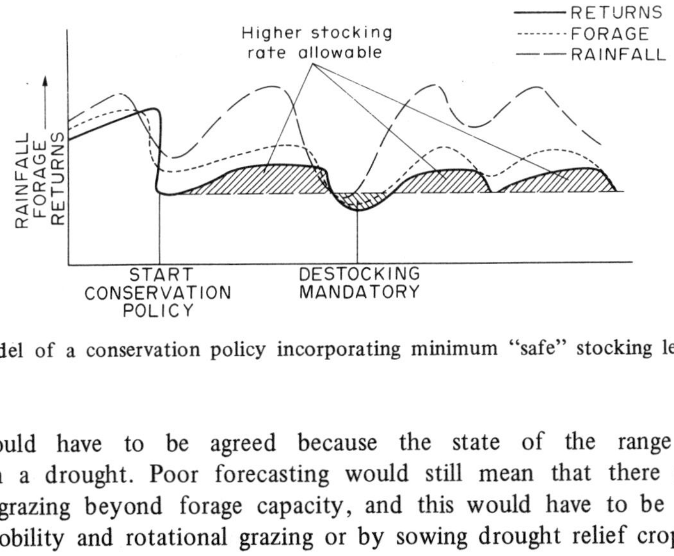

Fig. 18b. A model of a conservation policy incorporating minimum "safe" stocking levels (see text for explanation).

destocking would have to be agreed because the state of the range had seriously deteriorated in a drought. Poor forecasting would still mean that there might be some overshoot of grazing beyond forage capacity, and this would have to be catered for by encouraging mobility and rotational grazing or by sowing drought relief crops.

271. **"Sustained" pastoral production can only occur if there is careful monitoring of the grazing resource. Productive possibilities will remain highly variable, and systems of production must accommodate this. Sustained yield will depend on the ability to move herds rapidly from drought-stricken areas when the range deteriorates below a critical threshold.**

(b) Stabilizing yield at near present levels

272. **Many simple practices could reduce or virtually eliminate desertification, and might even reverse the trend of declining yield.** Perhaps the most important of these

would be the conferring of pastoralists with legal rights to their resources, either as individuals or as co-operatives. Given the security to manage their pastures, many pastoralists already know what they would do to improve their methods. Their practices might include rotational grazing, reducing their herds, less discriminating burning, and the maintenance of their own supplies of dry season or drought-relief forage.

(c) Investment in improved output

273. The methods and techniques for improving output are outlined in a separate component review (Anaya, 1976). Here it is only necessary to list the ecological principles that should underlie these techniques.

274. (i) *Water conservation.* Water is the prime constraint on biological production in dry environments, and dry ecosystems have evolved strategies to conserve and make use of as much of the scarce and erratic supply as they can. A more efficient crop can be expected if water conservation is further encouraged by certain land management techniques. Apart from the maintenance of stocking densities at or below recommended levels, the main techniques for this are: water spreading, contour ploughing, strip seeding of grasses, the maintenance of cover crops, and the installation of small dams.

275. (ii) *Soil conservation.* The soil not only helps to retain water, it is also a store of nutrients. Soil conservation is more vital in dry than in humid environments because of the vital role that the topsoil plays in both imbibing water and in holding nutrients. Investments in soil conservation include strip ploughing, the planting of cover crops of grasses and of wind breaks, the installation of small water and sediment-retaining dams, or of check dams in gullies, and the planting of quick-growing cover on devegetated areas and gullies.

276. (iii) *Afforestation.* It is probably true that the planting of forests is seldom economic in non-irrigated dry areas (Delwaulle, 1973b). Yields are simply too low to justify the expense. On the other hand there are situations when forestry can play a role other than as a provider of timber. In some cases shelter belts may be necessary, in others dunes may have to be prevented from encroaching on agricultural land, and in yet others a supply of nitrogenous forage may be required for fattening cattle or as a drought relief (Depierre *et al.*, 1975). Forestry can also play a role in the rehabilitation of very damaged areas which are enclosed for protection.

277. (iv) *Fencing.* Fencing of pastures is a costly method of ensuring that some areas are allowed to recover after particularly heavy use, that some forage can be kept as a drought reserve, or that rather less favourable vegetation is grazed, and perhaps thereby improved. Traditional pastoralists can often achieve similar ends by carefully herding their stock. Fences become necessary when labour becomes scarce or costly (see also section VI.B.4).

278. (v) *Well-digging.* If very carefully controlled, the multiplication of wells is a very important technique for distributing pressure on the range, and is one of the best investments (see section VI.B.4).

279. **Where investment can be attracted there are many ecologically sound ways in which it can be employed to improve production. These include soil and water conservation measures, the planting of wind breaks, fencing and well-digging.**

4. The Distribution of Pastoral Use

280. Dry pastures are poor, and their production very patchy. Only by allowing the use of different pastures at different times can production be sustained. There are two main strategies which can optimize production in this situation.

(a) Mobility

281. Mobility is important to many wild animals and human cultures for a number of reasons:

(i) In an environment where the level of plant productivity per acre is low, mobility is important to allow the intake of enough food for survival. Unrestricted movement allows the animal to move on when the best forage has been taken, leaving the remaining plants time to recover. Restrictions on movement force the animal to graze forage of poor quality and endanger the resilience of the range.

(ii) In a climate which varies seasonally, but in which there are also less predictable temporal and spatial variations in rainfall, mobility is vital to allow stock to move to where the rain has brought on the forage.

(iii) Different vegetation associations offer complementary types of forage: trees and shrubs offer less bulk but more nitrogen; some pastures offer salt, others a relief from too salty forage. Each niche needs to be available at a particular season or time in the breeding cycle.

282. Mobility is, of course, characteristic of most pastoral enterprises, not only among the nomads of the Old World, but also in the much more commercialized systems in Australia and the United States: cattle are shipped by train or truck from breeding areas to "store" areas and from there to "fattening" areas and can be rapidly moved away from areas affected by drought. Unlike the nomadic system, however, these commercial systems transfer cattle, but not people. "Livestock stratification" which is the policy of using different habitats for different stages in the animal production process has its dangers in that it may lead to the exploitation of one group of pastoralists in the chain by another, but it has clear ecological advantages.

283. Many problems of nomadic pastoralism can be attributed to enforced restrictions on movement. Although it may be difficult to prevent the agricultural encroachment onto pastoral territory to which many of these restrictions can be attributed, especially if it can be shown that agriculture in the new lands is productive and at a low risk, any policy that allows or encourages agricultural expansion should also make provision for the impact that it will have on displaced pastoralists.

284. **Mobility is an essential part of many land-use strategies in the dry zone. Ideas of settled land use imported from more humid zones should not thoughtlessly be imposed on the inhabitants of the deserts and their margins.**

(b) Rational distribution of pressure

285. Some of the more successful modern systems of dryland grazing are based on a multiplication of watering points. This allows the range to be grazed in some form of rotation: pastures can be used when their kind of forage is needed, and can be vacated

when they reach a critical condition. The system depends on the advanced technology needed to drill deep wells and sometimes on quite extensive fencing, but it has been introduced with some measure of success in Mongolia and in parts of Botswana in labour-intensive systems which do not need much fencing.

286. The principle of rational use of grazing lands, and the observation that much of the range was seldom used, underlay the policy of borehole drilling in much of colonial Africa. The outcome of this programme, however, is widely acknowledged to be responsible for much of the desertification that exists today. The FAO–SIDA mission to the Sahel (1974) saw the failure of the drilling programme as a function not of the original principle so much as of the concomitant failure to give the pastoralists themselves any responsibility for the pastures surrounding the new wells: they were not charged the full price of the water, and were not organized to manage the surrounding pastures in a rotation or to treat the network of boreholes as a whole; they moved away from one watering point as it became the centre of a large piosphere, and opened up another area that had been allowed to rest. Such practices already underlie some indigenous systems and should be encouraged. The "Hema" system in Syria is one such system of rotational grazing that has been allowed to lapse. In economics where labour costs are high, fencing can also be used to distribute grazing pressure.

287. **The exploitation of dry ecosystems is most efficient and poses lower risks where a maximum range of forage is utilized. The technical problems behind such exploitation are far less difficult to solve than the social ones.**

5. Matching Uses and Resources

288. The methods for estimating the pastoral resource and its carrying capacity for stock have been reviewed in Chapter V. Only the implications of these methods for policy will be stated here.

(a) Carrying capacity is difficult to estimate. A recent exercise in comparing estimates of carrying capacity made by four internationally acknowledged experts for an area in Ethiopia yielded figures which differed by more than ten times (Sandford, pers. comm.). Whereas detailed quantitative methods may be evolved in the near future for measuring the value of one paddock, the estimation of the carrying capacity for the whole of a range and, more important, estimating it for different cultural demands will probably be much more difficult.

(b) Given the responsibility for their own land, pastoralists may well be the best people to estimate its carrying capacity. Although their methods may be proved inadequate in some instances, their knowledge of the environment and its ecology give them a better basis for its management than is available to many outsiders, however well trained.

(c) Carrying capacity is an emotive concept. Any policy which demands that pastoralists reduce their herds will inevitably meet resistance unless its assumptions and methodology are clearly stated, and unless it is not seen to be too rigid.

289. **It is vital that carrying capacity figures be estimated and adhered to for all dry rangeland. If a centralized policy is necessary, it should be as rational as possible.**

C. RAINFED AGRICULTURE

290. Ecological principles apply less to agriculture than to pastoralism because of the technology that is employed to keep cultivated fields at an extremely simple and low successional stage (see Chapter III). The natural checks and balances of the ecosystem are replaced by cultivation, harvesting and pest-control. Some ecological principles, nonetheless, do have a bearing on the problems of rainfed agriculture; the most important of them concern the conservation of the soil and moisture reserves.

1. The Integrity of Those Parts of the Ecosystem that Remain

291. The shape of a slope is the result of a balance between the nature of the weathered mantle that underlies it, the plants that protect it and the intensity of the rainfall it has to withstand. If the natural vegetation is removed by cultivation, a new balance will be necessary, this being achieved by erosion of the weathered mantle. Careful cropping, by leaving the surface bare for a minimum period, can alleviate the erosional problem, but may not remove it. A widely applied technique is to terrace the slope, the intention being to reduce the angle and the soil loss in some places, and to trap the moving sediment in others. But, while terracing is very often an excellent technique, it is important to acknowledge that the slopes it creates are seldom adjusted to accommodate the effects of intense rainstorms, and these may therefore do greater damage to the terraces than to the original slope (e.g. Moldenauer and Wischmeier, 1960). Only if these destructive events are catered for with carefully constructed storm channels, and if damage is quickly made good, can the farming system hope to remain viable. Dumsday (1971) has shown that such labour-intensive practices mean that narrow-based terraces cannot produce a profit even over extended periods in modern capital-intensive Australian agriculture.

292. Although erosion above the "background" natural rate is undesirable, two factors should be borne in mind. First, background rates are already high in semi-arid areas, and if any attempt is made to halt all erosion, then the supply of nutrients to crops could be curtailed, and unstable depths of soil may accumulate and be in danger of initiating landslides. Second, if erosion rates are speeded up, weathering rates may also increase, and a new dynamic equilibrium may be achieved that may be no threat to production.

293. **The slope-soil-erosion system should be conserved as a dynamic process. If it is modified, as when slope angles are altered, greater investment may be necessary to maintain equilibrium.**

2. Working with the Ecosystem

294. Shifting agriculture should be seen as a useful, low cost method of maintaining soil fertility where investments are not available for fertilizers. Shifting agriculture needs room to manoeuvre, and cannot operate effectively when there are high human population densities. **If threatened by increasing population or encroachment by other land uses, then the system would have to be rescued with investments in more advanced agricultural technology for all the aspects of its operation.**

295. Temperate dryland farming makes use of a fallow to accumulate moisture, but in so doing increases the dangers of erosion. **The introduction of cover crops, notable forage**

crops which can fix nitrogen such as the medic species, are an intelligent use of locally occurring species.

3. The Timing of Agricultural Use

296. Figure 19 presents some alternative policies that might alleviate the long-term problems of desertification in individual small areas of rainfed agriculture. The lowest curve shows the effects of retrenchment. Withdrawal of land from agriculture should seriously be contemplated where land capability and agroclimatological surveys show that farming is too great a risk to be tolerated either by the farmer or by the community (e.g. Bernus, 1975). Surveys of the land and of the climate are able to quantify the risks that may be incurred, but do not dictate the acceptable level of risk. This is a matter for planners and individuals on the ground. The individual can, of course, only contemplate the abandonment of his fields if he is offered alternative land or employment. A policy of withdrawing land from cultivation would have to be a gradual and sensitive process.

297. The lower of the two central curves shows what could happen on low-risk land if new techniques were to be introduced. The pattern can be exemplified by the effects of

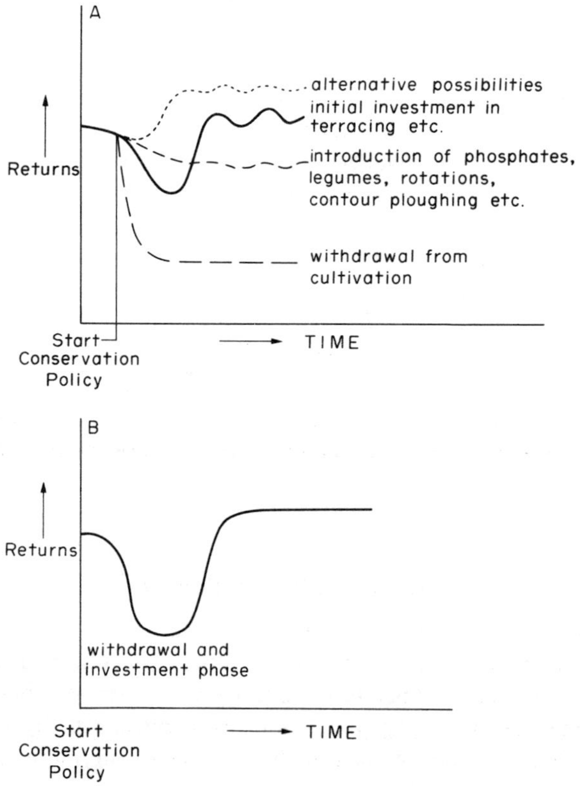

Fig. 19. The effects of conservation policies on patterns of dryland agriculture. A. Patterns for individual small areas. B. Pattern for a region extending from the desert to the wet margins of the semi-arid area.

introductory phosphatic fertilizers and leguminous crops to the wheat lands of South and West Australia. Some experimental farms using these techniques already exist in the Middle East and North Africa. The salvation of dryland agriculture in West Africa is seen by many to lie in the introduction of similar kinds of rotation, but with rather different break crops (Jones and Wild, 1975). These schemes may encounter marketing problems, but as a purely ecological strategy they have many advantages: the elimination of the bare fallow; the diversification of the economic base to include both a cereal and a wood or meat crop; and the low-cost fixation of nitrogen.

298. The upper of the two central curves illustrates the pattern of returns that would follow an investment in structures such as terraces, grassed waterways, contour ploughing, gully check dams, fences, etc. Although some of these investments may be large, most are very small. Delwaulle (1973b) has confirmed the findings of other workers in many parts of the World in his studies of sediment-conserving effects of cultivation techniques in West Africa: very simple and low cost modifications to current practices can reduce the loss of soil dramatically.

299. **The long-term strategy for agriculture in an area affected by desertification will depend on land capability. In poor land, retrenchment may be the only alternative, if repeated famines are not to recur. On better land there are many techniques that, with care, can be used to improve output.**

4. The Distribution of Agricultural Use

300. Few dryland fields can yield continuously. Fallowing is practised in most agricultural systems to conserve either nutrients or moisture. The area needed for a full cycle of activity is therefore much greater than the cropped area in any one season. **Unless investment is available to ensure supplies of fertilizer, or to protect the land against erosion, then such spatial strategies need to be protected.**

5. Land Capability

301. Land capability analysis is very important to any agricultural planning. The methodology has been developed most fully for agricultural land (e.g. Klingebiel and Montgomery, 1962; Moorman, 1974). **It is important that land systems and facets that are particularly vulnerable to erosion should not be cultivated. Agriculturalists have a less flexible strategy of land use than pastoralists so that their livelihood may have to be more carefully protected and planned.**

D. IRRIGATED AGRICULTURE

302. In a highly capitalized system such as irrigated agriculture where the ecosystem is vastly altered and bolstered by technology, ecological principles have little real meaning. This section merely extends some of the ideas of conservation outlined in section VI.A to the long-term problems of irrigation.

303. A possible conservation strategy for irrigated agriculture is shown in Fig. 20. The pattern of desertification followed in many irrigation schemes is shown by the solid-line curve. A somewhat similar, though not as exaggerated pattern has been experienced in at least parts of the Indus valley irrigation schemes in Pakistan (Karpov, 1964; Dorfman *et al.*, 1965; Michel, 1969). It is usually true that only when the problems of waterlogging

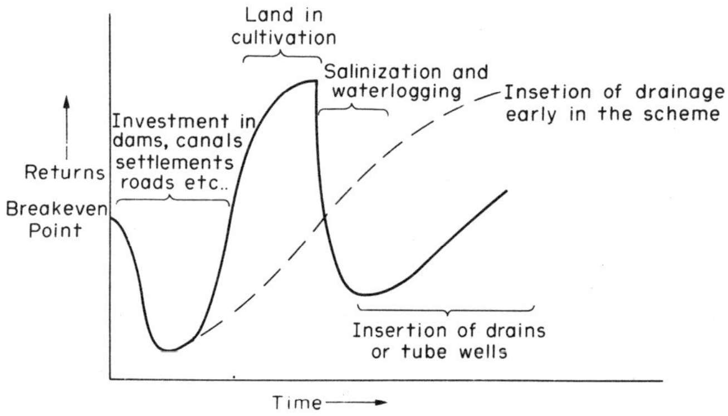

Fig. 20. The effects of conservation policy on irrigation agriculture.

and salinity suddenly appear on a wide scale and when production is seriously threatened, are programmes of drainage begun. The pious hope expressed by very many engineers that drainage works should be initiated at an early stage are illustrated in the broken-line curve. Once the canals are installed, the digging of drains or tube-wells should immediately be started, in anticipation of problems that will inevitably occur. The payoff would come quite quickly.

304. The detailed planning of drainage is a much more complex problem (see discussion following Dorfman *et al.*, 1965). In unconsolidated sediments drains may have to be so wide that a large proportion of valuable land may be consumed. Drains also present design problems if the slope of the surface is very slight; they become health hazards, are costly to clean of weeds, and they add to the costs of road building. Tube-wells, which are the alternative, themselves offer a number of options: should the wells be deep and widely spaced, or shallow and close together? Deep, widely spaced wells are expensive to construct, may have to bring up saline water from great depth in order to lower surface water-tables, and may require a complex administration to maintain; the loss of one deep well, say by the breakdown of a pump, may endanger a large area. Shallow wells can be built and maintained more cheaply by local labour and technicians and so can be capitalized by local farmers; they will usually raise sweet water which can be re-used for irrigation, but they may be difficult to control and so lead to a certain anarchy of pumping draw-down patterns. They will not be so useful in the reclamation of large areas from which population and local capital have already withdrawn.

305. **The prime principle in the management of irrigated agriculture is that water should be kept moving down through the soil to carry away the inevitable accumulation of salts. There are many alternative ways in which this can be achieved, each local situation demanding a different drainage strategy.**

E. OTHER LAND USES

1. Wood Collection

306. There are three strategies which conserve fuel resources:

(a) The development of alternative fuel resources. These include solar stoves, "gobar" or dung-gas generation, and the retailing of cheap kerosene or gas. The use of fossil fuels has already begun in some Arab countries which have rich oil and gas reserves, and it will be interesting to observe if the spread of their use will be followed by a recolonization of the desert with shrubs. "Gobar" gas is being pursued as an alternative fuel by the Indian Government and is said to have a great future and many advantages. Low-cost installations are being evolved. Solar stoves are being investigated by the UNEP Desertification Secretariat itself.
(b) The planting of fast-growing timber in wet sites, or even on irrigated sites. In some areas the evapotranspiration of water through the trees might be an adjunct to a policy for avoiding waterlogging and salinity. Planned timber plantations have been a part of the large irrigation schemes in Pakistan ever since their inception. Such plantations supply not only fuel but also good-quality construction timber. Their introduction is advocated for parts of the Sahel by Depierre *et al.* (1975).
(c) The protection from further pastoral or agricultural use of sites which have been badly eroded, and sites which provide windbreaks or amenity areas. Although the protected timber may not grow to a high quality, it should provide a sustained supply of firewood, if carefully managed (Depierre *et al.*, 1975). Crops of gum arabic and other forest products may be an additional benefit.

2. Urbanization

307. Two principal ecological policies apply to the management of urban peripheries:

(a) The planting of forests or the protection of reserves of native vegetation for amenity and to prevent dust blowing. In some cases grass can be sown and tended for the same purpose.
(b) The removal of waste waters to re-use schemes, to groundwater-recharge areas, or by remixing with trunk streams.

3. Mining and Industry

308. Similar principles apply to areas in which there is mining or industry:

(a) The avoidance of unnecessary ground disturbance by vehicles.
(b) The planting of areas bared by mining and on waste dumps. If these cannot be planted then alternatives such as the use of aerodynamically rough surfaces can prevent dust blowing (Marshall, pers. comm.).
(c) The disposal of noxious wastes, either by burial or by dilution.

4. Tourism

309. Tourists are attracted to dry areas by space, scenery and animals.

(a) Space

310. It is unlikely that dryland exploitation, however successful, will fill the deserts with people other than tourists, but the "perceptual carrying capacity" of dry areas is probably much lower than that of most other tourist areas except the open ocean, and **tourist authorities may need to restrict the use of deserts to a certain maximum number of people over any one period, to maintain their feeling of freedom.**

(b) Scenery

311. At a small scale the appearance of an area can be very much affected by tourists themselves, and tourists can in this respect also become a hazard to other land users. Roads and trails can become badly damaged by heavy vehicular or pedestrian use. **In particular popular sites' visitor routeways may need to be paved, and elsewhere traffic, especially by vehicle, may have to be restricted to only a few routes which can be monitored and serviced regularly.** Dryland sites vary in their physical capacity to sustain tourist traffic. Bare sand-dunes are virtually indestructible, but finer soils, particularly on steeper slopes, are very vulnerable. A land-capability approach may have to be adopted in heavily used areas. Litter can be a hazard to wild and domestic stock and is unsightly. **Tourist authorities need to have a litter policy.**

(c) Animals

312. If the tourist wants to see wildlife in the arid zone, game parks must be set aside for two main reasons. Wild animals sometimes become a problem to other land users: elephants damage trees, many wild herbivores harbour disease, and birds damage crops. It is not always the case that all wild herbivores compete with or are any real problem to domestic stock production. It has been shown in Australia, for example, that the red kangaroo eats quite different forage to sheep in good seasons. But the problems that wild animals can present, or are thought to present, to pastoralists or agriculturists means that there is usually a demand that they be confined to reserves. This is additionally necessary to protect the game from indiscriminate hunting.

313. Game reserves in dry climates have many attractions for the tourist, but present some problems (e.g. Jewell, 1969; Abel, 1976). The protection of wild animals from predators such as man can mean that the animal population can grow alarmingly, and there may be serious over-grazing. The protected game may then become a hazard to the pastoral or agricultural lands around. **The problems of managing both game and tourists themselves mean that a successful game park needs a considerable investment in labour, road, fencing and other services.**

314. Game parks have a second, more important function: they are useful to ecological research, for a relatively undisturbed ecosystem presents an easier case to study than one that is interfered with. Much more research is needed to elucidate the mechanisms of resilience and productivity in natural dry ecosystems. This kind of research is a final link in the chain that leads from the herdsman, through the extension officer to the applied ecologist and the modeller.

REFERENCES

Abel, N. O. J. (1976) Wildlife conservation problems in a semi-arid land, FAO Background Paper FO: AFC/WL/76/5.1, Feb. 1976, African For. Comm., Working Party on Wildlife Management National Parks, 5th session, 10 pp.

Abercrombie, F. D. (1974) Range development and management in Africa, A.I.D. and Office of Development Services, Bureau for Africa, 59 pp.

Achi, K. (1969) Salinization and water problems in the Algerian North East Sahara, in *The Careless Technology*, M. T. Farver and J. P. Milton eds., J. P. Stacey, London, pp. 276–287.

Acocks, J. P. H. (1975) Veld types of South Africa, *Mem. Bot. Surv. S. Africa,* Bot. Res. Inst., Dept. Agr. Techn. Serv., South Africa, no. 40, 128 pp.

Adam, J. G. (1967) Évolution de la végétation dans les sousparcelles protégées de l'UNESCO-IFAN à Atar, *Bull. I.F.A.N.* Sér. A (Dakar), **29**: 1, 92–106.

Allan, W. (1965) *The African Husbandman*, Oliver & Boyd, Edinburgh.

Allison, L. E. (1964) Salinity in relation to irrigation, *Adv. in Agronomy*, **16**: 139–180.

Alpay, D. N. (1974) Range management and animal husbandry practices in Afghanistan of demonstration and training in forest and range improvement project, FAO Project Working Doc., FO: SF/AFG/67/5.5, Rome, 67 pp.

Anaya-Garduno, M. (1976) *Technology and Desertification*, U.N.E.P.

Anway, J. C., Brittain, E. G., Hunt, H. W., Innis, G. S., Parton, W. J., Rosell, C. F. and Sauer, R. H. (1972) ELM: Version 1.0, *US/Intern. Biol. Progr. Grassland Biome'Tech. Rep.*, **156**: 1–285.

Arnold, G. W. (1960) The effect of the quantity and quality pasture available to sheep on their grazing behaviour, *Aust. J. Agric. Res.*, **11**: 1034–1043.

Aubréville, A. (1973) Rapport de la mission forestière anglo-française Nigeria-Niger (déc. 1936–fèv. 1937), *Bois et Forêts des Tropiques*, **148**: 3–26.

Ayyad, M. A. (1975) Systems analysis of Mediterranean desert ecosystems of Northern Egypt, Report No. 1: 1974 progress, Univ. Alexandria.

Ayyad, M. A. (1976) Systems analysis of Mediterranean desert ecosystems of Northern Egypt, Report No. 2: 1975 progress, Univ. Alexandria.

Baker, J. J. K. (1973) A background to the study of drought in East Africa, in *Drought in Africa*, D. Dalby and R. J. Harrison-Church eds., S.O.A.S., London, pp. 46–52.

Barker, S. and Lange, R. T. (1969a) Effects of moderate sheep stocking on plant populations of a Black/Oak Bluebush association, *Aust. J. Bot.*, **17**: 527–537.

Barker, S. and Lange, R. T. (1969b) Population ecology of *Atriplex* under sheep stocking, in *The Biology of Atriplex*, R. Jones ed., CSIRO, pp. 105–120.

Barth, F. (1964) *Nomads of South Persia*, Allen & Unwin, London.

Bartlett, H. (1956) Fire, primitive agriculture and grazing in the tropics, in *Man's Role in Changing the Face of the Earth*, W. L. Thomas, Jr. ed., pp. 692–720.

Barucha, F. R. and Shankanarayan, K. A. (1958) Effects of overgrazing on the grasslands of the western Ghats, India, *Ecology*, **39**: 152–153.

Beaumont, P. (1974) Land-use impacts and desertification in the Middle East, I.G.U. Working Group on Desertification in and around Arid Lands, Alice Springs, Field Conf., pp. 19–20.

Beaumont, P. and Atkinson, K. (1969) Soil erosion and conservation in northern Jordan, *Jl. Soil Water Conserv.*, **24**: 144–147.

Bentley, J. R. and Talbot, M. W. (1948) Annual plant vegetation of the California foothills as related to range management, *Ecology*, **29**: 72–79.

Benyon, P. R. (1976) The use of semi-arid lands for grazing with the accompanying risks of deterioration, erosion and desertification, SCOPE Project 5, Australian Contrib. to Internat. Workshop Indianapolis, Mar. 22–27, 1976, MS, 10 pp.

Bernus, E. (1975) Human geography in the Sahelian zone, in *The Sahel: Ecological Approaches to Land Use*, MAB Technical Notes UNESCO, Paris, pp. 67–74.

Bharadwaj, O. P. (1961) The arid zone of India and Pakistan, in *A History of Land-use in Arid Regions*, Arid Zone Research XVII, UNESCO, pp. 143–174.

Bille, J. C. (1974) Recherches écologiques sur une savane sahélienne du Ferlo septentrional, Sénégal: 1972, Année sèche au Sahel, *Terre et Vie,* **28**: 5–20.

Blair Rains, A. and McKay, A. D. (1968) The northern state lands, Botswana, Min. of Overseas Devel., Land Resources Division, Surbiton, 125 pp.

Blydenstein, J., Hungerford, C. R., Day, G. I. and Humphrey, R. R. (1959) Effect of domestic livestock exclusion on vegetation in the Sonoran Desert, *Ecology*, **38**: 522–526.

Boeck, E. (1974) Problèmes hydrogéologiques, in E.E.C. conférence élevage Sahel, Bruxelles, 13/14 juin, pp. 23–26.

Bonté, P. *et al.* (1976) (in press): *L'Occupation humain des écosystèmes pâtures tropicaux*, UNESCO, Paris.

Boserup, E. (1965) *The Conditions of Agricultural Growth: the Economics of Agrarian Change under Population Pressure*, Aldine, Chicago.

Boudet, G. (1972) Désertification de l'Afrique tropicale sèche, *Adansonia*, Sér. 2, **12**(4): 505–524.

Boudet, G. (1974) Rapport sur la situation pastorale dans les pays du Sahel, mimeo FAO/UNEP/IEMVT; réunion techn. sur l'aménagement écol. des zones arides et semi-arides d'Afrique et du Moyen Orient, 44 pp., FAO, Rome.

Boudet, G. (1975a) Pastures and livestock in the Sahel, in *The Sahel: Ecological Approaches to Land Use*, MAB Technical Notes UNESCO, Paris, pp. 29–34.

Boudet, G. (1975b) *Manuel sur les pâturages tropicaux et les cultures fouragères*, IEMVT, Maisons-Alfort, Paris, 254 pp.

Boudet, G. (1976) Les pâturages sahéliens – les dangers de dégradation et les possibilités de regénération – principes de gestion ameliorée des parcours sahéliens, FAO/IEMVT, Maisons-Alfort, Paris, 58 pp.

Boudet, G. and Coulibaly, M. (1975) Pâturages du Gourma et Seno Mango, in *Étude de l'évolution d'un système d'exploitation sahélien au Mali, Rapport de Campagne, 1975*, 13–61, Rep. Mali, Min. Devel., 112 pp.

Boudet, G. and Duverger, E. (1961) *Etude des pâturages naturels sahéliens, Le Hodh. Mauritanie*, Paris, Vigot, 160 pp.

Boudet, G., Lamarque, G., Lebrun, J. P. and Rivière, R. (1969) Pâturages naturels du Dallol Mauri (République du Niger), *Etude Agrostologique*, No. 26. IEMVT, Maisons-Alfort, France.

Bourlière, F. and Hadley, M. (1970) The ecology of tropical savannas, *Ann. Rev. Ecol. & Systematics*, **1**: 125–152.

Braun, H. (1973) Shifting cultivation in Africa (evaluation of questionnaires), in *Shifting Cultivation and Soil Conservation in Africa*, FAO, Rome, *Soils Bulletin* **24**: 21–36.

Bridges, K. (1971) Questions, Version 1, *US/Intern. Biol. Progr. Desert Biome Modelling Report Series*, **31**(11): 1–3.

Bromfield, A. R. (1974) The deposition of sulphur in dust in northern Nigeria, *J. agric. Sci. Camb.*, **83**: 423–425.

Brown, G. W., ed. (1968) *Desert Biology*, 2 vols., Academic Press, New York.

Brown, L. H. and Cochemé, J. (1969) *A Study of the Agroclimatology of the Highlands of Eastern Africa*, FAO, Frome, 330 pp.

Brown, T. C., O'Connell, P. F. and Hibbert, A. R. (1974) Chaparral conversion potential in Arizona, Part II: An economic analysis, USDA Forest Service, Res. Pap. RM-127, 28 pp.

Buckhouse, J. C. and Coltharp, G. B. (1968) Effects of simulated grazing on soil moisture content in mid-elevation reseeded rangeland, *Proc. Utah Acad. Sci., Arts, Letters*, **45**: 211–219.

Buffington, L. C. and Herbel, C. H. (1965) Vegetational changes on a semi desert grassland range from 1858 to 1963, *Ecol. Monogr.*, **35**: 139–164.

Burrows, W. H. (MS) *Vegetation Management Decisions in Queensland's Semi-arid Sheeplands*, 27 pp.

Butzer, K. W. (1974) Accelerated soil erosion: a problem of man–land relationships, in *Perspectives on Environment*, I. R. Manners and M. W. Mikesell eds., Amer. Assoc. of Geographers, Washington, D.C., pp. 57–78.

Capot-Rey, R. (1953) *Le Sahara Français*, Presses Univ. de France, Paris, 564 pp.

Chakela, Q. K. (1974) Studies of soil erosion and reservoir sedimentation in Lesotho, Dept. Phys. Geogr., Uppsala, Sweden, UNGI Rept. No. 34, pp. 479–495.

Chakela, Q. K. (1975) Erosion and sedimentation in some selected catchment areas in Lesotho: 1974/75, Internal Dept. Phys. Geog., Uppsala Univ., 16 pp.

Chakravarty, A. K. (1971) Grasslands of the Indian arid zone, 21st Int. Geogr. Congr., I.G.U., Symposium on Arid Zone, Calcutta, 63–68.

Chalk, A. T. (1963) Soil conservation in northern Nigeria and a suggested programme, USAID Rep. C-34.

Charley, J. L. and Cowling, S. L. (1968) Changes in soil nutrient status resulting from overgrazing and their consequences in plant communities of semi-arid areas, *Proc. ecol. Soc. Aust.* **3**: 28–38.

Charreau, C. (1972) Problèmes posés par l'utilization agricole des sols tropicaux par des cultures annuelles, *Agron. trop. sér. agron. gén. études tech.* no. 9.

Chepil, W. S. (1959) Wind erodibility of farm fields, *J. Soil Water Conserv.*, **14**: 214–219.

Chudleigh, P. D. (1971) *Pastoral Management in the West Darling Region of New South Wales*, Ph.D. thesis, Univ. N.S.W.

Clawson, M. (1963) Critical review of man's history in arid regions, in *Aridity and Man*, C. Hodges and P. C. Duisberg eds., A.A.A.S. Publn. No. 74, pp. 429–459.

Clos-Arceduc, M. (1956) Études sur photographes aériennes d'une formation végétale sahélienne: la brousse tigrée, *Bul. I.F.A.N.*, Sér. A, **18**: 677–684.

Cochemé, J. and Franquin, P. (1968) Étude agroclimatique dans une zone semi-aride en Afrique au sud du Sahara, WMO Tech. Note 86, Geneva.

Cointepas, J. P. (1965) Irrigation à l'eau salée et drainage en Tunisie, *Série Pédologie, Cahiers ORSTOM*, vol. III, fasc. 4, 299–305.

Condon, R. W. (1961) Soils and landform of the western division of New South Wales, *Soil Conserv. Jl.*, Jan. 1961, 1–16.

Condon, R. W. (1968) Estimation of grazing capacity on arid grazing lands, in *Land Evaluation*, G. A. Stewart ed., Macmillan, pp. 112–124.

Condon, R. W., Newman, J. G. and Cunningham, G. M. (1969) Soil erosion and pasture degeneration in central Australia, *Jl. of the Soil Conservation Service of New South Wales*, **25**: 47–92, 161–182, 225–250, 295–321.

Cooke, R. U. and Reeves, R. W. (1976) *Arroyos in the American South West*, O.U.P., London, 256 pp.

Cooke, R. U. and Warren, A. (1973) *Geomorphology in Deserts*, Batsford, London, 394 pp.

COWAR, (1976) *Arid Lands Irrigation in Developing Countries*, Acad. of Sci. Research & Technol. Cairo Symp. Proc., 386 pp.

Crider, F. J. (1955) Root-growth stoppage resulting from defoliation of grass, U.S. Dept. Agric. Tech. Bull. No. 1102, 1–23.

Crisp, M. D. (1975) *Long-term Change in Arid Zone Vegetation at Koonamore, South Australia*, Ph.D. thesis, Univ. Adelaide, June 1975.

Crowcroft, P. (1969) The sheep and the saltbush: the utilization of Australia's arid lands, in *The Careless Technology*, M. T. Farver and J. P. Milton eds., J. P. Stacey, London, pp. 742–752.

Curry-Lindahl, K. (1972) *Conservation for Survival. An Ecological Strategy*, London, Gollancz Ltd., 335 pp.

Dagg, M. and Pratt, M. A. C. (1962) Relation of stormflow to incident rainfall, *E. Afr. Agr. For. Jl.*, **27**: 31–35.

Daubenmire, R. (1968) Ecology of fire in grassland, *Adv. Ecol. Res.*, **5**: 209–266.

Delwaulle, J. C. (1973a) Désertification de l'Afrique au sud du Sahara, *Bois Forêts Tropiques (Paris)*, **149**: 3–20.

Delwaulle, J. C. (1973b) Résultats de six ans d'observations sur l'érosion au Niger, *Bois Forêts Tropiques*, **150**: 15–36.

Depierre, D. and Gillet, H. (1971) Désertification de la zone sahélienne du Tchad, *Bois Forêts Tropiques (Paris)*, **139**: 3–25.

Depierre, D., Gillet, H., Catinot, R. and Delwaulle, J. C. (1975) The role of the forester in land use planning in the Sahel, in MAB Technical Notes. *The Sahel: Ecological Approaches to Land Use*, UNESCO, Paris, pp. 41–53.

Despois, J. (1961) Development of land use in northern Africa (with references to Spain), in *A History of Land Use in Arid Regions*, L. D. Stamp ed., UNESCO, Paris, pp. 219–238.

Despois, J. and Raynal, R. (1967) *Géographie de l'Afrique du Nord-Ouest*, Payot (Paris), 563 pp.

De Vos, A. (1975) Africa, the devastated continent? *Monogr. Biol.* 26, 236 pp.

Dorfman, R., Revelle, R. and Thomas, H. (1965) Waterlogging and salinity in the Indus plain: some basic considerations, *Pakistan Devel. Rev.* **5**(3): 331–370.

Dumsday, R. G. (1971) Evaluation of soil conservation policies by systems analysis, in *Systems Analysis in Agricultural Management*, J. B. Dent and J. R. Anderson eds., John Wiley, Sydney, pp. 152–172.

Dunford, E. G. and Weitzman, S. (1955) Managing forests to control soil erosion, in *Water: The Yearbook of Agriculture*, U.S.D.A., pp. 235–242.

Dyer, R. A. (1955) Structural and physiological features of the vegetation of arid and semi-arid areas of the Union of South Africa, in *Plant Ecology*, Arid Zone Res. V, UNESCO, pp. 20–24.

Dyksterhuis, E. J. (1949) Condition and management of range land based on quantitative ecology, *J. Range Management*, **2**: 104–115.

Dyson-Hudson, N. (in press) The structure of East African herds and the future of East African herders (MS).

East Africa Royal Commission (1955) Report 1953–1955.

Eckholm, E. P. (1975) Desertification: A world problem, *Ambio* (Royal Swedish Acad. Sci.), **I**: 137–145.

El Gabaly, M. (1976) Problems and effects of irrigation in the Near East region, in *Arid Lands Irrigation in Developing Countries*, COWAR Symp., Acad. of Sci. Res. & Technol., Cairo, pp. 237–252.

Ellison, L. (1960) Influence of grazing on plant succession in rangelands, *Bot. Rev.*, **26**: 1–78.

Ellison, L., Croft, A. R. and Bailey, R. W. (1951) Indicators of condition and trend on high range-watersheds of the intermountain region, U.S. Dept. Agric., Handbook No. 19, 66 pp.

Emberger, L. and Lemée, G. (1962) Plant ecology, in *The Problems of the Arid Zone*, Arid Zone Res. XVIII, UNESCO.

Emmett, W. M. (1970) The hydraulics of overland flow on hill-slopes, U.S. Geol. Surv. Prof. Paper 662A, 68 pp.

Evenari, M., Shanan, L. and Tadmor, N. (1971) *The Negev, the Challenge of a Desert*, Harvard U.P., Cambridge, 345 pp.

FAO (1974) The ecological management of arid and semi-arid rangelands in Africa and the Near East: an international programme, Rept. of an expert consultation EMASAR, FAO Rept. AGPC:MISC/26, Rome, 52 pp.

FAO/SIDA (1974) Report on the Sahelian zone, FAO/SWE/TF117, FAO, Rome.

Fisher, I. H. (1974) *Resource Optimization in Arid Grazing Systems*, Ph.D. thesis, Univ. N.S.W., 377 pp.

Fitzgerald, K. (1968) The Ord River catchment regeneration project. Dealing with the problem, *Jl. Agric. W. Austral.*, March 1968.

Flannery, K. V. (1965) The ecology of early food production in Mesopotamia, *Science*, **147**: 1247–1256.

Flinders, J. J. and Hansen, R. M. (1975) Spring population responses of cottontails and jackrabbits to cattle grazing shortgrass prairie, *Jl. Range Management*, **28**: 290–293.

Flint, R. F. and Bond, G. (1968) Pleistocene sand ridges and pans in Western Rhodesia, *Geol. Soc. Amer. Bull.*, 79: 299–314.

Flohn, H. and Ketata, M. (1971) Investigations on the climatic conditions of the advancement of the Tunisian Sahara, WMO Tech. Note 116, 42 pp.

Floret, C. (1971) Recherches phytoécologiques entreprises par le CNRS sur le biôme 'zone aride' en Tunisie, 26 p. Roneo, CNRS, CEPE Doc. no. 57, Montpellier.

Floret, C. and Le Floc'h, E. (1972) Desertisation et ressources pastorales dans la Tunisie présaharienne, *Journées d'études sur la lutte contre la désertization*, 25–27 dec, 1972, Gabès; Min. Agric. units, 12 pp. Roneo.

Floret, C. and Le Floc'h, E. (1973) Production, sensibilité et évolution de la végétation et du milieu en Tunisie présaharienne, CNRS, CEPE, Montpellier, No. 71, 45 pp.

Forsling, C. L. (1931) A study of the influence of herbaceous plant cover on surface run-off and soil erosion in relation to grazing on the Wasatch Plateau in Utah, *U.S. Dept. Agric. Tech. Bull.*, **220**: 1–71.

Fournier, F. (1962) Carte du danger d'érosion en Afrique au sud du Sahara, CEE–CCTA.

Fournier, F. (1967) Research on soil erosion and soil conservation in Africa, *Sols Africains*, **12**: 53–96.

Fraser-Darling, F. (1960) Wildlife husbandry in Africa, *Sci. Amer.*, **203**: no. 5, 123–134.

Gallais, J. (1967) *Le Delta intérieur du Niger*, IFAN, Dakkar, 2 vols.

Garcia, M. (1974) La structure du troupeau bovin sahélien au Niger et en Haute-Volta après la sècheresse, *Revue trimestr. d'inf. techn. écon.*, CEBV, no. 8–9, pp. 4–13.

Garcia-Moya, E. and McKell, C. M. (1970) Contribution of shrubs to the nitrogen economy of a desert-wash plant community, *Ecology*, **51**: 81–88.

Gerasimov, I. P. (1968) Basic problems of the transformation of nature in central Asia, *Soviet Geography*, **9**: 444–458.

Gill, E. D. (1973) Geology and geomorphology of the Murray River, *Mem. Nat. Museum, Victoria*, **34**: 1–97.

Gillet, H. (1967) Essai d'évaluation de la biomasse végétale en zone sahélienne (végétation annuelle), *J. Agr. Trop. Bot. appl.*, **14**: 123–158.

Gillet, H. (1975) Plant cover and pastures in the Sahel, in MAB Technical Notes. *The Sahel: Ecological Approaches to Land Use,* UNESCO, Paris, pp. 21–27.

Gischler, C. E. (1975) Desert encroachment control, Rept. on visit to Khartoum, March 1975, Regional Office, Sci. Tech. Arab States, Cairo, Annex A, part I, 25 pp.

Glendening, G. (1952) Some quantitative data on the increase of mesquite and cactus on a desert grassland range in southern Arizona, *Ecology*, **33**: 319–328.

Goodall, D. W. (1970) Use of computer in the grazing management of semi-arid lands, *Proc. XIth Intern. Grassl. Congr.*, pp. 917–922. St. Lucia, Univ. Queensland Press.

Goodall, D. W. (1971) Building and testing ecosystem models, British Ecol. Soc. Symp. No. 12, pp. 173–194.

Goodall, D. W. (1974) Ecosystem modelling in the Desert Biome, in *Systems Analysis and Simulation in Ecology*, vol. III, B. Patten ed., Academic Press, New York and London.

Gould, P. R. (1965) Wheat on Kilimanjaro: the perception of choice within game and learning model frameworks, *General Systems*, pp. 157–166.

Graetz, R. D. (1973) Biological characteristics of Australian Acacia and Chenopodiaceous shrublands relevant to their pastoral use, in *Arid Shrublands*, D. N. Hyder ed., Soc. Range. Management, Denver, Colo., pp. 33–39.

Granier, P. (1975) *Rapport du service d'agrostologie 1974–75. Note sur l'introduction des techniques d'amélioration de la productivité de l'élévage en zone Sahélienne. Niger.* IEMVT, Maisons-Alfort, Paris, 37 pp.

Grove, A. T. (1973) Desertification in the African environment, in *Drought in Africa*, D. Dalby and R. J. Harrison-Churon eds., School for Oriental and African Studies, London, pp. 33–45.

Hall, E. A. A., Specht, R. L. and Eardley, C. M. (1964) Regeneration of the vegetation on Koonamore Vegetation Reserve, 1926–1962, *Aust. J. Bot.*, **12**: 205–264.

Halwagy, R. (1962a) The incidence of the biotic factor in northern Sudan, *Oikos*, **13**: 97–117.

Halwagy, R. (1962b) The impact of man on semi-desert vegetation in the Sudan, *Jl. Ecol.*, **50**: 263–273.

Hansen, R. M. and Reid, L. D. (1975) Diet overlap of deer, elk, and cattle in southern Colorado, *Jl. Range Management*, **28**: 43–47.

Hare, F. K. (1976) *Climate and Desertification*: UNEP Report, 2nd Draft.

Harris, D. R. (1969) Agricultural systems, ecosystems and the origins of agriculture, in *The Domestication and Exploitation of Plants and Animals*, P. J. Ucko and G. N. Dimbleby eds., Duckworth, London, pp. 3–15.

Haskell, P. T. (1969) Locust control: ecological problems and international pests, in *The Careless Technology*, M. T. Farver and J. P. Milton eds., J. P. Stacey, London, pp. 499–526.

Hastings, J. R. (1959) Vegetation change and arroyo cutting in south-east Arizona, *Arizona Acad. Sci. Journal*, **1**: 60–67.

Hastings, J. R. and Turner, R. M. (1965) *The Changing Mile: an Ecological Study of Vegetation Change with Time in the Lower Mile of an Arid and Semi-Arid Region*, Tucson, Univ. Arizona, 317 pp.

Hay, J. (1969) Salt Cedar and salinity on the Upper Rio Grande, in *The Careless Technology*, M. T. Farver and J. P. Milton eds., J. P. Stacey, London, pp. 288–300.

Heady, H. F. (1960a) Range management in the semi-arid tropics of East Africa according to principles developed in temperate climates, *Proc. 8th Int. Grassld. Congr.*, Reading, pp. 223–226.

Heady, H. F. (1960b) *Range Management in East Africa*, Govnt. Printer, Nairobi, 125 pp.

Heady, H. F. (1969) Ecological consequences of Bedouin settlement in Saudi Arabia, in *The Careless Technology*, M. T. Farver and J. P. Milton eds., J. P. Stacey, London, pp. 683–693.

Heady, H. F. (1973) Structure and function of climax, in *Arid Shrublands*, D. N. Hyder ed., Soc. Range Management, Denver, Colo., pp. 73–80.

Heady, H. F. (1975) *Rangeland Management*, McGraw-Hill, N.Y., 460 pp.

Heathcote, R. L. (1975) *Australia*, Longmans, London, 246 pp.

Hefley, H. M. and Sidwell, R. (1945) Geological and ecological observations of some High Plains dunes, *Amer. Jl. Sci.*, **243**: 361–376.

Held, R. B. and Clawson, M. (1965) *Soil Conservation in Perspective,* Johns Hopkins, Univ. Press for Resources for the Future Inc., Baltimore, 344 pp.

Hibbert, A. R., Davis, E. A. and Scholl, D. G. (1974) Chaparral conversion potential in Arizona. Part I: water yield response and effects on other resources, USDA Forest Service Res. Paper RM.126, 36 pp.

Holling, C. S. (1973) Resilience and stability of ecological systems, *Ann. Rev. Ecol. Syst.*, **4**: 1–24.

Holmgren, R. C. (1973) The Desert Experimental Range: description, history and program, in *Arid Shrublands*, D. N. Hyder ed., Soc. for Range Management, Denver, Colo., pp. 18–23.

Hooper, P. T., Sallaway, M. M., Latz, P. K., Maconochie, J. R., Hyde, K. W. and Corbett, L. K. (1973) Ayers Rock–Mt. Olga National Park Environmental Study, 1972. Arid Zone Res. Inst., Dept. Northern Terr., Alice Springs. Land Conserv. Series No. 2, 52 pp.

Hopkins, B. (MS) Fire in Tropical Savannas.

Hotes, F. L. and Pearson, E. A. (1976) Effects of irrigation on water quality in *Arid Lands Irrigation in Developing Countries* (COWAR Symp.), Acad. of Sci. Res & Technol., Cairo, pp. 102–143.

Hugot, P. (1965) *Le Tchad*, Nouvelles Edits. Latines, Paris, 155 pp.

Humphrey, R. R. and Lister, P. B. (1941) Native vegetation as a criterion for determining correct range management and run-off characteristics of grazing lands, *Jl. Forestry*, **9**: 837–842.

Humphrey, R. R. and Mehrhoff, L. A. (1958) Vegetation changes on a southern Arizona grassland range, *Ecology*, **39**: 720–726.

Hunting Technical Services Ltd. (1961) *Ghulam Mohammed Barrage Command Vol. 1, Physical, Soil and Agricultural Investigations*, West Pakistan Water & Power Development Authority, Lahore, 159 pp.

Hunting Technical Services Ltd. (1974) Development Plan, Southern Darfur Land-use Planning Survey, Min. of Agr. Food & Nat. Res., Sudan.

Hurault, J. (1966) Étude photo-aérienne de la tendance à la remobilisation des sables éoliens sur la rive nord du Lac Tchad, *Rev. Inst. Français de Pétrole*, **21**: 1837–1846.

Hutchinson, K. J. and King, K. L. (1970) Sheep numbers and soil arthropods, *Search*, **1**: 41–42.

Hyde, K. W. (1975) *Management of Pastoral Lands*, A Report on Churchill Fellowship Studies in South Africa and the U.S.A., Jan–June, 1975.

Ingebo, P. A. and Hibbert, A. R. (1974) Run-off and erosion after brush suppression on the natural drainage watersheds in central Arizona, USDA Forest Service Research Note RM.275, 7 pp.

Jacobsen, T. and Adams, R. M. (1958) Salt and silt in ancient Mesopotamian agriculture, *Science*, **128**: 1251–1258.

Jacques-Félix, H. (1956) Écologie des herbages en Afrique intertropicale, *Agronomie Trop.*, **11**: 217–233.

Jenik, J. and Hall, J. B. (1966) The ecological effects of the Harmattan wind in the Djebobo Massif (Togo Mountains, Ghana), *Jl. Ecol.*, **54**: 767–779.

Jewell, P. A. (1974) Problems of wildlife conservation and tourist development in East Africa, *J. Sth. Afr. Wildl. Mgmt. Assoc.*, **4**: 59–62.

Johnson, D. L. (1973) Jabal Al-Akhdar, Cyrenaica: an historical geography of settlement and livelihood, Univ. of Chicago, Dept. of Geog., Research Paper, 148, 240 pp.

Johnson, D. L. (1975) The status of pastoral nomadism in the Sahelian zone, in *The Sahel: Ecological Approaches to Land Use*, MAB Technical Notes, UNESCO, Paris, pp. 75–88.

Johnson, W. M. (1956) The effect of grazing intensity on plant composition, vigor, and growth of pine-bunch grass ranges in central Colorado, *Ecology*, **37**: 790–798.

Jones, M. J. and Wild, A. (1975) *Soils of the West African Savanna*, Tech. Communic. No. 55, Commonwealth Bur. of Soils, Harpenden, England, 246 pp.

Karpov, A. V. (1964) Indus Valley – West Pakistan's lifeline – discussions, *Proc. Amer. Soc. Civ. Engineers J. Hydrol. Div.* 90 HY1, pp. 371–390.

Kassas, M. and Girgis, W. A. (1965) The units of a desert ecosystem, *Jl. Ecol.*, **53**: 715–728.

Kates, R. W., Johnson, D. L. and Johnson Haring, K. (1976) *Demographic, Social and Behavioural Review*, U.N.E.P., 83 pp.

Klingebiel, A. A. and Montgomery, P. H. (1962) *Land Capability Classification*, Agri. Handbook No. 210, Washington D.C. USDA Soil Cons. Service, 1962.

Kovda, V. A. (1961) Land use development in the arid regions of the Russian plain the Caucasus and Central Asia, in *A History of Land Use in Arid Regions*, L. D. Stamp ed., UNESCO, Paris, pp. 175–218.

Lacey, J. A. and Sallaway, M. M. (1975) Management policies: Ayers Rock–Mt. Olga. Some aspects of the formulation of a planning policy for the management of the Ayers Rock–Mt. Olga National Heritage Area Symposium Managing Terrestrial Ecosystems, Brisbane, in *Proc. Ecol. Soc. Austr.*, 9, 256–266.

Lamprey, H. (1976) Survey of desertification in Kordofan Province, Sudan, UNEP, Nairobi.

Land Resources Division (1972) *Land Resources of North East Nigeria,* Land Resources Division, Directorate of Overseas Surveys, Tolworth, U.K. *Vol. 3, Land Systems*, by M. G. Bauden, D. M. Carroll and P. Tuley, 466 pp; *Vol. 4, Present and Potential Land Use*, by P. N. de Leeuw, A. Lesslie and P. Tuley, 106 pp.

Langbein, W. B. and Schumm, S. A. (1958) Yield of sediment in relation to mean annual precipitation, *Trans. Amer. Geophys. Union*, **39**: 1076–1084.

Langdale-Brown, I., Osmaston, H. A. and Wilson, J. G. (1964) *The Vegetation of Uganda and Its Bearing on Land Use*, Govnt. Printer, Entebbe, Uganda.

Lange, R. T. (1969) The piosphere: sheep track and dung patterns, *Jl. Range Management*, **22**: 396–400.

Lattimore, O. (1951) *Inner Frontiers of China*, Amer. Geog. Soc. Res. Series 21, New York, 585 pp.

Lee, R. B. and DeVore, I. (1968) *Man the Hunter*, Aldine, Chicago.

Leeuw, P. N. de (1966) The role of savanna in nomadic pastoralism: some observations from western Bornu, Nigeria, *Neth. J. agric. Sci.*, **13**: 178–189.

Le Houérou, H. N. (1958) Écologie, phytosociologie et productivité de l'olivier en Tunise méridionale, *Bull. de la Carte Phytogéographie (CNRS), Paris*, Sér. B, **4**: 7–22.

Le Houérou, H. N. (1959) Recherches écologiques et floristiques sur la végétation de la Tunisie méridionale, *Mém. Hors de série de l'Inst. Rech. Sahar. Alger.*

Le Houérou, H. N. (1970) North Africa: past, present, future, in *Arid Lands in Transition*, H. E. Dregne ed., Amer. Assoc. Adv. Sci., Washington, D.C., pp. 227–277.

Le Houérou, H. N. (1974) Deterioration of the ecological equilibrium in the arid zones of North Africa, Spec. Publ. 39, Agron. Res. organ. Volvani Centre Bet. Dagan, pp. 54–57.

Le Houérou, H. N. (1975) The nature and causes of desertification: *Proc. IGU Cambridge Symposium on Desertification*, pp. 9–27.

Leigh, J. H. (1974) Diet selection and the effects of grazing on the composition and structure of arid and semi-arid vegetation, in *Studies of the Australian Arid Zone II. Animal Production*, A. D. Wilson ed., CSIRO, pp. 102–126.

Leigh, J. H. and Noble, J. C. (1972) *Riverine Plain of New South Wales, Its Pastoral and Irrigation Development*, CSIRO Div. of Plant Industry, Canberra, 63 pp.

Lendon, C. (MS) *The Development of Range Condition Assessment Methodology for Central Australian Rangelands*, 17 pp.

Lendon, C. and Lamacraft, R. R. (MS) *Standards for Testing and Assessing Range Condition in Central Australia*, 20 pp.

Lericollais, A. (1970) La définition d'un terrain Sob, en pays Serer, *Études rurales*, **37–39**: 113–128.

Long, G. (1974) Bases écologiques de l'aménagement du territoire et de la gestion des ressources naturelles biologiques dans les zones arides de l'Afrique du Nord, *Proc. Symp. Rech. Ecol. Relatives au Développement des Zones Arides (desserts médit.) à Précip. Hivernales*, Israel–France, Spec. pub. 39, pp. 17–33.

Long, G. A. and Perry, R. A. (1975) *Iran: Basic Principles and Main Conclusions related to Research and Development for Arid and Semi-arid Iran*, Draft summary report, Unesco, Paris, 10 pp.

Low, B. S., Birk, E. M., Lendon, C. and Low, W. A. (1973) Community utilization by cattle and kangaroos in mulga near Alice Springs, N.T., *Tropical Grasslands*, **7**: 149–156.

Lusby, G. C. (1965) Causes of variations in run-off and sediment yield from small drainage basins in W. Colorado, US Dept. Agr., Agric. Res. Service, Misc. Pub. no. 970, pp. 94–98.

Lyford, F. P. and Qashu, H. K. (1969) Infiltration rates as affected by desert vegetation, *Water Resour. Res.*, **5**: 1373–1376.

Lynch, J. J. (1974) Behaviour of sheep and cattle in the more arid areas of Australia, in *Studies of the Australian Arid Zone II. Animal Production*, A. D. Wilson ed., CSIRO, pp. 37–49.

MacArthur, R. M. and Connell, J. H. (1966) *The Biology of Populations*, Wiley, New York.

McGinnies, W. G., McComb, A. L. and Fletcher, J. E. (1963) Of watersheds and forests in the arid west, in *Aridity and Man*, C. Hodge and P. C. Duisberg eds., AAAS Publn, no. 74, pp. 277–307.

McKell, C. M. and Goodin, J. R. (1973) United States shrublands in perspective, in *Arid Shrublands*, D. N. Hyder ed., Soc. for Range Management, Denver, Colo., pp. 12–18.

Mabbutt, J. A. (1968) Review of concepts of land classification: in *Land Evaluation*, G. A. Stewart ed., pp. 11–28.

Mabbutt, J. A. (1969) Landforms of arid Australia, in *Arid Lands of Australia*, R. O. Slatyer and R. A. Perry eds., Austral. Nat. Univ. Press, Canberra, pp. 11–32.

Mabbutt, J. A. (1974) Desert types of Australia and their inherent vulnerability, *Proc. IGU Working Group on Desertification in and around Arid Lands*, Field Conf., Alice Springs, pp. 5–6.

Mabbutt, J. A. (1976) Physiographic setting as an indication of inherent resistance to desertification, in *Problems in the Development and Conservation of Desert and Semi-desert Lands*, 23rd Int. Geogr. Congr., Univ. N.S.W., Working Group on Desertification in and around Arid Lands, Ashkabad, Pre-Congr. Symposium, K26, pp. 189–197.

Main, A. R., Shield, D. J. W. and Waring, H. (1959) Recent studies in marsupial ecology, *Monog. Biology*, **8**: 315–333.

Mann, H. S. (1975) Measuring desert encroachment in the Rajasthan desert region, *Proc. IGU Cambridge Symposium on Desertification*, p. 39.

Marlowe, B. J. (1958) A survey of marsupials of New South Wales, *CSIRO Wildlife Res.*, **3**: 71–114.

Marshall, J. K. (1970) Assessing the protective role of shrub-dominated rangeland vegetation against soil erosion by wind, *Proc. XIth Intern. Grassld. Congr.*, pp. 19–23.

Marshall, J. K. (1971) Drag measurements in roughness arrays of varying density and distribution, *Agric. Meteor.*, **8**: 269–292.

Marshall, J. K. (1972) Principles of soil erosion and its prevention, in *The Use of Trees and Shrubs in the Dry Country of Australia*, Forestry & Timber Bureau, Canberra, pp. 90–107.

Marshall, J. K. (1973) Drought, land use and soil erosion, in *The Environmental, Economic and Social Significance of Drought*, J. V. Lovett ed., Angus and Robertson, pp. 55–77.

Mensching, H. and Ibrahim, F. N. (1973) Problems of desertification in the northern Saharan boundary zone, the steppe region of the Maghrib and the Sahelian zone, *Proc. IGU Cambridge Symposium on Desertification*, pp. 65–67.

Michel, A. A. (1969) The impact of modern irrigation technology in the Indus and Helmand basins of south-west Asia, in *The Careless Technology*, M. T. Farver and J. P. Milton eds., Stacey, London, pp. 257–275.

Mohammed, G. (1965) Rejoinder to paper by Dorfman *et al.*: Waterlogging and salinity in the Indus Plain, some basic considerations, *Pakistan Devel. Rev.* **5**: 393–407.

Moldenhauer, W. C. and Wischmeier, W. H. (1960) Soil and water losses and infiltration rates on Ida silt-loam as influenced by cropping systems, tillage practices and rainfall characteristics, *Proc. Soil Sci. Soc. Amer.* **24**: 409–413.

Monod, Th. and Toupet, Ch. (1961) Land use in the Sahara–Sahel region, *Arid Zone Res.* **17**: 239–253.

Moore, C. W. E. (1973a) Effect of exploitation on the vegetation of north-western New South Wales, *Proc. Arid Zone Confer.*, Broken Hill, pp. 6–38 to 6–39.

Moore, C. W. E. (1973b) Regeneration of productive range in north-western New South Wales, *Proc. Arid Zone Confer.*, Broken Hill, pp. 6–39 to 6–41.

Moore, R. M. (1962) Effects of the sheep industry on Australian vegetation, in *The Simple Fleece*, A. Barnard ed., Melbourne, pp. 170–180.

Moore, R. M. and Biddiscombe, E. F. (1964) The effects of grazing on grasslands, in *Grasses and Grasslands*, C. Barnard ed., Macmillan, London, pp. 221–235.

Moorman, F. R. (1974) Classification of land for its use capability and conservation requirements, *FAO Soils Bull.* **24**: 230–234.

Morel, R. and Quantin, P. (1964) *Agron. trop. (Paris)*, **19**: 105–136.

Mortimore, M. J. and Ologe, O. (1975) A study of systems stability in a Sudan savanna community in Nigeria, *Proc. IGU Cambridge Symposium on Desertification,* pp. 68–75.

Moss, R. P. (1969) The ecological background to land-use studies in tropical Africa, with special reference to the West, in *Environment and Land Use in Africa*, M. F. Thomas and G. W. Whittington eds., Methuen, pp. 193–238.

Mott, G. O. (1960) Grazing pressure and measurement of pasture production, *Proc. 8th Int. Grassland Congr.*, Reading, pp. 606–611.

Mouttapa (1973) Soil aspects in the practice of shifting cultivation in Africa and the need for a common approach to soil and land resource evaluation, in *Shifting Cultivation and Soil Conservation in Africa*, FAO Soils.

Muller, C. H. (1953) The association of desert annuals with shrubs, *Amer. Jl. Bot.*, **40**: 53–60.

Myers, K. (1962) A survey of myxomatosis and rabbit infestation trends in the eastern Rivernina, New South Wales, 1951–1960, *CSIRO Wildlife Res.*, **7**: 1–12.

Myers, K. (1970) The rabbit in Australia, *Proc. Adv. Study Inst. Dynamics Numbers Popul.* (Oosterbeek), pp. 478–506.

Myers, K. and Parker, B. S. (1975) A study of the biology of the wild rabbit in climatically different regions in eastern Australia. VI. Changes in numbers and distribution related to climate and land systems in semi-arid north-western New South Wales, *Aust. Wildl. Res.*, **2**: 11–32.

Myers, K. and Poole, W. E. (1963) A study of the biology of the wild rabbit, *Oryctolagus cuniculus* (L.), in confined populations. IV. The effects of rabbit grazing on sown pastures, *J. Ecol.*, **51**: 435–451.

National Academy of Sciences (1974) *More Water for Arid Lands. Promising Technologies and Research Opportunities*, Washington D.C., 153 pp.

Newman, J. C. and Condon, R. W. (1969) Land use and present condition, in *Arid Lands of Australia*, R. O. Slatyer and R. A. Perry eds., ANU Press, Canberra, pp. 105–129.

Newsome, A. E. (1964) Anoestrus in the red kangaroo, *Megaleia rufa*, in central Australia, *Aust. J. Zool.*, **12**: 9–17.

Nir, D. and Klein, M. (1974) Gully erosion induced by changes in land use in a semi-arid terrain (Nahal Shiqma, Israel), *Zeit. Geomorph.*, Supplementband 21, Jerusalem–Elat Symposium on Geomorphic Processes in Arid Environments, pp. 191–201.

Noble, I. R. (1975) *Computer Simulations of Sheep Grazing in the Arid Zone,* Ph.D. thesis, Univ. Adelaide.

Novikoff, G. *et al.* eds. (1973) Tunisian Presaharan Project, Progress Rept. no. 1 (1972–73), US/IBP Desert Biome, Utah State Univ., Logan.

Novikoff, G. (1974) The desertisation of range lands and cereal cultivate lands in Presaharan Tunisia. A statement on some possible methods of control. Secretariat for Internat. Ecology, Sweden, Rept. No. 3.

Noy-Meir, I. (1973a) Desert ecosystems: environment and producers, *Ann. Rev. Ecol. & Systematics*, **4**: 25–51.

Noy-Meir, I. (1973b) Desert ecosystems: higher trophic levels, *Ann. Rev. Ecol. & Systematics*, **4**: 195–214.

Noy-Meir, I. (1975) Stability of grazing systems: an application of predator–prey graphs, *J. Ecol.*, **63**: 459–481.

Noy-Meir, I. and Harpaz, Y. (1976) Nitrogen cycling in annual pastures and crops in a semi-arid region, *Proc. Symp. on Cycling of Mineral Nutrients in Agric. Systems*, Amsterdam, Elsevier.

Nye, P. H. and Greenland, D. J. (1960) The soil under shifting cultivation, Comm. Bur. Soils Tech. Comm. No. 51.

Ogden, P. R. (1973) Biological and environmental constraints for livestock production on arid shrublands, in *Arid Shrublands*, D. N. Hyder ed., Soc. for Range Management, Denver, Colo., pp. 89–91.

Osborn, T. G. B., Wood, J. G. and Paltridge, T. B. (1932) On the growth and reaction to grazing of the perennial salt bush *Atriplex vesicarium*, an ecological study of the biotic factor, *Proc. Linn. Soc. NSW*, **57**: 377–402.

Otterman, J. (1974) Baring high-albedo soils by overgrazing: a hypothesised desertification mechanism, *Science*, **86**: 531–533.

Packer, P. E. (1951) An approach to watershed protection criteria, *Jl. Forestry*, **49**(9): 639–644.

Packer, P. E. (1953) Effects of trampling disturbance on watershed condition, run-off and erosion, *Jl. Forestry*, **51**: 28–31.

Panahi, Z. (1976) *Modelling as a Technique for Optimizing Transhumance in the Pastoral Industry of Iran*, M.Sc. Thesis, Utah State Univ., pp. vi + 93 + Appendices.

Parker, K. W. (1954) Application of ecology in the determination of range condition and trend, *Jl. Range Management*, **7**: 14–23.

Payne, A. L., Kubicki, A. and Wilcox, D. G. (1974) *Range Condition Guides for the West Kimberley Area, Western Australia*, W. Austr. Dept. Agric., 141 pp.

Pearse, C. K. (1970) Range deterioration in the Middle East, *Proc. XIth Int. Grassld. Congr.*, Surfers Paradise, pp. 26–30.

Pearson, H. A. (1975) Herbage disappearance and grazing capacity determinations of Southern Pine Bluestem Range, *Jl. Range Management*, **28**: 71–73.

Pearson, L. C. (1965) Primary production in grazed and ungrazed desert communities of eastern Idaho, *Ecology*, **46**: 278–286.

Peck, A. J. (1975) Effects of land use on salt distribution in the soil, in *Ecological Studies. Analysis and Synthesis, vol. 15. Plants in Saline Environments*, A. Poljakoff-Mayber and J. Gale eds., Springer-Verlag, Berlin, pp. 77–90.

Pels, S. and Stannard, M. E. (1976) Environmental changes due to irrigation development in semi-arid parts of New South Wales, Australia, in *Arid Lands Irrigation in Developing Countries*, COWAR Symp. Acad. of Sci. Res. and Technol., Cairo, pp. 144–166.

Pereira, H. C. (1961) Land-use hydrology in Africa, *Proc. Inter-Afr. Conf. on Hydrol.*, Nairobi, CCTA, pp. 45–54.

Pereira, H. C. *et al.* (1962) Grazing control in semi-arid ranchland, *E. Afr. Agr. For. Jl.*, **27**: 42–63.

Perrin, R. M. S. and Mitchell, C. W. (1969) An appraisal of physiographic units for predicting site conditions in arid areas, Military Engineering Experimental Establishment, Christchurch, Hampshire, U.K., MEXE Rept. No. 111, 2 vols.

Perry, R. A. (1967) Integrated surveys of pastoral areas, Paper given to 2nd Internat. Seminar on Integr. Surveys of Natural Grazing Areas, 17th–22nd April, Delft, The Netherlands, ITC–UNESCO Centre for Integrated Studies, pp. 5–20.

Perry, R. A. (1968) Australia's arid rangelands, *Ann. Arid Zone*, **7**: 243–249.

Perry, R. A. (MS) Range condition and trend techniques, CSIRO, 17 pp.

Peyre de Fabrègues, B. (1971) Rapport sur l'évolution des pâturages sahéliens au sud Tunisie, Étud. Agrost. No. 32, IEMVT, Maisons-Alfort, France.

Poupon, H. and Bille, J. C. (1974) Recherches écologiques sur une savane sahélienne du Ferlo septentrional, Sénégal: Influence de la sècheresse de l'année 1972–1973 sur la strate ligneuse, *Terre et Vie*, **28**: 49–75.

Pratt, D. J. (1969) Management of arid rangeland in Kenya, *J. Brit. Grassld. Soc.*, **24**: 151–157.

Quezel, P. (1965) *La végétation du Sahara, du Tchad a la Mauritanie*, Stuttgart; Paris, 1 vol., 333 pp.

Quinn, J. A. and Hervey, D. R. (1970) Trampling losses and travel by cattle on Sandhills Range, *Jl. Range Management*, **23** 50–55.

Rapp, A. (1974) A review of desertization in Africa – water, vegetation and man, Secretariat for Internat. Ecology, Sweden Rep. 1, 77 pp.

Rapp, A., Rust, D. H. M., Christiansson, C. and Berry, L. (1972) Soil erosion and sedimentation in four catchments near Dodoma, Tanzania, *Geografiska Annaler*, **54A**: 255–318.

Rauzi, F., Fly, C. L. and Dyksterhuis, E. J. (1968) Water intake on mid-continental rangelands as influenced by soil plant cover, US Dept. Agr. Tech. Bull. 1390, 58 pp.

Reynolds, H. G. and Bohning, J. W. (1956) Effects of burning on a desert grass–shrub range in southern Arizona, *Ecology*, **37**: 769–777.

Richards, L. A. (1954) Diagnosis and improvement of saline and alkali soils, US Dept. Agric. Handbk. No. 60.

Rijks, D. A. (1971) Water use by irrigated cotton in the Sudan; (3) Bowen ratio and advective energy, *Jl. appl. Ecol.*, **8**: 643–663.

Ross, M. A. (1973) GAME – A model of growth and soil moisture extraction of *Eragrostis eriopoda*, contribution to Water in Rangelands Symposium held at Alice Springs, August 1973 15 pp. 15 pp.

Rowe, P. B. and Reimann, L. F. (1961) Water use by brush, grass-forb vegetation, *Jl. Forestry*, **59**: 175–181.

Sampson, A. W. (1944) Effect of chaparral burning on soil erosion and on soil–moisture relations, *Ecology*, **25**: 171–191.

Sandford, S. (1976) *Pastoralism under Pressure*, MS.

Sayer, J. A. (1976) Wildlife conservation in the Sudano–Sahelian zone, FAO Background Paper FO:AFC/WL:76/5.2 African Forestry Comm. Working Party on Wildlife Management and National Parks, 5th session, 5 pp.

Schmidt-Nielson, K. (1964) *Desert Animals. Physiological Problems of Heat and Water*, Oxford U.P., London, 277 pp.

Schumm, S. A. (1969) A geomorphic approach to erosion control in semi-arid regions, *Trans. Amer. Soc. Agric. Engineers*, **12**: 60–68.

SEDES (1975) Recueil statistique de la production animale, Répub. Française, Min. Coop., 1201 pp.

SEDES (1976) Éléments de statistique pour une analyse du développement socio-économique dans les six pays du sahel, 266 pp.

Seifert, H. S. H. (1975) Animal production and health in the Sahelian zone, in *The Sahel: Ecological Approaches to Land Use*, MAB Technical Notes, UNESCO, Paris, pp. 55–60.

Shiflet, T. N. (1973) Range sites and soils in the United States, in *Arid Shrublands*, D. N. Hyder ed., Soc. for Range Management, Denver, Colo., pp. 26–33.

Shreve, F. (1929) Changes in desert vegetation, *Ecology*, **10**: 4, 364–373.

Simms, D. H. (1970) *The Soil Conservation Service*, Pall Mall, London, 238 pp.

Slatyer, R. O. (1962) Climate of the Alice Springs area, in *General Reports on Lands of the Alice Springs Area, Northern Territory, 1956–1957*, CSIRO Aust. Land Res. Series No. 6, pp. 109–128.

Slatyer, R. O. (1973a) Structure and function of Australian arid shrublands, in *Arid Shrublands*, D. N. Hyder ed., Soc. for Range Management, Denver, Colo., pp. 66–73.

Slatyer, R. O., ed. (1973b) Plant response to climatic factors, *Proc. Uppsala Symposium*, UNESCO, Paris.

Smith, C. C. (1940) The effect of overgrazing and erosion upon the biota of the mixed-grass prairie of Oklahoma, *Ecology*, **21** 381–397.

Smith, D. D. and Wischmeier, W. H. (1962) Rainfall erosion, *Adv. in Agron.*, **14**: 109–148.

Stamey, W. L. and Smith, R. M. (1964) A conservation definition of erosion tolerance, *Soil Sci.*, **97**: 183–186.

Stanhill, G. (1965) The concept of potential evapotranspiration in arid zone agriculture, *Proc. Montpellier Symp., Method Plant Eco-Phys.*, UNESCO, Paris, pp. 109–116.

Stebbing, E. P. (1953) *The Creeping Desert in the Sudan and Elsewhere in Africa.* Khartoum, Sudan Gov.

Stewart, G., Cottam, W. P. and Hutchings, S. S. (1940) Influence of unrestricted grazing on northern salt desert plant associations in western Utah, *J. Agr. Res.*, **60**: 289–316.

Stoddart, L. A. and Smith, A. D. (1943) *Range Management*, McGraw-Hill, 547 pp.

Swift, J. (1973) Disaster and a Sahelian nomad economy, in *Drought in Africa*, D. Dalby and R. J. Harrison-Church eds., SOAS, London, pp. 71–78.

Swift, J. (1974) The "causes" of the Sahel disaster, Seminar on Points of Controversy in the Study of Drought in Africa, Sept. 30, 1974, Int. African Inst., London. Mimeograph.

Swift, J. (1975) Two years B.P., the political ecology of drought and desertisation in Africa, Cambr. Mtg. Desertification. IGU, 22–26 Sept. 1975, pp. 85–88.

Tadmor, N. H., Evenari, M. and Shanan, L. (1972) Primary production of pasture plants as function of water use, in *Eco-physiological Foundation of Ecosystem Productivity in Arid Zone*, Publishing House "Nauka", Leningrad, pp. 151–158.

Tadmor, N. H., Noy-Meir I., Safreil, U., Seligman, N., Goldman, A., Katznelson, I., Eyal, E. and Benjamin, R. (1974) A simulation model of a semi-arid Mediterranean grazing system. Typescript, 88 pp.

Talbot, L. M. (1969) Ecological consequences of rangeland development in Masailand, East Africa, in *The Careless Technology*, M. T. Farrer and J. P. Milton eds., J. P. Stacey, London, pp. 694–711.

Task Committee on Preparation of Sedimentation Manual (1970) Chapter IV. Sediment Sources and Sediment Yields, *J. Hydrol. Div. Proc. Amer. Soc. Civ. Eng.*, **96**: 1283–1329.

Temple, R. S. and Thomas, M. (1973) La sècheresse au Sahel – une catastrophe pour le bétail, *Rev. Mond. Zoot.*, FAO, (8), 1–7.

Trlica, M. J. and Cook, C. W. (1971) Defoliation effects on the carbohydrate reserves of desert range species, *Jl. Range Management,* **24**: 418–425.

Trombie, J. M., Renard, K. G. and Thatcher, A. P. (1974) Infiltration for three rangeland soil–vegetation complexes, *Jl. Range Management,* **27**: 318–321.

Tueller, P. R. (1973) Secondary succession, disclimax and range condition standards in desert shrub vegetation, in *Arid Shrublands*, D. N. Hyder ed., Soc. for Range Management, Denver, Colo., pp. 57–66.

Tueller, P. R., and Blackburn, W. H. (1974) Condition and trend of the Big Sagebrush/Needle and thread habitat type in Nevada, *Jl. Range Management*, **27**: 36–40.

U.S. Department of Agriculture (1951) *Soil Survey Manual*, US Dept. Agric. Handbook No. 18, 503 pp.

Uvarov, B. P. (1962) Development of arid lands and its ecological effects on their insect fauna, in *The Problems of the Arid Zone*. Arid Zone Research XVIII, UNESCO, Paris, pp. 235–248.

Vaché-Grandet, G. (1959) L'erg du Trarza, notes de géomorphologie dunaire, *Inst. de Rech. Sahariennes, Travaux*, **18**: 161–172.

Van Keulen, H. (1975) *Simulation of Water Use and Herbage Growth in Arid Regions*, Simulation Monographs, Pudoc, Wageningen.

Verstappen, H. Th. (1968) On the origin of longitudinal (seif) dunes, *Zeit. für Geom. N.F.*, **12**: 200–220.

Vink, A. P. A. (1975) *Land Use in Advancing Agriculture*, Springer-Verlag, New York, 394 pp.

Vita-Finzi, C. (1969) *The Mediterranean Valleys: Geological Changes in Historical Time*, Cambridge U.P., Cambridge, 140 pp.

Vuillaume, G. (1969) Analyse quantitative du role du milieu physico-climatique sur le ruissellement et l'érosion à l'issue de bassins de quelques hectares en zone sahélienne (Bassin de Kountkouzout, Niger), *Série Hydrologie, Cahiers ORSTOM*, vol. VI, No. 4, 87–132.

Walker, B. H. (1974) Ecological considerations in the management of semi-arid ecosystems in South-Central Africa, *Proc. 1st Int. Congr. Ecol.*, The Hague, pp. 124–129.

Wallén, C. C. (1967) Aridity definitions and their applicability, *Geografiska Ann.*, **49**: 367–384.

Warren, A. (1967) East Africa, in *Africa in Transition,* B. W. Hodder and D. R. Harris eds., Methuen, London, pp. 163–217.

Warren, A. (1970) Dune trends and their implications in the central Sudan, *Zeitschrift für Geomorph.* (Stuttgart), Supplementband, **10**: 154–180.

Warren, A. (1975) Changes in the desert boundary in the Negev-Sinai area, Cambr. Mtg. Desertification, IGU, 22–26 Sept. 1975, p. 44.

Warren, A. (1976) (in press) Geomorphology and sedimentology of the Nebraska Sand Hills, *Jl. Geology*.

Warren, A. and Goldsmith, F. B. (1974) *Conservation in Practice*, J. Wiley & Sons, London, 512 pp.

Warren, J. F. (1975) The scalds of western New South Wales – A form of water erosion, *Australian Geographer*, **9**: 282–292.

Watson, J. A. L. and Gay, F. J. (1970) The role of grass-eating termites in the degradation of a mulga ecosystem, *Search*, **1**: 43.

Weaver, J. E. (1950) Effects of different intensities of grazing on depth and quantity of roots of grasses, *Jl. Range Management*, **3**: 100–113.

West, O. (1969) The ecological impact of the introduction of domestic cattle into wildlife and tsetse areas of Rhodesia, in *The Careless Technology*, M. T. Farver and J. P. Milton eds., J. P. Stacey, London, pp. 712–725.

Western, D. (1974) The environment and ecology of pastoralists in arid savannahs, Paper read to SSRC Symp. on the Future of Traditional Primitive Societies, Cambridge, Dec. 1974.

Whitman, W. C. (1971) Influence of grazing on the microclimate of mixed grass prairie, in *Plant Morphogenesis as the Basis for Scientific Management*, US Dept. Agric. Misc. Publ. 1271, pp. 207–218.

Whittaker, R. H. (1970) *Communities and Ecosystems*, Macmillan, London, 158 pp.

Whyte, R. O. (1961) Evolution of land use in south-western Asia, in *A History of Land Use in Arid Regions*, L. D. Stamp ed., UNESCO, Paris, pp. 57–118.

Wilcott, J. C. (1973) *A Seed Demography Model for Finding Optimal Strategies for Desert Annuals*, Ph.D. dissertation, Utah State University, 137 pp.

Wilcox, D. G. and McKinnon, E. A. (1974) *A Report on the Condition of the Gascoyne Catchment*, Dept. Agric. W. Austral./Dept. Lands & Surveys W. Austral., 10 chapters.

Wilkin, D. C. and Norton, B. E. (1974) Resource management, US/Intern. Biol. Progr. Desert Biome, RM 74–65: 1–7.

Williams, M. (1976) (MS) The perception of the soil erosion hazard in South Australia, Paper given to Hazard Symposium, Canberra, May 1976.

Williams, O. B. (1961) Principles underlying the improvement of dryland country, *Wool Technology and Sheep Breeding*, **viii**: June 1961, 51–58.

Williams, O. B. (1969) Studies in the ecology of the Riverine Plain. V. Plant density response of species in a *Danthonia caespitosa* grassland to 16 years of grazing by merino sheep, *Aust. J. Bot.*, **17**: 255–268.

Wilson, A. D., Leigh, J. H., Hindley, N. L. and Mulham, W. E. (1975) Comparison of the diets of goats and sheep on a *Casuarina cristata* – *Heterodendrum oleifolium* woodland community in western New South Wales, *Austrl. Jl. Exper. Agric. Anim. Husb.*, **15**: 45–53.

Wilson, J. G. (1962) The vegetation of Karamoja District, Northern Province of Uganda, Uganda Protectorate, Dept. Agric., Memoirs Res. DN. series 2: Vegetation, no. 5, 163 pp.

Winkworth, R. E. (1970) The soil water regime of an arid grassland (*Eragrostis eriopoda*, Benth.) community in central Australia, *Agr. Meteorol.*, **7**: 387–399.

Wright, H. A. (1974) Range burning, *J. Range Management*, **27**: 5–11.

Yaalo, D. H. and Ganor, E. (1975) Rates of aeolian dust accretion in Mediterranean and desert-fringe environments of Israel, *Israel J. Earth Sci.*, **24.**

Young, J. A. and Evans, R. A. (1974) Population dynamics of green rabbitbrush in disturbed big sagebrush communities, *Jl. Range Management*, **27**: 127–132.

Zeuner, F. E. (1963) *A History of Domesticated Animals*, Hutchinson, London.

POPULATION, SOCIETY AND DESERTIFICATION

ROBERT W. KATES, DOUGLAS L. JOHNSON and KIRSTEN JOHNSON HARING

Graduate School of Geography, Clark University, Worcester

with the assistance of
L. Berry, M. Bowden, T. G. Carroll, C. Carter,
A. Dennis, S. W. Sawyer and F. Vogel Robott

CONTENTS

LIST OF ILLUSTRATIONS

LIST OF TABLES

ACKNOWLEDGEMENTS

The assistance provided by many colleagues and friends deserves special recognition. While the conclusions presented in the final report remain our responsibility, they have been informed and improved by their observations. Numerous individuals have reviewed earlier drafts of this report and contributed helpful and thought-provoking comments, suggestions, and criticisms. Among those who have contributed in this way are D. Brokensha, J. Caldwell, H. Dregne, M. Horowitz, M. Kassas, J. Mabbutt, M. Nag, J. Newmann, S. Sandford, T. Scudder, C. Swift, J. Walls, and A. Warren. During field visits to desertification-prone areas, a wide range of experience and insight was made available by resident scholars and officials. Our gratitude is expressed to I. Adly, D. Amiran, M. Beglibaev, J. Ben-David, M. Clor, J. Cooke, R. Daly, J. Doughrameji, R. Dutton, F. Fennick, M. Gebre-Medhin, D. Hopcroft, G. Hundt, J. Joury, R. Ketchum, G. Kressel, C. Lendon, S. P. Malhotra, H. S. Mann, A. Mascarenhas, A. Mehretu, R. Millington, L. Ngutter, M. Obote, R. A. Perry, M. P. Petrov, M. L. Purohit, J. Schechter, A. Schinhal, A. Schmueli, R. Slayter, C. Toupet, C. G. Widstrand, G. Williams, O. Williams, and M. Zehni.

In addition, twelve profiles of change, as well as other written material, were contributed by M. Bowden, R. Brooks, D. Campbell, B. Davidson, R. Faulkingham, R. Hansis, R. L. Heathcote, K. Hewitt, A. Mascarenhas, F. Vogel Roboff, S. Smith, J. Whitney, A. Whyte, M. Williams, A. Wilson, and B. Wisner. While not published as part of this report, these supplementary materials have had a major influence in shaping our thinking about the human dimensions of desertification. Various of the United Nations agencies, in particular FAO, UNESCO, and WHO, made available their personnel and extensive document collections for consultation. Their support was always unstinting, generous, and appreciated.

Two of the diagrams in the text were first published elsewhere. Figure 3 appeared in Michael Williams, *The Making of the South Australia Landscape* (London: Academic Press, 1974, p. 271). A minor error in the yield figures of the original figure was discovered too late for the appropriate corrections to be made in this version of the report. Figure 5 depicts the growth and decline of population in the Diyala basin; it is an adaptation and redrawing of data first presented in Robert McC. Adams, *Land Behind Baghdad* (Chicago: University of Chicago Press, 1965, p. 115). Finally, Fig. 4 was compiled from an array of statistical material made available by S. P. Malhotra, some of which is contained in his unpublished papers cited in the bibliography.

A concluding word of thanks is due to R. Baril, C. Gediman, J. McGrath, and P. Monahan who worked on various drafts of the manuscript.

I. INTRODUCTION

1. The hot deserts of the world, the areas of little or no vegetation, are natural phenomena. Desertification, the aggravation or intensification of such conditions, is a human phenomenon, arising most commonly from society's search for secure livelihoods in dry environments. In most instances this search proves successful. In others it involves destructive processes in which the productive base deteriorates and the social system is imperilled. Unlike drought, which is usually a short-term diminution in available moisture (1–5 years), the physical processes involved in desertification are long-term, chronic, and pervasive: dune and sand encroachment, degradation of vegetative cover and resources, soil erosion, and, where irrigation is used, waterlogging and salinization. Although distinct from short-term fluctuations, these long-term processes are intimately affected by them. Wind, storms, and drought greatly accelerate chronic destructive conditions.

2. The simple definition of desertification, which views it as the spread or intensification of desert-like conditions, provides us with a gauge to measure this destructive process, but tells us nothing about its causes, its dynamics, or its human consequences. Rapp (1974) has attributed desertification to a combination of drought and poor land use. However, the origins of desertification appear to be much more complex.

3. We see desertification as arising from interaction (Fig. 1) between the three major livelihood systems of the dry lands (agriculture-based, animal-based, and urban-based) and a combination of fluctuations in the natural environment and changes in human social systems at varying scales. Often this interactive process acquires a synergistic quality, as in the instance of the Mendoza oasis, where the very remedies for desertification become part of the hazard's causal mechanism. While each livelihood system possesses different characteristics and is exposed to varying degrees of vulnerability to desertification, in dry lands, in addition to fluctuations in climate, two major processes of human change

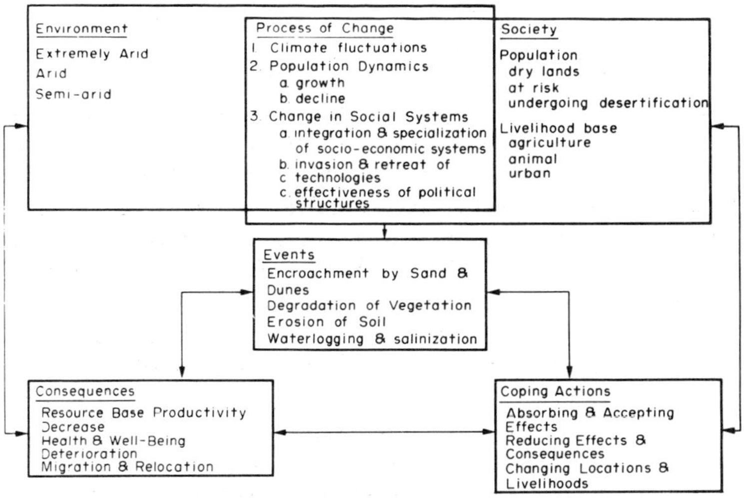

Fig. 1. Coping with desertification hazard

characterize the causal component of this interaction: (1) population dynamics is an important part of the process, for both growth and decline in population appear to cause desertification; and (2) three types of social change also contribute to desertification: integration into wider socio-economic systems alters the dynamics of local livelihood systems; the invasion of new and the retreat of old technologies truncates the evolution of indigenous expertise; and the fluctuating strength and effectiveness of governments drastically affect stability and survival in the dry margins. Out of these interactive processes develop the events that lead to desertification. In turn, these events give rise to desertification's salient consequences: deterioration in health and well-being and migration. And finally, societies respond to desertification with a variety of coping mechanisms whereby its effects are absorbed or accepted, its consequences are reduced, or its pressures are translated into change in livelihood or location.

4. Seventy-eight million people live in regions that have experienced severe or very severe desertification; of this number, 50 million are directly affected. In addition, the pressures produced by desertification affect a larger, but indeterminate, population as demands are placed on their resources for the support of less fortuitously located populations. For those populations directly affected, the human consequences of desertification are primarily twofold: the threat to the health and well-being of individuals and groups placed at risk by desertification and the problems and opportunities that arise from desertification-induced migration and resettlement. The threats to health and well-being arising from desertification are not easily identifiable as to cause and effect, but are reflected in the inadequacy of diet, the decrease in nutritional status, and accompanying vulnerability to disease that characterize declining resource productivity. Migration and relocation occur when livelihoods can no longer be sustained within weakened production systems and this entails both new opportunities and additional health, psycho-social, and economic problems.

5. Human societies are not merely passive recipients of the harmful effects of desertification. All peoples possess coping strategies enabling them to bear, reduce, or modify the consequences of natural hazards. These strategies may be deeply ingrained in the social fabric or they may be recently adopted innovations. Although they play a crucial role in people's survival, these coping strategies are not yet fully understood or, in many cases, even recognized. Adjustments to short-term fluctuations such as drought are relatively easy to document and this fact is reflected in a substantial body of literature dealing with human response to drought (e.g. Berry and Kates, 1972; Glantz, 1976; Heathcote, 1969; Kates, 1972). Desertification, on the other hand, has had little comprehensive study (the work of LeHouérou, 1968, and Rapp, 1974, comprising notable exceptions) and is a long-term process difficult to detect in short-term field study. Further, the more successful of society's mechanisms for coping with a long-term process such as desertification may be so embedded in everyday livelihood activity that they are no longer easily distinguishable as discrete coping strategies.

6. In spite of the existence of coping actions by societies affected by desertification hazard, much room is left for additional, especially international, effort. However, these efforts cannot succeed without a clear understanding of how desertification is both cause and consequence of change in the human occupance of dry lands. Isolated from the context of regional, national, and global change, an understanding of the desertification process will prove to be illusory. The demographic, social, and behavioural aspects of desertification can only be seen in the perspective of continuing change in the lives and

livelihoods of the 628 million people (1974) who live and work in the dry lands of the world. For these peoples, desertification serves to accelerate changes already taking place in their lives, and, in turn, is exacerbated by them.

7. The will and wisdom of many nations will now concentrate on ways to combat desertification and ameliorate its effects. Such efforts, if they are to prove practicable and successful, need to be rooted in a basic understanding of the interaction of natural phenomena with social process, of the dynamics of how humans cope with environmental hazards. This review provides a framework for understanding these dynamics. It sets forth the numbers and patterns of human occupance of dry lands, a simple typology of life and livelihood systems, within which we will trace both the causes of and the processes of change instituted by desertification. In this wider context, we can examine the vulnerability of peoples and livelihood systems to desertification, the various modes of coping with the hazard that exist in different livelihoods, and the implications for action that flow from these understandings.

II. POPULATION AND LIVELIHOODS AT RISK

8. No assessment of desertification and the prospects for ameliorating its effects is complete without an understanding of the human populations placed at risk by environmental and social change. This section explores the basic demography of different dryland countries, examines the major livelihood adaptations to aridity, compares the relative vulnerability of these systems to desertification processes, and estimates the magnitude of the population placed at risk by desertification.

A. POPULATIONS OF THE WORLD'S DRY LANDS[1] AND REGIONS UNDERGOING DESERTIFICATION

9. Approximately 14 per cent of the world's population, 628 million people, live in the dry lands. Of these, the majority, 72 per cent, live within semi-arid zones, 27 per cent inhabit the arid zone, and 1 per cent live in the extremely arid zone. Rural population densities vary according to the productivity of the micro-environment, but they generally remain below 1 per square kilometre in the extremely arid zones, below 5 per square kilometre in the arid zones and below 10 per square kilometre in the semi-arid zones (see Table 3). However, densities are considerably greater wherever irrigation agriculture is practised.

10. Table 1 contains population estimates by livelihood groups for the dry lands. Since demographic data are of doubtful accuracy for many of the countries, the figures should be read and used with caution. Even more problematic are livelihood estimates for nomadic populations which can vary by a factor of from 2 to 8 depending on the criteria used (US, CDC, 1973, versus Berg, 1975). Another problem is the urban population estimate, which appears to be too high. Some national censuses accord urban status to places with as few as 500 inhabitants, while others use a much higher threshold (Davis, 1969). A more accurate livelihood analysis might shift up to one-third of the urban population into the agricultural category. On a world scale, the result of this modification is a dry world population that is 21 per cent urban, 72 per cent agricultural, and 7 per cent animal-based. Despite the many necessary qualifications, this estimate can serve as a basis for examining the relative vulnerability of dryland people to desertification.

11. Of the 628 million people inhabiting the world's dry lands, approximately 78 million presently live in areas undergoing severe or very severe desertification. These figures are summarized in Table 2. According to Dregne (n.d.) severe desertification means that resource-base productivity is threatened, while very severe desertification indicates that the deterioration is so acute that it is economically irreversible. Not all of the 78 million inhabitants of these areas are directly affected by desertification, since occupation or economic status may provide a shield against the hazard. Nonetheless, it seems reasonable to assume that a majority of this population experiences some of desertification's adverse effects.

[1] Dry lands is here used as a comprehensive term for Meigs's (1953) threefold classification of the world's rainfall-deficient regions into extremely arid, arid, and semi-arid.

TABLE 1. *Estimates of dryland[a] populations by region[b] and livelihood group (in thousands)*

Region	Dry lands Total population[c]	Livelihood populations in dry lands: Urban based	Agriculture based	Animal based
Mediterranean Basin	106,800	42,000 (39%)	60,000 (57%)	4,200 (4%)
Sub-Saharan Africa	75,500	11,700 (15%)	46,800 (62%)	17,000 (23%)
Asia and the Pacific	378,000	106,800 (28%)	260,400 (69%)	10,300 (3%)
Americas	68,100	33,700 (50%)	29,300 (43%)	5,100 (7%)
	628,400	194,200 (31%)	397,100 (63%)	37,100 (6%)

[a]Meigs's classification (1953) including extremely arid, arid, and semi-arid areas.
[b]Groupings as designated by UNEP Governing Council for regional meetings.
[c]Total world population was estimated to be 3.86 billion in 1974.

12. How many people make up this majority is difficult to define precisely. Using a different approach, this review estimated that 50 million people are exposed to desertification. Therefore, an overall vulnerability figure can be expected to lie somewhere between 50 and 78 million. In areal terms the problem also is significant, for Dregne's categories cover an area of 30 million square kilometres, or 22 per cent of the dry lands as defined by Meigs (1953). The populations inhabiting these areas account for 15 per cent of the total dryland population. Significantly, this population is differentially exposed to desertification. Animal-based populations are only 6 per cent of the total population of dry lands, yet they represent 22 per cent of the population threatened by desertification.

TABLE 2. *Estimates of populations and livelihoods resident in areas recently undergoing severe desertification[a] (in thousands)*

Region	Total population	Urban based	Agriculture based	Animal based	Area (km²)
Mediterranean Basin	9,820	2,995 (31%)	5,900 (60%)	925 (9%)	1,320,000
Sub-Saharan Africa	16,165	3,072 (19%)	6,014 (37%)	7,079 (44%)	6,850,000
Asia and the Pacific	28,482	7,740 (27%)	14,311 (54%)	6,431 (19%)	4,361,000
Americas	24,079	7,683 (32%)	13,417 (56%)	2,979 (12%)	17,545,000
	78,546	21,490 (27%)	39,642 (51%)	17,414 (22%)	30,076,000

[a]As estimated by H. Dregne (includes both severe and very severe categories).

B. DRYLAND LIVELIHOOD SYSTEMS AND THEIR RELATIVE VULNERABILITY TO DESERTIFICATION

13. Faced with the need to derive sustenance from dryland environments that are hazard-prone, harsh, and capable of great interannual fluctuation in productivity, people have evolved an array of livelihood systems. These livelihoods not only enable survival, but also produce usable and desired products, many of which are exported to more humid regions.

14. Within each major livelihood system, agriculture-based, animal-based, and urban-based (see Fig. 2), further subdivisions can be distinguished, each possessing its own demographic characteristics, technology, social organization, and behavioural traits. These livelihood systems do not operate in isolation; rather they are complexly interlinked and interdependent. Most culture groups do not rely on just one exploitive system, but instead utilize a dynamic mix or complex of techniques. Over a period of time complementary

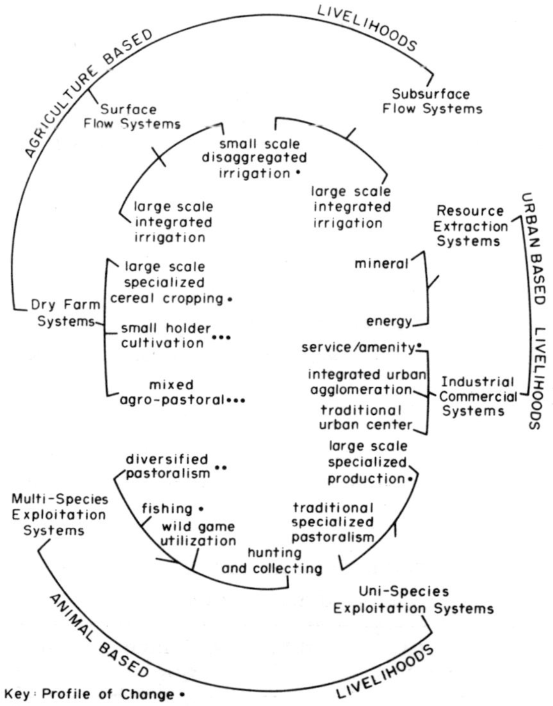

Fig. 2. Livelihood systems in dry lands

sets of activities have evolved that spread the risks inherent in dryland environments, maximize productivity to the extent that available technology permits, and meet cultural and economic ends.

15. At the same time, these adaptations are not without their problems, for as local systems become increasingly integrated into national and international market economies, they become exposed to forces operating beyond their boundaries. In today's world economy where prices, marketing structures, and political events integrate the most disparate groups, it can be said that no culture stands alone, timeless, changeless and self-sufficient.

16. These relations between the world market economy and the three basic livelihood systems are mediated by the concrete conditions operating in different nation states. Therefore, a generalized vunerability assessment based solely on livelihood systems is at best too general to be useful in specific cases, and at worst misleading. However, the world market economy is such that it is composed of two basic categories of nation states: developed industrialized countries, and developing countries. If this criterion is taken into consideration along with that of livelihood systems, then certain basic parameters of vulnerability to desertification can be outlined. Additionally, patterns of land ownership and the distribution of wealth come to bear upon the differential vulnerability of groups within a society. In the following pages an attempt will be made to incorporate these elements in an assessment of livelihood system vulnerability.

1. Agriculture-based Livelihoods

17. Two basic types of agricultural livelihood exist. One relies on dry farming (rainfall dependent) techniques and the other employs irrigation technology of varying degrees of sophistication and at differing scales. Together these livelihoods account for 5 out of every 7 people inhabitating the dry world. Here are found some of the highest birth and death rates among dryland peoples, with population generally growing at a rate of 3 per cent per annum. It is mostly from this sector that the ranks of urban migrants are swelled.

18. Dryland farming, the most prevalent livelihood in the semi-arid zones, is characterized by a range of strategies and techniques enabling farmers to cope with a high degree of rainfall variability. Except in so far as they relate to people's vulnerability to desertification, a full description of these strategies is beyond the scope of this review. (See the companion review on Ecological Change and Desertification for full coverage of this material; Warren and Maizels, 1976.) In this regard, two basic categories of dry farming strategies are of interest. The first category is smallholder cultivation which largely emphasizes strategies that permit moisture maximization within field systems. The second category involves strategies combining cultivation with gathering and/or animal husbandry. Both moisture maximization and mixed agro-pastoralism provide the livelihood with needed flexibility and diversity enabling maximum security within uncertain rainfall régimes.

19. Moisture maximization, particularly in non-industrialized developing settings, involves an array of diverse techniques that include: careful seasonal plantings of fast-maturing varieties, scattered plantings of drought-resistant crops, careful weeding and mulching practices, the use of moist bottomlands, and the construction of terraces, bunds and dams to impound run-off water. In terms of desertification, it can be said that most of these moisture maximization techniques are either benign, or actually conservative of soil and soil fertility. For example, the primary objective of ephemeral stream dams and hillslope terraces is to trap run-off from infrequent rainstorms. In the process, these structures also act as a means to impound soil and organic material carried by the run-off, thus providing fields with a renewable source of fertility. This technique not only enables cropping under circumstances which otherwise would not be appropriate to cultivation, but, in fact, also can support relatively high population densities. West (1970), for example, estimates that semi-terracing contributed substantially to pre-Conquest densities of up to 90 persons per square kilometre in Central Mexico.

20. Moisture maximization techniques such as terracing are labour intensive and therefore vulnerable to losses in manpower. As traditional societies increasingly become tied to an outside market, economy terrace extension and maintenance can suffer. Reports from sites as widely scattered as Mexico (Johnson Haring, 1976a), Kenya (O'Keefe, 1975), and Yemen (Dow, 1975) document a process of terrace degradation that is linked to labour out-migration. The full ramifications of this type of desertification are unknown, and its effect on socio-economic structures awaits investigation.

21. Most dry farming in developing countries combines cultivation with gathering, craft-manufacture, and animal husbandry. In most cases, these activities provide a crucial complement to an uncertain agricultural situation, thereby forming an integrated agro-pastoral production system. Mixed agro-pastoralism can take numerous forms. Let us examine two of these in order to assess their dynamics and consequent vulnerability to desertification.

22. In the first case, traditional mixed agro-pastoralism (Faulkingham, 1976; Johnson Haring, 1976b; Malhotra, 1976b; Wisner, 1976) combines the gamut of semi-arid resource-use activities in one household productive unit. Changes occurring in the larger national or international setting often render some of these activities economically obsolete. Handicraft manufacture, for example, is frequently replaced by industrialization. Handicrafts provide a key source of supplementary cash income for many dryland agriculturalists; therefore, when they are undercut a family must resort to other options. Increasingly they resort to labour out-migration. This, in turn, leads to a decline in field maintenance, and consequently in local productivity, thus further weakening the local system. The deterioration of field maintenance, as in the case of terrace systems, can lead to serious desertification. In other cases labour out-migration merely entails a deintensification of local cultivation with no irreversible damage to the productive base.

23. In the second case, mixed agro-pastoralism operates within a system composed of separate but integrated parts held together by a system of unequal interdependence in which restrictive land tenure patterns play a crucial role. This is the case where large livestock ranches and peasant share-croppers (e.g. Brooks, 1976) enter into an arrangement whereby landowners provide subsistence plots in exchange for the labour needed in the cattle raising operations. This system is prone to desertification when an expanding population is concentrated on marginal lands while the bulk of the land is devoted to extensive grazing. Overcultivation, accelerated soil erosion, and deforestation are the inevitable consequence of this arrangement.

24. Commercial dry farming in the industrialized world copes with rainfall uncertainty by means of strategies and support systems (e.g. government subsidies, insurance programmes) unavailable to peasant cultivators in the developing world. In this sense, their stability is assured in the short term (Bowden, 1976). The consolidation of farms into large corporate units using capital intensive production is frequently an important aspect of this stability. Capital and technology seem, at present, to be able to cope with problems of soil erosion and dune encroachment (e.g. Heathcote and Williams, 1976). Williams's review (n.d.) of the history of soil erosion in farming areas of South Australia suggests that periods of declining resource productivity were followed by the adoption of technical innovations that would allow recovery at higher levels of productivity (Fig. 3). However, the long-term success of such short-term stabilized systems is unclear. Escape from the pressures of the local physical environment are made at the expense of a dependence on market economy over which the producer has little or no control. In the

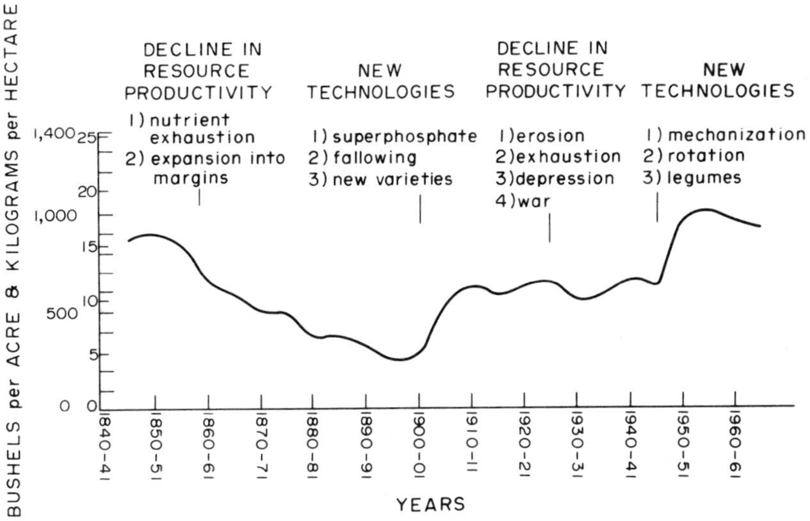

Fig. 3. Resource productivity and technological adjustments in commercial farming, South Australia

Australian Mallee between the wars, for example, profit margins were slowly whittled away as operation costs rose and unexpected market fluctuations changed anticipated profit into oversupply and sudden debt. The response was to seek greater economies of scale achieved through the use of more labour-saving technology with higher energy cost. Such changes may only delay a day of reckoning. In this sense, short-term success may initiate a process that ultimately increases long-term vulnerability.

25. The second principal agricultural livelihood in dry lands is based on irrigation. Irrigation often provides more secure support, but not necessarily greater resistance to desertification, than does dry farming. Irrigation results in the growth of dense population concentrations in the major riverine valleys of the arid lands. This growth is not without its own special problems, especially salinization, siltation, and periodic destructive flooding, but the regional disparities in population distribution within dryland countries can be explained largely on the basis of the high agricultural productivity potential of both modern and traditional irrigation systems.

26. Several basic types of irrigation systems can be identified. On the one hand are the systems that rely on surface flow. These include traditional systems based on annual floods inundating floodplain basins (either natural or man-made) which then serve as settling points for silt-rich alluvium. Success in this system depends in large part on the regularity of the annual flood. In the case of the Nile, an elaborate civilization and population developed without requiring massive investments in canals and other hydraulic works because the annual flood was extremely regular in both time and quantity. In most other instances, stream régimes were more irregular, requiring integrated canal systems to smooth out system irregularities and to bring water to areas normally beyond the reach of agricultural operations (Jacobsen, 1960; Walters, 1970; Adams, 1965). The development of urban civilizations is associated with such irrigation settings.

27. This system often, at least in its initial stages, resulted in desertification, since the rise of cities encouraged a concentration of population and an intensification of production close to the urban area with associated land abandonment in more peripheral

districts (Adams, 1974). The growth of salinization and siltation difficulties also characterized this era (Jacobsen and Adams, 1958). Continued growth of population in these rich alluvial environments has sparked interest in converting all such annual systems to perennial multi-crop enterprises. It is this drive that underlies the High Dam at Aswan and the Euphrates Dam in Syria. Yet all of these activities remain subject to desertification processes inherent in the system and, since they are all dependent on water derived from humid areas outside the arid zone, they all remain exposed to political and climatic events that might reduce the volume of water available. Moreover, salinity and waterlogging continue to plague modern irrigation systems. Often, it is inadequate provision of drainage facilities, as in the Mussayib project (van Aart, 1971), that leads to decreased productivity which can only be restored at great expense. Problems of salinization and waterlogging are not limited to developing countries. Warren and Maizels (1976) review examples of serious deterioration in irrigation systems of industrial nations. Where political boundaries are involved, as in the cases of the Colorado River flowing through the United States and Mexico, this problem can be especially intractable. Over the years, as agricultural use of the upper Colorado Basin intensifies and the salt content of the river steadily increases, the downstream users in Mexico have had a particularly difficult time in exerting control over the quality of the water they receive.

28. Somewhat less vulnerable to desertification are the point irrigation systems that tap groundwater. This generalization is especially applicable to the underground water collection galleries found in many parts of the Middle East (*qanat* in Iraq; *aflag* in Oman; *foggara* in North Africa). However, outflows of labour to economic opportunity elsewhere in the country may reduce the labour supply needed to keep the system functioning. Shallow groundwater wells or agricultural systems using date palms to tap near-surface groundwater are also less vulnerable as long as overexpansion of the system is not encouraged. Deep bore wells may be productive in the short run, especially the central privot systems that have replaced extensive ranching operations in Nebraska and provide a high technology alternative to the traditional oasis gardens of Kufra oasis in Libya (Splinter, 1976). Such systems are vulnerable to desertification wherever groundwater reserves are imprecisely known. Moreover, their dependence on high energy fossil fuels is a considerable risk in an era of rising energy costs and uncertain long-term supply. Deep bore wells may be productive in the short-run, but watertable depletion makes them prone to their own peculiar form of desertification.

2. Animal-based Livelihoods

29. A detailed review of animal-based livelihood systems is beyond the scope of this paper. However, a few important characteristics should be noted. The most significant feature of traditional animal-based livelihoods is their emphasis on mobility. It is by movement to seasonal pastures at different locations that the pastoralist is able to balance pasture and water through an annual cycle. To accomplish the same purpose, modern commercial ranching or game harvesting schemes usually must also incorporate large areas, but at a lower stocking rate. Western (1974), in comparing the food chain efficiency of various pastoral management strategies in the Amboseli grassland, found that Masai had a food chain efficiency of 12 per cent compared to the 6.5 per cent figure of commercial ranching operations. This is, in part, a reflection of the advantage obtained by the pastoralist's cattle since they are brought into contact with easily digestible young plant

growth for a longer portion of the year. While pastoral movement takes place for a number of other reasons, including escape from disease, access to markets, acquisition of farm products and religious devotion, a major factor is an ecologically astute assessment of seasonally variable pasture quality and productivity. Hunters (Lee, 1969) exhibit analogous patterns of movement that are linked closely to seasonally available water supplies, to the location of forage, and to animal movements that are related to available moisture. While fishing populations are more immediately restricted to the locale of their resource base, they are found wherever water is sufficient to support fish. Fishing people are important protein producers in dry lands (UN, FAO, 1975), and will move over great distances to take advantage of the fishing potential created by new dams (Lagler, 1973).

30. A second major strategy employed by animal-based populations is to diversify their livelihood systems. This is done in a number of ways. Pastoralists often herd a mixture of animal species in order to maximize the productive potential of the various ecological niches available to them (Western, 1974; Johnson, 1973). Hunters most commonly exploit a variety of game species, while fishing cultures (with only rare exceptions) rely on a mixed catch. Only under the most extreme environmental conditions where only the hardiest species survive, do pastoralists concentrate on one species. Commercial livestock ranching, which exploits a rigidly bounded space with herds of one species, stands in contrast to these principles of traditional animal-based exploitation régimes.

31. Thus the combination of mobility, flexibility, and diversity characteristic of traditional livelihood systems reduces their vulnerability to desertification and lessens their contribution to desertification processes.

32. Hunting populations are so small on the global scale that their contributions to desertification are insignificant.

33. Fishing populations are minor contributors to desertification; however, they are dramatically affected by its consequences. The case of the Mohanas of Lake Manchar in Pakistan, as documented by Hewitt (1976), illustrates the risk run by a fishing culture when its basic resource is diminished. As Lake Manchar shrinks, and finally dries completely, even such basic tools as the Mohanas' boats are threatened. Should present conditions worsen, this basic resource would be damaged beyond repair. Compounding the problem is the fishing population's small size and its relatively insignificant political power in contrast to competing water resource consumers. Just as water diversion for irrigation may permanently endanger Lake Manchar's status as a fishery resource, so, too, the migration of fishing populations to newly created dam sites is often resisted because of its potential for overloading the local fishing capacity (Henderson, 1974) to the detriment of the indigenous inhabitants. Moreover, the dams and irrigation systems developed to combat desertification, while opening up new fishing potential, can cause disproportionately greater alterations in the aquatic ecology of downstream areas. In the case of the Aswan Dam, the decline of productivity in both the coastal marshes and the deep sea fishing grounds may have exceeded gains in Lake Nasser (George, 1972).

34. The role of pastoral communities in desertification is not as clear cut. In one sense, pastoralists are somewhat less vulnerable to desertification than are agriculturalists. This is a consequence of their superior mobility, which permits accumulated capital reserves, in the form of animals, to be shifted to less affected regions. Possession of herds containing a number of species may ensure the survival of at least some animals in severe drought and permits a shift in herd composition to larger numbers of hardier species should an acute drought phase trigger a chronic decrease in productivity.

35. Yet if nomads are ecologically less vulnerable than sedentarists, they are frequently more vulnerable politically. Colonial boundaries often truncated ecological zones that, from a pastoral perspective, constitute one grazing system. Thus, the bedouin of northern Arabia found their grazing lands split among several countries, and analogous conditions can be found elsewhere (Cossins, 1971; Melamid, 1965). Governments controlled by sedentary groups of different ethnicity also serve to increase the marginality of pastoral groups, since distance from the centre, spatially and politically, often delays the input of aid until deleterious conditions are well advanced. Furthermore, economic returns to investment in agriculture often show a higher net return, prompting a greater flow of capital into non-pastoral areas (Western, 1974).

36. Nomadic populations have often been blamed for their presumed tendency to promote overgrazing. Conventional wisdom contends that a growth in human population is paralleled by an increase in animal numbers, inevitably accelerating pressure on the land and initiating a long-term downward spiral in productivity (EMASAR, 1975). Even a small steady growth in nomadic population should, therefore, result in more animals on the range since only in this way, in the absence of other employment, can a larger population be supported. Larger size herds per family are not required to postulate overstocking. However, many nomads settle spontaneously, reducing the pastoral population needing support from the animal economy. Moreover, herd composition often changes toward species that meet the food preferences of the growing sedentary population; such changes increase the pressure on certain fodder species while decreasing the pressure on others. The overall impact of these variables is being investigated quantitatively in southern Tunisia (UNESCO, MAB, 1975). It is also possible to find rangeland deterioration in specific areas at least partly as a result of increases in livestock. LeHouérou (1968) has reviewed the evidence for portions of the North African steppe and argues that a substantial decrease in primary productivity has taken place there. Peyre de Fabrègues (1971) points out the impact of excessive animal concentrations around watering points. Heady (1972) has described the island of degraded vegetation that rapidly develops around a newly sedentarized pastoral community, an effect that has been widely noted in the literature (Toupet, 1975; Shamekh, 1975; Darling and Farvar, 1972). Overgrazing pressures are as likely to come from an expansion of herds kept by settled agriculturalists or by the confinement of nomads to more restricted and marginal rangeland, as they are likely to stem from mismanagement by the pastoral community. For example, loss of rangeland and dry season fodder reserves in the Awash Valley to an expanding agriculture (van Lier, n.d.) and to hydroelectric power development accounts for much of the stress experienced by Afar pastoralists (Cossins, 1972). Political impotence leaves the pastoralist increasingly vulnerable to desertification and doubly maligned, as the victim is blamed for a problem not solely his responsibility.

37. Acute and rapid desertification is also found in commerical ranching. This frequently results when persistent wet periods coincide with unfavourable market prices. Herd growth stimulated by abundant pasture is only worthwhile if the price of meat remains high enough to make movement to market profitable. With depressed prices, ranchers are encouraged to retain animals on pastures that may only be temporarily productive. The experience documented in the Gascoyne case study (Williams and Suijdendorp, 1976) suggests that reversion to more normal rainfall patterns will be accompanied by both degradation of the rangeland and extensive animal losses. An excess of animals produced by precisely these conditions threatens much of interior Australia at the present time.

Commercial ranching schemes inserted into regions dominated by traditional pastoral systems can also lead to desertification. In these cases pressures are more apt to appear outside the fenced rangeland as traditional pastoral groups find their grazing areas restricted. On occasion partial implementation of ranching schemes that attempt to reshape traditional pastoral practices has resulted in desertification due to overstocking within fenced areas (Talbot, 1972). Even within well-managed livestock schemes in industrialized countries desertification can occur unless great care is taken to control grazing, and in the USSR it has been found that pasture rotation systems that in certain respects approximate the mobility principle of traditional pastoralists provide the best way to maximize productivity without rangeland degradation (Nechayeva, 1976).

3. Urban-based Livelihoods

38. Between 20 and 30 per cent of the population of the dry lands live in urban centres. However, it is difficult to estimate the proportion of this urban-based percentage that actually engage in non-agricultural production. Nonetheless, it is clear that the productive activities centred in dryland cities are qualitatively different from those of their hinterland.

39. Historically, dryland cities have served as commercial and administrative centres, have grown up around mines, have developed as ports and amenity centres, and, to a degree, have industrialized. Except in so far as manufacturing may play a lesser role than other functions, dryland cities are not fundamentally different from their counterparts in more humid environments. It is not in urban functions but in the interaction between dryland urban growth, with its associated consumption needs, and the arid environmental setting that the source of these cities' distinctiveness, as well as their possible vulnerability to desertification, is to be found.

40. In this regard it is possible to distinguish between two types of urban/arid environment interactions. The first type of interaction is *intrinsic*; that is, it relates to problems that might arise from any concentration of population in the arid lands: the need for fuel, shelter, water and food which may not be available in sufficient quantities in the immediate hinterland of the urban centre. In the developed world these needs are taken care of because the dryland city operates as a specialized component of a larger, modern system of exchange. In the developing world, where production, technology and supply networks have not reached this level of development, the needs of urban dwellers often must be satisfied by exploiting the local hinterland. Therefore, the need for fuel and firewood can often lead to serious deforestation around urban centres (Warren and Maizels, 1976). Cattle concentrations near cities can also devegetate large areas leading to soil erosion and the formation of localized dust bowls (Berry and Kates, 1972). In turn, devegetation and soil erosion can threaten limited water supplies by increasing reservoir infilling.

41. The second type of interaction arises from the *indirect* effects of drought or desertification taking place in the urban centre's hinterland. In this case the city becomes the recipient of a sudden, heavy influx of rural population. Except in the case of drought refugees, it is difficult to disaggregate rural-urban migration induced by desertification from that brought about by other pressures and temptations upon the rural sector. Nonetheless, urban centres located within regions experiencing desertification can expect a continued flow of rural migrants.

42. This migration has positive and negative aspects. For one, it is likely to entail the improvement in the lot of individual migrant families. For example, evidence from studies done in Lima, Peru, indicates that migrants are better off there than if they had stayed in the rural sector (Mangin, 1970). Another example shows that wages three times as high in Cairo as in the Delta underlie the movement of fellahin into urban employment (Abu-Lughod, 1961). Once a rural-urban flow is established, permanent links, complete with modest monetary remissions to the rural sector, often develop. There is some evidence indicating that this, in turn, may contribute to the underdevelopment of the peasant rural sector (Nolan, 1975). It should also be noted that larger urban populations create larger markets for rural produce and this allows farmers to make monetary savings.

43. The negative effects of heavy urban immigration in developing countries are obvious. Urban services, often inadequate to begin with, are further overburdened. The intrinsic problems of urban/arid environment interaction are exacerbated by burgeoning numbers of city dwellers. Water consumption rises dramatically leading, as in the case of Mexico City, to the diversion of agricultural water supplies for urban needs (Fox, 1965). This triggers desertification in the hinterland which, in turn, generates a new source of migrants. Similarly, as the demand for fuel increases in urban centres the radius of deforestation rings extends dramatically, and at the same time the cost of charcoal and wood rises for a population that can ill afford it.

44. The spectacular growth of some dryland cities in the developed world exhibits a different pattern. Amenities such as sun and mild winters attract large numbers of new residents to these centres (e.g. Wilson, 1976). In addition, the growth of service industries provides a healthy economic basis. Under these conditions, the dryland city can afford to import a large portion of its necessities. Except for the destruction of native vegetation in order to make way for suburban development, and for strains placed upon groundwater supplies, these cities appear to be viable in at least the short term.

C. SUMMARY

45. Approximately 14 per cent of the world's population, or 628 million people, live in dry lands. Twenty-two per cent of this dry realm is exposed to severe or very severe desertification and 15 per cent of the dryland population inhabits these vulnerable areas. Thus, between 50 and 78 million people are directly threatened by desertification. However, vulnerability to the hazard varies considerably; economic status or urban residence shields many, while some livelihoods, such as the animal-based (who are 6 per cent of the dryland population but 22 per cent of the risk-prone population), are disproportionately exposed to risk.

46. All agriculture-based livelihood systems in the dry lands are potentially vulnerable to desertification. The nature and degree of this vulnerability depends on the interrelation between intrinsic features of the livelihood system and large-scale societal and demographic factors. In developing countries desertification is a predictable outcome in peasant dry farm systems when these are circumscribed within marginal lands, when supplementary income-generating activities are rendered obsolete, or when labour out-migration leads to a decline in field system maintenance. Due to its ability to invest capital in conservation measures, commercial dry farming seems in the short-term to have escaped the pressures characteristic of its peasant counterparts. Irrigation agriculture in both developed and developing nations suffers from problems of salinization, waterlogging, and potential

groundwater depletion. Vulnerability to this type of desertification will diminish for those schemes provided with adequate drainage and for individual users able to bear the burden of salinization control measures. However, vulnerability remains high for those unable to do so.

47. A balanced assessment of the relations among pastoralism, overgrazing and desertification awaits the collection of more comprehensive data. In the interim, it can be said that environmental degradation occurs when drought, the constriction of grazing lands due to political factors, the extension of cultivation, or sedentarization lead to a concentration of herds in specific locations. With few exceptions, pastoral nomads constitute socially and politically vulnerable groups within present day nation states. It is difficult, if not impossible, to isolate these external pressures from an assessment of pastoral land use and their threatened productive base.

48. Urban centres in the dry lands of developing countries are vulnerable to desertification, both as the result of intrinsic patterns, and as the result of rapid population growth. On the other hand, urban centres in the arid parts of the developed world appear to have transcended a dependence on immediate hinterland resources (except, in some cases, water) and seem able to prosper in spite of population growth and environmental limitations.

III. THE SOCIAL CAUSES AND CONSEQUENCES OF DESERTIFICATION

49. The human experience of desertification is often obscured in the gross figures of animals lost, refugees and migrants created, ill-health experienced, trauma and stress endured, and inoculations given. Grinding and debilitating, the struggle for survival under harsh conditions is often a desperate, silent business. Frequently the sense of frustration engendered by cumulative events outside the individual's control colours the experience, while the promise of a bright alternative future in urban or technologically sophisticated modern systems often proves elusive. Considerable adaptive powers reside in the human systems exposed to desertification, and so successful are these coping devices that it frequently requires an extreme series of events to highlight the problem. Yet cause and consequence of desertification remain inextricably linked, if not always overtly observed. Their mutual interactions and interdependencies can be illustrated by a case example for a large-scale, integrated, irrigation-based livelihood system.

A. DESERTIFICATION AS A SYNERGISTIC PROCESS: DROUGHT, DESERTIFICATION AND SOCIETY IN MENDOZA

50. At time scales significant to people, desertification is a complex phenomenon in which one or more important processes of social change interact with natural fluctuations in climate and ecosystems. While the evidence is not conclusive, there is reason to suspect that desertification may be a synergistic process in which the existence of arid conditions breeds increasingly arid conditions, while the very remedial measures undertaken to reduce desertification's impact may lead to a long-term worsening of society's ability to cope with both natural and societal change. Moreover, not only does desertification have cumulative features, but also it is a selective hazard taking its heaviest toll on socio-economically marginal populations. Thus, one overwhelming feature of the desertification process in their lives is loss in their basis of sustenance and in their health and well-being.

51. The Mendoza oasis illustrates in its historical evolution and contemporary condition the complex, synergistic character of desertification. Located in the rain shadow of the eastern slopes of the Andes, Mendoza receives insufficient precipitation to support agricultural activities and instead must rely on irrigation water supplied by the district's exotic streams. It is, therefore, vulnerable to fluctuations in regional climate and weather, as well as variation in world market prices, inequities and complications in water rights and land tenure practices, price fluctuations due to overspecialization in grape monoculture, and urban encroachment on agricultural land. These forces have resulted in serious problems of waterlogging, salinization, and a differential diminution of social health and well-being.

52. Periodic shortfalls in precipitation, especially when they continue for successive years, while not in themselves responsible for desertification, cause severe problems for many oasis farmers (Hansis, 1976). Location at a distance from the main distribution canals increases vulnerability. This spatial disadvantage is reinforced by the legal structure of water rights which gives some properties priority rights to water (Morris, 1969). This insecurity in water availability leads to damaging misapplications of water in both

favoured and unfavoured sites. Salinization is frequently the result and is particularly acute for those farmers located at the base of a slope or downslope from heavy irrigation use. In such sites, soils often become waterlogged as the water applied to upslope areas accumulates in the groundwater table. The marginal farmer becomes doubly disadvantaged in such situations, because, located far from primary canals and thus exposed to crop loss due to moisture deficiencies, many are also threatened by the subterranean movement of excess water from their more fortunately located and endowed neighbours.

53. No factor is more important in contributing to desertification in Mendoza than is the land tenure system. Viticulture is the primary agricultural activity and large numbers of Mendoza farmers work under a sharecropping (*contratista*) system. While the owner of a property supplies the capital equipment and farm land, the *contratista* provides the labour. The contract generally runs for 8 years. During this time, the *contratista* must clear the land, plant it in vines, cultivate and harvest the crop, and pay 82 per cent of the yield as rent. At the end of the contract period the land reverts to the owner just as the vines reach maximum production and the sharecropper must look for employment elsewhere. Only if conditions have been unusually favourable will the *contratista*, who finds both profit and capital reduced during drought years, be able to save enough on this arrangement to purchase his own land.

54. The *contratista* system contributes to desertification in a number of ways. Because a *contratista* holds the land for only a short time, his ability and willingness to invest in capital improvements is minimal. Yet these capital improvements are often essential to reducing the risk of salinization. Improved distribution systems such as sprinklers or increased drainage are seldom found on *contratista* holdings. It is no accident that the most saline soils in the area are those in the oldest part of the oasis where the *contratista* system is most pervasive (Morris, 1969).

55. Moreover, soils in most of the Mendoza oasis are light and of loose structure. This exposes them easily to wind deflation and to rilling as a result of surface irrigation water flow. To some extent both wind and water erosion go hand in hand with human use of the agricultural potential of the oasis and are unavoidable. But much of their ill effect could be minimized if irrigation were conducted by more capital intensive sprinkler or drip irrigation techniques. For the *contratista* the capital to shift to such practices is nonexistent; anything that can be saved must be hoarded against the risk of future crop failures, depressed market prices or the possibility of acquiring one's own farm. The *contratista's* unwillingness to put capital into improvement of a short-term holding is matched by the landowner's reluctance to invest in upgrading land quality. For the landlord, the land primarily represents a long-term hedge against inflation (Morris, 1969) and, because the tenant and his labour are already present, there is little incentive to make *contratista* labour more productive, or the exploitation system more efficient, by investing in capital improvements. Indeed, the landowner can expect to make a profit even on desertified land, since the physical expansion of Mendoza city into nearby *contratista* districts makes the land valuable for non-agricultural purposes.

56. Specialization and integration into the global economy compounds the problems of the Mendoza oasis. Originally developed as a diversified agricultural oasis that produced its own food, while shipping alfalfa-fattened cattle over the mountains to Santiago, Mendozan agriculture became increasingly specialized following the arrival of the railroad from Buenos Aires in 1884. Economic and environmental considerations favoured a concentration on wine and table grapes, and a flood of Italian and French immigrants

poured into the district. A prosperous global economy stimulated rapid expansion, but sudden drops in prices, especially during the thirties, visited economic hardship upon monocultural grape growers and increased the difficulties of the small farmers and *contratista* labourers of the oasis. Locked into a rigid land tenure system, they were unable to diversify production on their holdings. While some farmers replaced a portion of their vines with olives, the *viña baja* system of closely spaced vines made intercultivation difficult. Dependence on flow irrigation and an antiquated system of water allocation made water distribution too sporadic to permit the development of vegetable cultivation in the initially developed areas in the northern part of the oasis. Diversification into vegetables based on wells brought prosperity to the larger farms in the southern part of the oasis, but only served to increase the marginality and vulnerability of farmers and *contratista* located closer to the expanding urban zone of Mendoza.

57. The vulnerability of economically and socially marginal people to desertification in Mendoza mirrors the experience of other regions. However, unlike other examples of desertification causes and consequences cited in this section, population growth is not a prime factor in desertification in Mendoza. Rather, antiquated land tenure systems, differential access to technology, and exposure to market fluctuations appear to be prime casual factors. These, together with population dynamics, constitute the social causes of desertification.

B. THE SOCIAL CAUSES OF DESERTIFICATION

1. Population Dynamics

58. The most important process of social change associated with desertification is demographic, the change in the numbers of people dependent for sustenance on the dry lands and in the numbers capable of providing such sustenance. In the conventional wisdom, this is described primarily as overpopulation, or as exceeding some implied carrying capacity of dry lands, at a fixed level of living and technology, beyond which desertification occurs.

59. The physical linkages between population numbers and desertification are well defined. Concentrations of people requiring firewood lead to the destruction of trees and shrubs, the destabilization of dunes, the degrading of vegetation and the exposure of soils to wind and water erosion. Overgrazing, arising from increases in herd size that originate in population increases or concentrations, plays a similar role, as does the cultivation of marginal lands to feed an increasing population. Finally, groundwater depletion is directly related to the amount of use, and waterlogging and salinization often accompany the decrease in fallow associated with more intensive demands on the subsistence base.

60. There are, however, many compensating mechanisms that can lead to population increase with little increase in ecosystem damage. Intensification of production, new livelihoods, migration, and technological change may all mitigate the inevitability of the physical processes. Overpopulation is an oversimplification that ignores the considerable variability in population density found in similar regions, the differences in population processes and trends world-wide, the ability of some societies and ecosystems to absorb considerable population increase with minimal ecosystem damage, and the ways in which a decline in productive population can also lead to desertification.

(a) Population density

61. Aside from their climate and sparse vegetation, there is little that is shared by all the dry lands. Population densities and trends not only reflect the particular socio-economic setting of dry lands, but also the national setting in which they lie. This setting ranges from being a part of some of the world's richest countries to the poorest; from being closely integrated into the world economic system to being isolated and landlocked; from highly specialized, technological societies to near subsistence economies. It is not surprising that the carrying capacities of dry lands cannot be inferred from their current population densities.

62. Indeed, a comparison of estimates of such densities for the rural population of the dry lands as shown in Table 3 is evidence for the variety found in the deserts and desert margins of the world. Even when comparisons are limited to comparable zones and levels of economic development, the differences are highly significant. Densities range between less than one person to 10 people per square kilometre in arid zones, and between less than one person to 100 or more per square kilometre in the semi-arid zones. The greatest range is among the developing countries, with the poorest or least developed exhibiting densities similar to the industrialized nations. In both cases, these reflect the sparseness of settlement and isolation of arid areas.

63. The extraordinary range of population densities accommodated within broadly similar climatic zones should give pause to facile conclusions (Eckholm, 1976) regarding the carrying capacities of dry lands. Production systems have evolved in countries like India or Nigeria capable of sustaining much higher human densities than are found in much of the Sahelian–Sudanic zone or the Middle East. In the close-settled zone of Kano Province, northern Nigeria, population densities average 180 per square kilometre, while portions of semi-arid Rajasthan in India reach densities of 150–200 per square kilometre. Despite such striking examples of success in supporting large populations in areas that elsewhere are lightly populated and/or prone to desertification, there are few comparative studies of the indigenous human ecology of homoclimatic areas. Such studies constitute a first step in explaining these apparent variations. Until this is done, inferences as to inevitable desertification by population growth seem unwarranted.

TABLE 3. *Dryland rural population densities by number of countries and economic development*

Zone and economic development		Number of countries with zonal density/km²					
		< 1	1-10	10-25	26-50	51-100	> 100
Extremely arid:	*N*						
Poorest nations	(7)	3	3	1	–	–	–
Other developing	(12)	5	6	–	1	–	–
Industrial	(2)	2	–	–	–	–	–
Arid:							
Poorest nations	(13)	5	7	1	–	–	–
Other developing	(30)	3	15	8	3	–	1
Industrial	(5)	2	2	–	1	–	–
Semi-arid							
Poorest nations	(10)	1	5	4	–	–	–
Other developing	(31)	2	12	5	6	5	1
Industrial	(5)	2	3	–	–	–	–

64. At the same time, the broad range of population densities conceal a major emergent problem of underpopulation, in that a decline in agriculturally active people may imperil the continuity of an indigenous productive system or the viability of a settlement. In many developing countries, indigenous systems are declining locally in productivity as a result of the temporary or permanent migration of the most active members of the work force. Such migration may be in response to the attractions of higher wage returns (as of Yemenis to Saudi Arabia; Dow, 1975; Shamekh, 1975), or they may arise from the desperate condition of drought (e.g., Faulkingham, 1976; Brooks, 1976) or declining productivity due to desertification, reduction in fallow and soil nutrients, and overgrazing. The initial impact of the out-movement of active males may well be favourable, when through their remittances new resources are brought to an impoverished area. But there are increasing signs that a critical point is reached in which the resource base itself becomes imperilled by the lack of workers needed for the maintenance and operation of labour-intensive cropping systems (e.g. Nolan, 1975). The literature is ambiguous and contradictory when attempts are made to assess whether the migrant and his larger social system is better or worse for the change, and strong opinions are held on either side. The main loss in such situations is the decrease in production from the local resource base that often accompanies such migration, but whether these production systems could survive in any case in the face of changing economic relationships is unclear. Certainly little attention is devoted to their problems and potentials. This neglect is especially tragic if one assumes (as seems likely) that population growth will continue in most semi-arid regions in the decades ahead; under such conditions of depopulation many indigenous production systems will no longer be able to make their accustomed contribution. For those countries with only limited productive opportunities, losses of this type are particularly serious.

65. In the industrialized countries, agricultural production remains high in the face of continued out-migration and decreases in the number of agriculturally active persons. What are imperilled in these situations are the urban-based livelihoods in the commercial and service centres of the lowest level of settlement. In Australia and North America, for example, the smaller settlements of the dry lands face a continuous and unfavourable struggle for viability and survival (Heathcote and Williams, 1976).

(b) Population trends

66. Data with which to analyse the demographic characteristics of dryland populations are scanty, for only 20 of 58 dryland countries have published a census in the last decade. Moreover, the data that do exist are often of questionable accuracy. Still, some figures and trends, however tentative, allow us to make an interim assessment. From among the 58 countries with considerable dry lands, five patterns of demographic change or scenarios for the future emerge. These types are associated with level of economic development, sparseness of rural population and their isolation, and degree of control exercised in settlement policy. Table 4 describes the component rates for each pattern.

67. A significant number of the world's dryland population live in isolated countries that have experienced only a minimum of socio-economic change and integration. The urban and rural population of dry Chad illustrates many of the population trends characteristic of this situation. The traditionally high birth rates remain greater than 40 per thousand, with the death rates between 25 and 30 per thousand (Reyna and

TABLE 4. *Population trends in dryland countries*

		Trend component				
	Pattern (N = 58)	Births	Deaths	Immigration	Out-migration	Overall growth
I.	Isolated, least developed nations (22)					
	Urban	High	Moderate	Moderate	Negligible	Moderate
	Rural	High	High	Negligible	Moderate	Low
II.	Developing nations, uncontrolled settlement (28)					
	Urban	High	Low	High	Negligible	High
	Rural	High	Moderate	Negligible	High	Low
III.	Developing nations, controlled settlement (3)					
	Urban	Moderate	Low	Low	Negligible	Moderate
	Rural	High	Moderate	Negligible	Low	Moderate
IV.	Industrialized nations, uncontrolled settlement (4)					
	Urban	Low	Low	High	Negligible	High
	Rural	Low	Low	Negligible	Moderate	Negligible
V.	Industrialized nations, controlled settlement (1)					
	Urban	High	Low	Moderate	Negligible	High
	Rural	High	Low	Negligible	Moderate	Moderate

Bouquet, 1975). Migration from the rural dry areas, an influential component of many dryland growth trends, is comparatively moderate. Under normal conditions, most of this movement is of a seasonal nature and limited to adult males who migrate to urban centres either within the dry area or in the humid regions where economic opportunities may be greater.

68. Yet even in this state of development, rural out-migration to the capital and other key cities is occurring. In the capital city of Chad, N'Djamena (population 100,000) at the fringes of the semi-arid area, annual population increase due to inmigration was found to be around 4–5 per cent in addition to its 1.5 per cent natural growth rate. However, given the recent civil disturbances in the north of the country it is uncertain how large a percentage of this increase originated from dryland populations. Under drought conditions out-migration greatly accelerates and temporary migration across international borders is also significant.

69. In countries such as Chad, present trends indicate moderate urban growth in small cities and low rural increase. The critical component in this prognosis is the death rate; this can be anticipated to fall in the next two decades with improved health care delivery to children and reduction in infant mortality patterns (Caldwell, 1975).

70. In the midst of development, the dryland population growth trends of Mexico illustrate many of the demographic characteristics of people undergoing rapid

socio-economic transformation. Death rates have been lowered, especially in the urban centres where they are just 8 per thousand, while birth rates remain high, 47 per thousand (CONAZA, n.d.). The rural population, experiencing relative environmental and economic impoverishment, is migrating in accelerating numbers to urban centres both humid and dry. The result is that dryland cities have been growing at an annual rate of 6.0 per cent for the past 20 years with 20 to 30 per cent of the inhabitants coming from the countryside. These cities now account for half of the dryland population (MacGregor and Valverde, 1975). The flight from rural areas to the cities has also had the effect of dampening the rural growth rate considerably to between 0.7 per cent and 1.3 per cent in some areas. The lowered rural growth rate has the effect of reducing total dryland population growth rates below the national average (2.5 per cent compared to 3.2 per cent), although with the urban component it is still high.

71. Although many dryland developing countries have a similar pattern, several exceptions exist. India, with many of the same component trends as Mexico, has not experienced the same degree of out-migration from its arid areas. This is probably due to the absence of strong economic alternatives elsewhere and the success of rural irrigation projects such as the Rajasthan canal project (Singh, Singh, and Singh, 1971). The net effect has been an extremely rapid increase in rural densities. For example, in the Rajasthan arid area, densities have doubled over the past four decades and the growth rate of the arid areas is slightly higher than that of the national average (Fig. 4). While many farmers appear to believe that conditions are worse now than in the past, most observers have concluded that desertification has not increased commensurate with the areas's higher population densities.

72. Very little precise data is available for many dryland populations and the Peoples Republic of China is no exception (e.g. Chen, 1973). Nevertheless, it is probably the leading example of controlled settlement in the developing world. A successful attempt to regulate the movement of people, particularly the rural to urban migration pattern, by minimizing regional economic disparities produces a population trend that is qualitatively different from other developing countries (Tien, 1973). This policy was first introduced in 1958 and is characteristic of most countries with planned settlement policies. The primary effect of these policies is to reduce rural out-migration rather than to reverse the process.

73. Australia and other industrialized countries differ strikingly in their demographic trends. Birth and death rates are low (20 and 9 per thousand respectively) and the

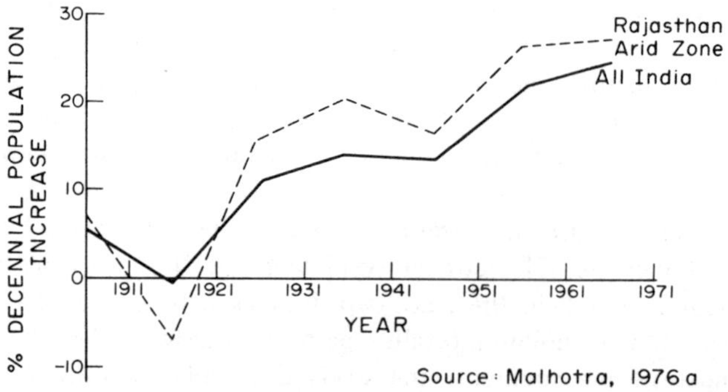

Fig. 4. Decennial increase in population in Rajasthan arid area and in all India

natural growth rate is only 1.1 per cent (Australia, Commonwealth Bureau of Census and Statistics, 1973). Dryland population increase and decrease is strongly affected by migration. There is intense migration to the larger urban centres from both the dry and humid rural areas. The decreasing influence of communal ties, the expansion of communication networks, and the ease of mobility characteristic of an industrialized nation, make the social and economic attraction of the cities more powerful than in any of the other four demographic groups. This movement is encouraged by an economic shift from agriculture to extractive industries, light manufacturing, and tertiary activities. In many rural areas, this combination of low birth rates and out-migration results in a negligible, even negative, rural growth rate. Because of mechanized farming, the rural population of Australia has decreased in both dry and wet areas not just as a percentage of the total, but also in absolute numbers.

74. In the Soviet Union, an industrialized nation with controlled settlement, the demographic trends of the dry lands reflect the government's policy of developing the economic and population base of these regions. Here urban growth has been averaging over 5 per cent and the rural growth rate is over 3 per cent per year (Harris, 1971). It is anticipated that with industrial development, the dry land's growth rates will decrease and conform more to the national norm. At this scale, the 1970 Census indicates that low birth rates and regulated migration have effectively reduced urban growth to just over 3 per cent per year while rural numbers remain constant.

75. In summary, population trends indicate that most of the world's dry lands people are rapidly expanding their numbers, but at rates lower than national averages with major exceptions being India, Pakistan, the USSR, and the United States. This less rapid growth is a product of the lower growth rate of dryland populations who have yet to experience rapid reductions in death rates and of the out-migration of rural people to cities which are often outside the arid zone. In many countries this process is only beginning, and, to the extent that population pressure is related to desertification, the peaks are yet to be experienced. In the wealthiest nations, lower rural population growth rates ameliorate the worst effects of the process. But rapid growth or urban centres and the money economy can also encourage rural depopulation and the loss of labour required to maintain traditional livelihoods in developing countries and the viability of smaller settlements in industrialized countries. These trends are often moderated in both industrialized and developing nations when planned settlement is employed. Overall, in terms of the relationship of demographic change to desertification, the critical components are the expected future reduction in death rates in the isolated, least-developed countries, and the degree to which national governments are able and willing to regulate rural to urban migration everywhere else.

2. Change in Social and Political Structures

76. The societal forces that interact with the physical environment to produce desertification are themselves constantly changing. Just as the dynamics of human populations, in the ebb and flow of their demographic structure and spatial location, are both cause and consequence of desertification, so too are changes in global social, political and economic structures important variables. Because these cause and effect relationships are often indirect, feedback in the system is complex and causes are both multiple and several stages removed from consequences. This makes it difficult to distinguish between

general processes of change and desertification. Yet desertification changes at a local scale can often be traced back to forces and processes set in motion in distant places, while the problems of desertification (and, in more dramatic fashion, drought) in semi-arid regions set in motion events whose ramifications ultimately impinge upon the resources and consciences of the denizens of more favoured climes.

77. For much of the dry world it is a shift from relatively closed and isolated to increasingly open, interlocked, and interdependent systems that has significant influence on desertification causes and serious impact on the well-being of communities exposed to desertification. In particular, changes in the social order, varying rates of technological growth, and fluctuations in the authority and wisdom of central governments all intersect, at differing scales, with the livelihood practices of dryland peoples, to produce or compound the problem of desertification. Some ramifications of these processes, and their impact on the health and well-being of individuals and groups, are outlined in these and subsequent sections.

(a) Integration, specialization and constriction of livelihood systems

78. Dry lands were the locus of some of the great civilizations of antiquity. As such, they became centres of regional and even continental integration of power, trade, and production. Periods of power and prosperity have fluctuated with periods of poverty and isolation. Typical of these fluctuations are the 2½ cycles of population growth and decline over 4000 years that have been documented for the Diyala floodplain of Iraq (Adams, 1965; see Fig. 5). The causes of these great fluctuations are complex, but they involve dramatic changes in the social order occurring at a time when an over-extended irrigation system was struggling with waterlogging, siltation, and salinization.

79. We live in a new interdependent world where these historical centres have been relegated to the periphery. The links of this interdependence were forged by the sixteenth-century European expansion into Africa, Asia, and the New World and reconstituted in a somewhat different form by the nineteenth-century imperial powers. This undertaking destroyed or dismantled many non-Western civilizations, decimated or

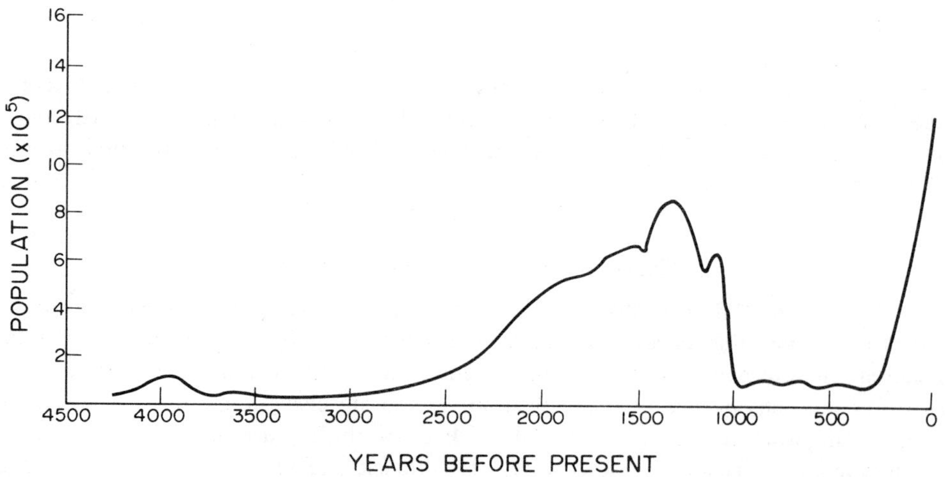

Fig. 5. Population fluctuations in the Diyala floodplain

enslaved populations, and, in Saharan Africa, reoriented an entire system of trans-continental trade to serve new North American and European ends. The wealth generated in the process of transforming old centres into new peripheries flowed to the growing European core, fuelling an industrial revolution which, in its turn, created new integrative bonds.

80. This interdependence is still with us, albeit clothed in a twentieth-century, post-colonial garb. The basic trend is toward wider coordination into a world system of production and exchange. Isolated and self-sufficient systems are increasingly becoming the exception rather than the rule. Systems tied to a world market with specialized functions are the new norm.

81. These world-scale processes also operate within nations as dryland regions increasingly become peripheral to more humid centres of power and production. Nonetheless, they continue to provide needed raw materials and foodstuffs. Thus, milk moves nearly 500 kilometres from the Rajasthan desert to the markets of Delhi, and beef moves 7500 kilometres from the central desert of Australia to the markets of Tokyo.

82. This economic and political integration and specialization is further complicated by the selective constriction and isolation of certain livelihood systems. For example, the often artificial borders of the now independent, former colonial states place serious new obstacles in the path of the long seasonal migrations characteristic of animal-based livelihoods at the same time that traditional grazing lands have been occupied by agriculturalists.

83. The implications of this historical evolution for an understanding of desertification are manifold. As traditional societies inhabiting what until recently had been peripheral dry lands undergo a process of change, and old livelihood strategies become unviable options, both the natural and the cultural components of their ecosystem deteriorate. O'Keefe (1975) and Wisner (1976) illustrate the unfolding of this process at the village level in Kenya. When additional stresses, such as drought, are placed upon the system, it becomes hard for it to recover on the basis of purely endogenous resources. Yet, due to its still peripheral status, outside resources do not readily flow in upon demand, thus further hampering recovery. Growing integration into a world system has meant relief in some cases, and, over the long run, perhaps new opportunity. However, to date, relief has generally been limited to palliative efforts and we have yet to see the new interdependence of the world translate itself into positive development for these threatened arid ecosystems.

84. Similar implications hold even in wealthy countries. The arid lands of the wealthy countries play little or no subsistence role for their own peoples, only serving distant export markets through capital-intensive productive systems. The problems are not nearly as dramatic except for some marginal peoples of industrialized centres, e.g., Australian aborigines and native North Americans. Nevertheless, because ecosystem management becomes subject to the vagaries of distant markets, economic demands may be out of step with climatic potential and the basic system inevitably leads to the loss of smaller communities and populations.

(b) The invasion of new and retreat of old technology

85. There is no such thing as a non-technological society. All peoples possess knowledge and equipment enabling them to produce the means of their subsistence. Some societies, however, possess a powerful and modern technology while others possess

technology that has not changed substantially since the neolithic. This unequal development of different societies' productive forces has meant that there has been a retreat of the indigenous technologies and a growing hegemony of Western science and technology. This process unfolds along two characteristic paths. The first occurs when indigenous technologies atrophy, land use changes, and a technological vacuum is created. Entire spheres of productive activity then disappear, resulting in a loss to productive use of significant portions of the ecosystem. In arid and semi-arid lands this may entail desertification. For example, once ancient water-collecting systems are first neglected and then abandoned, the investment in labour and capital required to bring them back into production is too great and they may disappear forever.

86. The second path along which indigenous systems retreat involves their direct competition with modern technology. In these instances modern productive enterprises compete successfully with traditional systems, enabling the former to gain control of scarce resources: water, land, and labour. This, in turn, may lead to the wholesale abandonment of traditional resource-use techniques. In these areas where the competition is not with agriculture but with urban spread, the trend has accelerated. Iraq, for example, has lost almost half of her date palms to urban-based livelihood systems in the past 15 years (Fahmi, 1976).

87. The transition from indigenous to modern technologies has not always brought about the desired levels of social well-being that an increase in productivity might suggest. The substitution of modern for indigenous technology, linked as it is to rates of structurally unequal development, has proven an inadequate vehicle for integrated social change. Punjabi farmers were among the major beneficiaries of the Rajasthan irrigation canal, but the nomadic inhabitants became worse off with a shrinking grazing area (Malhotra, *et al.*, 1976). Moreover, the new technology is often available only selectively to members of the traditional community. Those owning land, possessing capital, or having access to education and information are most likely to be in a position to benefit from the new introductions. Not only do the resulting introductions often increase existing social disparities, but also they tend to eliminate labour-intensive coping mechanisms found in more traditional technology and supported by indigenous value systems. The consequence is often increased pressure on the most marginal resources, partial – and frequently destructive – integration of technology into local productive systems, and the breakdown of existing, often supportive, social arrangements.

88. But something more than well-being is lost in this unequal contest. The natural evolution of indigenous technologies is truncated by the importation of exotic techniques bearing no organic relation to people's ecological setting or to their culture history. What often confronts them is the invasion of modern technology, rather than its rational integration into the productive fabric of the society. Since industrial technology, and the management practices associated with it, are generally outside the experience of practitioners of the traditional system, or run counter to their objectives, values, and customary wisdom, adoption of such innovations is often retarded or is rejected outright. Careful attention to the fit between the traditional and the modern, and to the identification of appropriate technologies, is an essential prerequisite in the struggle against desertification.

89. By truncating the evolution of indigenous technologies, the world also loses much accumulated folk wisdom. These indigenous technologies are not simply anthropological anachronisms, but contain valid, time-tested principles built on detailed local knowledge.

With the new opportunities and threats that arise from a wider world integration, specialization and constriction, they need to be given opportunity to evolve and change. Unfortunately, they tend to disappear. For example, there is probably no better way to extract the diffuse and marginal resources of arid lands than pastoral migration. But most efforts at change seek to sedentarize pastoralists and to limit them to fixed ranch locations, instead of seeking to modernize pastoralism (using radio, trucks, etc.) in order to capitalize on its mobility. The destruction rather than the evolution of pastoralism can lead either to a new loss of productivity or to overgrazing of a more constricted area, both of which are forms of desertification.

(c) Political structures and desertification

90. Ironically, central governments are responsible both for resisting desertification and, generally inadvertently, for stimulating its spread. This ambiguous situation can be illustrated by several examples, with irrigation civilizations being a particularly poignant case. While it is unlikely that an integrated centralized state was an essential causal mechanism in bringing complex irrigation systems into existence as is suggested by some (Wittfogel, 1957), it is certain that, once established, co-operative activities such as canal cleaning and maintenance require considerable centralized direction if an integrated system is to be maintained.

91. The Kara-Kum canal in Soviet Central Asia is an example of the way central governments can organize resources and overcome scale problems to bring water to areas that once were seasonal pastures (Petrov, 1976). However, historical experience shows that the system created is utterly dependent for success upon the continuity of governmental resources, attention, and authority. Any decline in authority is quickly followed by problems of canal maintenance, difficulties in water distribution, and a diminished ability to support existing populations (Adams, 1965; Fernea, 1970).

92. Run-off farmers in semi-arid areas are also vulnerable to changes in governmental effectiveness. Ecologically astute systems (Kassas, 1972; Evenari *et al.*, 1971) that make use of extremely limited and sporadic rainfall, they depend on the security provided by central governments to resist the potential depredations of their nomadic neighbours. When the government's fortunes decline, these marginal agricultural regions slip into eclipse. Such episodic collapses have resulted in the desertification of large tracts of semi-arid and arid lands, as evidenced by the fluctuating fortunes of the Diyala floodplain (Fig. 5).

93. But just as central governments can increase productivity by providing support to irrigation and run-off systems, so too can their demands for increased productivity, combined with technological difficulties, contribute to desertification. Salinization appeared by 4000 B.P. in the Tigris and Euphrates floodplain as a result of over irrigation and poor drainage (Jacobsen and Adams, 1958), while decreases in the fallow cycle may well have been the mechanism that instituted desertification in much of Mesopotamia (Gibson, 1974). Similarly, government efforts to promote agricultural development in desert oases have often dramatically increased both salinization and sand-dune encroachment (Meckelein, 1976). In the pastoral sphere, governments have done much to develop water resources. While intended for use by the pastoral community, these well and surface flow collections facilities have frequently become growth points for a settled agro-pastoral population. The resulting intensification of grazing pressures frequently

contributes to wind and water-borne soil erosion, to impoverishment of vegetation resources, to increasing administrative and legal difficulties for mobile pastoralists (Ahmed, 1973), and to the spread of desertification. Indeed, there is reason to fear that an inevitable corollary of development may be the enhancement of desertification, unless fundamental changes in priorities occur, for a recent simulation of the effects of the aid programmes proposed for recovery in development of the drought-afflicted Sahelian region showed that the result to be expected is rapid and complete desertification of pastoral areas (Picardi and Seifert, 1976).

94. In summary, central government policy and practice are intimately associated with desertification. Its record over both long and short time scales is a chequered one and, like so many other facets of exploitation of the humanized earth, both good and bad consequences have resulted. In many cases the security and central direction provided by governments has been essential to the use of dry lands. But the very conditions that encourage exploitation of the dry margins may also promote the spread and intensification of desertification, thus contributing to the synergistic character of the problem. No special aura of success and rationality is attached to central bureaucratic structures and they should be employed only with caution and with a healthy respect for their limitations as well as their potentials.

C. THE SOCIAL CONSEQUENCES OF DESERTIFICATION

95. The consequences of desertification are manifold and extend from decreases in the productivity of the resource base to the emotional and physical traumas visited upon individuals and societies. However, discussion here is limited to those that effect, or are manifest in, social practices and structures. Two levels of consequences are identified: those that are highlighted in the direct experience of individuals and small groups, and those that have a far-reaching effect on large numbers of people. Within this latter category two particular consequences, migration and changes in health and well-being, seem to be most important.

1. The Human Experience of Desertification

96. An understanding of the full implications of desertification hinges, in part, on the ability to view the process at an immediate human scale. The family household or community occupying an area undergoing desertification may have to respond to short or long-term ecosystem change, to the movement and growth (or depletion) of animal and human populations, to changing relations in the total socio-economic system, and to the obsolescence of old and invasion of new technologies. The nature of this response will increase or decrease the family or community members' vulnerability to the desertification process. In all cases the family or community response is linked to a series of desertification-induced pressures which constitute the immediate consequences of desertification. These consequences are felt in their most acute forms by those marginal peoples least able to bear them.

97. Drawing upon data contained in the Profiles of Change (available in the supplementary materials) this section reviews the consequences of desertification as experienced by selected vulnerable peoples. Because the dynamics of productivity loss are

analysed by Warren and Maizels (1976), the aim of this brief section is to highlight the effects of desertification upon human livelihood and sustenance.

98. The consequences of desertification upon agriculture-based populations range from loss of their resource-base productivity to imperilled health and forced migration. The short case studies tell of:

. . . peasant farmers in Central Mexico whose deteriorated terraces retain less field moisture. This leads to lower yields and often to crop failure in three out of five years;

. . . sharecroppers in north-east Brazil whose holdings are affected by soil erosion and declining fertility; when their households' ability to produce enough food to cover basic needs and to survive drought is threatened, impoverished farmers stream into urban areas;

. . . peasants in Rajasthan who must contend with overgrazed pastures, deforestation, soil erosion, and salinization of irrigated lands; growing numbers force cultivation of marginal lands, thus increasing vulnerability to drought as well as contributing to the degradation of their resource base;

. . . Argentinian sharecroppers and smallholders whose lands are being waterlogged and salinized and who cannot afford the capital investment necessary to reclaim them; faced with declining yields, as well as inflation and uncertain market conditions, these skilled farmers often must give up their holdings and migrate to low-paying jobs in the cities;

. . . commune dwellers in the People's Republic of China who must mobilize large quantities of labour in order to combat dune encroachment. In one county, peasants established a tree belt 5 kilometres long and ½ kilometre wide in order to hold down drifting sands; they constructed sand dams totaling 35 kilometres in length in order to expand their fields; and they drilled thousands of wells in order to combat drought;

. . . Cape Verdean peasants who underwent famines due to a deteriorating resource base and social inequities and who now, following independence, face the task of rebuilding a degraded resource base on the strength of mass mobilization and self-rule.

99. Desertification also entails pressures upon the livelihood patterns of animal-based peoples. Often these pressures threaten an entire way of life and compel sedentarization. Thus:

. . . Fulani nomads find it increasingly difficult to reconstitute their herds after drought and 25 per cent of those who arrived at one refugee centre during the last Sahelian drought are now considering non-pastoral work because they lack sufficient livestock to resume pastoralism;

. . . Iranian settled nomads find their agricultural resource base inadequate to sustain their needs and health as evidenced by low standards of living and a high infant mortality rate;

. . . Pakistani fishing peoples are faced with the prospect of a lake drying up due to drought and to upstream irrigation water withdrawals which deprive them of their livelihood and contaminate their water supplies.

100. By drawing upon outside resources, urban-based populations can sometimes buffer the immediate effects of desertification. Nevertheless, rapidly growing urban concentrations often place strains on their immediate hinterlands. This can occur in both large- and small-scale concentrations. For example:

. . . newly settled *ujamaa* villages in Tanzania must contend with devegetation and diminishing soil fertility in the environs of their settlement; this implies that scarce resources will have to be allocated for afforestation projects and for fertilizer;

. . . so far, Tucson, USA, has been able to provide for its water needs from ground water supplies. However, if demand for residential and industrial water keeps increasing, the city may be faced with water resource depletion and the need to divert water away from its highly specialized commercial agriculture.

101. The experience of desertification is a widespread one in all dryland societies. In industrialized and urban livelihoods the resources available to combat desertification are greater and the possibilities for transcending its consequences are more numerous. For families and livelihoods on the margins of the modern world, or for those isolated by tradition, social inequalities, ignorance, and indifference from access to the resources and organization required to resist desertification, the experience of productivity loss is often bleak and barren. For them diminished health, social stress, stagnation, migration, and the abandonment of time-honoured ways of life constitute the human experience of desertification and mirror its social consequences.

2. The Social Consequences of Desertification

102. The experience of desertification at an individual, human scale is often a subtle process that accumulates slowly and is often difficult to distinguish from other changes that are taking place. This is especially true when large groups are considered, and the difficulty is compounded by both lack of data and insufficient focus in previous work on the desertification process itself. Nonetheless, both in terms of causing movements of people and changes in health and well-being, desertification has a significant, if at times ambiguous, impact on the structure, sustenance, and stability of social systems.

(a) Migration as a response to desertification

103. Mobility within local rural areas and out-migration to settle in urban areas are adjustments to environmental and political conditions in dry lands that have existed for centuries in both nomadic and settled livelihoods (Lee and DeVore, 1968; Mitchell, 1969; McGee, 1973; Mangin, 1970; Redfield, 1956; and Wolf, 1966). In this light, migration in response to the long-term decrease in productivity associated with desertification can be seen as an adjustment that is a continuous extension and expansion of an established adaptation.

104. Except for the sudden impulse given to migration by prolonged drought, migrants

from desertifying districts pass unnoticed in the larger flow of population from rural to urban centres. Initially, migration represents a short-term adjustment to a temporally limited period of adverse conditions. Migrants in this situation exploit urban resources only during crisis (Graves and Graves, 1974) and envision returning to their village setting once environmental conditions improve. Only when drought continues for several consecutive years, often exposing underlying desertification or raising doubts in refugees' minds about future resource base productivity, do migrants plan to remain in, or to seek alternative livelihood options. This apparently accounts for the large number of those interviewed who indicated unwillingness to return to their original Sahelian habitats even if some governmental assistance were provided (Laya, 1975).

105. Longer-term adaptations generally involve the establishment of more or less permanent ties between two economic systems, one rural and one urban, and the periodic movement of labour between them. The initial result of this linkage is to improve the total productivity of the rural system in question. Nolan (1975) indicates that the situation of the Bassari community in eastern Senegal was improved by monetary remissions from migrants engaged in urban employment. While these cash returns seldom are very large (Campbell, 1976b; Nolan, 1975), they are essential to largely subsistence communities if taxes are to be paid and consumer goods purchased. As long as the labour consumed by urban employment is small and the migrants return to provide needed communal labour during crucial periods of the agricultural year, the results are an improvement or stabilization of existing living standards. But when labour migration begins to compete with the labour required to keep the traditional agricultural system functioning, conditions rapidly worsen in the traditional sector. At the same time, for those migrants who choose to remain in urban employment for extended periods, the change frequently involves an improvement in economic potential and living standard. The resulting network of kin living in rural and urban areas serves as a communications network for the flow of people, information, and materials, and provides a ready-made point of contact in the urban milieu for those individuals who wish to follow their kinsmen into an alternative life style.

106. This network is particularly important when desertification pressures become intense, since a supportive network of relationships exists to assist a more rapid and large-scale movement away from afflicted districts. Yet the very success of urban migrants often undercuts the viability of the traditional sector. As soon as too many migrants stay away too frequently for too long, the productivity of the rural resource base decreases as the labour available is no longer sufficient to maintain yields. Migrants in this situation become both cause and consequence of desertification. An example, among many, is provided by the Otomi Indians of central Mexico who, in order to seek wage employment outside their communities, are forced to neglect their terraced fields. The unmaintained terraces erode and fail to provide the moisture-retaining capacity needed to produce dependable yields under erratic rainfall conditions. In good years, the average family holding yields a 6-month supply of maize. In bad years it yields nothing. Farmers are often even unable to harvest enough fodder to keep their livestock through the winter dry season. This, in turn, places a heavier burden on the already overgrazed communal pasture land. Consequently, precious animals weaken and die during the winter months, further threatening the family's survival-hedge. The old pattern of life appears inadequate to maintain Otomi households, and thus accelerates the out-migration of much needed labour power (Johnson Haring, 1976b).

107. The search for economic improvement in urban areas thus involves more and more individuals. As social bonds between migrant and home base weaken, those left behind are increasingly marginalized and a debilitating retrogressive spiral can be instituted. What initially appeared to be a stable dual economy becomes a prescription for decline at the margins. Physical and social desertification march hand in hand, often spurred on by drought, and increased migration often represents the individual's and the group's only option. The motive underlying this spontaneous migration is economic; both the image and the reality of improved well-being attract the migrant. It seems likely that resettlement and development schemes designed to combat desertification will be most attractive to potential migrants and will have the best prospect for success if they place their highest priority on the economic benefits that will accrue to participants. Similarly, the flow of migrants can be channelled most judiciously and their integration into new livelihoods can be accomplished with least stress if migrants are provided with needed skills via relevant educational programmes.

(b) The health and well-being of individuals and groups

108. In regions undergoing desertification, many sedentary farmers, nomadic pastoralists, and recent urban migrants are caught in a web of inadequate diet, disease, poor sanitation and inadequate health care services. They suffer physical debilitation, demoralization, and emotional stress as a result. This traumatized physical and emotional state hampers the adaptive abilities needed for successful change. Yet in experiencing these calamities, desertification-afflicted communities and individuals are not appreciably different than other social groups involved in rapid and traumatic change. Similar problems emerge and there is little that is distinctively and uniquely related to desertification. Only when desertification, drought, and disease coincide are the chronic problems of desertification-prone populations different in scale from those experiencing more general pressures associated with development and culture change.

109. If there is one health aspect of desertification that is popularly recognized as a serious problem, it is malnutrition and, ultimately, death by starvation. Certainly the loss of productivity characteristic of desertification leads logically to the conclusion that death by starvation is inevitable for a population without alternative life-support options. However, the evidence to support this belief is lacking, and most existing information is related to drought-induced famine rather than desertification. Malnutrition was found among migrants to refugee camps during the Sahelian drought, but levels were not appreciably greater than those of individuals who did not move to camps (Seaman, *et al.*, 1973), perhaps because famine is essentially over before action can be taken to cope with it (Gebre-Medhin, 1974). Nomads exhibited higher levels of malnutrition than migrants from sedentary populations (Greene, 1975; Sebai, 1969; Omolalu, 1976), but this differential is as easily correlated with distance from relief outlets as to generally lower nutritional levels or to greater ecological vulnerability.

110. Yet crop failure, famine, and disease are related, as are the intensifying features of drought years in desertifying regions. During the recent drought in the Sahel, Hausa cultivators may have suffered up to a 70 per cent loss of their normal harvest. Hard pressed farmers resorted to cash cropping in limited well-watered areas, wage-labour, out-migration, and foraging in order to feed their families. In spite of this, a food crisis ensued. In part, this was due to the failure of cash crops on which peasants depended to

provide capital for essential purchases. Other factors may have included the calorie and protein deficiencies that frequently develop when migrants move to areas where traditional foods are not available and new substitutes are often nutritionally inadequate. In addition, as new technologies are introduced, folk use of important resources such as famine foods frequently diminishes; in desertifying settings such foods may no longer survive to nourish a drought-affected population.

111. Despite these nutritional problems, drought-related deaths in the Sahel were less numerous than originally feared precisely because the region's integration into the global economic and political system permitted a rather rapid and effective response. Where mortality did increase, it was disease related and affected the young, aged, infirm, and pregnant as opposed to the entire population. This pattern of drought, famine, and disease in areas exposed to desertification is closely paralleled among peasant cultivators in semi-arid Kenya who must operate within a constricted resource base. Their surplus product in good rainfall years is insufficient to carry them through the bad; this places the vulnerable members of the population, the very young and the old, in jeopardy from malnutrition and gastrointestinal diseases (Wisner, 1976). But it is important to emphasize that desertification's impact on nutrition is indirect, and that no evidence presently exists correlating malnutrition and desertification.

112. Analogous difficulties obscure the relationship between desertification and infectious diseases. Only trachoma exhibits an apparent distributional relationship between aridity and incidence of the disease (Tarizzo, 1976), but this is more apparent than real. Its control has been achieved outside the dry world and only isolation and insufficient resources prevent trachoma's demise within it. Other diseases, such as leishmaniasis, schistosomiasis, and trypanomiasis, are associated with attempts to make dry environments more productive (Ormerod, 1976), but few would forego these projects to avoid disease. Rather, they would try to minimize the disease potential. Dust exacerbated respiratory diseases are likely to be somewhat more frequent where desertification is increasing, but these do not seem to be significant problems. Sanitary problems associated with sedentarization and migration to urban concentrations are hardly unique to the desertification process. There appears to be no diseases or related health problems so closely related to desertification that they qualify as fellow travellers which require singling out for specific attention. This conclusion does not in any way eliminate the need and responsibility to focus international attention on these and other health problems related to development and human well-being. But it does suggest that there are few health problems uniquely linked to desertification and that the resources available to combat desertification *per se* should be targeted in other areas.

113. Certain longer term changes affecting the health and well-being of social groups should also be noted. Loss of a privileged social position placed most pastoral groups in a powerless position *vis-à-vis* governmental authority in the pre-drought era and increased the sense of helplessness engendered when environmental adversity arrived to compound political and economic constraints. Many farmers most seriously affected by the drought were also economically marginal, many having migrated into agriculturally vulnerable areas when their own agricultural land was pre-empted for commercial production. Use of increasingly marginal resources invariably involved greater distance between and less social interaction with compatriots. As resource productivity decreased, social bonds that once were crucial to individual and group viability were shattered because neither the resources nor the coincidence in time and space were sufficient to maintain them. Thus social

networks that normally supported individuals were already in decline in many areas before the Sahelian drought struck and coping mechanisms were in an atrophied state. Moreover, vulnerability increased in direct proportion to the breakdown in social networks, the decrease in livelihood interdependence, and the increasing tendency to concentrate on specialized production. All these factors contributed to a lack of social, economic, and political well-being which was preset for disaster when drought exposed basic desertification pressures.

114. Yet despite the multitudinous stresses brought to bear on individuals and groups, basic adaptability remained. Many nomads stressed in this way showed considerable willingness to engage directly in agricultural development projects such as Tin Aicha in Mali (Smith, 1976). Farmers were able to engage in wage labour migration without abandoning their commitment to the traditional village scene (Faulkingham, 1976). Moreover, nomadic groups have shown considerable ability to bring their production activities more closely into line with modern organizational concepts. Mongolia's experience indicates that stress can be moderated in a development programme if attention is paid to traditional patterns (Aubin, 1967; Humphreys, 1976), while traditional *hima* systems of grazing resource control have been found suitable for development in a co-operative context (Draz, 1974). The record, therefore, is ambiguous. Severe stress is apparent in many areas, yet it is difficult to isolate those stresses stemming from desertification itself rather than from drought or social and economic change. And evidence for successful adaptation is also present. Little work has been done on the psychological characteristics of much of the non-Western world (an exception is Swift and Asuni, 1975) and full elucidation of the problem awaits further research.

D. SUMMARY

115. Desertification is a complex process, and is the product of the interaction of social change with fluctuations in climate and ecosystems. The social processes are usually those associated with modernization and integration of social systems, population shifts, changes in the geopolity and political economy of regions and countries, and the sometimes careless replacement of traditional techniques by industrialized technologies.

116. Natural fluctuations have always occurred in the climates of the dry lands. Both natural ecosystems and older livelihood systems have worked in response to this inherent variability. When these fluctuations are modest and of short duration, their impact on desertification is minor. Climate becomes an active contributor to desertification when abnormally persistent dry or wet periods occur. Drought can lead to desertification by decreasing vegetative productivity and delaying its recovery rates. Ironically, exceptionally wet fluctuations may also promote desertification by encouraging extension of agricultural and pastoral settlement into areas generally too dry to utilize. Wet periods may also encourage an intensified pattern of land use. In both cases, the processes can lead to desertification when the climate reverts to its drier, less productive norm.

117. The most dramatic process of social change is demographic, the rapid change in the number of people utilizing and dependent on the resources of the dry lands. Often imbalances caused by population growth or concentration place increased pressure on the local resource base. When this occurs, the intensified utilization of the resource base without compensating increases in productivity can lead to soil erosion, degradation of

plant cover, groundwater depletion, waterlogging, salinization, elimination of fallow periods, and other phenomena that cause long-term deterioration.

118. Rural depopulation is also a potential contributor to desertification. Out-migration of economically active rural people may increase levels of living in the short term. But in the long run, competition for labour between rural and urban spheres tends to undermine the viability of rural livelihoods which require intensive labour or specialized knowledge. As a consequence, the productivity of their resource base is likely to decline.

119. To the extent that population growth exacerbates desertification, a prolonged worsening of conditions can be anticipated in many rural and most urban areas of the developing world. For industrialized nations the threat posed by population growth is centred on the larger urban areas. Where depopulation contributes to the process, it appears that selected rural areas of developing nations are most threatened by desertification.

120. Demographic trends indicate that population imbalances can be expected to increase in the near future. In the dry rural areas of developing countries, the imbalances will be the product of a steady reduction in death rates without a concomitant reduction in birth rates. Urban populations will grow at even higher rates because of both immigration and local growth. These rates of growth, particularly the urban component, are expected to accelerate in the future as the death rate is further reduced in the countryside and as people move off the land and migrate into urban centres in increasing numbers. Only countries with controlled settlement policies appear able to moderate the rural to urban flows of population by emphasizing the economic viability and social well-being of the countryside.

121. A study of population densities indicates that great variation exists, even among peoples who inhabit homologous climatic regions. The densities reflect the national settings of the dry lands and the great variation in the productivity of the different livelihood complexes employed. The extraordinary range of population densities accommodated within broadly similar climatic zones should discourage facile conclusions about fixed carrying capacities in the dry lands. Conversely, the findings encourage optimism about the flexible and dynamic ability of dry lands to support human populations.

122. A more complex process, and one which has received less careful attention, is the impact that world integration and specialization has had on the traditional livelihood practices and adaptive techniques of dryland peoples. Frequently on the periphery of the world economic system, they are vulnerable to external market forces over which they have little or no control. Often modernization has brought more problems than benefits to these peoples as traditional cultural ecological systems lose their viability and no alternatives take their place. When additional stresses, such as drought, are placed on these systems, it becomes more difficult for them to recover on the basis of their own resources. Yet due to its peripheral status, outside resources are not fully available to aid in recovery. In extreme cases, the productive base may be so degraded that people are forced to abandon their traditional livelihood and habitat entirely.

123. Desertification is also exacerbated by rapid but often incomplete diffusion of unsuitable technologies into traditional societies. Unfortunately, arbitrary implantation of exogenous technologies has triggered processes that have led to long-term loss in the productivity of many dryland areas. In addition, the introduction of such technologies truncates the balanced evolution of indigenous systems of production and desertification

prevention that contain valid, time-tested principles of dryland resource management. The constriction of both livelihood systems and indigenous expertise with proven viability is a serious contributor to the desertification process and a severe handicap in efforts to combat it.

124. The health impact of desertification is more ambiguous. The nutritional mal-effects of decreasing resource base productivity seem straightforward, but little evidence of adjustment to such conditions is presently available. Most existing information relates to drought, an exacerbating factor in desertification, rather than to desertification *per se*. Other diseases also fail to exhibit a close linkage to desertification, being, instead, more directly related to general changes in environment or location or both. The psycho-social stresses experienced by desertification-prone populations are similar to those encountered by most individuals and groups exposed to sudden and rapid change. Their trauma is neither unique nor extreme; indeed, there is evidence to suggest that, provided considerable latitude is available for free choice and for preservation of desired social norms and objectives, the practitioners of traditional livelihoods are capable of adapting successfully to altered conditions.

125. A different type of ambiguity surrounds the role of migration in desertification. In the short term, migration appears to have positive effects; income is increased, extra-regional employment cushions the group from the worst features of local environmental perturbations, and outlets for potentially surplus population are created. Paradoxically, the very success of migrants in adapting to economic opportunity threatens the survival of indigenous livelihood systems. This occurs when labour is shifted into urban and commercial sectors of the economy and the traditional livelihood can no longer maintain basic productivity. The corollary of successful migration often is a weakening of social ties with the home region, a decline in the productivity of the traditional system and an increased marginalization of those who remain behind.

126. A similar array of positive and negative consequences flow from government actions. Often essential to the support of agricultural livelihoods in dry lands, governmental activities are also simultaneously responsible for much desertification. Thus, the government that provides the security, technology, and organization to extend irrigation into new districts often spreads salinization or, by diverting former grazing land to agricultural use, is responsible for increasing desertification pressures on adjacent rangelands. The role of governments in desertification is a chequered one and the very governmental factors that promote use of dry lands often cause their desertification as well.

127. Each of these natural fluctuations and social processes has been identified where desertification occurs. But rarely does one process alone appear to be an adequate explanation for the decline in the productivity of a dry land's resource base. Rather their complex combination and interaction is responsible for most of the world's desertification. Efforts to understand the process and to combat it will be more effective if this causal complexity is recognized.

IV. COPING WITH DESERTIFICATION

128. Desertification is inherent to the human use of dry lands and alternating periods of intensification or decline of desert-like conditions are evidenced clearly in their long history. There are many ways in which dryland peoples have learned to cope with desertification, and new techniques are under development. Four principal modes of coping with desertification may be discerned.

A. ABSORBING DESERTIFICATION EFFECTS

129. Human livelihood systems are able to absorb some significant degree of desertification with little or no harmful effect. This occurs because ways of coping with desertification are deeply embedded in the everyday functions of livelihood systems. But because they are part and parcel of a way of life, they are difficult to isolate.

130. Sometimes the reduction of hazard occurs incidentally as a byproduct of some other functional purpose. For example, the practice of alternate-year-fallow grazing in irrigated areas of Iraq was basic to animal maintenance in an essentially agriculture-based livelihood. Employment of at least half the arable land in fallow fodder also served, incidentally, to reduce waterlogging and salinization (Gibson, 1974). Abandonment of this practice causes a rapid rise in the watertable with a consequent increase in saline soil levels unless large investments are made in drainage.

131. The most common adaptations are found in the myriad ways in which the culture of dryland livelihood systems copes with threats to, and fluctuations in, its resource productivity. Such everyday practices range from the basic principles of mobility inherent in pastoralism to the tree worship rituals of Rajasthani women that ensure the regular watering of important but exotic desert species. The latter indicates the difficulty in identifying the functionality of resource practice. In the view of the forester, it is a practice that provides shade and species reproduction; in the view of the practitioner, it is a measure of devotion.

132. Taken all together, these different mechanisms provide a livelihood system with a capacity to absorb desertification without much harm. Unfortunately, we have no quantitative estimates of the degree to which a society can suffer the productivity loss of desertification with little or no ill effect. From other hazards, especially drought, we know that short-term losses of up to 10 per cent can be easily coped with.

B. ACCEPTING DESERTIFICATION EFFECTS

133. Regardless of the built-in capacity of a system to absorb desertification with little or no harm, such capacity can eventually be exceeded and an awareness of desertification ensues when stress becomes apparent. We know of only one study to date which seeks to identify the perception of desertification by farmers, an opinion survey of 74 farmers in a 10-km wide transect across the 300 mm isohyet in Rajasthan (Malhotra, 1976c). Their perception of fluctuations in traditional wind patterns, degradation of vegetation, soil erosion, and salinization that have taken place over the span of one generation is not only

a measure of the severity of these processes, but is also in accord with current scientific observations.

134. When such awareness develops, the initial response is to *accept* the consequence of desertification. Individuals and societies bear their losses in health, wealth and productivity when they occur, or share them with kin, organized insurance, or community, national and international relief. However, except where acute disaster intersects with desertification, the slow chronic decrease in productivity may get little or no organized attention, leaving the burden to fall on the individuals, communities and regions concerned.

C. REDUCING DESERTIFICATION CONSEQUENCES

135. A continued decline in productivity cannot be sustained over a long period of time. Individuals and societies act to *reduce* desertification either by modifying or preventing the physical processes or by diminishing the consequences and loss potential.

136. The rich repertoire of techniques to counter dune and sand encroachment, to reduce vegetation degradation and limit the consequent soil erosion, and to prevent waterlogging and salinization is described in the companion review on technology (Anaya-Garduño, 1976). What that review does not address are the social obstacles to adoption of new technologies or developing indigenous areas. Three major obstacles stand in the way of adopting new technologies or furthering the evolution of older ones: difference in perceptions, productive goals, and scale of social organization.

137. No better example of differential perceptions of the desertification problem exists than in comparing the image of erosion held by a technician with that of a traditional dryland farmer. For the technician erosion generally is an unmitigated evil. Whenever deep gullies develop on a denuded hillside the process is seen as one that irrevocably reduces the productivity of that ecological setting. The peasant farmer often does not view erosion as a problem; rather, to him it appears to be a desirable situation replete with opportunities for increasing productivity. Gullies can be blocked by low stone walls and both soil and water concentrated behind the dam (Kirkby, 1973). The fields produced in this manner are more productive, have greater fertility, and, by retaining more moisture, resist drought longer than do cultivated plots on neighbouring hillsides. Similarly, the development of alluvial soils in larger valleys often is enhanced by erosion and richer and flatter fields are developed in the process. It is obvious that the solutions to gullying and erosion proposed by and acceptable to dryland farmers and modern technicians are likely to be very different and may well be mutually incompatible.

138. Even where desertification and the technique for coping with it are similarly appraised, difference in acceptance may be great because of different goals or capabilities for change. Plant cover, an integral component of dune stabilization, often serves a dual function in people's livelihood systems as a source of firewood, forage, and food. In times of stress a tension occurs between people's needs and the sound management principles detailed in the companion review on ecological change (Warren and Maizels, 1976). Other adjustments, such as enclosures or the barring of cultivation in areas perceived as marginal, while serving sound ecological ends, may also imply heavy individual and social costs due to the fact that these measures entail a short- or long-term loss of access to a crucial resource base for needy populations.

139. Finally, to be efficacious new techniques require certain scales of application. For

example, since wind is capable of transporting enormous quantities of sand or soil (McGinnies, McComb and Fletcher, 1963), combating sand encroachment can be a laborious and costly proposition, often beyond the means of a single family or community. Therefore, successful large-scale efforts based on local self-reliance entail a supportive national structure, as is the case in the People's Republic of China (Whitney, 1976). In other cases, efforts on a smaller scale may be undertaken by single entrepreneurs only with the aid of capital and energy-intensive equipment (Heathcote and Williams, 1976).

140. Efforts can be taken not only to reduce the physical processes of desertification, but also its consequences. In coping with the consequences of reduced nutritional intake, people can do less work, resort to famine foods, seek outside relief, or develop hostile, atomistic, but self-preserving, interaction patterns. While each of these alternatives is widely resorted to in times of crisis, none is a viable long-term solution since this would entail intolerable levels of pathology and dependency.

141. A much more common response to declining yields and the lowered nutritional intake that these entail is a resort to seasonal labour or handicraft manufacture. In many dryland livelihood complexes, these two alternatives are already worked into the productive system (e.g., Johnson Haring, 1976a). However, when desertification lowers the productivity of farming or husbandry, people must rely on the other options open to them. In extreme cases, this may mean that people lose control over their productive base and become impoverished craftspeople or landless wage labourers. In wealthy countries, in response to declining yields, there is a continuing enlargement of the productive unit. This, in turn, requires a high capital investment to operate the enlarged unit without more labour. While this enables the entrepreneur to maintain net economic return in the face of declining productivity, it is seldom an option that is available to the bulk of a population exposed to desertification.

D. CHANGES IN USE OR LOCATION

142. Finally, more important to coping with desertification than to other hazards are *changes* of livelihood or location or both. Historically, the responses of people to desertification have involved such changes in contrast to direct efforts to combat desertification. Thus, herds were reduced or moved, nomads without animals became sedentarized, and long-fallows were practised to allow for natural recovery; alternately, and less frequently, efforts were increased to intensify production through irrigation.

143. The most drastic of all changes is permanent abandonment. Whenever people abandon land, they turn to urban or rural wage labour, small or large commercial ventures, or attempt to enter the bureaucracy. This social mobility is as much a basic response to the unpredictability of an arid ecosystem as the livelihood movements described in previous sections of this report. Whether or not this mobility represents an adjustment that is beneficial to the individual or the social group very much depends on the circumstances where it takes place; there is no one generalization that will cover the range of this adjustment. Pastoral nomads who have lost their herds, and who now subsist in what were to be temporary food distribution centres, represent a case where change has led to a (at least temporary) dead end. On the other hand, peasants who are forced to leave a countryside that no longer provides a basis for their sustenance can find work, and often prosper, in cities.

144. In any society, the particular mix of coping modes is related to the wealth and technology available to reduce desertification as well as to the alternative livelihoods and locations available to the people undergoing desertification. The profiles of response (Fig. 6) present an impression of the relative effort in terms of both social costs of adjustment and the consequences suffered for four different livelihood systems. No monetary value is assigned to the cost axis, although for many elements of that component such values could be developed. The social effort required to cope with desertification can be measured in terms of crop losses sustained, relief aid given, person-hours of administrative attention consumed, insurance policies purchased and so on. Far more difficult to quantify are the costs associated with relocating in new settings, the stress placed on individuals when traditional social support mechanisms no longer function, or the strains experienced when adjustment to new livelihoods must rapidly be undertaken.

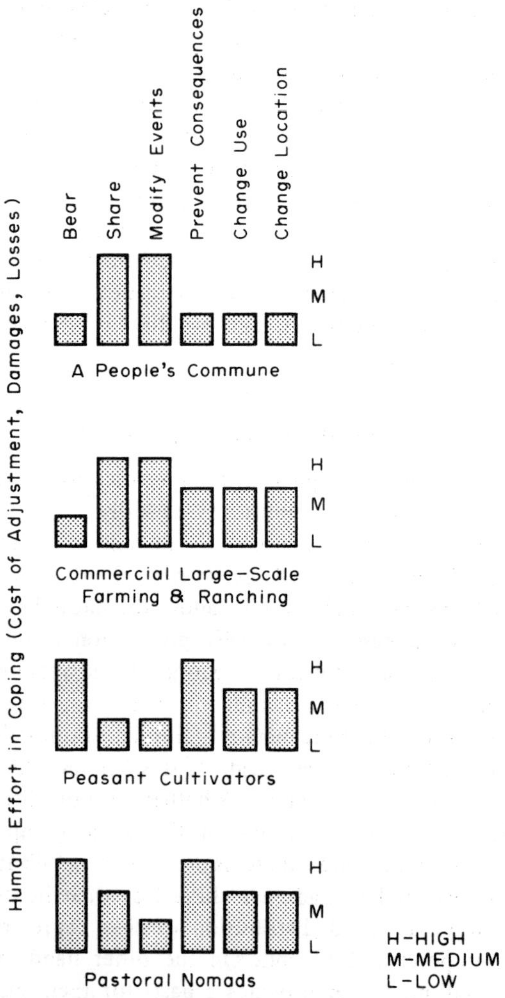

Fig. 6. Adjustments to desertification: examples from four different livelihood systems

145. Thus, the scale of social costs is selective and impressionistic as there are no basic economic studies of desertification. However, they are partly based on two analogous and related sets of studies: drought adjustment and soil erosion. From studies of agricultural drought adjustment in Australia and Tanzania (Burton, Kates and White, forthcoming), we know that the ongoing costs of reducing drought vulnerability and the average residual drought losses are about equal, total average costs per person at risk being about $1.50 (U.S.) in Tanzania and $43.00 (U.S.) in Australia. While the Australian costs and losses are 30 times greater than those of Tanzania in absolute terms, the reverse is true in relative terms (per cent of per capita GNP). This reflects the greater seriousness of drought in developing countries. Desertification costs and losses are likely to evidence a similar pattern.

146. The second set of studies relates to soil erosion. In the most comprehensive review to date and available only for a developed country, the United States, Held and Clawson found in 1965 that the U.S. had spent some 9 billion dollars on soil conservation over a 30-year period. Using a variety of sources, they found that neither the physical accomplishment of that expenditure nor its economic benefits (if any) were assessable. At best, they could conclude that only 20–60 per cent of the needed effort to reduce soil erosion in the U.S. from 1930s levels had been accomplished. As to economic payoff, they concluded that in the absence of further subsidy there was little profit potential in soil conservation. Indeed, in many cases there would be losses to the farmers concerned.

147. The absence of comprehensive cost and loss data would imply a need to encourage such studies as preliminary to international efforts to combat desertification. But as in many such assessments, the costs will be more easily defined than the benefits (the losses to be averted). This recurrent problem of economic evaluation is the most serious deficiency in desertification study, given its poorly examined, slow, cumulative, and long-term effects on productivity.

E. SUMMARY

148. People cannot have been passive in the face of desertification and still have survived in the dry lands, the oldest of humankind's habitats. Ways of coping with desertification are built into existing livelihood systems, into the expectation and acceptance of fluctuating fortune, into a wide range of techniques to combat desertification and reduce its consequences, and into the willingness to move or to adapt to new livelihoods.

149. Thus all livelihood systems have a limited capacity to absorb some measure of desertification without suffering significant consequences. Cultural adaptations, part and parcel of everyday life, serve to buffer slowly evolving losses of productivity. And given the fluctuations in well-being characteristic of dryland livelihoods, productivity losses, even when apparently harmful, may be accepted for long periods of time.

150. But in the long run, continued loss of productivity and threats to life and livelihood evoke active responses, some to combat the physical processes of desertification, others to provide an alternative focus on places or livelihoods. Historically, changing land use, building irrigation works, reducing or moving herds, instituting long fallows, and allowing for natural recovery were more common or effective responses than human

initiative to restore productivity or reverse desertification. The mix of coping modes utilized in a livelihood system is related to available wealth and technology. Today technological advances in methods of prevention and restoration, and increases in the scale of social organization, make possible, but do not ensure, a broadened range of human response.

V. MAJOR FINDINGS AND IMPLICATIONS FOR ACTION

151. Desertification poses a serious threat to the livelihoods of an estimated 78 million people living in severely or very severely desertified areas; 50 million of this total are directly affected by the loss of resource base productivity. Most of these are numbered among the poorest peoples or are inhabitants of the poorer nations. To a degree, they are victims of an environmental hazard not of their own making, and one that will increase considerably over the next two decades. Thus, it is fitting that an international forum be developed to examine this threat, to attend to the plight of its victims, and to organize assistance and preventive action.

152. But while the numbers and areas affected may be large, they are the aggregate of many different environmental and human conditions. Desertification is a complex environmental hazard and vulnerability to it is highly selective. Its physical processes are varied, its causes are multiple, and its consequences are subtle and long-term. Desertification is a human phenomenon, more a social, economic and political problem than a technical one.

153. This complexity is the major finding of our review. It should caution those who would expand on the popular myth of a world-wide problem of advancing deserts, those who would propose universal actions or facile prescriptions. It should inform the international community concerned with desertification of the need to be selective in choice of problems to be addressed, to be flexible in terms of measures proposed, and to be sustained in terms of the efforts undertaken.

A. ASSIST THE MOST VULNERABLE

154. **The limited resources of nations and the international community need to be directed towards reducing the threat for the most vulnerable peoples. These groups should be the target of assistance because desertification's impact on them is proportionately greater than it is on more secure peoples and because their resources are insufficient to provide a flexible, multi-option response to desertification. Among nations, the most vulnerable are developing nations undergoing more or less rapid change in population, technology, and land use and who are primarily dependent on dryland resources. Within nations, the most vulnerable peoples are those dependent for their livelihoods on agriculturally marginal lands, and those who by virtue of location, ethnicity or class are peripheral to centres of power, assistance, and investment.**

155. The 50 million people directly undergoing desertification are not uniformly distributed across the dry lands nor are they simply chance victims of natural settings or climatic fluctuations. They are rarely found in industrialized countries such as Australia, the Union of Soviet Socialist Republics or the United States, where impressive progress has been observed in overcoming earlier problems. Nor, for the most part, are they in the extremely arid zone, in the mineral-rich nations, or in countries in which population, land use, and technology are changing slowly.

156. The greatest vulnerability to desertification occurs selectively within nations. The reduction in productivity of dryland resources characteristic of desertification affects mostly those whose diets are already inadequate or whose land and water resources are

currently limited. Such constraints may overcome the ecological advantage of certain livelihoods. For example, animal-based livelihoods, which by virtue of their mobility are ecologically least prone to desertification, may become vulnerable as their relative access to resources is increasingly constrained. Conversely, the intense desertification that accompanies urbanization, or the waterlogging and salinization that follows from irrigation, may be offset by the diverse opportunities cities provide or the heavy investment devoted to irrigation.

157. **The implications of this vulnerability analysis run counter to proposals that would address desertification problems in all of the dry lands or whenever the physical signs of desertification are observed. Instead, it would urge the international community to direct assistance to the most vulnerable nations, and would encourage nations to direct their attention to their most vulnerable peoples rather than to the most desertified areas. To do so, a more detailed vulnerability analysis, as sensitive to the social, economic, and political contexts of people and their livelihoods as to measures of physical process, is the first step needed in order to target international and national efforts to combat desertification.**

B. ADDRESS THE ROOT CAUSES

158. **Desertification is a human phenomenon. It inevitably arises from the need or desire to subsist upon, improve, or exploit the land and water resources of the dry lands. While some degradation in the natural environment is surely acceptable, when the immediate benefits outweigh the costs, much of it threatens the long-term well-being of the people whose livelihoods are dependent on these resources. Such desertification usually arises from two sets of imbalances in the human use of dry lands: (1) those related to population growth, decline and movement, and (2) those related to the introduction or abuse of technologies, or new land uses and economic ties. These imbalances are further exacerbated and intensified when they coincide with detrimental climatic fluctuations. If efforts to combat desertification are to be more than palliatives, they need to be addressed to the root causes of these imbalances.**

159. Such a task is not a simple one. The causes of these imbalances are multiple and contradictory. Population decline as well as growth can lead to desertification. New technologies can combat desertification or induce it. The major impetus for desertification may stem from the livelihood practices within a region or from policies and practices adopted in centres far from the dry lands. Finally, the most careful plans and practices may prove inappropriate in the face of short-term natural fluctuations of climate and vegetation.

1. Population Growth, Decline, and Movement

160. In a global context, the notion of a generalized carrying capacity for dry lands, which when exceeded leads to desertification, seems chimerical. Population densities and trends vary widely in similar climatological zones and the tendency to equate desertification with numbers of people or animals at best oversimplifies and thereby obscures the origins of desertification. At its worst, it seeks to shift the responsibility from desertification onto its victims. Indeed, an important finding of this demographic review is how selective loss of the agriculturally active population can serve to encourage desertification.

161. **Regions with a potential for desertification can anticipate further population**

growth due to the fact that their birth rates remain high despite improvement in reducing death rates. In developing countries, rural population will increase by 2 per cent to 3 per cent per year. This is a growth rate which can be absorbed in very few dry rural or urban areas without causing environmental disruption and human problems. Yet, given the values of many peoples and their governments, sudden change to a low birth/low death pattern is unlikely, and many dry lands will be forced to accommodate high growth rates for many years. This indicates that for many regions more desertification, not less, is the likely immediate prospect.

162. For many of the rural people, the vast majority of whose lives are sustained by agriculture, life in the countryside will remain economically and socially viable. However, millions of others will migrate to urban centres in both dry lands and beyond. Where this out-migration is composed largely of labouring males, the loss of agriculturally active population may well result in a loss of productivity in the rural environment. Terraces no longer maintained may collapse, leading to soil erosion; in the absence of herders, animals may be sold; loss of specialized labour may result in lower yields from orchard crops and abandonment of farm land. Conversely, these effects may be compensated by remittances that lead to productive investment, by the substitution of gas or oil for charcoal as a fuel source, and by opportunities to supplement domestically produced foodstuffs.

163. Related to this issue of the loss of labour in agricultural societies are restrictions on the movement of pastoralists and fishing peoples. In the interests of national sovereignty, there are often rigid controls placed on population movement from one territory to another. Pastoralists and fishing peoples have experienced particular problems in this regard, since the national boundaries inherited from the colonial era often cut across natural ecosystems and ethnic and livelihood regions. The importance of free movement across such lines becomes particularly apparent during droughts when it becomes difficult, if not impossible, to escape to districts that traditionally were available.

164. **Movement and migration policies should be placed at the centre of efforts to combat desertification in preference to population control policies (an exception being South Asia). It should be recognized that continued outflow of population from arid lands will take place in most developing countries. Attention should focus on how best to prepare and facilitate such migration, where appropriate, particularly by youth, and effort is needed to encourage remittances for productive investment in their home areas. The cause of such movements lies in the disparities between dry lands and the centre, disparities which must be addressed by efforts to create viable social and economic alternatives in the area of origin. In this regard, the experience of some planned economies as well as the methods of compensatory subsidy support found in many market countries should be carefully studied. International agreements should be negotiated to facilitate movement by pastoralists and fishermen, particularly in periods of stress.**

2. New Technologies, Land Uses and Economic Ties

165. Integration into a world market economy frequently has brought more problems than benefits to countries and regions subject to desertification. These areas are often located within the underdeveloped periphery of the world system, or the under-developed region of a nation-state. Thus, they are vulnerable to external market forces over which they have little or no control. Historically, the needs and problems of the developed, highly industrialized countries or regions at the centre of the world market system have

dictated the path and pace of development in the periphery. This development entails the modernization, i.e. destruction, of traditional, self-sufficient livelihood systems and, in this sense, plays an important role in the dynamics of desertification. **As traditional societies in peripheral arid regions undergo a process of disintegration, and as old productive strategies and exchange patterns become unviable options, both the natural and the cultural components of the ecosystem deteriorate. When additional stresses, such as drought, are placed upon the system, it becomes hard for it to recover on the basis of purely indigenous resources.** Yet, due to its still peripheral status, outside resources do not readily flow in upon demand, further hampering recovery. In extreme cases, the productive base may have become so degraded that people are forced to abandon their old livelihoods and habitats.

166. It is not clear that this is a historically necessary pattern. With the introduction of new technologies, indigenous ones are in retreat. But paradoxically, the loss of these systems comes at a time when there is growing recognition of their inherent original rationality and ecological wisdom. Acknowledgement of this characteristic of traditional systems is not based solely on appreciation of their longevity. Rather, the ability to use marginal resources in ways that are ecologically astute and that are efficient in energy terms seems especially important. Any system capable of reducing or avoiding fossil fuel inputs represents a substantial saving to the total socio-economic system.

167. Not only does the introduction of new technologies truncate the evolution of indigenous systems, but the former seldom fulfil their own promise. There seems to be an inevitable and endemic lag in the transfer process in which only central functions of the new technology are transferred without the necessary complements. For example, irrigation water is introduced without sufficient drainage leading to salinization, or mechanized ploughing is promoted without the benefit of conservation farming resulting in the loss of topsoil due to aeolian action.

168. New land uses or economic ties have a similar effect. Export crops with cash returns capture the available resources of the agricultural enterprise and reduce the traditional "normal surplus" of food production. This often creates famine where before there were shortages in slightly less-favourable years. Single crop productive systems replace traditional systems without the complementary handicraft or collecting activities that enriched the traditional livelihoods. Industrialized countries are also not immune to similar problems for the use of their arid lands becomes subject to distant market forces or competitive production.

169. **Several implications follow from these observations. Development projects in dry lands should be stringently assessed for their impact on desertification. Indigenous technologies and productive systems require sympathetic study to identify opportunities to support their evolution. New technologies and land uses need critical appraisal to identify needed complements without which desertification and its consequences would be increased. In general, maintaining the viability of diversified livelihood systems should take precedence over the optimization of any single product system.**

C. COPE WITH THE CONSEQUENCES

170. The major consequences of desertification are the loss of productivity of the resource base, the threat to the health and well-being of people at risk, and the problems to populations that arise from desertification-induced migration and resettlement.

1. Loss of Productivity

171. **A long-term chronic decline in productivity of the resource base is the most serious manifestation of desertification, yet its direct social and behavioural effects are poorly understood.** It is clear, however, that societies possess strategies that help them cope with the losses in productivity which are caused by desertification. Some of these strategies serve to stay or reverse the deterioration of the resource base, some develop new patterns of use of existing resources, while others involve total changes in livelihood and location. Although most societies adopt a range of these coping strategies, certain characteristic modes of coping can be distinguished, depending on societal wealth and level of organization. For example, agriculture-based societies possessing sufficient capital and organization transform dryland use by the introduction of irrigation, or they enlarge the scale of operations by increasing the inputs of land and capital. Individual peasant agriculturalists, who characteristically have access to neither land nor capital, commonly resort to labour out-migration or to handicraft manufacture using locally available materials. Pastoralists may abandon nomadism altogether and become farmers or urban workers.

172. **In general, and with the exception of the wealthy industrial nations, conditions favourable to individual efforts directed at land restoration do not exist among the populations vulnerable to desertification. Programmes aimed at stemming loss of productivity and restoration of land in the developing world will necessarily involve capital input from sources outside the affected areas. These investments yield maximum social benefits if they are used in such a manner as to complement existing coping strategies and thereby reinforce the viability of rural life. Therefore, land reclamation and rural infrastructure projects which emphasize the use of local materials and local labour, as well as focus on developing collective solutions to collective needs, should be given priority over other alternatives.**

2. Health and Well-Being

173. **The health and well-being of individuals and livelihood systems are adversely affected by desertification, but there is little evidence that directly links health problems to increasing aridity. The health consequences of desertification are invariably indirect and, hence, difficult to measure. Lower nutritional intake, however, is a direct consequence of desertification and, for vulnerable populations, it may have serious repercussions on health.**

174. Diseases associated with arid areas, such as trachoma or pulmonary infirmities, are serious problems that persist in arid areas, requiring changes in land use, water availability, and health care to reduce them. Except in so far as wind-borne dust may increase the incidence of eye infections, respiratory irritation, and associated diseases, evidence for a direct causal linkage between desertification and disease is lacking.

175. Most diseases commonly associated with desertification are in fact a consequence of drought or human migration. In the case of the latter, the first contact health problems are of the kind to be expected whenever an isolated population is brought into close association with otherwise alien diseases. For these public health problems, and for sanitary difficulties linked to population concentration, knowledge of the appropriate

remedial measures is well understood by national officials and international agencies. Drought-induced health problems are also well-known, the signs of nutritional deficiency are generally unambiguous, and the unequal pattern of mortality that strikes the aged, the infirm, and the young is well established.

176. Additional health problems associated with desertification are those related to psycho-social stresses. These are most serious in acute situations such as drought or resettlement involving new environmental settings or an unfamiliar livelihood. There is little evidence for similar pressures being present in chronic desertification; indeed, where such change is gradual there is reason to believe that the adaptive capacity of culture and livelihood will facilitate the development of effective coping mechanisms.

177. **Specific attention to the health problems of scattered and marginal populations is justified on both general development and humanitarian grounds. This can best be accomplished by adopting strategies which place the primary responsibility for the initial stages of health care on trained indigenous individuals and which selectively use modern transportation and communications technology to move serious cases to centralized medical facilities. It is important for planners to take into account the probability that the initial effect of this health care improvement will most likely be a reduction in the death rates and a consequent increase in population.**

3. Migration, Relocation and Livelihood Change

178. **The major response of livelihood systems to chronic desertification is migration and/or change in basic livelihood patterns.** The role that such shifts in location and livelihood play in traditional systems varies. For some, migration is the only alternative; for others, seasonal mobility is a fundamental part of long-term adaptation; for still others, it is a positive response to new opportunities. Because population growth can be expected to continue in dry lands for the foreseeable future, provision for the accommodation of some of this increase outside desertification-prone districts will have to be made. Most of these migrants will be individuals or family units rather than entire social groups and assistance provided should prepare participants at both ends of the migration continuum to cope with the often rapid required changes.

179. Action programmes supportive of migration's adaptive elements should take into account the following three areas: (1) Education developed to combat desertification must incorporate relevance to local concerns as its paramount objective. In order to be effective, education must, therefore, focus on techniques for increasing the productivity of local resource systems, as well as develop programmes that prepare individuals with skills that enable them to find employment outside desertified areas. (2) In all cases, it is essential to incorporate the people most affected by desertification in meaningful decision-making roles. The best measure of the desirability of an anti-desertification project should be the willingness of local people to participate in it. Formal education and propaganda are inadequate as substitutes for genuine participation. (3) Resettlement where necessary can be carried out in ways that minimize psycho-social stress if the following principles are used. Economic viability is the crucial variable in determining the success of a project; other considerations are secondary. Projects of a design and scale that maintain cultural and social integrity, as well as maximize local participation in decision-making, will minimize the negative effects of relocation.

180. **To facilitate migration and to reduce its burdens, there is need to develop in the short-run meaningful forms of local participation, in the medium-run selective economically-viable resettlement opportunities, and in the long-run effective work-related educational programmes.**

BIBLIOGRAPHY

Aart, R. van (1971) *Aspects of Land Use Planning in the Lower Mesopotamian Plain: A Problem Analysis*. Technical Report No. 33. Abu Ghraib, Iraq: Institute for Applied Research on Natural Resources.

Abu-Lughod, Janet (1961) Migrant adjustment to city life: the Egyptian case, *American Journal of Sociology,* **lxvii**: 22–32.

Adams, Robert McC. (1965) *Land Behind Baghdad: A History of Settlement on the Diyala Plains.* Chicago: University of Chicago Press.

Adams, Robert McC. (1974) Historic patterns of Mesopotamian agriculture, in *Irrigation's Impact on Society,* Theodore E. Downing and McGuire Gibson, eds., Tucson: University of Arizona Press, pp. 1–6.

Ahmed, Abdel Ghaffar M. (1973) Nomadic competition in the Funj Area, *Sudan Notes and Records,* **liv**: 43–56.

Anaya-Garduño, M. (1976) *Desertification and Technology.* Preliminary Draft Report prepared for the United Nations Conference on Desertification.

Australia. Commonwealth Bureau of Census and Statistics (1973) *Demography*, Canberra: Commonwealth Bureau of Statistics.

Benyoussef, J. *et al.* (1974) Health effects of rural-urban migration in development countries: Senegal, *Social Science and Medicine,* **viii**: 243–258.

Berg, Elliott (1975) *The Recent Economic Evolution of the Sahel.* Ann Arbor: University of Michigan, Centre for Research on Economic Development.

Berry, Leonard and Robert W. Kates (1972) Views on environmental problems in East Africa, *The African Review,* **ii**: No. 3, 299–314.

Bowden, Martyn J. (1976) No dustbowl in the future: saying nay to doomsday in the North American Great Plains. MS prepared for Supplementary Materials to *Population, Society and Desertification*.

Brooks, Reuben H. (1976) The peasant farmer of the Sertão of Taua, Ceara, Northeast Brazil, in *Profiles of Change*, Supplementary Materials to *Population, Society and Desertification.*

Burton, Ian, Robert W. Kates and Gilbert F. White (forthcoming) *The Environment as Hazard.* New York: Oxford.

Caldwell, John C. (1975) *The Sahelian Drought and its Demographic Implications.* OLC Paper No. 8. Washington, D.C.: Overseas Liaison Committee, American Council on Education.

Campbell, David (1976a) A nomadic family in Niger, in *Profiles of Change,* Supplementary Materials to *Population, Society and Desertification.*

Campbell, David (1976b) *Strategies for Coping with Drought in the Sahel: A Study of Recent Population Movements in the Department of Maradi, Niger.* Ph.D. Dissertation, Clark University, Graduate School of Geography, Worcester, Massachusetts, U.S.A.

Chen, Cheng-Sian (1973) Population growth and urbanization in China 1953–1970, *Geographical Review*, **lxiii**: 55–72.

CONAZA (n.d.) Report on Mexico's arid lands. Unpublished MS. Mexico D.F.: Comision Nacional de las Zonas Aridas.

Cossins, Noel J. (1971) *Pastoralism Under Pressure: A Study of the Somali Clans of the Jijugga Area of Ethiopia.* Addis Ababa.

Cossins, Noel J. (1972) *No Way to Live: A Study of Afar Clans of the North-East Rangelands Carried Out for the Livestock and Meat Board of the Imperial Ethiopian Government.* Addis Ababa: North-East Rangelands Development Project.

Darling, F. Fraser and Mary A. Farvar (1972) Ecological consequences of sedentarization of nomads, in *The Careless Technology: Ecology and International Development*, M. Taghi Farvar and John P. Milton, eds., Garden City, New York: Natural History Press, pp. 671–682.

Davis, Kingsley (1969) *World Urbanization,* 1950–1970. *I: Basic Data for Cities, Countries and Regions.* Population Monograph Series No. 4. Berkeley, California: University of California, Institute of International Studies.

Dow, Michael (1975) *Trip Report: Visit to Sana'a, Yemen Arab Republic, December 2nd–8th, 1975.* Mimeo.

Draz, O. (1974) *Report to the Government of the Syrian Arab Republic on Range Management and Fodder Development.* United Nations Development Programme No. Ta 3292. Rome: United Nations, Food and Agriculture Organization.

Dregne, Harold (n.d.) Desertification: a worldwide phenomenon. MS.

Eckholm, Erik P. (1976) *Losing Ground*. New York: W. W. Norton.

EMASAR (1975) *The Ecological Management of Arid and Semi-Arid Rangelands in Africa and the Near and Middle East*. Rome: United Nations, Food and Agriculture Organization.

Evenari, Michael, Leslie Shanan and Naphtali Tadmor (1971) *The Negev: The Challenge of a Desert*. Cambridge, Mass.: Harvard University Press.

Fahmi, Adnan S. M. (1976) Personal communication, Acting Director Palm and Dates Research Centre, Baghdad, Iraq, May 9, 1976.

Faulkingham, Ralph H. (1976) Alasan and the drought: a case study from Niger, in *Profiles of Change*, Supplementary Materials to *Population, Society, and Desertification*.

Fernea, Robert A. (1970) *Shaykh and Effendi: Changing Patterns of Authority among the El-Shabana of Southern Iraq*. Cambridge, Mass.: Harvard University Press.

Fox, David J. (1965) Man-water relationships in Metropolitan Mexico, *Geographical Review*, **lv**: 523–545.

Gebre-Medhin, Mehari (1974) Famine in Ethiopia, *Ethiopian Medical Journal*, **xii**: 105–107.

George, Carl J. (1972) The role of the Aswan Dam in changing the fisheries of the Southeastern Mediterranean, in *The Careless Technology: Ecology and International Development*, M. Taghi Farvar and John P. Milton, eds., Garden City, New York: Natural History Press, pp. 160–178.

Gibson, McQuire (1974) Violation of fallow and engineered disaster in Mesopotamian civilization, in *Irrigation's Impact on Society*, Theodore E. Downing and McQuire Gibson, eds., Tucson: University of Arizona Press, pp. 7–19.

Glantz, Michael H. Ed. (1976) *The Politics of Natural Disaster: The Case of the Sahel Drought*. New York: Praeger.

Graves, Nancy B. and Theodore D. Graves (1974) Adaptive strategies in urban migration, *Annual Review of Anthropology*, **iii**: 117–151.

Greene, Mark (1975) Impact of the Sahelian drought in Mauritania, *African Environment*, **i**: No. 2, 11–21.

Hansis, Richard A. (1976) Viticulture in the Mendoza oasis, Argentina, in *Profiles of Change*, Supplementary Materials to *Population, Society, and Desertification*.

Harris, Chauncey, D. (1971) Urbanization and population growth in the Soviet Union, 1959–1970, *Geographical Review*, **lxi**: 102–124.

Heady, Harold F. (1972) Ecological consequences of Bedouin settlement in Saudi Arabia, in *The Careless Technology: Ecology and International Development*. M. Taghi Farvar and John P. Milton, eds., Garden City, New Jersey: Natural History Press, pp. 694–711.

Heathcote, R. L. (1969) Drought in Australia: a problem in perception, *Geographical Review*, **lix**: 175–194.

Heathcote, R. L. and M. Williams (1976) The Mallee of South Australia, in *Profiles of Change*, Supplementary Materials to *Population, Society, and Desertification*.

Held, R. Burnell and Marion Clawson (1965) *Soil Conservation in Perspective*. Baltimore: Johns Hopkins Press for Resources for the Future, 1965.

Henderson, H. Francis (1974) *Current State of the Fish Stocks of Lake Kossou*. Rome: United Nations, Food and Agriculture Organization.

Hewitt, Ken (1976) Problems of aridity in Pakistan, in *Profiles of Change*, Supplementary Materials to *Population, Society, and Desertification*.

Horowitz, Michael M. (1972) Ethnic boundary maintenance among pastoralists and farmers in the Western Sudan (Niger), in *Perspectives on Nomadism*, William Irons and Neville Dyson-Hudson, eds., Leiden: E. J. Brill, pp. 105–114.

Jacobsen, Thorkild (1960) The waters of Ur, *Iraq*, **xxii**: 174–185.

Jacobsen, Thorkild and Robert M. Adams (1958) Salt and silt in Ancient Mesopotamian agriculture, *Science*, **cxxviii**: No. 3334, 1251–1258.

Johnson, Douglas L. (1973) *Jabal al-Adkhdar, Cyrenaica: An Historical Geography of Settlement and Livelihood*. Chicago: University of Chicago, Department of Geography, Research Paper No. 148.

Johnson Haring, Kirsten (1976a) The Otomi of the Mezquital Valley, Hidalgo, Mexico, in *Profiles of Change*, Supplementary Materials to *Population, Society, and Desertification*.

Johnson Haring, Kirsten (1976b) *Do As the Land Requests: A Study of Otomi Resource-Use on the Eve of Irrigation*. Ph.D. dissertation, Clark University, Graduate School of Geography, Worcester, Massachusetts, U.S.A.

Kassas, M. (1972) A brief history of land-use in Mareotis Region, Egypt, *Minerva Biologica*, **i**: No. 4, 167–174.

Kates, Robert (1972) The hazard of drought, in *Patterns and Perspectives in Environmental Science*, Washington, D.C.: National Science Foundation, pp. 218–221.

Kirkby, Anne V. T. (1973) *The Use of Land and Water Resources in the Past and Present Valley of Oaxaca, Mexico. Prehistory and Human Ecology of the Valley of Oaxaca*, Vol. 1. Ann Arbor: Memoirs of the Museum of Anthropology, University of Michigan No. 5.

Lagler, Karl F. (1973) *A Review of the Lake Kossou Fishery Development Project, Ivory Coast.* Rome: United Nations, Food and Agriculture Organization.

Laya, Diuldé (1975) Interviews with farmers and livestock-owners in the Sahel, *African Environment*, i: No. 2, 49–93.

Lee, Richard B. (1969) !Kung bushman subsistence: an input-output analysis, in *Environment and Culture*, Andrew P. Vayda, ed., Garden City, New York: Natural History Press, pp. 47–79.

Lee, Richard and Irving DeVore (1968) *Man the Hunter*. Chicago: Aldine.

LeHouérou, H. N. (1968) La désertisation du Sahara septentrional et des steppes limitrophes (Libya, Tunisie, Algerie), *Annales Algériennes de Géographie*, 3me Année, No. 6, 5–30.

Lickiss, J. Norelle (1975) Health problems of urban aborigines: with special reference to the aboriginal people of Sydney, *Social Science and Medicine*, ix: 313–318.

Lier, R. A. J. van (n.d.) *Report Concerning the FAO/UNDP Project for Strengthening the Awash Valley Authority (Ethiopia 11)*. Rome: United Nations, Food and Agriculture Organization.

McGee, T. G. (1973) Peasants in the cities: a paradox, a paradox, a most ingenious paradox, *Human Organization*, xxxii: No. 2, 135–142.

McGinnies, W. G., A. L. McComb and J. E. Fletcher (1963) Role of watersheds and forests in the arid west, in *Aridity and Man*, C. Hodge and P. C. Duisberg, eds., Washington, D.C.: American Association for the Advancement of Science, pp. 277–307.

MacGregor, Maria T. G. de and Carmen Valverde V. (1975) Evolution of the urban population in the arid zones of Mexico, 1900–1970, *Geographical Review*, lxv: 214–228.

Malhotra, S. P. (1976a) Demographic, social and behavioural attributes and their impact on desertification in arid zones of Rajasthan. Unpublished MS. Jodpur: Central Arid Zone Research Institute.

Malhotra, S. P. (1976b) A case study among Bishonis, Chirai Village, Jodhpur District, in *Profiles of Change*, Supplementary Materials to *Population, Society, and Desertification*.

Malhotra, S. P. (1976c) Opinion survey on desertification, a case study. Unpublished MS. Jodhpur: Central Arid Zone Research Institute.

Malhotra, S. P., L. P. Bharara and F. C. Patwa (1976d) Introduction of irrigation in arid Rajasthan: a case study, in *Problems in the Development and Conservation of Desert and Semidesert Lands*. Twenty-Third International Geographical Congress, Working Group on Desertification in and around Arid Lands, Pre-Congress Symposium K-26, Ashkhabad, USSR, July 20–26, pp. 115–118.

Mangin, William, ed. (1970) *Peasants in Cities: Readings in the Anthropology of Urbanization*. Boston: Houghton Mifflin.

Meckelein, W. (1976) Desertification caused by land reclamation in deserts: the example of the New Valley, Egypt, in *Problems in the Development and Conservation of Desert and Semidesert Lands*. Twenty-Third International Geographical Congress, Working Group on Desertification in and around Arid Lands, Pre-Congress Symposium K-26, Ashkhabad, USSR, July 20–26, pp. 151–153.

Meigs, Peveril (1953) World distribution of arid and semi-arid homoclimates, in UNESCO, *Arid Zone Hydrology*. Arid Zone Research No. 1. Paris: United Nations, Educational, Scientific and Cultural Organization, pp. 203–209.

Melamid, Alexander (1965) Political boundaries and nomadic grazing, *Geographical Review*, lv: 287–290.

Mitchell, J. Clyde (1969) Structural plurality, urbanization and labour circulation in Southern Rhodesia, in *Migration*, John A. Jackson ed., London: Cambridge University Press, pp. 156–180.

Morris, Arthur S. (1969) The development of the irrigation economy of Mendoza, Argentina, *Annals Association of American Geographers*, lix: 97–115.

Nechayeva, N. T. (1976) Influences of anthropogenic factors on the ecosystems of the deserts of Central Asia, in *Problems in the Development and Conservation of Desert and Semidesert Lands*. Twenty-Third International Geographical Congress, Working Group on Desertification in and around Arid Lands, Pre-Congress Symposium K-26, Ashkhabad, USSR, July 20–26, pp. 125–129.

Nolan, Riall (1975) Labour migration and the Bassari: a case of retrograde development, *Man*, x: No. 4, 571–588.

O'Keefe, Philip (1975) Gakarara: a study in the development of underdevelopment, Occasional Paper No. 6, Disaster Research Unit, University of Bradford.

Omolalu, A. (1976) Personal communication, June 2, 1976.

Ormerod, W. E. (1976) Ecological effect of control of African Trypanosomiasis, *Science*, cxci: No. 4229 (27 February), 815–821.

Petrov, M. P. (1976) *Deserts of the World*. New York: Halsted.

Peyre de Fabregues, B. (1973) *Synthèse des études de la zone de modernisation pastorale du Niger: amelioration de l'exploitation pastoral*. Paris: Institut d'étude médecine véterinaire des pays tropicals.

Picardi, Anthony C. and William W. Seifert (1976) A tragedy of the Commons in the Sahel, *Technology Review*, lxxviii: No. 6, 42–51.

Rapp, Anders (1974) *A Review of Desertification in Africa: Water, Vegetation, and Man*. Stockholm: Secretariat for International Ecology, Sweden.

Redfield, Robert (1956) *Peasant Society and Culture*. Chicago: The University of Chicago Press.

Reyna, A. P. and Christian Bouquet (1975) Chad, in *Population Growth and Socioeconomic Change in West Africa*, J. Caldwell ed., New York and London: Columbia University Press for the Population Council, New York, pp. 565–581.

Seaman, J, J. Rivers, J. Holt and J. Murlis (1973) An inquiry into the drought situation in Upper Volta, *The Lancet*, No. 7832 (October 6), pp. 774–778.

Sebai, Zohair A. (1969) *The Health of the Bedouin Family in a Changing Arabia*. Ph.D. dissertation. Baltimore, Maryland: School of Hygiene and Public Health, The Johns Hopkins University.

Shamekh, Ahmed A. (1975) *Spatial Patterns of Bedouin Settlement in al-Qasim Region Saudi Arabia*. Lexington, Ky: University of Kentucky, Department of Geography.

Singh, J., R. L. Singh, and K. N. Singh (1971) Rajasthan, in *India: A Regional Geography*, R. L. Singh, ed., Varanasi: National Geographical Society of India, pp. 7–82.

Smith, Susan E. (1976) Personal communication, July 14, 1976. International Division, Africa Programs, American Friends Service Committee, Philadelphia, Pennsylvania, USA.

Splinter, William E. (1976) Center-pivot irrigation, *Scientific American*, ccxxxiv: No. 6, 90–99.

Swift, C. R. and T. Asuni (1975) *Mental Health and Disease in Africa: With Special Reference to Africa South of the Sahara*. Edinburgh, London and New York: Churchill Livingstone.

Talbot, Lee M. (1972) Ecological consequences of rangeland development in Masailand, East Africa, in *The Careless Technology: Ecology and International Development*, M. Taghi Farvar and John P. Milton, eds., Garden City, New York: Natural History Press, pp. 694–711.

Tarizzo, Mario (1976) Control of trachoma, *World Health*, February–March, pp. 10–14.

Tien, H. Yuan (1973) *China's Population Struggle: Demographic Decisions of the People's Republic, 1949–1969*. Columbus, Ohio: Ohio State University Press.

Toupet, Charles (1975) *La sedentarisation des nomades en Mauritanie centrale sahelienne*. Doctorat d'Etat et Lettres Thesis (University of Paris VII). Dakar.

United Nations. Educational, Scientific, and Cultural Organization. Man and the Biosphere (MAB) (1975) *Regional Meeting on the Establishment of Co-operative Programmes of Interdisciplinary Ecological Research, Training and Rangeland Management for Arid and Semi-Arid Zones of Northern Africa*. MAB Report Series No. 30. Paris: UNESCO.

United Nations. Food and Agriculture Organization (FAO) (1975) *Report of the Consultation on Fisheries Problems in the Sahelian Zone*. CIFA Occasional Paper No. 4. Rome: Committee for Inland Fisheries, FAO.

United States. Public Health Service. Center for Disease Control (CDC) (1973) *Nutritional Surveillance in Drought Affected Areas of West Africa (Mali, Mauritania, Niger, Upper Volta)*. Atlanta: Center for Disease Control. Mimeo.

Walters, Stanley D. (1970) *Water for Larsa: An Old Babylonian Archive Dealing with Irrigation*. New Haven and London: Yale University Press.

Warren, Andrew and J. K. Maizels (1976) *Ecological Change and Desertification*. Preliminary Report Prepared for the United Nations Conference on Desertification.

West, Robert (1970) Population densities and agricultural practices in pre-Columbian Mexico, with emphasis on semi-terracing, *Verhandlungen des XXXVIII Internationalen Amerikanisten-kongresses*. Munchen: Klaus Renner. Vol. 2, pp. 361–369.

Western, David (1974) The environment and ecology of pastoralists in arid savannas. Paper read at the SSRC Symposium on the Future of Traditional "Primitive" Societies. Cambridge.

Whitney, Joe (1976) Chungwei County, Ninghsia-Hui Autonomous Region, China, in *Profiles of Change*, Supplementary Materials to *Population, Society, and Desertification*.

Williams, Michael (n.d.) The perception of the hazard of soil erosion in South Australia: a review. Mimeo.

Williams, O. B. and H. Suijdendorp (1976) Draft of Gascoyne catchment: case study of desertification. Canberra, Australia: CSIRO.

Wilson, Andrew (1976) Tucson, Arizona, U.S.A.: a case study in desertification, in *Profiles of Change*, Supplementary Materials to *Population, Society, and Desertification*.

Wisner, B. (1976) The peasants of Tharaka Division, Meru District Kenya, in *Profiles of Change*, Supplementary Materials to *Population, Society, and Desertification*.

Wolf Eric R. (1966) *Peasants*. New Jersey: Prentice-Hall.

TECHNOLOGY AND DESERTIFICATION

MANUEL ANAYA GARDUÑO

CONTENTS

INTRODUCTION

1. Desertification is the impoverishment of arid, semi-arid, and some subhumid ecosystems by the impact of man's activities. It is the process of change in these ecosystems that leads to reduced productivity of desirable plants, alterations in the biomass and in the diversity of life forms, accelerated soil degradation, and increased hazards for human occupancy. Desertification is the result of land abuse.

2. While climatic variability in time and space has occurred and will occur, the world climate appears to have remained fairly constant during the past two thousand years or more (Hare, 1976). That being the case, the modern phenomenon of desertification must be ascribed to man's interventions, not to increasing aridity. This does not imply, however, that protracted droughts have not accelerated the process of desertification. On the contrary, droughts, in conjunction with man's activities, have had a major adverse impact on the land and, more importantly, on the people who wrest their livelihood from it (Warren, 1976; Kates and Johnson, 1976).

3. The role that the application of technological innovation has played, on the one hand, in reducing the threat of desertification or restoring desertified land to a more productive status and, on the other, in the hastening of desertification needs to be defined if a balanced assessment of the value of technology in combating desertification is to be made. Certainly there are numerous examples of both a bettering and a worsening of the situations, as will be shown later. Current popular emphasis is on the manner in which the misuse of technology has been responsible for many environmental problems of which desertification is but one. Quite properly, the "quick technological fix" has become discredited, but carefully planned technological introductions can be, and have been, highly beneficial.

4. Technology, for the purpose of this review, refers to man's inventions. They may be as simple as rotational grazing of pastoral lands or as complex as a computer system using soil sensors to automate the applications of irrigation water. Whether technology is called simple, intermediate or advanced depends upon one's frame of reference, but "simple" is not necessarily to be equated with "traditional" or "primitive". Appropriate technology could, from the standpoint of its complexity, be simple or advanced or intermediate; any one technology may be appropriate in one set of circumstances and inappropriate in another.

Constraints on the Application of Technology

5. There is no shortage of potential technological solutions to desertification problems. With no limit to financial resources, any harmful situation associated with desertification of natural resources can be remedied. Whether the results are worth the cost is another matter, as is the question of whether the technology selected is appropriate to the condition for which a remedy is sought.

6. Technological solutions which call for complicated mechanical or electrical systems obviously require a high degree of technical knowledge to maintain the systems in operating condition. Additionally, spare components must be readily available in order to

keep the equipment operating. These two considerations – technical knowledge and spare parts – commonly are ignored by purchasers or recipients of equipment.

7. The availability of trained people to use new techniques effectively, of local sources of materials, of foreign exchange to pay for imported items, of adequate transport and communications to support and administer projects, and of energy supplies to power equipment are among the constraints that affect the utilization of new technologies. The admonition of Schumacher (1973) that small is beautiful serves as a reminder that the best technology is the smallest (and simplest) that will do the job. The smallest capable of being effective, however, is not necessarily simple in operation and maintenance. A drip irrigation system is best devised to give almost complete control of water applications in the field; it is small, but it is far from simple to operate and maintain.

8. The fact that simple also is beautiful should not obscure the reality that advanced technologies, when appropriate and properly applied can be socially acceptable, environmentally sound, and economically viable. The ability to recognize and eliminate constraints will determine whether its use is good or bad.

9. Developing nations with limited financial resources initially require technologies that are generally low cost, easy to use, small scale, based on locally available resources, labour intensive, culturally fitting, and environmentally acceptable. Most of the technologies described in this report fit those needs if they happen to be culturally fitting. Those technologies that require a high level of technical competence to operate and maintain, are costly, require materials that are hard to get or are subject to other constraints identified in Tables 2–6.

The Use and Misuse of Technology

10. Desertification is a man-made phenomenon caused by exceeding the limits of the land's resilience. Any use of land by man imposes stresses on the system; the role of technology is to minimize those stresses while maintaining or increasing productivity. That this can be done was demonstrated two millenia ago by the Romans who constructed terraces in North Africa to control water run-off from the slopes and, in the process, made that region the breadbasket of the empire. The unhappy sequel to that endeavour illustrates the hazards associated with the application of some technological innovations. After maintenance of the terraces was abandoned following the breakup of the empire, soil erosion became more destructive than before the works were built. To bring the Roman experience into the twentieth century, terracing has been reintroduced into Tunisia and, in conjunction with reforestation of steep slopes, is restoring the land to its earlier productivity.

Cultivated Land

11. Land degradation through wind erosion is the to-be-expected consequence of extending cultivation into marginal climatic areas. Nations around the world (China, USSR, United States and Argentina, as examples) seem to have gone through the same experience over and over again. The sequence of events begins during a relatively moist period in the dry areas, when new land is penetrated, a crop is planted, and – frequently

– a good yield is obtained. Prosperity may last for a few years but then the next inevitable drought occurs, crops fail, and the new land is abandoned. Despite the history of drought-induced crop failures, land-hungry cultivators will again crop the land during the next rainy period, and the cycle of cropping-abandonment-cropping begins anew. Hope is eternal among dryland cultivators and pastoralists that droughts will not recur, but the hope is never realized. What has been realized, however, is improvement in land management technology (wind-breaks, strip cropping, crop-fallow systems, stubble mulching) that has permitted the successful extension of cultivation into drier areas than were previously possible to use on a sustained basis. Technological advances have been responsible for changing the Great American Desert of the nineteenth century into the wheat empire of twentieth century United States and Canada. The progressive change from periodic land abuse in the 1800s and early 1900s in the Great Plains of North America to reasonably satisfactory conservation of the land in the middle 1900s was a slow one. Many ideas for technological improvement were tested and discarded because they failed to meet local needs. Research, teaching, extension, a viable agricultural industry, and a favourable economic climate combined – but only slowly – to make success possible. The first land broken out in the Great Plains was ploughed with simple ox-drawn implements. In time, advances were made that led, ultimately, to the present-day machines that have revolutionized agriculture. While this process of gradual development can be accelerated, it probably cannot be discarded.

Grazing Lands

12. Pastoralism was not responsible for severe desertification in earlier centuries because periodic droughts and diseases kept livestock and human numbers down. All of that changed with the dramatic development and utilization of disease control during the past few decades. Techniques of vaccination, drug treatment, and sanitation enabled people and animals to live longer, which, in turn, increased pressure on limited land resources and led to accelerated desertification with lowered productivity; it is ironic that the humanitarian extension of health services should have inadvertently accelerated desertification.

13. In nations with extensive grazing lands and low human densities, desertification was a natural result of an exploitation attitude and a feeling that resources were limitless. When pastoralists came to realize that resources were finite, they still had time to change to land-conserving and drought-coping management techniques such as reduced stocking rates, the introduction of improved breeds of livestock, deferred grazing, rotational grazing, revegetation and the provision of well-distributed watering points. This has been the case in the developed countries. These same technological improvements are, in principle, equally valuable for the developing nations, as has been proven in Botswana and Mexico, among other places. Poorer nations have problems, however, which hamper application. These are low educational and economic levels, absence of effective research and extension services, absence of a strong industrial or commercial base, lands of marginal productivity, unstable government, and subsistence patterns of living. The result is that there is little room for innovation because the penalty of failure is too great. One innovation alone, that of simple rotational grazing, would very likely increase long-time productivity significantly, but the results would not be felt for a few years under average conditions. During those few years, outside support would be needed or adjustment to a

lower standard of living would have to be made. The latter would be unacceptable to most people. Some kind of insurance against risk is essential if new practices are to be adopted by poor people.

Irrigated Land

14. Salinization and waterlogging are the twin curses of irrigated agriculture. Salinization may be caused by the use of too little water, by applying saline waters, or by irrigating naturally saline soils. The usual causes, however, are waterlogging arising from canal seepage, overirrigation, and poor internal drainage. As watertables rise, salts accumulate on the soil surface when soil water evaporates, leaving precipitates behind.

15. History is replete with instances where immense works have been constructed to store and distribute irrigation water without consideration of drainage of the irrigated land. Then, as drainage problems gave rise to salt deposition and desertification occurred, costly efforts were made to remedy the situation. The only known method of salinity control is to wash excess salts from the soil by applying more irrigation water than the crop requires. Good drainage is indispensable for salinity control and effective drainage systems call for careful design, installation, and maintenance. Only in recent years have open drainage ditches, with all their problems of weeds, mosquitos, hazard to children, land wastage, and unsightliness, been replaced by underground drain pipes. A technological advance of this type can be labour-intensive or capital-intensive since hand labour or machines can dig the ditch, lay the pipe, and refill the ditch equally well, with proper supervision. Construction of open ditches likewise can be either labour-intensive or capital-intensive. Soil salinization has adversely affected crop production and human livelihood in virtually all of the major irrigated areas of the world. The Indus Valley in Pakistan, the Tigris–Euphrates Valley in Iraq, the Imperial Valley in the United States, the Mexicali Valley in Mexico, the Nile Valley in Egypt, and the Kura–Araks Valley in the USSR are all affected by salinity to the point where severe desertification has occurred in at least part of those areas. The ubiquity of salinization problems in irrigated areas has persuaded many people that permanent control of irrigation is impossible. And certainly it is not possible unless technological solutions are made use of.

16. One example of a successful attempt to control waterlogging and salinity comes from Pakistan. There, a massive programme of soil reclamation was initiated in the 1960s with Salinity Control and Reclamation Project No. 1 (SCARP I), which covers about one million acres in the Punjab. About 45 per cent of the waterlogged land was reclaimed in the first 9 years through the use of tubewells which served to lower the watertable while supplying irrigation water. Subsequent progress has been slower, presumably due to the development of soil alkalinization as salinity was reduced. Yields of wheat, a major crop, have doubled and the gross value of crop and livestock production has increased 2.5 times what it was before reclamation was begun. Reclamation alone does not account for the improvement; fertilizers, good quality seeds, and better land management have also played an important role.

Dependence on Technology

17. Salinity control in irrigated areas illustrates a basic element involved in the introduction of at least some technologies. Once reliance is placed on a technological

solution to one problem, it becomes necessary to introduce additional technologies in order to cope with problems arising from the technology used to solve the first problem.

18. An excellent illustration of this phenomenon can be found in the Nile Valley of Egypt. Soil salinity problems south of the Delta (the Delta is in a special position) were, by and large, of minor importance when the Nile was an uncontrolled river that flooded the valley annually. Floods caused much human misery when they were exceptionally violent, but they had the great merit of flushing salts out of the soil regularly and of depositing fertile sediment on the fields. That combination of effects enabled irrigation to be practiced for millennia without interruption, and it proved that irrigation – under certain circumstances – could truly be permanent. With the construction of the Aswan High Dam, however, things changed. First, there was no more annual flushing of salts from the soil; second, little sediment was carried in the water because it had settled out in Lake Nasser; and third, waterlogging became a greater problem because the availability of irrigation water throughout the year led to overirrigation. Thus, control of the variable flow in the river and its flooding problem resulted in greater salinity, waterlogging and fertility problems, all of which required technological solutions, some simple, some complicated. Where these solutions have been applied, the technologies have proved their value. Since no one is interested in destroying the dam and letting the river run wild again, the nation's agriculture is on a technological course that probably could not be reversed even if that were desirable. And that means ever more advanced technologies will be needed to cope with ever arising problems. Planners should recognize this fact of life.

Extensive and Intensive Land Use

19. A word needs to be said about intensive and extensive use of land. Either type of land use can lead to desertification, just as productivity can be high with either type. Management is the key to whether there will be improvement or degradation of the land resource. If population pressures are low and productivity per unit of area is acceptable, then extensive land use is appropriate if the management is such that environmental degradation does not occur. Once the decision is made to increase yields per unit area, some degree of intensification of management practices becomes essential, and that demands technological advances. Old or traditional technologies are not necessarily better or worse than modern technologies, but traditional ones probably are not adequate to meet new and higher productivity demands. The Chinese may be the most efficient food producers among nations using traditional production methods, yet they are relying increasingly on commercial fertilizers to help meet their production requirements.

Old Versus New Technologies

20. Perhaps one of the best examples of where an old and simple technology is inferior to a new and more complicated one is the case of water harvesting. A water harvesting system called run-off agriculture was devised in the Negev Desert two thousand or so years ago. The system was an effective one for the purpose for which it was designed because it permitted crop production in an area so arid that the land is virtually barren of vegetation. However, it was also an environmentally damaging technique because it was based on acceleration run-off and erosion in the upper reaches of catchments. The water and silt were collected in the valleys (wadis) where agriculture was practised.

Modern use of the same technology accomplishes the same purpose but without the damaging erosion. The difference lies in the availability of materials, (plastic, asphalt, chemicals) and machines, and in knowledge of physics, chemistry and biology.

Evaluation of Technologies

21. The selection and application of technologies to combat desertification in different countries will depend upon the education and motivation of the people, availability of well-prepared staff, government policies, the size of the areas to be reclaimed, the level of investment, and the time dedicated to reclaiming specific areas of degraded land. The solution to desertification problems requires the immediate implementation of short-term and long-term programmes, with special emphasis on more productive areas where the rate of degradation in productivity is slow and where relatively low investments produce high returns. Because there are so many different ecological, socio-economic and political conditions in the arid and semi-arid areas, it is obvious that there are no universal technological strategies for combating desertification. Each situation must be considered individually and appropriate adjustments made in general land-use plans.

22. Developed and developing countries provide examples of successful programmes in controlling desertification. Effective watershed management, salinity control, drainage installation, soil and water conservation, range management, sand-dune fixation, afforestation, and the establishment of natural reserves and national parks have contributed to the recovery of degraded areas. Conversely, there are also instances where land degradation is still faster than land reclamation due to lack of (a) economic and technical resources; (b) the organization of concerted effort; and, (c) application of appropriate technologies already known.

23. Traditional technologies developed by local people should be considered before establishing new land-use systems. Desirably, traditional technologies can be complemented, if necessary, with intermediate and transitional modern technologies which will permit a gradual change from extensive to intensive systems of production.

24. Combating desertification requires the organized application of available and appropriate technologies. The gap between technical-scientific knowledge and its application to the solution of problems may be bridged through education and training. Research and training centres are necessary to generate and disseminate information in different socio-economic and ecological condition.

25. Programmes of agricultural extension, including economic and sociological aspects, should be reinforced in order to diffuse appropriate technologies. The development unit should be an area belonging to a single community and of a size that is not too large to control and not so small that it is unrepresentative of the community. Development should be based upon a plan of action in which local people, as well as technicians and decision-makers, participate.

26. As underdeveloped countries are characterized by abundant labour and scarce capital, the use of intermediate technology should be considered. That level of technology has the merit of requiring only low to moderate investments in the small-scale labour-intensive application of appropriate technologies for rural communities, and of permitting a gradual change from extensive to intensive systems of production.

27. It must be pointed out that technologies by themselves are not sufficient to halt desertification and to increase land productivity. It is necessary to have an infrastructure

TABLE 1. *Physical and biological processes that cause desertification – problems and solutions*

Factor	Problems	Causes	Solutions
Water	Scarcity	Low precipitation Poor and erratic distribution of rainfall Mismanagement of irrigation water Over-exploitation of groundwater and surface reservoirs Uncontrolled water losses by evaporation	Improved water supplies, water conservation
	Mismanagement in dryland agriculture	Poor and erratic rainfall distribution Uncontrolled run-off	Run-off management, plant management, soil conservation
	Irrigation mismanagement	Unknown consumptive use Drainage system deficiency	Improved irrigation methods, drainage systems, salinity control, dependable water supplies
		Deficient land levelling Inadequate water distribution Imprecise water measurements Poor irrigation methods	
	Floods	Poor and erratic rainfall distribution Uncontrolled run-off	Flood control
Soil	Erosion (by water and wind)	Reduction of vegetation cover Uncontrolled run-off Sedimentation and silting Soil structure degradation Improper tillage practices Strong winds Soil profile depth diminishment Loss of fertile surface soil Reduction of water-holding capacity	Soil conservation, soil moisture conservation, plant establishment, plant conservation, fertilization

	Soil salinity waterlogging	Canal seepage Waterlogging Poor water quality Deficient leaching practices Irrigation water mismanagement Deficient drainage systems Floods	Irrigation water salinity control, dependable water supplies Soil salinity control, drainage, plant establishment, plant conservation, run-off management, canal lining
Plants	Reduction in vegetative productivity	Land clearance Plant mismanagement Over-cultivation Overgrazing Invasion of undesirable plant species Uncontrolled wood gathering Forest over-cutting Fire (uncontrolled burning) Drought	Irrigation, run-off management, water supplies, soil conservation, plant establishment, plant management
Animals	Reduction in animal productivity	Water scarcity Deficient fodder crops and food reserves Health and nutrition Overpopulation	Water supplies, water conservation, range management, livestock management, soil conservation, plant productivity, improved breeds, plant conservation, control of pests, game ranching
Energy	Scarcity and misuse of fuel	Uncontrolled wood gathering Misuse of the available energy	Afforestation, solar energy, wind energy, energy conservation, biogas

and, more important, to have the enthusiastic support of the people through a real understanding of the new technologies and their purpose, along with an effective degree of organization.

Desertification Factors

28. This review of technology and desertification describes, evaluates and discusses appropriate technologies to combat desertification in different land-use systems, with the following purposes: (a) to improve productivity by prevention measures; (b) to halt and reverse desertification by corrective measures; and, (c) to reclaim desertified areas. The analysis of technologies will be done according to the principal factors (water, soil, plant, animal, and energy) of the different land-use systems. The intent is to consider a great diversity of conditions and select the most appropriate technologies for different ecological, climatological, and socio-economic conditions.

29. Table 1 summarizes the problems responsible for desertification, causes and solutions according to the five principal factors. More detailed analyses of solutions are given in Tables 2 to 6.

I. TECHNOLOGY AND LAND USE

A. RURAL LAND USE

Irrigation

30. Irrigated areas cover 200 million hectares (ha) or 14 per cent of the total area of cultivated land in the world. It has been estimated that the total area irrigated in the year 1800 was 8 million ha. By the end of the nineteenth century about 40 million ha were under irrigation, and by 1950 the total had reached about 100 million ha. At the present time, the more than 200 million ha under irrigation are distributed as follows: Asia 140 million ha; United States 26 million; USSR nearly 13 million; Europe 12 million; Africa about 10 million; and Australia and New Zealand over one million ha. At the end of the present century the total area under irrigation may well approach 300 million ha. Drainage and salinity problems will increase more than proportionally, and solutions to these problems must be found. In general, the more severe the aridity, the more intense the salinity problem. Integral watershed management is essential to reduce sediment in irrigation waters and thereby to stop and reverse land degradation in the uplands.

31. According to Kaczmarek (1976), irrigated agriculture is the largest and most wasteful user of water. Better methods of irrigation are therefore urgently required to increase water-use efficiency and to rectify already widespread damage from excessive and improper water application.

Rainfed Agriculture

32. A large part of the semi-arid lands – how much is not known precisely – is devoted to rainfed crop production. Rainfed agriculture is the main source of the world's great cereal grains, employs many workers and is frequently characterized by low technological level, high risk, low production per unit area, and low living standards for the farmers who depend on this land-use system.

33. Three grain crops (wheat, rice and maize) are, with rice, the most important basic foods. Wheat, sorghum, barley and millet originated in semi-arid regions and are ecologically tied to that type of climate.

34. High rates of population growth have forced man to expand his tilled acreage in order to procure more food. The areas under cultivation continue to expand. In North Africa and the Near East, the area under cereals increased from 10.6 million ha in the 1950s to 11.7 million ha in 1970 and in the Sahel–Sudan region from 16.3 million ha in 1960 to 17.2 million ha in 1970. Beginning in the plains and fertile river valleys, cultivation has gradually expanded onto the hillsides and into rangelands.

35. Attempts to grow crops in marginal areas is one of the main reasons for desertification. Expansion into unsuitable climatic areas has been achieved at the expense of the pastoral territories and has accentuated a decline in grazing resources. Further, the poor results (cereal yields of 700–800 kg/ha) do not compensate for the disastrous ecological effects of cultivating such areas. Technical progress and increased production,

which are vital factors in development, require the stabilization, demarcation and improvement of areas to be cultivated instead of the indiscriminate opening up of new lands.

36. Unfortunately, traditional rainfed farming systems are usually characterized by low production and high risk due to the uncertainty of rainfall in amount distribution and intensity. Some of the causes of their low productivity are: (a) high economic risk related to input; (b) availability of inputs; (c) marketing problems; (d) low educational level of farmers; (e) inadequate systems of production; (f) deficient technical assistance; (g) lack of trained scientists in this field; (h) reluctance of farmers to adopt technologies which require more inputs; (i) the practice of extensive instead of intensive agriculture; and, (j) traditional types of energy used (manpower and animal draught).

Rangeland

37. Rangeland provides fodder for cattle, sheep, pigs, goats, buffaloes, horses, mules, camels and wild animals. Unfortunately, the productivity of much rangeland has been reduced due to human and livestock pressures reinforced by natural hazards. Desertification has made its appearance in soil and plant degradation and in the replacement of fodder plants by undesirable species. In the arid and semi-arid regions, animal husbandry covers larger areas than crop productivity. The more intensively cultivated areas must be used for forage cropping and stall-feeding practices. It is also imperative to maintain an equilibrium among water resources, numbers of people and livestock, soil and plant productivity and energy sources in order to sustain productive rangeland ecosystems.

Forestry

38. Trees and shrubs are a vital part of a healthy biosphere. They serve to reduce wind velocity, assist in reducing wind erosion, help to stabilize sand dunes, provide control of gullying, reduce soil loss in the upper parts of watershed, provide fuel for fire and wood for agricultural tools, railroads, and houses, provide browse for animals and improve the appearance of arid landscapes, providing shade and recreational sites. Most of the now degraded semi-arid and subhumid regions have suffered from the uncontrolled removal of tress and shrubs, unregulated burning, uncontrolled grazing and shifting cultivation.

39. The proper protection and management of forests call for research and training in tree planting methods, for extension education and for programmes for developing forest resources.

40. Cutting wood for fuel has reached devastating levels around many urban centres. Perhaps no other technology would so successfully combat desertification as the development of socially and economically acceptable substitutes for wood as a fuel.

B. URBAN AND INDUSTRIAL USES

Transport Routes and Mining

41. Land is readily degraded along transport routes and around the sites of mineral exploitation. Roadside erosion has increased dramatically in recent times with its

accompaniment of landslides and local flooding. The successful control of land degradation around roads and mines requires special water and soil conservation technologies, plus legislation and socio-economic support. The restoration of strip-mined land is very difficult in semi-arid regions and almost impossible without irrigation in arid regions.

42. Although capital-intensive commercial ranching reduces the number and size of rural settlements, worldwide population growth acts to spur the establishment and expansion of towns and cities. When settlements are located in the dry lands, they will develop desertified perimeters unless strict conservation measures are introduced to prevent, for example, uncontrolled wood collection and localized overgrazing. The protection of good farmland near settlements is demanded from both economic and aesthetic perspectives.

C. CLASSIFICATION AND EVALUATION OF TECHNOLOGIES

Water

43. The most important methods for managing water resources are listed in Table 2. The present times give access to a great number of ancient and modern techniques which have been developed under a great diversity of socio-economic and ecological conditions. As space prevents listing all of them, only the best known are included in Table 2, the purpose of which is to present an orderly system of classification and evaluation.

44. Problems generated by mismanagement of water resources in different land-use systems can be solved by proper application of: (a) dependable water supplies; (b) water conservation; (c) improved irrigation methods; (d) salinity control; (e) run-off management; and, (f) flood control. Each of these categories represents a group of technologies which can improve productivity, halt desertification, reverse the process and reclaim extremely degraded areas. These technologies may be applied in one or more of seven different land-use systems. Most of them have the objective of improving productivity, some of them may help to halt desertification, and a few may be useful in reversing the process.

45. Generally speaking, water management technologies as listed in Table 2 present only slight desertification potential with these exceptions: watershed conservation, salinity and drainage control, and run-off farming (which have a medium-to-high potential for land degradation). Most technologies are used in semi-intensive or intensive systems of production, characterized by high yields and good returns; their costs range from medium to high.

Recycling and Re-using Water

46. The re-use of water can greatly lower the pressure on water resources. Waste water can be used for irrigation, for industry and for recharge of groundwater and in special cases, properly treated waste water has been used for municipal drinking supplies.

47. Processes are available to obtain potable water from used domestic and industrial sources. Most of them, however, require large investments in capital, equipment, power and chemicals. The cost of such water is relatively high, but may be lower than the cost

TABLE 2. *Classification and evaluation of water management technologies and their effect on desertification and productivity applied to seven land use systems*

Management method	Technology	Land-use system	Objective or purpose	Desertification potential	Production system	Cost	Level of technology
Sources of supplies	Dams	S,I	A,B	High	Intensive	High	Complex
	Horizontal wells	S,I	A	Slight	Semi-intensive	Medium	Interm–complex
	Qanats	S,I	A	Slight	Semi-intensive	Medium	Intermediate
	Dug wells	S,I	A	None	Semi-intensive	Low	Simple
	Boreholes	S,I	A	High	Extensive	Low	Simple
	Springs	S,I	A	Slight	Semi-intensive	Low	Intermediate
	Desalinization	S,I	A	None	Intensive	High	Complex
	Recycling sewage water	I	A	Slight	Semi-intensive	Medium	Complex
	Solar distillation	S	A	None	Semi-intensive	High	Simple–complex
	Rain collection from roofs	S	A	None	Semi-intensive	Low	Simple
	Dew and fog collection	S	A	None	Semi-intensive	Medium	Simple–complex
	Rainfall augmentation	S,RA,R,F	A	None	Semi-intensive	High	Complex
Conservation	Groundwater recharge	I,RA,S,R,F	A	None	Semi-intensive	Medium	Interm–complex
	Reservoirs, cisterns	S,I	A	None	Semi-intensive	Low	Simple
	Sand taps	S,R,	A	None	Semi-intensive	Medium	Intermediate
	Rubber bags	S,R	A	None	Intensive	High	Interm–complex
	Ponds	S,R	A	None	Semi-intensive	Low to High	Interm–complex
	Seepage control	I	A	Medium	Semi-intensive	Medium	Interm–complex
	Evaporation suppression	S,I,RA,R,F	A	None	Extensive	High	Interm–complex

Irrigation	Flooding	I	A	Slight	Semi-intensive	Low	Simple
	Furrows	I	A	Slight	Semi-intensive	Low	Simple–interm
	Overhead or sprinkler	I	A	Slight	Intensive	High	Interm–complex
	Drip or trickle	I	A	Medium	Intensive	High	Complex
	Subirrigation	I	A	High	Semi-intensive	High	Complex
	Innovative						
	centre-pivot sprinkler	I	A	None	Intensive	High	Complex
	pitcher	I	A	None	Intensive	Low	Simple
	Leaching	I,RA,	C,B,A,D	Slight	Intensive	Medium	Complex
	Supplementary irrigation	I,RA,R	A	Slight	Semi-intensive	Low to High	Interm–complex
Salinity and alkalinity control	Physical	I	C,B,A,D	Medium	Semi-intensive	Medium	Interm–complex
	Chemical	I	C,B,A,D	None	Semi-intensive	Medium	Interm–complex
	Combination of methods	I	C,B,A,D	Slight	Intensive	High	Complex
Run-off	Waterspreading	RA,R,F	A,B	Slight	Semi-intensive	Low	Simple–complex
	Water harvesting	RA,R,F	A,B	Slight	Semi-intensive	Low	Simple–complex
	Run-off farming	RA,R,F	A,B	Medium to High	Semi-intensive	Low	Simple–complex
	Flooding	RA,R,F	A,B	Slight	Semi-intensive	Low	Simple–complex
Flood control	Channels	S,F,R,RA,I	B,A	Slight	Intensive	Medium	Complex
	Levees	S,F,R,RA,I	B,A	Slight	Intensive	Medium	Interm–complex
	Diversion	S,F,R,RA,I	B,A	Slight	Intensive	Medium	Simple–complex
	Land treatment measures	S,F,R,RA,I	B,A	Medium	Intensive	Medium	Simple–complex

Land-use systems: irrigation agriculture (I); rainfed agriculture (RA); rangeland (R); forestry (F); mining (M); transport routes (T); settlements (S).

Objective or purpose: A – To improve productivity; B – To halt desertification; C – To reverse the process; D – to reclaim extremely degraded areas.

of developing alternative sources of water, though not – in most cases – if treatment is needed to remove dissolved mineral salts.

48. One important advantage of water re-use is that, it can, if properly managed, reduce by several-fold the demand on water from original sources. Recycling 50 per cent of waste water in effect doubles the water supply. Although the technology for re-use is now available, economic considerations will probably limit its application to specialized locations or purposes. As acceptance groups and fresh water supplies diminish, widespread recycling of waste water into the potable supply system becomes even more likely.

Desalinization

49. Desalinization techniques are already known and currently used to produce sweet water for human and animal consumption. In the least expensive cases (nuclear energy powered plants) the cost ranges from US 10 cents per cubic metre to ten times that much. For the time being, this is still too expensive for agricultural purposes and can be economically used to produce only luxury cash crops such as out-of-season vegetables, an application that has, of course, little to do with desertification. In the not-too-distant future, however, water desalinization offers more favourable prospects although it will require very large plants powered by nuclear or solar energy which only rich countries will probably be able to afford.

Water Conservation

50. With respect to surface reservoirs, it is important to apply appropriate technologies to control sedimentation and silting because they reduce the storage capacity and provoke land degradation in the upstream part of the watershed. Seepage control measures should be included to reduce water losses. Evaporation suppression techniques may be applied to increase the reservoir storage efficiency but the technology is not yet sufficiently well developed. Technologies for improving efficiency of long-distance water transport systems need to be further developed.

Underground Water

51. Underground water is a valuable resource. Its exploitation requires an understanding of the recharge capacity and the water volumes to be extracted for different purposes (municipal, agricultural, industrial, etc). It is imperative to maintain a rational equilibrium between groundwater resources and their exploitation and avoid the indiscriminate over-exploitation which may lead to desertification.

Small Reservoirs

52. Due to the scarce and erratic rainfall of the arid zones, small reservoirs for human and livestock consumption may be needed. There are several ways of designing such reservoirs (cisterns, galvanized tanks, sand traps, and rubber bags, etc.), some simple and relatively inexpensive, others costly.

Watering Points for Livestock

53. The establishment of new watering points in areas already damaged by overgrazing should be avoided until the pastures have recovered. In certain cases it is possible to plan pasture rotation by temporarily closing off or prohibiting access to certain watering points, or by developing new watering points. Procedures should be established to determine the lowest cost combination of catchment area, its efficiency and storage size, water demand schedules and rainfall patterns by surveying the local situation.

Irrigation Methods

54. Water application to the soil for plant production through an irrigation method intends to produce the highest yields with the minimum amount of water and without deteriorating the soil system.

55. Flooding and furrow methods of irrigation have been extensively and intensively used by farmers all over the world; they do not require a high level of technology. Yet they also have disadvantages: (a) both are surface irrigation methods which generally present low to medium water use efficiency because they demand high volumes as water flows and becomes distributed by gravity; (b) when misapplied, they may accelerate the potential for salinity and waterlogging; salts will be deposited in the soil profile when the leaching and drainage systems are not properly designed; and, (c) excess water will raise the watertable and wash out nutrients.

56. Overhead or sprinkler methods show an improved water-use efficiency because they distribute water with greater accuracy, more uniformity and control watering penetration at specified depths. If a proper salt balance is maintained in the soil profile, salinity hazards may be avoided. These are expensive methods, requiring considerable technical knowledge, and are more likely to be used in intensive systems of production.

57. The drip or trickle method has the highest water-use efficiency because water is delivered in small amounts at the precise sites where it will be consumed and may be applied according to the water requirements of the plant. It requires a high level of technology and is expensive and is therefore applied only in intensive systems of production. Salt accumulating in the soil is a potential problem.

58. Subirrigation methods offer high efficiency if properly controlled. They are applied only in very localized areas, are related to intensive systems of production, are expensive and require advanced technical knowhow. They have a high potential salinity hazard.

59. Other innovations (centre-pivot sprinkler, fixed-orifice sprayer, gated pipes) have been developed, each with its own desertification potential. With proper research and technical assistance, however, such problems as excess water, under-irrigation, salinity and waterlogging may be avoided. Most such innovative systems are expensive.

60. Salt balance in the soil profile may be achieved by properly designed leaching procedures. Some of the above-mentioned irrigation methods (flooding, furrow, sprinkler and several innovations methods) are clearly capable of meeting leaching requirements. Hand drip and subirrigation and some innovations methods such as pitcher irrigation are limited in their leaching capacities.

Irrigation with Saline Water

61. The use of saline water for irrigation has been limited in the absence of sufficient research into: the breeding and selection of plants that can use water with high salt content, the relationship between saline water and the physiological stress by different irrigation, fertilization, soil aeration, leaching, hormonal, chemical and physical treatments.

62. A tremendous amount of saline water (i.e. 75 per cent of global water resources is available in the world from the sea, lagoons, lakes, and some underground reserves. However, such water is not easily usable for plant growth. At present, use of saline water is confined to places where soils are highly permeable, salt tolerant plants are grown, and a large supply of water is available for leaching purposes.

63. Another way of utilizing water of high salt content is by mixing it with fresh water to reduce the salt concentration. This procedure will increase the total amount of water available for irrigation but will also lower the quality of the water, so the following points must be taken into account: (a) the increase in yields due to the increase in water supply should be at least as large as the potential decrease in yields caused by increased salinity; (b) when the decrease in water quality through mixing requires leaching, the additional water required for leaching should be added to the calculations; (c) the use of good quality irrigation water to dilute drainage water so that the latter may be re-used and thus increase the total supply is a questionable practice which is useful only when leaching is provided by rainfall.

Rainfed Agriculture

64. As dryland systems (rainfed agriculture, rangeland, and forestry) will depend upon rainfall patterns, an optimum use of this natural resource must be achieved. One of the most promising methods to improve land productivity and control desertification consists in run-off management.

65. Most rainfed areas in the dry lands will need to increase the application of technologies for the improved utilization of run-off to increase water infiltration and to improve plant production; otherwise, uncontrolled run-off may cause erosion and desertification.

66. Four steps may be taken to improve the productivity of rainfed lands: (a) control erosion; (b) optimize rainfall by water harvesting techniques; (c) improve soil fertility and, (d) select appropriate plant species adapted to local conditions. Techniques to accomplish the first three are available and generally simple to apply; the fourth is only matter of selection if the proper seeds are available.

Flood Control

67. A system of flood control normally consists of a variety of structural and non-structural measures to reduce flood damage. Effective flood control is achieved by a combination of measures.

68. The selection and combination of structural and non-structural measures depend upon the social, political and economic characteristics of the countries concerned. When rivers cross national boundaries, international co-operation, exchange of data, and mutual assistance in flood control and warning are required.

69. Along with flood control, sediment control is very important. Excess sediment causes reduction in reservoir storage, plugging of canals, ditches, and drains, and pollution of water supplies.

D. SOILS

70. The problems produced by man's mismanagement of soil resources in different land-use systems can be solved by applying: (a) soil conservation; (b) soil moisture conservation; (c) salinity control; (d) drainage; and, (e) measures to improve soil productivity. Each category represents a group of technologies, as shown in Table 3, which displays their distinctive characteristics.

71. It should be emphasized that of 31 technologies described in Table 3, all of them have as an objective the halting of desertification. Soil management will thus have a significant impact in controlling desertification.

72. The most important technologies related to soil management are: revegetation, dune stabilization, sedimentation control, gully control, conservation on transport routes, terracing, land levelling, mulching, tillage practices, salinity control, drainage, and organic fertilization. Most of these technologies have several purposes (to improve productivity, to halt desertification, to reclaim extremely degraded areas). In developing countries it is necessary to develop and apply a concept of intermediate technology, which means that to increase productivity and halt desertification, it is necessary to pass from extensive systems to semi-intensive and intensive systems of production by using technologies which work without requiring high or complicated inputs.

73. The potential of technologies for desertification must not be forgotten. Misapplied technologies, such as injudicious tillage practices or uncontrolled land clearance, help to further degradation, as has happened in several countries.

Soil Conservation

74. Soil erosion is as old as the land. A detailed understanding of the mechanisms involved and the development technologies to prevent or control it is, however, very new. Several authors (White ,1976; Barry, 1971) have estimated a total annual loss of dissolved solids carried to the ocean by streams at about 3000 million metric tonnes. In this century, most countries have established national conservation programmes, but their results indicate only a slowing down of the rate of destruction in different land-use systems, not a halting or reversal of destruction.

75. Land capability must be assessed before applying a soil conservation programme. The arid and semi-arid zones are in regions which show variations in customs, land tenure, economy, climate, soil, plant cover and other variables.

76. When uncontrolled run-off prevails in dry lands it represents a potential source of desertification. Programmes for run-off management are essential to improve the land's productivity and to prevent and correct desertification problems.

77. As wind and water erosion are major causes of desertification, soil conservation measures should be developed in both short-term and long-term perspectives and at local, national, regional and international levels with the organized participation of decision-makers, scientists, technicians, extensionists, villagers and farmers. Appropriate technologies must be applied to prevent and correct degradation and to reclaim productive

TABLE 3. *Classification and evaluation of soil management technologies and their effect on desertification and productivity applied to several land-use systems*

Management method	Technology	Land-use system	Objective or purpose	Desertification potential	Production system	Cost	Level of technology
Soil conservation	Sediment control	I,RA,R,T,M,S	B,C,D,A	Slight	Semi-intensive	High	Interm–complex
	Gully control	I,RA,R,F,T,M,S	B,D,C,A	Slight	Extensive	Low to High	Simple–complex
	Transport routes conservation	I,RA,R,F,T,M,S	B,A	High	Intensive	High	Interm–complex
	Soil stabilizers	S,T,I,M,RA,R,F	B,A	None	Intensive	High	Complex
	Revegetation	R,F,RA,I,T,M,S	B,C,D,A	None	Extensive	Low to High	Simple–complex
	Mechanical practices	I,RA,R,F,T,M,S	B,C,D,A	Medium	Semi-intensive	Medium to High	Interm–complex
	Dune stabilization	I,S,T,RA,R,F,M	B,C,D,A	Medium	Extensive	Medium	Interm–complex
	Tillage practices	I,RA,R,F,T,M,S	A,B,C,D	High	Intensive	High	Simple–complex
	Land levelling	I,RA,R,T,S	B,C,D,A	Medium	Intensive	High	Interm–complex
	Strip cropping	I,RA,R	A,B	Slight	Semi-intensive	Low	Intermediate
Soil moisture	Tillage practices	I,RA,R	A,B	Slight	Semi-intensive	Medium	Simple–complex
	deep ploughing	I,RA,R,F	A,B,C,D	High	Intensive	High	Intermediate
	ripping	I,RA,R,F	A,B,C,D	Medium	Semi-intensive	Medium	Intermediate
	range pitting	R	B,A	Slight	Extensive	Low	Intermediate
	Tied ridge	RA,R	B,A	Slight	Extensive	Low	Simple–interm
	Mulching	I,RA,T,M,S	B,A	None	Semi-intensive	Low to High	Simple–interm
	Stubble mulching	I,RA	B,A	None	Semi-intensive	Medium	Simple–interm

Salinity control	Physical	I,RA,R	B,C,A,D	Slight	Semi-intensive	High	Simple–complex
	deep ploughing	I,RA,R	B,C,A,D	Slight	Semi-intensive	High	Intermediate
	subsoiling	I,RA,R	B,C,A,D	Slight	Semi-intensive	High	Interm–complex
	Sanding	I,RA,R	B,C,A,D	Slight	Semi-intensive	High	Intermediate
	Biological	I,RA	B,A	None	Semi-intensive	Medium	Simple–complex
	Chemical	I,RA,R	B,C,A,D	Slight	Intensive	High	Interm–complex
	Combination of methods	I,RA,R	B,C,A,D	Slight	Intensive	High	Simple–complex
Drainage	Open channel or ditches	I,RA,R	B,C,A,D	Slight	Semi-intensive	High	Simple–complex
	Pipe	I,RA,R	B,C,A,D	Slight	Semi-intensive	High	Simple–complex
	Hole	I,RA,R	B,C,A,D	Slight	Semi-intensive	High	Simple–complex
	Combination pipe-mole	I,RA,R	B,C,A,D	Slight	Semi-intensive	High	Interm–complex
	Interceptor system and pumping	I,RA,R,S	B,A	Slight	Semi-intensive	High	Complex
Fertilization	Organic	I,RA,R	B,A,C,D	None	Semi-intensive	Medium	Simple–interm
	Chemical	I,RA,R	A,B	Slight	Semi-intensive	High	Simple–interm

Land use irrigation agriculture (I): rainfed agriculture (RA); rangeland (R); forestry (F); transport routes (T); settlements (S).

Objective or purpose: A – To improve productivity; B – To control deserts (improve conditions: halt sand dune encroachment; C – To reverse the process: D – To reclaim extremely degraded areas.

soil in all land-use systems. Here it should be remembered that prevention is vastly cheaper than cure.

78. Measures to reduce flood and sedimentation hazards include: (a) trenching, furrowing or pitting soil surfaces; (b) revegetation – vegetation can be restored on depleted and eroded areas by contour trenching and reseeding; (c) water spreading, consisting of dike systems constructed to divert flood flows from gullies and arroyos and spread them over adjacent lands.

79. Gullies will continue to appear when new areas are opened to dry farming on sloping land without adequate soil and water conservation. Some ancient civilizations that cultivated steep slopes susceptible to erosion developed an ability to control gullies by building terraces, some of which are still protecting soil because care has been taken to maintain them generation after generation.

Soil Conservation on Transport Routes

80. Megahan (1976) describes four basic principles for reducing erosion: (a) minimize the amount of disturbance caused by road construction (by controlling total mileage and by reducing the area of disturbance on the roads that are built); (b) avoid construction in highly erodible areas; (c) minimize erosion on areas disturbed by road construction through suitable soil conservation practices; (d) minimize the off-site impact of erosion. Prevention rather than control of land deterioration seems to be the most effective method of reducing erosional effects; prevention may also avoid irreparable damage or costly repairs which may exceed original construction costs. The first and second principles mentioned above are designed to prevent erosion rather than to control it.

81. The entire road development process, beginning with broad land-use planning, should include location, design, construction, maintenance and closure. Road construction programmes should include multidisciplinary teams of soil scientists, geologists, engineers, land-use planners, hydrologists and agronomists.

82. When considering the economic details of roadside erosion control, the cost of the selected treatment should be evaluated not only with reference to the road itself, but also with reference to adjoining land. The successful control of roadside erosion requires special attention to the choice of soil conservation technologies.

83. Slope stabilization measures include revegetation and other practices to control surface erosion on road cuts and on dumping and borrow areas. Soil stabilization should be carried out immediately, preferably during construction. A wide variety of plants (grasses, shrubs and trees) can be used for stabilizing slopes, alone or in combination.

Revegetation

84. As a measure to control land degradation, revegetation is directed toward: (a) control of wind erosion and wind speed; (b) control of water run-off and its consequent erosion; (c) aesthetic improvement; (d) shade; (e) food production; (f) production of other plant products; and, (g) control of sediment around ponds.

85. Where there is possibility of irrigation, improved land-utilization systems are necessary for more rational exploitation of the local potential under the given environmental conditions.

86. The natural vegetation of arid and semi-arid zones consists mainly of ephemerals

and drought-resistant grasses and perennial shrubs which, if properly managed, could be a source of forage in critical periods of the annual grazing cycle and during droughts.

Sand-dune Fixation

87. There are several large-scale experiments carried out in some countries of the Middle East in which seedlings are protected from burial or wind scouring by erecting micro-windbreaks (fences) made from dead or living plant materials, or by soil stabilizers (principally oil derivatives).

88. Sand-dune afforestation programme should be considered since they would provide: (a) protection against wind erosion for roads, industrial areas, settlements, water reservoirs and agricultural areas; (b) an increase in the water supply by increasing deep percolation; (c) grazing benefits with the introduction of more palatable species; (d) recreational areas and wildlife conservation; (e) wood production (for fuel, timber and pulp). The cost/benefit ratio could be high because this kind of afforestation represents a multiple purpose activity.

Tillage Practices: Implements for Soil and Water Conservation

89. The proper choice and use of tillage methods through efficient methods through efficient implements is a primary criterion for preventing land degradation under dry land-use systems. One major agricultural advance in the last 25 years has been obtained in the design, construction, and use of tillage implements that leave crop residues at the soil surface. This helps in controlling water erosion and run-off, increasing water-use efficiency by suppressing evaporation and improving the water-holding capacity of soil.

90. Governments should establish national programmes for the inventory of agricultural implements which have been designed by farmers and complement them with the design and large-scale construction of hand tools, animal traction and machines. Implements should be made available to small farmers who cannot afford to purchase them.

Soil Management (Rainfed Agriculture)

91. Soil management under arid and semi-arid conditions requires maintaining a cloddy and residue-mulched surface and carrying out all tillage and planting operations across the slope of the land (contour culture), forming a series of parallel ridges and furrows perpendicular to the slope and reducing the flow of water downhill.

92. Kassas (1970) mentions that dryland farming in Mareotis depends on water redistribution with the help of small-scale terracing, man-made hillocks and contour ploughing in a series of limestone bars located between Alexandria and the Libyan border where many small farms on the basal areas of these bars can be planted to olive or barley. The coastal sand dunes are at present planted with figs.

93. In the semi-arid tropical environment of Hyderabad, India, well-maintained systems of bunds have a very definite effect on decreasing soil erosion on a watershed basis. The cultivators are often advised by technicians of the Soil Conservation Service to establish certain plant species on the bunds and are provided with grasses to be established on the beds of waterways so as to avoid soil erosion. In India, the All-India Co-ordinated

Research Project for Dryland Agriculture, started in 1970, provides an excellent example of an integrated approach towards research and action on the problems of semi-arid agriculture.

The Fallow System

94. This system is applied in the lower rainfall (less than 400 mm) areas of North America, Latin America, Australia, the Near East and North Africa and consists in cropping in alternate years and fallowing between crops. The amount of moisture stored during fallowing is often no more than 20 per cent of the rain received, which together with that falling during the crop season, usually assures a satisfactory crop. In addition, the nutrient content is also enhanced.

Furrow Basins

95. The stability of agriculture systems relies on the bulk of the soil being held in place at all times. If rainfall infiltrates the root zone, there will be no surface run-off and no water erosion. With furrow basins, land is always maintained in ridges laid out across the slope; the furrows are dammed into a series of rectangular depressions. The ridges and the cross-ties can be constructed using handpower, animal traction or machinery.

96. This method is followed by farmers in many countries. The implement used is merely a shovel for making basins in furrows. This is done after the first cultivation and when the plant has grown somewhat.

Soil and Water Conservation (Pastoral Lands)

97. Rangeland management requires among other things that soil be conserved from erosion and moisture utilized to its fullest extent.

98. As overgrazing is the main cause of soil and plant degradation on rangelands, it is obvious that proper maintenance of the vegetative cover is essential. Anything which improves existing conditions for plant growth will lead to protection against desertification and an improvement in land productivity. Prevention techniques are the key to soil erosion problems.

99. Species selection and seeding operations for pastoral lands should take into account: (a) adaptability to soil and climate; (b) vegetative cycle; (c) expected yields; (d) physiological response to grazing; (e) nutritive value; (f) palatability; (g) compatibility in mixtures of species, and, (h) forage production to meet seasonal deficits.

Mechanical Practices

100. Terracing has been practised since the dawn of agriculture. In many parts of the world (Philippines, China, Japan, Peru, Mexico) ancient terraces are still productive thanks to constant maintenance by villagers. When well constructed, protection terraces can assure the productivity of land with slopes as high as of 60–70 per cent.

101. Terracing is expensive. There are a number of ways in which terraces can be constructed: (a) local people can build them by hand, receiving from the government technical assistance and some plant materials to cover surfaces; (b) the government

constructs the terraces and local people maintain them; (c) the farmers apply for a survey of the land; the government constructs the terraces, gives technical assistance and charges 25–50 per cent of the cost; (d) the farmers construct their own terraces and maintain them. (e) Construction can be carried out by private contractors. Terraces once constructed will require permanent maintenance to keep them productive. Otherwise they quickly deteriorate.

Fertilization

102. Fertilization comes in various types, ancient and modern, including organic substances, fluids, mixtures, green manuring and newly developed types which should be made economically available to farmers.

103. Soil productivity can be maintained through proper soil and water management combined with optimum levels of fertilization. Crop rotation represents another method to preserve soil fertility and maintain a proper vegetation cover in drylands.

104. Rangeland fertilization may increase profits in an animal industry, but the response of pasture to this practice will be a function of soil moisture availability and plant management. The maintenance of rangeland productivity is a long-term operation and in most areas must overcome a number of deficiencies, such as a lack of seeds and plant varieties.

Organic Fertilizers

105. Organic materials (plant, animal and human wastes) have been applied to soils since the dawn of agriculture. They are the most widely used fertilizers in less developed countries and they often represent a higher economic potential than the chemical fertilizers characterized by high prices and shortages. However, organic materials have not received enough attention from either the technical or economic point of view. These important resources need much more attention from researchers, technicians, extensionists, governments, and farmers in order to optimize their rational use.

Soil Salinity and Waterlogging

106. Both under-irrigation and over-irrigation are harmful. Inadequate water supply fails to meet evapotranspiration and leaching requirements and results in salinization. Enough water must be supplied to satisfy both leaching and consumption requirements. The main course of the failure of crop production under irrigation are the combined and related effects of excessive salt accumulation in the root zone.

107. The quality of the water must also be taken into account in any irrigation scheme. Quality will have a bearing on the type of soil that can be irrigated, the crops that can be grown, and the management techniques and irrigation methods that can be used.

108. Waterlogging is an increasing problem produced by irrigation practices which raise the watertable and cause salts to accumulate in the soil profile. Over-irrigation is the main cause of waterlogging. Seepage control and drainage systems should be installed where necessary to prevent and reduce salt accumulation in the root zone.

109. It is estimated (Israelsen and Hansen, 1965) that from 20 to 30 per cent of the

irrigated lands in arid and semi-arid regions need improved drainage if they are to continue as productive sources of crops. Drainage requirements arise because: (a) man cannot precisely control the water used for irrigation; (b) it is difficult to apply water in the exact amounts needed by plants; normally, too much rather than too little water is applied; and, (c) in arid areas, excess water must be applied in order to leach accumulated salts from the root zone (Donnan and Schwab, 1974).

110. Salinity and waterlogging are the principal problems to be found in irrigated lands although they may also occur elsewhere. Both problems call for reclamation techniques which are more expensive than preventive measures.

Salinity Control

111. Reclamation of saline and soda soils can be achieved by physical, biological, chemical or hydrotechnical methods or a combination of such measures. In areas where saline or saline soil are used for cropping, more salt-tolerant crop varieties must be selected because of their lower leaching requirements and the consequent reduction in the amount of water for irrigation. Also, crops of low evapotranspiration reduce leaching requirements.

112. Reclamation is undertaken to reduce salt concentration in the root zone to a level compatible with crop growth and to lower the exchangeable sodium level where necessary. In order to obtain a good salt balance near the root zone, an efficient drainage system is required.

Drainage

113. Adequate drainage facilities are often not provided until problems become acute. A lack of proper drainage systems has been the cause of the failure of many irrigation districts. Under certain conditions, surface drainage may be enough, but often subsurface drainage (ditch or tile) has to be provided. If canal seepage is considerable, an intercepter drain may be effective, as has been shown in Australia, Iraq, and Pakistan, Syria, and the United States.

114. Data from twelve countries indicate that clay pipe is still the most commonly used drainage material, but plastics are fast gaining ground because of low initial investment costs and advantages in transport and handling. Smooth plastic pipe, which was the first competitor to clay pipe, is being replaced by a corrugated type, because the latter has greater mechanical strength against deformation, and is also lighter.

115. The watertable should be maintained at a depth depending on the hazard of upward salt movement with recommendations indicating between 1 m and 2 m. Most countries suggest drain spacing of 60–80 m from the point of view of costs. However, recommendations vary from 15 m to 10 m.

116. Leaching requires proper internal drainage, which is essential for the achievement of permanent irrigation agriculture. Combinations of surface and subsurface drainage systems are well documented but not often employed. Surface drainage plus levelling allows adequate removal of run-off and prevents waterlogging; its cost is low and it could easily be applied in most countries.

E. PLANTS

117. Plants (agricultural, pasture, tree and native) sustain both human and animal populations. Their productivity depends on nature's influence and man's management. Erratic and badly distributed rainfall may lead to fluctuating production and subsequent famine and scarcity.

118. The problems involved with plant management in different land-use systems may be overcome by these activities: (a) plant establishment; (b) plant conservation, and, (c) plant production (agricultural, rangeland, and forestry). Each of these activities represents a group of technologies, as can be seen in Table 4.

119. The most promising technologies to halt desertification, to reverse it, or to reclaim deteriorated areas are these: revegetation, rotational grazings, deferred grazing, rotation pasture, crop rotation, wood fuel control, windbreaks, shelter belts and fencing.

Reduction of the Vegetative Cover

120. Most of the now degraded arid and semi-arid regions have suffered from forest removal, bush fires, uncontrolled grazing, shifting cultivation, etc. Through the devastation of forests, man eliminates his best ally for the control of soil erosion.

Shifting of the Vegetative Cover

121. Shifting cultivation is a farming system which consists of the rotation of fields cleared for cultivation, with short periods of cropping alternating with long periods of natural fallow. Slash-and-burn methods are used to clear vegetation and fertility is maintained by allowing the vegetation to regenerate. The primary implements are the axe, machete or panga, and hoe, sometimes supplemented with the digging stick in Africa. Once the principal tool in America and Asia, the digging stick is now supplemented by other implements. Shifting cultivation implies that soil fertility must be restored by fallow before crops can be grown again. Nowadays, the length of the fallow period is increasingly reduced as a result of excessive population growth, and its consequent pressures.

Rangeland

122. Rangelands provide fodder for both domestic and wild animals. Mismanagement of either domestic stock or wildlife gives rise to overgrazing which is the main cause of soil and plant degradation in rangelands. Other causes include insufficient watering points, gully encroachment, replacement of fodder plants by undesirable species, invasion of grazing land by cropping systems and alteration in the hydrologic cycle.

123. Other difficulties are generated by permitting animal concentration around watering sites, in lanes or gates between pasture, and in shady areas. The latter, even though small, can become a source of erosion and desertification. It will be impossible to improve rangelands where livestock numbers cannot be controlled.

TABLE 4. *Classification and evaluation of plant management technologies and their effect on desertification and productivity applied to several land-use systems*

Management method	Technology	Land-use system	Objective or purpose	Deserti-fication potential	Production system	Cost	Level of technology
Plant establish-ment	Seeding methods	I,RA,F,R,T,S,M	B,A	Slight	Extensive	Low to High	Intermediate
	Planting techniques	I,RA,R,F,M,T,S	B,A	None	Semi-intensive	Low to High	Intermediate
	Reseeding	R,T,M,S,F	B,A	None	Extensive	Low to High	Simple–interm
	Afforestation	I,RA,R,F,S,T,M	B,A	None	Extensive	Low to High	Intermediate
Conservation	Water efficient plants	I,RA,R,F,T,S,M	A,B	None	Semi-intensive	Medium	Interm–complex
	Genetic technology	I,RA,R,F	A	None	Intensive	High	Complex
	Transpiration suppression	I	A	None	Intensive	High	Complex
	Controlled environments	I,S	A	None	Intensive	High	Complex
	Revegetation	I,RA,R,F,T,M,S	B,C,D,A	None	Intensive	Low to High	Simple–complex
	Control of undesirable vegetation	I,RA,R,F,T,M,S	A,B	Slight	Extensive	Low to Medium	Simple–interm
	Cutting control	F,R	B,A	Slight	Extensive	Medium	Simple–complex
	Fencing	R,F,RA,I,T,M,S	B,C,A	Slight	Extensive	High	Simple–complex
	Fire control	R,F,RA,I,S	A	Slight	Semi-intensive	Low to High	Simple–complex
	Plant disease control	I,RA,R,F	A,B	Slight	Semi-intensive	Medium	Complex
	Insect and rodent control	I,RA,R,F,S	A	Slight	Semi-intensive	Medium	Intermediate

Agriculture	Basic crops	I,RA	A	Slight	Semi-intensive	Medium	Simple
	Crop rotation	I,RA	B,A	None	Semi-intensive	Medium	Intermediate
	Multiple cropping	I,RA	A	Medium	Semi-intensive	Medium	Simple–complex
	Monoculture	I,RA	A	High	Semi-intensive	Medium	Simple
	Vegetables	I,RA	A	Slight	Intensive	High	Interm–complex
	Fruit trees	I,RA	A,B	Slight	Semi-intensive	High	Simple–complex
	Industrial crops	I,RA,F	A	Slight	Intensive	High	Interm–complex
Rangeland	Fodder plants	R	A,B,C,D	None	Extensive	Medium	Simple–complex
	Rotational grazing	R	B,A	High	Extensive	Low	Interm–complex
	Deferred grazing	R	B,A	Medium	Extensive	Low	Interm–complex
	Rotational pasture	R	B,A	Medium	Extensive	Medium	Interm–complex
	Pastoral land	R	B,A	High	Extensive	Low	Simple–complex
Forestry	Wood gathering control	F	B,A	High	Semi-intensive	Low	Simple–complex
	Wood fuel	F,S,R,RA	B,A,C,D	None	Extensive	Medium	Simple–complex
	Windbreaks and shelterbelts	I,RA,R,S,T,M	B,A,C,D	None	Extensive	Low to High	Interm–complex
	Native vegetation	R,RA,F	A	Medium	Extensive	Low	Simple–complex

Land use: irrigation agriculture (I); rainfed agriculture (RA); rangeland (R); forestry (F); mining (M); transport routes (T); settlements (S).

Objective or purpose: A – To improve productivity; B – To halt desertification; C – To reverse process; D – To reclaim extremely degraded areas.

Forestry

124. Excess woodcutting and gathering and over-exploitation of wood fuel cause serious land degradation due to the elimination of the vegetative cover which protects the soil against erosion.

125. Many cities and villages in dry lands are surrounded by belts, sometimes up to a radius of 60–100 km, of land denuded of trees and wood and (Le Houérou, 1975; Makhijani and Poole, 1975) increasing land degradation.

126. The Energy Survey of India Committee reports that about 120 million tonnes of wood, 20 million tonnes of dry dung, and 30 million tonnes of vegetable wastes are burned in that country every year. It is estimated that the use of wood and other organic materials in the rural areas of many countries is over one tonne per capita per year (Makhijani and Poole, 1975).

127. According to Makhijani and Poole (1975) Latin America uses 25 million Btu per person per year of wood for cooking and heating. Tropical Africa has the smallest input of energy into farm work. In Gambia about 55 per cent of wood fuel is used for cooking, 35 per cent for heating, and 10 per cent for other uses.

128. Data from India and Gambia indicate an annual wood fuel use of 5–7 million Btu per person per year; approximate data on the use of petroleum gases for cooking in India are of 1–1.5 million Btu per capita and of 1 of 2 million Btu for gas stoves.

129. In countries where wood is used as fuel and where no control exists on the wood gathering system, millions of hectares have been degraded due to the indiscriminate elimination of forest cover.

130. Pimentel (1976) estimates that 25 per cent of the world's energy (including wood) is used in the food system, an amount that will have to be increased three times to meet the food needs of the 7 billion human beings in the year 2000.

131. When vegetative cover is reduced, it may well be replaced by bare land or by undesirable species, and gradually the loss of fertile topsoil by wind creates erosion and results finally in unproductive and desertified land.

Agricultural Solutions

132. Rainfed agriculture occupies 86 per cent of the world's cultivated area, represents the major part of the agrarian economy of less developed countries, and will require much more attention through long-term plans of action because such systems produce most of the basic cereal crops and this land use occurs in ecosystems which are fragile because of variability of climatological conditions. Research should be encouraged on the development of varieties of basic, industrial and forage crops with drought-resistant characteristics and adapted to arid and semi-arid environments.

Selecting and Managing Crops to Use Water More Efficiently

133. A breeding programme aimed at developing cultivable varieties for agricultural use under arid conditions can be approached (a) by selecting wild plants that now survive in desert conditions and developing them as food or cash crops; or, (b) by selecting individual water-efficient plants from varieties of already-domesticated crops such as

barley, sunflower, melons, sorghum, millet and beans, and using them in breeding programmes.

134. The following kinds of plants seem to offer promise: (a) plants that grow in cool seasons when evaporation rates are lower; (b) rapid-growing plants (but without greatly increased water requirements) that shorten the time in which water is lost by transpiration and evaporation; (c) high-yielding plants that require no appreciable increase in water supply; (d) plants with low transpiration losses, and, (e) plants that can tolerate low-quality (e.g. saline) water. Reflectant materials have been tried on artichokes in Israel with considerable success.

Cereal Production

135. In the arid and semi-arid regions of North Africa, special dry farming methods consist in ploughing and sowing in fields divided into a number of rectangular strips. The ploughing is a mere scratching of the surface. On heavy soils the sowing plough is harnessed to a pair of oxen, horses or mules. On light soils, one animal is sufficient. Farmers who are also herdsmen use less grain for sowing, from 25 to 100 kg/ha of wheat and barley. Barley is planted in the driest areas, since it requires less water and ripens earlier than wheat. Harvesting is done in the second half of April or early May. Some straw is left on the ground to be used by the cattle (Despois, 1961).

Fruit Trees in Tunisia

136. Olive trees are planted in areas with a mean annual rainfall of only 170–250 mm. Soils are sandy and condensation of water vapour takes place in the soil system. Trees are planted at intervals of 24 m. Each tree occupies 576 m^2 and the density per hectare is 17 or 18. Almonds, apricots and peaches are planted about 12 m apart. Vines are planted 5–6 m apart. The land must be ploughed at least twice and must have two surface dressings. For the first ploughing, wooden swing ploughs are used; for the second, a horizontal blade penetrates the shallow soil to a depth of 7–8 cm.

Multiple Cropping

137. Multiple cropping offers an opportunity for small farmers to significantly increase the quantity and variety of food supplies. It also favours a third dimension to agricultural production – time. In many cases it can be carried out best by small growers with underemployed family labour.

138. Multiple cropping comprises a package of technologies. In some favoured regions, it may be a clear choice; in more marginal areas, traditional practices might better be used.

139. The goals of multiple cropping must be established within the context of national conditions and the stage of development of the nation. Further scientific studies could help to establish more general guidelines, while a better exchange of national studies and experience could provide insights for local application.

140. Although thousands of years old, multiple cropping has considerable potential for helping meet contemporary problems; however, much more research will be needed.

Climate, soil, available water, and the characteristics of different plant varieties play their roles in making multiple cropping possible and in determining systems.

141. Mixed cropping under rainfed conditions is extensively practised by farmers with these objectives: (a) due to high risks in this type of agriculture, they try to assure that at least one or two crops will survive for subsistance purposes; (b) the total production per unit is usually higher with multiple cropping than with an individual crop; (c) multiple cropping is more efficient in the use of the two principal sources of energy (manpower and animal traction); (d) with multiple cropping labour utilization is more continuous; (e) it provides more efficient weed control due to intensive hand cultivation.

Shifting Agriculture

142. This system restores nutrients to the soil, playing a role like manure, fertilizers, legumes and rotation. If intermixture of crops is practised, the cover protects the soil against erosion. Shifting agriculture can be carried on with a minimum of implements. It appears to give a higher output per man-hour than most other peasant agricultural systems.

143. The disadvantages of this system are that it works well only under rain-forest conditions, and it usually gives way before more intensive systems.

Rangeland

144. The productivity of pastures can be maintained or restored by the development and introduction of drought-resistant, high-yielding plants, supplemented by control of herd size and with rotational and deferred grazing in accordance with pasture availability and soil and water conservation measures.

145. Seeding of rangelands may be necessary where desirable native vegetation is absent. The need for additional forage on ranchlands and the necessity to match species with habitat determine the varieties to be seeded. Selection and seeding should be carried out in accordance with the: (a) adaptability of plant varieties to soil and climate; (b) vegetative cycles; (c) expected yields; (d) physiological response of plants to grazing; (e) their nutritive value; (f) palatability; (g) the compatibility of mixtures of species; and, (h) the need for forage production to meet seasonal deficiencies.

146. In places with less than 350 mm of rainfall, possibilities for reseeding should be explored.

Revegetation for Fodder Production

147. Palatability must be evaluated as a function of the order in which species are eaten, of the time spent by animals in grazing them and of the quantity of matter consumed. Generally indices of palatability vary during the year (Forti, 1971).

148. There exist several methods to reduce excessive run-off, encourage water infiltration and to increase plant production on native range.

Fencing, Natural and Artificial Recovery

149. Fencing protects ecosystems from such deteriorating agents as man and livestock. Natural ecosystems usually respond to fencing by increasing biological activity and by improving soil characteristics. Recovery rates will depend on vegetation stability and resilience and on the stage and amount of deterioration. As a rule, biological recovery will be slower the more arid the conditions.

150. The use of barbed wire for fencing was reported for as early as 1884 (Simpson, cited by Bhymaya *et al.*, 1966). Another enclosure method consists in electric fencing comprising a single wire attached to an automobile-type battery. Angle iron posts and barbed wire are the commonest fencing materials throughout the world. Other materials and methods of fencing consist of angle iron posts with woven wire, stone and barbed wire, wooden posts with barbed wire, ditches and mound, corewall and thorns, stone-wall and cactus fencing (live fencing).

151. A biological recovery system might be considered as a grazing reserve during drought periods when animals could be admitted on a temporary basis.

152. *Semi-natural recovery*. Artificial manipulation of soil, rainfall, and plant cover through scarification, terracing, water spreading, fertilization, reseeding and other techniques may help natural recovery and speed up. These techniques can save many years (Le Houérou, 1975).

153. *Artificial recovery*. Here the ecosystem is transformed through management and species introduction (establishment of drought-resistant fodder shrubs and trees).

154. Several fodder shrubs and trees have been successfully established and economically exploited in arid areas with less than 150 mm of annual precipitation in the Mediterranean, Mexico, South Africa, the United States and Northeast Brazil.

Control of Woody Vegetation and Undesirable Plants

155. Moving, crushing and cutting are used to eliminate weedy vegetation, sometimes complemented by the use of fire. However, many species may respond vigorously after these measures; then a follow-up treatment is required, usually accompanied by reseeding with desirable plants. Some plants are poisonous to domestic livestock and need to be eliminated. A great variety of herbicides are already known for the control of undesirable vegetation. They must be evaluated as to their harmful effects on animal life, including man.

Forestry

156. The forest is a vital part of a healthy biosphere. Its uses by human beings are numerous. It can serve as a windbreak against wind velocity and for the reduction of eolic erosion, stabilization of sand dunes, control of soil erosion in gullies, reduction of sediments in the upper part of watersheds, provision of fuel for fire and wood for an immense number of other purposes including agricultural tools. Forest also serves as a source of browse and changes the physiognomy of arid landscapes, providing recreational sites. A forest may also be cropped.

157. Much more needs to be done than at present for forestry protection and management. Any plan for forestry development must take into account tree planting

methods, forest research, extension services, legislation, and rural programmes for the development of this land use.

Control of Wood Gathering

158. Openshaw, cited by Makhijani and Poole (1975), estimates that the cost of establishing village woodlots to supply fuel, excluding the cost of land and local labour, would be about US$20 per hectare and for large-scale plantations, including the cost of labour, about US$200 per hectare. Many villagers could co-operate in tree planting programmes to establish wood fuel plantations provided they were shown the basic skills. At the present time, village woodlots would appear to be greatly needed.

159. If every hectare had an annual yield of 15 tons of wood, equivalent to cooking fuel for 40–50 people, then with a value of only US$5 per tonne, the annual output per hectare would be worth US$75 and would do much to ensure the stability of the ecosystem (Makhijani and Poole, 1975). Premature cutting in the village woodlot must be avoided and a plan for equitable distribution of the wood should be developed.

160. Alternate fuels such as gas or petroleum may be distributed to villages. Algeria and Libya have already begun regional programmes for distributing bottled gas (Le Houérou, 1975). The suggestion has been made that petroleum pellets be used for portable heaters or petroleum combined with sawdust for the same purpose.

161. Biogasification (see below) provides an alternative source of energy which can substitute for wood fuel. Solar cookers are becoming more popular in arid and semi-arid zones even though their cost is still high and women are reluctant to cook in the sunshine. There is an immediate need to develop cheap, light non-toxic solar cookers that retain heat after the sun has gone down.

Shelter Belts and Windbreaks

162. More research is needed on the choice of species, whether local or introduced, to be used as shelter belts, and on their effects on crop production under irrigation and rainfed agriculture, on the evaluation of their costs and cost/benefit ratio, and on the products derived from shelter belts. In rural areas, shelter belts may be established by providing farmers with tree seedlings either free or at nominal prices and putting at their disposal technical assistance. An example of a government policy to encourage farmers to plant trees can be seen in Libya where since 1952 the private sector has planted over 48 million trees covering an area of 30,000 ha. The Government of Sudan has adopted the policy that 50 per cent of any agricultural area must be reserved as tree plots. In this way, people have plenty of wood for fuel and other uses, and they refrain from cutting the natural protective forests. Seedlings are supplied to citizens to encourage them to plant trees on their farms.

163. The prevention of uncontrolled wood fuel gathering requires the co-operation of local people as well as energy sources alternative to wood and manure fuels such as economical solar heaters, wind-energy generators and biogasification plants.

164. In Tunisia, coastal and dune fixation totals 30,000 ha, now completely stabilized and supporting a healthy coniferous forest (*Pinus pinea* and *P. pinaster*).

165. Once identified, areas affected by sand dune encroachment can be treated by tree planting, fencing, grass sowing and soil stabilizers, with the participation of villagers in integral rural development programmes.

Controlled Environments

166. Controlled environments dedicated to flower, vegetable and fruit production are one of the modern ways of developing arid and semi-arid regions. Off-season agriculture produces cash crops with investments of US$220,000 or more per hectare.

167. In Aravah in Israel, vegetables and fruits are grown for export by air to Europe, especially flowers, peppers and tomatoes (Dalrymple, 1973). This type of agriculture requires high technology and investments and much manpower. Multiple cropping means a very intensive agricultural system under controlled environmental conditions. It has long been practised in American and European green houses with higher outputs than those obtained under field conditions – up to 145 metric tonnes of tomatoes per hectare. Several European governments have stimulated greenhouse production through low interest loans.

168. In Kuwait and Abu Dhabi, greenhouses for food production use desalinized sea water for hydroponic or trickle culture production, with high water-use efficiency. These activities provide a source of employment in rural areas and can be productive where other forms of agriculture are not possible. In Abu Dhabi, the power for desalinization is provided by heat exchangers attached to the exhaust and water jackets of diesel engines which drive electric generators.

169. The costs are quite variable, depending mainly on the type of structure, the nature and extent of environmental control equipment, and auxiliary structures for grading, packing and offices. The risks in these production systems are related to market fluctuations. Plastic structures (low tunnels or plastic row covers) are used for crop production in some places. More research is needed principally in developing countries to improve methods of construction, heating, watering and plant breeding. In spite of the savings gained from shifting away from traditional glass structures, more research is required to reduce costs.

170. Controlled environments are expensive, but may find a place where food cannot be produced at reasonable cost under natural environmental conditions.

F. LIVESTOCK

171. According to UNESCO (1975), half of the world's cattle, one-third of its sheep and two-thirds of its goats are contained in the arid and semi-arid zones. Bio-products, principally meat and milk, are the main sources of food to people living in pastoral lands.

172. The problems related with animal production may be solved under the following headings: (a) biological improvement; (b) animal type; and, (c) animal use, as shown in Table 5.

Animal Health, Nutrition and Rangeland Management

173. Some excellent livestock are not used because of the prevalence of animal diseases for which control methods are not available. Methods of disease control should be applied

TABLE 5. *Classification and evaluation of animal management technologies and their effect on desertification and productivity applied to several land-use systems*

Management method	Technology	Land-use system	Objective or purpose	Desertification potential	Production system	Cost	Level of technology
Biological	Health and nutrition	I,RA,R,F,S	A,B	None	Semi-intensive	Medium	Interm–complex
	Species introduction	I,RA,R,F,S	A,B	Slight	Semi-intensive	Medium	Interm–complex
	Genetic technology	I,RA,R,F,S	A	None	Semi-intensive	Medium	Complex
	Feeding places	I,RA,R,F,S	A	None	Intensive	Medium to High	Simple–complex
	Insect and rodent control	I,RA,R,F,T,M,S	A,B	Slight	Extensive	Medium	Simple–interm
	Shade	I,RA,R,F,S	A	None	Semi-intensive	Low to High	Simple
Type	Domestic	I,RA,R,F,M,T,S	A,B,C,D	Medium	Extensive	Low	Intermediate
	Wild	R,F	A	High	Extensive	Low	Intermediate
Uses	Traction	I,RA,R,F,M,T,S	A,B,C,D	Slight	Extensive	Low	Simple
	Carrying	I,RA,R,F,M,T,S	AB	Slight	Extensive	Low	Simple
	Transport	I,RA,R,F,M,T,S	A	Slight	Extensive	Low	Simple
	Bio-products	I,RA,R,F,M,T,S	A,B	None	Semi-intensive	Medium	Intermediate

Land use; irrigation agriculture (I); rainfed agriculture (RA); rangeland (R); forestry (F); mining (M); transport routes (T); settlements (S).

Objective or purpose: A – To improve productivity; B – To halt desertification; C – To reverse the process; D – To reclaim extremely degraded areas.

with full cognizance of their broader ecological effects and should be complemented by improvement in feeding, management, and breeding.

174. Animal movements and activities indicate that sheep spend more time grazing and travelling than do cattle. Animals frequently pull plants from the soil, tear developing grass stems from their sheaths, and sever pieces of plants which they do not eat.

175. In the arid zone, one sheep needs about 2 ha. The average production is 0.7–1.0 units of feed or 2–3 kg of dry matter per millimetre of rainfall (Le Houérou, 1975).

Systems of Pastoralism

176. Pastoral nomadism characterizes the great arid belt of the Old World, from the Atlantic shores of the Sahara to the steppes of Mongolia (Grigg, 1974). In Nigeria over four-fifths of all cattle are owned by the pastoral Fulani. Pastoral nomads rely for their subsistence upon their herds, thus the larger the herd the greater the chance of survival. Mixed herds also provide insurance against drought. Nomadic herdsmen depend upon natural vegetation for fodder and neither sow pastures nor storage forage. At present, however, the balance between land productivity and stocking rates has been altered unfavourably in many places due to human population growth.

177. Rangeland management requires among other things the conservation of soil from erosion and the fullest possible utilization of moisture. Watering points, such as holes and bores, should be avoided in erosion-susceptible soil. It has been shown in many areas that the development of water supplies without grazing control or co-operation from villagers leads to the rapid destruction of vegetative cover by overgrazing. Watering points govern the movements and distribution of grazing animals and represent potential hazards if they are not adequately spaced.

Solutions

178. According to FAO (1974), efforts to improve the health of livestock should take these factors into account: (a) emphasis should be placed on maintaining a diversity of species and seeking genetic improvement through selection in indigenous stock rather than through the introduction of exotic blood; (b) improvement programmes should be direct at developing, for each major range environment, appropriate combinations of the best adapted livestock; (c) veterinarians should give more attention to debilitating diseases and and conditions which impair reproductive and metabolic efficiency.

179. The basic control mechanism on grazing lands is obviously the maintenance of vegetative cover. Provision of harvested forage to feed cattle during periods of poor vegetative growth is the key to the control of erosion by wind and water. Feeding livestock during the periods of scarcity can follow several strategies: (a) the establishment of plantations of fodder shrubs and trees as feed reserves; (b) the production of irrigated fodder crops when possible; (c) the production of concentrates of high value; and, (d) the utilization of deferred grazing systems.

Livestock Management by Nomads

180. Survival of nomads and their livestock depends on a combination of quick and ingenious decision-making, risk management, and other fallback activities that can be resorted to in the event of catastrophe.

(a) Herd diversification is one strategy based on the fact that different species have different grazing habits and therefore occupy different niches in the ecological community.
(b) Loaning animals and sharing herds is another strategy and also helps build up social contacts within an area. If catastrophe hits, owners in one area will always have some stock on loan or on a sharing basis with associates somewhere else.
(c) Movement of herds is an obvious strategy for survival, and pastoral nomads show all kinds of migration patterns varying from simple, seasonal, short-distance transhumance to long-distance disaster migrations.

181. Generally, rangeland is characterized by extensive systems with low productivity. The productivity could be considerably increased by giving much more attention to the more productive areas, if there are any, where intensive systems reinforced by irrigation or other water supplies could be established, yielding better returns than extensive management in larger areas and allowing their recovery from overgrazing.

182. FAO (1974) describes a variety of pastoral types: (a) cultivators who combine cropping with livestock, with the latter retained in relatively close proximity to the cultivated areas; (b) cultivators who combine cropping with livestock, but whose herds on flocks migrate over long distances (often under the control of paid herdsmen or shepherds) during at least part of the year; (c) pastoralists who, seasonally at least, have strong links with cultivators and who consequently have access to stubble grazing for their livestock and grain food for themselves; (d) pastoralists more directly dependent on their livestock for subsistence, distinguishing between those who are more or less nomadic and those who are more or less sedentary; (e) people (even merchants or professions) who own livestock which are grazed on other people's land (or state land) under a variety of arrangements.

183. The question of what kind of unit is likely to make the most efficient use of vegetation and other resources should always receive the most careful consideration. For instance, among ex-pastoralists whose diet is predominantly milk, the substitution of grain foods for a portion of the milk diet offers the prospect of reducing overgrazing without having to reduce the number of people.

184. FAO (1974) makes this recommendation: assuming that an average family needs 20 livestock units when on a normal 70 per cent milk diet (at 3 livestock units of 450 kg bodyweight per human adult), the substitution of grain for half of the milk ration allows the size of herd to be reduced to 10 livestock units, thereby permitting about twice as many people to live in the area at the same standard. This works if people accept a change of diet and if grain is cheap and available.

185. Another approach is to finish and fatten animals elsewhere, leaving mainly female stock on the range. This can reduce the size of the subsistence herd. The most useful strategies, however, are those which remove animals permanently from the overused range.

186. The average yearly demand on rangeland can be calculated according to the following assumptions: 1 sheep = 1 goat = 0.33 cow G = 0.2 camel = 0.5 donkey = 300

units of feed (UF) per year = 300 kg of barley = 210 units of starch = 495,000 kcal (1 unit of feed = 1650 kcal). This implies that the arid regions of North Africa can support 31 million sheep. Animal consumption is estimated at 180 units of feed per year per sheep or 540 kg of dry matter per sheep in a region of 250–300 mm of annual precipitation and allowing 1.7 ha/sheep (Le Houérou, 1975).

187. A great number of different grazing systems can be devised. One employs a grazing reserve which is used only during the most adverse time of the year (frequently the last 3 months of the dry season) and which is allowed to recover during the subsequent rainy season. This procedure has been implemented in the past by some pastoral groups in Africa and the Near East and is considered as a rotational grazing method (FAO, 1974).

188. South Australia has developed an advanced settled system of animal husbandry adapted to the climate and soils of a Mediterranean-type zone. This system is stable, is increasing the level of soil fertility and has controlled soil erosion. The policies that have been developed for these regions lead the world in the management of stable livestock production.

Complementarity and Integration with Other Land Uses

189. Some groups keep livestock not primarily for the direct subsistence they offer but because by tradition they are livestock people and have access to extensive grazing areas that are too dry for cultivation and can only be utilized by migratory herds. Much of the subsistence of such groups is derived from grain foods, either grown in the higher rainfall areas of their own territory or obtained from neighbouring agriculturists in exchange for livestock.

190. In semi-desert regions, where rainfall is patchy in the extreme, free movement over very large areas is essential for survival, and opportunities for limiting movement are slight (FAO, 1974).

191. Tewari (1976) describes the pastoral activities in Jaisalmer Desert mentioning that 70 per cent of animal stock are in pastures being grazed beyond their carrying capacity. Some 400,000 adult animal units require 1.12 million tonnes of dry matter, provoking an annual deficit of 200,000 tonnes. The scarcity may be overcome by increasing the present area by one fifth but it is also necessary to maintain control of animal pressure.

192. In several countries a growing gap between production of and demand for meal and milk will have to be met through imports and/or regional long-term programmes to halt desertification and improve the productivity of land. Some semi-intensive systems of animal husbandry could be established by integrating animal production and agriculture through the use of intermediate technologies according to the conditions of the migratory animal producer and the sedentary farmer.

193. According to FAO (1960), rangeland problems can be overcome by: (a) supplying adequate watering sites, and if possible, moving them from time to time; (b) fencing or otherwise excluding animals from threatened portions of the pasture; (c) avoiding the placement of gates or lanes on vulnerable sites and moving them when problems arise; (d) providing wind barriers to protect agricultural land; (e) applying corrective measures promptly to potentially vulnerable areas; (f) the application of manure or commercial fertilizers; (g) the use of contour furrowing to reduce run-off and increase rainfall

efficiency; (h) systems of rotational grazing to maintain desirable plant species and increase production.

194. These principles of control apply everywhere, but the relative utility of each varies with the climate, soil and land-use conditions. Usually not one but a combination of methods will be most effective for rangeland management.

G. ENERGY

195. Problems generated by the mismanagement of energy may be overcome by analysing the energy sources and the uses to which the energy will be put, as is shown in Table 6. The most important sources of energy are: manpower, animal draught, chemical, mechanical, hydrothermal, geothermal, wind, solar hydraulic and biogasification. In some cases they represent double or multiple-purpose technologies.

196. Energy is used for: pumping systems, domestic purposes, the generation of electricity, traction and transport. People might find energy usually available to them reduced or chronically lacking often because the broader setting in which they live is characterized by underdevelopment. Even when available, energy is sometimes misapplied or irrationally exploited. The most immediate result is that mechanization cannot be applied and full advantage taken of the potential productivity of the area in question.

197. Before starting any development programme, energy sources, such as oil and coal resources, water potential to produce energy, solar energy, windpower, biogasification, animal draught and manpower, should be evaluated. Energy availability is related to the intensity of production systems in general. The less energy available the more extensive and less productive is the agricultural system which can be applied.

Sources of Energy

198. As most developing countries show deficiencies in conventional energy, it is necessary to consider alternative sources. Makhijani and Poole (1975) say that together with wood, animal dung and crop residues, human and animal labour provide all the energy available to the vast peasant populations who live in Asia, Africa and Latin America.

199. Ingenuity will be required to increase energy at low cost in the developing world. Conventional technologies for supplying fuel and electricity for industry and agriculture have been widely applied. Other sources of energy have not been so intensively studied or evaluated, and these would include solar energy, wind power and the use of basic raw materials for producing fuel and fertilizers in biogas plants. One of the parameters for evaluating living standards is energy consumption per inhabitant.

Energy Management Technologies

200. Land-use systems under rainfed agriculture, rangeland and forestry conditions have utilized wind, animal and manpower since early history. These sources continue to be used in the developing countries for: pumping systems, land preparation, seeding and harvesting, as well as for domestic purposes, the generation of electricity, traction, and transport.

TABLE 6. *Classification and evaluation of energy management technologies and their effect on desertification and productivity applied to several land-use systems*

Management method	Technology	Land-use system	Objective or purpose	Desertification potential	Production system	Cost	Level of technology
Sources	Animal	I,RA,R,F,T,M,S	A,B,C,D,	Slight	Extensive	Low	Simple–interm
	Man	I,RA,R,F,T,M,S	A,B,C,D	High	Extensive	Low	Simple
	Mechanical	I,RA,R,F,T,M,S	A,B,C,D	Slight	Semi-intensive	Medium	Simple–complex
	Wind	I,RA,R,F,M,T,S	A,B	Medium	Extensive	Low	Intermediate
	Solar	I,RA,R,F,M,T,S	A,B	None	Semi-intensive	Low to High	Simple–complex
	Biogasification	I,RA,R,F,M,S	A,B	None	Semi-intensive	Medium	Intermediate
Uses	Pumping systems	I,S	A,B	Slight	Semi-intensive	Medium	Simple–complex
	Domestic uses	S	A	None	Intensive	High	Simple–complex
	Traction	I,RA,R,F,M,T,S	A,B,C,D	Slight	Semi-intensive	Low to High	Intermediate
	Transport	I,RA,R,F,M,T,S	A,B	High	Semi-intensive	High	Simple
	Carrying	I,RA,R,F,M,T,S	A,B	Slight	Semi-intensive	Medium	Simple
	Combination of conventional and non-conventional energy sources	I,R,A,R,F,M,T,S	A,B,C,D	Medium	Semi-intensive	Low to High	Intermediate

Land use: irrigation agriculture (I); rainfed agriculture (RA); forestry (F); mining (M); transport routes (T); rangeland (R); settlements (S).

Objective or purpose: A – To improve productivity; B – To halt desertification; C – To reverse the process; D – To reclaim extremely degraded areas.

Mechanization

201. Agricultural mechanization is inherently complex with its physical and biological limitations. Soils, climatic factors and crop characteristics determine the conditions under which machines must operate. For success, appropriate agricultural and mechanical technology must be employed within the existing social system to form an economically attractive package.

202. Development of a local farm equipment industry provides employment, reduces dependence on imports, saves foreign exchange, and facilitates the supply of parts and service. These are the advantages when agricultural tools and animal draft implements are designed and fabricated in the country where they are to be used. Unfortunately, very little effort has been made to develop indigenous farm equipment industries (Stout and Downing, 1975).

203. Hand and animal power will predominate in many countries for decades to come. A mechanization policy should therefore include attention to the improvement of hand tools and animal implements as well as the breeding and rearing of draught animals.

204. Recent developments in the design and production of machinery have made dryland farming systems more stable and productive under arid and semi-arid conditions. Practices to conserve and utilize soil moisture to the fullest extent, proper tillage and seeding methods and the application of organic or chemical fertilizers will help to achieve maximum production. As conditions vary from area to area, appropriate systems of agricultural production must be developed to meet specific conditions (FAO, 1971).

205. Peak labour demands (seeding and harvesting times) can be reduced or eliminated by selective mechanization of agricultural operations. Increased sources of energy can be found in more draught animals, stronger draught animals, or in fuels such as oil, biogas, wood and coal.

206. Animal draught has the advantage of leaving intact the well rooted plants which hold the soil in place on hillsides and provide shade and sometimes fodder for animal life.

Water Pumping Systems

207. Water pumping systems may use manpower, animal-powered pumps or motor-powered systems. Their costs include the overhead costs of initial investment, depreciation and perhaps interest charges, and the running costs of operating the pumping equipment (Watt, 1975).

208. Man-powered pumps are extensively used in developing countries. They are usually small and portable, are sometimes locally made or provided by governments. Wind power is widely used for pumping water.

Solar Energy

209. According to Makhijani and Poole (1975), the energy from the sun that falls on the earth is the Third World's most abundant energy resource. It is estimated that the solar energy falling on Saudi Arabia each year is equivalent to the world's entire proved reserves of coal, oil and natural gas. The solar constant is 1360+ watts per square metre. A number of countries, USSR, France, Iraq, Egypt, Turkey, Mexico, Japan, Israel, the United States and Kuwait, for example, are applying, at various stages of development, technologies for using solar energy.

210. Some applications of solar energy are: (a) the conversion of solar radiation to electricity, by using solar cells, for the pumping of irrigation water; (b) the heating of water for domestic purposes using flat, oriented glass collectors is under commercial development in Israel, Australia, Mexico, and other countries and could be extensively used in areas where deforestation is taking place; (c) solar dryers which heat air instead of water are being used for drying grains, and could be complemented by small electric grinders fed by electric motors (300 watts); (d) Duffie and Beckman (1976) estimate that solar heating and cooling of buildings, with energy collected by flat-plate collectors, will be one of the first large-scale application of solar energy; (e) other uses include solar evaporation for salt production, solar distillation of potable water for human or animal consumption, solar refrigeration for food preservation, sea water desalinization, and solar cookers.

211. As fuel costs rise, solar energy becomes more competitive, especially as its technology improves. Lusting (1974) calculates that by the year 2000 solar energy could economically provide in the United States up to 35 per cent of total building heating and cooling load, 30 per cent of what is now provided by natural gas, 10 per cent of what is provided by liquid fuel and 20 per cent of electric energy requirements. Building heating could be in use by the public within 5 years, building cooling systems in 6–10 years, synthetic fuels from organic materials in 5–8 years, and electricity production in 10–15 years. Such developments would provide additional energy for agricultural production.

212. The first international meeting on solar energy in modern times was the UNESCO symposium of 1954. The international Solar Energy Society was founded 22 years ago and meets annually. Regional and international co-operation for the use of solar energy is under development.

213. The large-scale use of solar technologies in the arid and semi-arid areas of the World would have a minimum effect on the biosphere and would serve as a tremendous asset for the development of rural communities, which represent about 80 per cent of the population in developing countries.

214. At the present time, there are several rural energy centres in countries of America, Asia and Africa which are demonstrating alternative and integral uses of several sources of energy (an energy mix) under a concept of complementarity among solar energy, wind power and biogas from organic wastes and manure. The objective is a supply of energy to meet the requirements of villages of about 1000 people mainly for lighting, cooking and pumping water (Chatel, 1976).

Biogasification

215. Biogasification is a method of converting biological material into fuel. Dung and crop residues can be converted into methane, a gaseous fuel akin to natural gas, by the action of anaerobic bacteria. The residue is high in nitrogen, phosphorus and potassium, and can be used as a fertilizer.

216. The use of this gas as a source of energy for villages is now the object of a special study by the United States National Academy of Sciences. Thousands of small biogas plants are used on farms in India and Taiwan. Costs vary from US$0.20 to US$0.70 per million Btu depending on the state and availability of manure. Bottled gas

may provide fuel for farm machines, for cooking and for generating electricity. The residuum is a good organic fertilizer (Makhijani and Poole, 1975).

217. A biogas plant to convert dung to methane and organic fertilizer with a capacity of 10^9 Btu of biogas per year would cost about US$10,000 (Makhijani and Poole, 1975). Its energy could supply 400 million Btu for cooking and 500–550 million Btu for operating the compressor and pumps of the biogas plant. It would also supply 3 tonnes of fixed nitrogen, 1 tonne of phosphorus (P_2O_5), and 1 tonne of potassium oxide per year.

H. CONCLUSIONS

218. Present scientific and technological knowledge should be applied for the optimum management of available resources, especially where resources are scarce. Impediments to the application of knowledge should be removed.

219. As rainfed agriculture represents a major factor in the agrarian economy of the developed countries and is the major source of cereal crops, programmes should be launched to transform low-yield practices to medium- and high-yield methods through appropriate technologies related to soil, plant and rainfall management.

220. Governments should establish large-scale national programmes for evaluating, designing and constructing agricultural implements to be used by hand, animal traction or machines. Implements should be made available free to small farmers who cannot afford to purchase them.

221. Traditional technologies should be considered before establishing new land-use systems, and these may be complemented with intermediate and transitional modern technologies with the idea of changing gradually from extensive to intensive systems of production.

222. Since underdeveloped countries are characterized by abundant labour and scarce capital, the concept of intermediate technology should be fully considered, as it involves low to medium investment in the application of appropriate technologies which would transform extensive to intensive systems of production. The growing socio-economic gap between the irrigated areas and the rainfed lands should be reduced.

223. There are many current examples of successful programmes in developed and developing countries in controlling desertification. However, their diffusion has been limited due to lack of funds and trained staff.

II. TECHNOLOGICAL APPLICATIONS

A. WATER

1. Water Supplies

Vertical wells

224. *Characteristics:*
Land-use system: Settlements, irrigation agriculture
Desertification potential: Medium
Production system: Semi-intensive to intensive
Cost: Low to high
Level of technology: Simple to complex

Description:

225. Wells dug in rock or soil or in the sands and gravels of dry river beds are found in all countries. They often afford the only permanent water for homes or villages in the dry lands and in cases in the desert, and may support animals as well as a human population, especially in the dry season.

226. The term "well" is often used for a groundwater structure excavated by hand, and such a well usually ranges from about 0.9 to 1.8 m in diameter, whereas a borehole, or bored well, is drilled by a drilling machine and is usually 15 to 20 cm or more.

227. In most cases, local people have been accustomed for many generations to dig their own wells, and these vary greatly in construction from place to place according to the nature of the enclosing rocks and the level of technology. The water is often raised in rudimentary devices such as skin containers with leather thongs, usually by hand and sometimes with the aid of stock, either hauling directly or working mechanical gear. Often the shafts are of small diameter, about 50 cm, just enough for one man to work in. Others are immense crumbling excavations reaching down to the watertable, as in the gypsum wells of northern Kenya. Still others follow a zigzag course downwards along the weathered joints of solid rocks, and the water is passed up hand over hand in small buckets, while yet others are great tunnel-shaped excavations in loose sand 10 m or more in depth, in which diggers are not infrequently trapped.

228. The character of rainfall in arid climates presents both advantages and difficulties in storage, for, whereas the great majority of rainy days produces small amounts of rain, there occurs practically every year a small number of days with rainfall of great intensity and considerable amounts, often compressed within a few hours. The same pattern would hold good for extremely arid areas where the intervals between heavy rainfalls can be spaced over years.

Evaluation:

229. Modern drilling and pumping equipment can reach down to far greater depths and yield more abundant water supplies, but it must be kept in mind that modern powered systems represent potential hazards that, if not controlled will rapidly deplete the underground water. This may lead to desertification once the aquifers are empty, and people start migrating to more favoured locations.

Horizontal wells

230. *Characteristics:*

Land-use system: Settlements, irrigation agriculture
Desertification potential: Slight
Production system: Semi-intensive
Cost: Medium
Level of technology: Intermediate to complex

Description:

231. A horizontal well is a cased spring as can be seen in Fig. 1. A horizontal boring is used to drill a hole and install a steel pipecasing into a mountain or hillside to tap a trapped water supply. Good sites for horizontal wells are dike formations, impervious tilted clay, or rock walls that form a natural dam (National Academy of Sciences, 1974).

232. No pumps are needed on horizontal wells and the costs of maintenance are low. The drilling equipment is simple and portable and operational costs are insignificant. The flow can be controlled, and the system supplies water of good quality.

Qanats

233. *Characteristics:*

Land-use system: Settlements, irrigation agriculture
Desertification potential: Slight
Production system: Semi-intensive
Cost: Medium
Level of technology: Intermediate

Description:

234. A Qanat system is composed of three essential parts. One or more vertical head wells dug into the water-bearing layers of an alluvial fan collect the water. A gently downward-sloping underground horizontal tunnel leads the water from the head wells to a lower point at the surface (in part, the tunnel acts as a subsurface drain to collect water). A series of vertical shafts between the ground surface and the tunnel are used for ventilation and removal of excavated debris (National Academy of Sciences, 1974).

235. The physical basis of qanats lies in the combination of mountains receiving a considerable and quite stable amount of rainfall and, therefore, groundwater recharge, and nearby plainland with good agricultural soil but an inadequate supply of water (see Fig. 2).

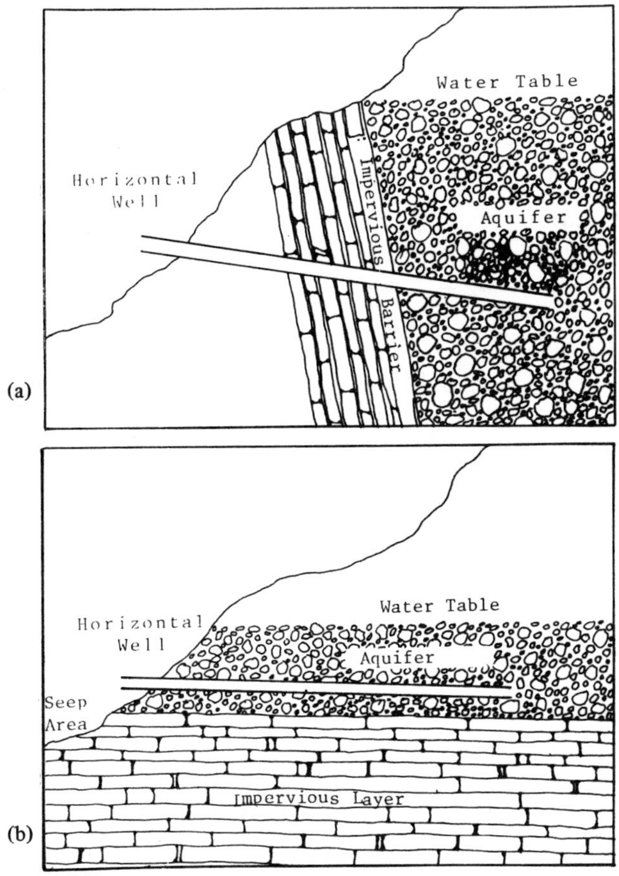

Fig. 1. Horizontal wells.
(a) Dike spring formation. (b) Contact type of spring formation.

Evaluation:

236. Because qanats deliver water without pumps and pumping costs, they can be used where pumped wells are too expensive to operate. They prevent evaporation losses, although seepage losses may be considerable.

Springs

237. *Characteristics:*

Land-use system: Settlements, irrigation agriculture
Desertification potential: Slight
Production system: Semi-intensive
Cost: Low
Level of technology: Intermediate

Description:

238. Springs appear as small water holes or wet spots at the foot of hills or along riverbanks. There are two general requirements necessary in the development of a spring

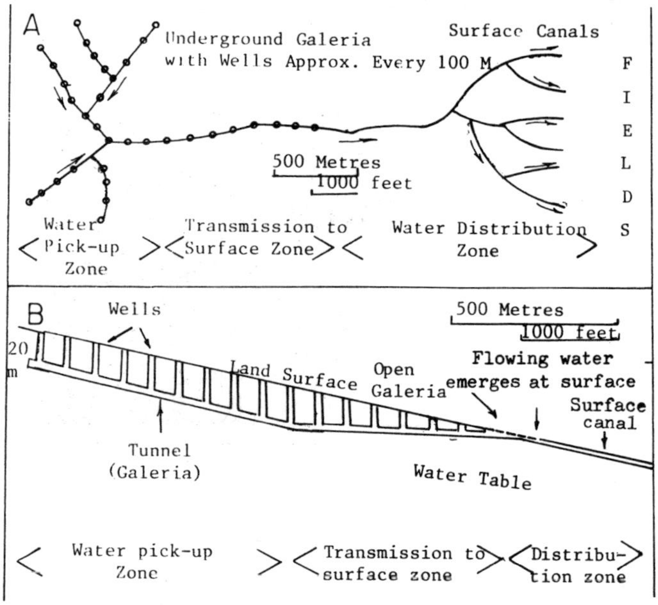

Fig. 2. Typical galeria (tunnel) irrigation system in Tehuacán, Mexico. (A) Plan, (B) Cross-section.

to be used as a source of domestic water: (a) selection of a spring with adequate capacity and quality for its intended use throughout the year; (b) protection of the sanitary quality of the spring.

239. The features of a spring encasement are the following: (a) open-bottom, watertight basin intercepting the source which extends to bed-rock or a system of collection pipes and a storage tank; (b) a cover which prevents the entrance of surface drainage or debris into the storage tank; (c) provision for the cleanout and emptying of the tank contents; (d) provision for overflow, and (e) a connection to the distribution system or auxiliary supply (U.S. Public Health Service, 1963).

240. Certain precautionary measures will help to ensure developed spring water of a consistently high quality; (a) provide for the removal of surface drainage from the site. A surface drainage ditch located at least 16 m away on the uphill side of the spring will prevent contaminating material from entering the source; (b) construct a suitable fence at least 35 m from the water source to prevent entry of livestock. The drainage ditch should be inside the fence at all points uphill from the source; (c) provide for access to the tank for maintenance, but prevent removal of the cover by a suitable locking device; (d) monitor the quality of the spring water through periodic checks for contamination. A marked increase in turbidity after a storm is a good indication that surface run-off is reaching the spring.

Evaluation:

241. Gravity springs in particular are subject to contamination, so sanitary precautions should be taken. An illustration of spring protection is found in Fig. 3.

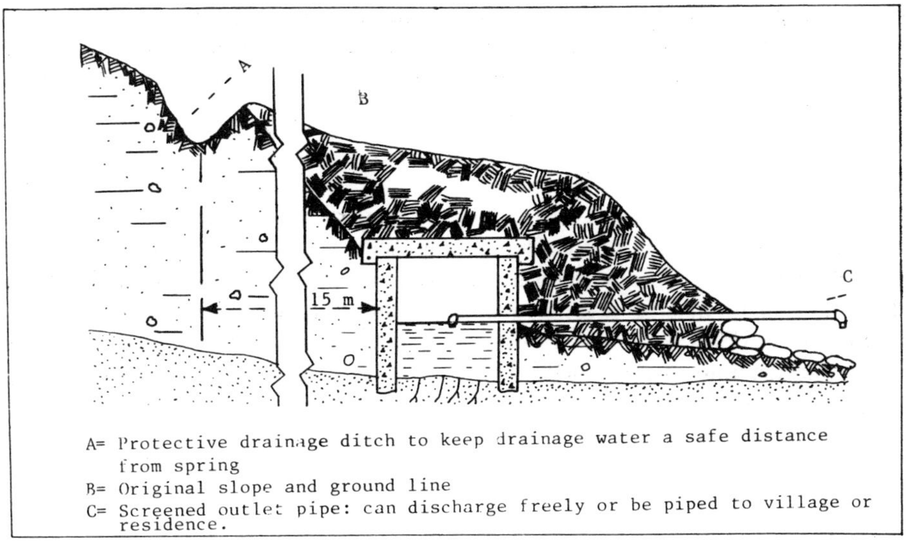

Fig. 3. Spring protection. (Wagner and Lanoix, 1959).
A. Protective drainage ditch to keep drainage water a safe distance from spring.
B. Original slope and ground line.
C. Screened outlet pipe: can discharge freely or be piped to village or residence.

Wind power for water pumping

242. *Characteristics:*

Land-use system: Rangeland, settlements, irrigation agriculture
Desertification potential: Medium
Production system: Extensive
Cost: Low
Level of technology: Intermediate

Description:

243. Wind power is extensively used for pumping water, and there are many different devices extensively used in Australia, Africa, the United States, Mexico, India and in other places. The most useful for West Africa, where a gentle wind régime is characteristic, are the low-speed multi-blade and the Cretan Sail wind wheels. The power available from the wind can be expressed as: $P = 0.006AV^3$ where A = swept area in m^2, V = wind velocity in m/sec and P = power in rated watts. The power from a wind wheel depends on the square of the linear dimension; the weight and cost depend on the cube; to double the size of the wind wheel is therefore to quadruple its power and to multiply its cost by a factor of 8 (Watt, 1975).

244. Wind wheels must be designed to start and work in low wind speeds and be equipped with a governing device to prevent their destruction during high winds. Many commercial windwheel pumps have power take-off gears so that pumps may be worked by animals, diesel engines or manpower during periods of no wind. They must also have storage tanks to contain the pumped water for use when needed (see Fig. 4).

245. Wind power is cheap but limited in output and variable due to shifts in the weather. The best results for pumping water are obtained at depths lower than 60 m. The daily volume will be rather low, averaging a few dozen cubic metres, which limits the use of windmills to small watering points.

Evaluation:

246. According to Baudelaire (1972), for a technical amortization period of 40 years, the development of watering points from springs with a reservoir of 150 m^3 gave a water cost of US\$0.07/m^3. For development of watering points from an artificial reservoir including its construction (a compacted earth dam made with 20,000 m^3 of earth) the water costs were US\$0.18/m^3; and for development from drilled or bored wells with a technical amortization period, with a reservoir or 150 m^3 the water costs were US\$0.21/m^3.

Desalting

247. *Characteristics:*

Land-use system: Settlements, irrigation agriculture
Desertification potential: None
Production system: Intensive
Cost: High
Level of technology: Complex

Description:

248. The many schemes for desalinization of water may be classified as thermal, mechanical, electrical, and chemical. Thermal and mechanical schemes include those involving phase changes of the water and have energy requirements which are independent of the initial salinity of the water being processed. Chemical and electrical methods are sensitive to the initial salinity of the water being processed and require appreciably less energy for the purification of brackish water than for seawater. It should be noted that at present the thermal and mechanical methods are cheapest for desalting seawater while chemical and electrical methods are cheapest for brackish water. However, since thermal and mechanical methods are far from achieving optimum power consumption, declines in fuel or power costs should render them attractive for saline water conservation.

249. Desalinized water would not help in the struggle against desertification unless the water is more wisely used than conventional irrigation water. The oldest and still most widely used method of purifying saline water is by distillation.

250. Undoubtedly the most significant benefit that could occur in relation to water would derive from the discovery of a cheap process for producing fresh from saline or mineralized waters, including of course these of the sea or salt lakes. With good assurance that world population will double within the next 40 or 50 years, even lands that today are well supplied with water will find, as is indeed the case today, that their use of water will need to be restricted, and the price of water as a commodity will rise. Accordingly, very large sums are currently being spent on research into demineralization, and several large plants have been constructed to serve areas having special needs. The cost of desalinized water is currently too high to permit its use for irrigation.

Fig. 4. A windmill in Alice Springs, Australia.

Evaluation:

251. Desalinization techniques are well known and currently used to produce sweet water for human and animal consumption. However, the cost of such water is still too high for agricultural purposes. In the least expensive cases (power from nuclear energy) the cost is about US\$0.10/m^3 and it can rise to tenfold that amount. As one needs at least ½ m^3 to produce 1 kg of dry matter, the cost of 1 g of dry matter is at least US\$0.05/kg, and the water can therefore, only be used to produce luxury products such as out-of-season vegetables, all of which has little to do with desertification. But in the near future, water desalinization is a promising prospect (at US\$0.05/m^3 although it will require very large plants powered by nuclear energy which calls for huge investments that only oil- or mineral-rich countries will be able to afford. It will be of little help to the landlocked countries of the Sahel (Le Houérou, 1975).

252. Desalting of brackish surface and ground waters by osmosis, ion exchange and electrodialysis is both economically and technically feasible for drinking water but not so for agriculture.

Re-using water

253. *Characteristics:*

Land-use system: Settlements, irrigation agriculture
Desertification potential: None
Production system: Intensive
Cost: Medium
Level of technology: Complex

Description:

254. The agricultural use of municipal waste water for irrigation is especially attractive where agricultural lands are located close to cities, because the nutrients in sewage would otherwise go to waste. Mexico City, for one example, delivers sewage as irrigation water to an area of about 40,000 ha in the Mezquital Valley.

255. Industrial waste water may also be used for irrigation, but it may require treatment when the industrial process adds chemicals detrimental to plant growth or public health. Municipal waste water from secondary treatment plants can be used for industrial cooling or other purposes which do not have requirements for high water quality (National Academy of Sciences, 1974).

256. In Windhoek, a metropolitan area of 84,000 people meets its water needs by treating and recycling into the potable water supply 4 million litres per day of its sewage, which represents one-third of the total daily supply (National Academy of Sciences, 1974).

Evaluation:

257. In any scheme for the re-use of water, certain constituents of waste water have to be considered: pathogenic bacteria, viruses, parasite eggs, heavy metals, salts and nitrates. The cost and difficulty of re-using water depends on the treatment process needed.

258. Electrodialysis and reverse osmosis show promise for removing many kinds of dissolved impurities, but better antifouling techniques and membranes that require less

pre-treatment are needed. Improved biological processes are needed for removing ammonia and nitrates in secondary effluents as are new low-cost specific ion exchanges for removing mineral salts.

Solar distillation

259. *Characteristics:*
Land-use system: Settlements
Desertification potential: None
Production system: Extensive
Cost: Low
Level of technology: Simple

Description:
260. In solar distillation the sun's radiation passes through a transparent cover onto brine water, and the vapour condenses on the cover which is arranged to collect and store it. This principle may be applied in either simple or sophisticated ways (see Fig. 5).
261. Water can be obtained from soil by a simple distillation technique. The soil is covered with plastic film formed and held in a conical shape by a rock placed in the centre. Water collects on the under side of the plastic, runs to the point of the cone, and drops into the container. A yield of 1.5 litres per day of potable water was obtained from a single survival still (Jackson and Van Bavel, 1965).

Evaluation:
262. The simple survival still uses solar energy to distill drinking water from soil and plant materials and may be used as a supplementary water supply.

Rain collection from roofs

263. *Characteristics:*
Land-use system: Settlements
Desertification potential: None
Production system: Semi-intensive
Cost: Low
Level of technology: Simple

Description:
264. There are arid and semi-arid zones where no perennial or seasonal river or groundwater supplies exist. One way to obtain drinking water is to catch the rain from the roofs or from smoothed slopes and store it in cisterns or in galvanized steel tanks lined with polyvinylchloride as can be seen in Fig. 6.

Evaluation:
265. This is a cheap method for collecting rain and is extensively used for supplementary water supply.

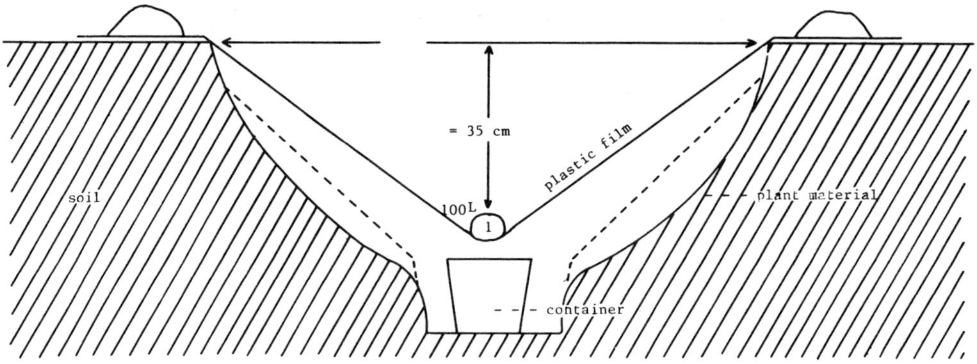

Fig. 5. Diagram of a survival still.

Fog collection

266. *Characteristics:*
Land-use system: Settlements
Desertification potential: None
Production system: Semi-intensive
Cost: Medium
Level of technology: Simple–complex

Description:

267. The principal arid lands of South America lie in the coastal deserts of Peru and Chile, and in the Monte and Patagonian deserts of Argentina. Brazil, Colombia and Venezuela contain other arid regions.

268. Henrich, mentioned by Cornejo (1970), estimated a water content of 0.1 to 5.8 g/m^3 of fog. As fog is present for several hours during the 120 days of winter in Peru, water condensation on the leaves of trees can be considerable. He has calculated 5.0×10^4 litres of water per square metre, and assumes that if the foggy mass of air has a velocity of 3.6 km/hr the amount of condensed water will be 1.8 litres per square metre per hour.

Evaluation:

269. On the coasts of South America some people use a fog catcher or "camanchaca". It consists of an instrument 2.5 m high with a series of nylon strings 1 mm in diameter and 1.2 m long. The density of strings is of 320 per metre. The yield of water is about 18 litres per day.

2. Water Conservation

Groundwater recharge

270. *Characteristics:*
Land-use system: Irrigation agriculture, rainfed agriculture, settlements, rangeland, forestry

Fig. 6. Galvanized steel lined with polyvinylchloride using roof catchment for drinking water in Australia.

Desertification potential: None
Production system: Semi-intensive
Cost: Medium
Level of technology: Intermediate to complex

Description:

271. In the arid and semi-arid zones, stream flows should be saved for use as groundwater by diverting them into basins, pits, furrows and recharge wells. A substantial amount of the rainfall run-off which is usually lost as river flow may be used to recharge scarce groundwater reserves. The aquifers may be supplemented from controlled run-off, seepage from canals, tiled channels, ponds and tanks, influent drainage from rivers and deep percolation from irrigated fields (Dakshinamurti *et al.*, 1973). Massive and intensive programmes of groundwater are urgently needed.

272. Groundwater recharge can also be achieved by applying soil conservation measures (contour bunding and gully plugging) and artificial measures (basin method, furrow or ditch method and flooding). In coastal areas, the pumping rate should be less than the average recharge in order to permit groundwater flow in the seaward direction and to prevent saline intrusions into aquifers (Dakshinamurti *et al.*, 1973).

Evaluation:

273. Optimum development and rational management are basic to the sustained use of groundwater supplies. It is necessary to establish an equilibrium between recharge and pumping and avoid indiscriminate over-exploitation which may lead to desertification.

Reservoirs

(a) Cisterns

274. *Characteristics:*

Land-use system: Settlements, irrigation agriculture
Desertification potential: None
Production system: Semi-intensive
Cost: Low
Level of technology: Simple

Description:

275. Cisterns have been constructed for thousands of years to catch the rain and store it for various purposes. There are various types and cross-sections.

276. One type of cistern catches rainwater, another floodwater (the latter are called "harable" by the Bedouins and "maagurah" in Hebrew). A typical floodwater cistern will have an area of about 100 m^2 and 5–7 m of height, giving a storage capacity of about 500–700 m^3. Groups of them provide water for small villages. Figures 7 and 8 show typical cross-sections of cisterns and the entry of a floodwater cistern located in the arid Negev in Israel at the geological contact between a hard limestone overlying a soft chalk of much reduced permeability. Low-cost construction of floodwater cisterns requires a proper location where contact between a hard cap-rock and a soft, easily workable

Fig. 7. Cistern constructed near Beersheba, Israel, for drinking and stockwater.

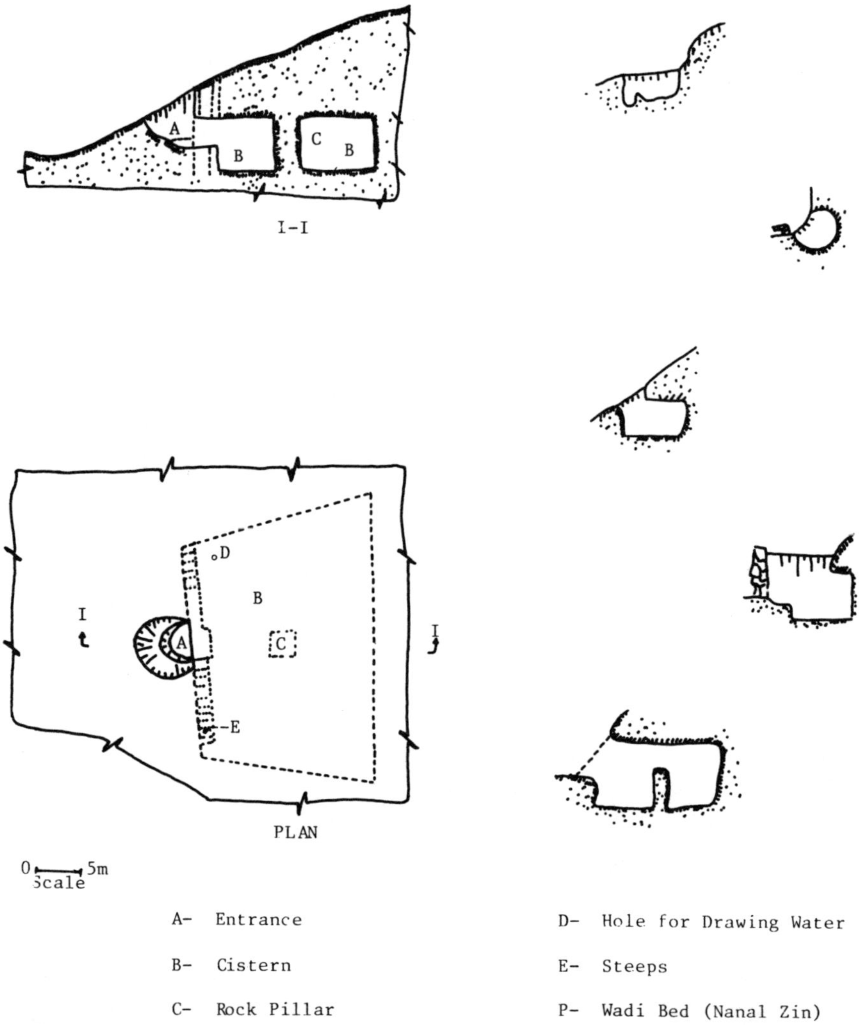

Fig. 8. Cross-section and ground plan of cisterns and typical cross-sections in Advat, Israel (Evenary, Shanan, Tadmor, 1974).
A. Entrance
B. Cistern
C. Rock pillar
D. Hole for drawing water
E. Steeps
P. Wadi bed (Nanal Zin)

reservoir rock is present (Hills, 1966). In Nabataean times both rain and floodwater cisterns were constructed in large numbers to serve the caravans crossing arid regions with 90–100 mm of annual rainfall.

277. Cisterns may be located underneath the interior courtyard of a building or next to a wadi bed or on the lower part of a slope, and have been used for nomadic and sedentary occupation of arid lands.

Evaluation:

278. Cisterns have an advantage over open reservoirs, because they prevent evaporation losses and can keep water clean and free from pollution.

279. Societies at higher economic levels utilize canals, deep wells, aqueducts and pipelines for water supply purposes. The source of water is usually from springs, rivers, dams, small reservoirs or bores. Operation and maintenance require a permanent administration. Modern cities usually involve an average daily water consumption of 200 to 700 litres per capita, while some nomadic groups use as little as three litres per capita.

(b) Rainwater catchments filled with sand

280. *Characteristics:*
Land-use system: Settlements, rangeland
Desertification potential: None
Production system: Semi-intensive
Cost: Medium
Level of technology: Intermediate

Description:

281. Such reservoirs are filled with sand and loose rock. A water-storage tank is shown in Fig. 9. The sand reduces evaporation and filters the water as it enters and is withdrawn, making it suitable for drinking.

Evaluation:

282. This system could be applied as a supplementary water supply for small communities located in arid and semi-arid zones.

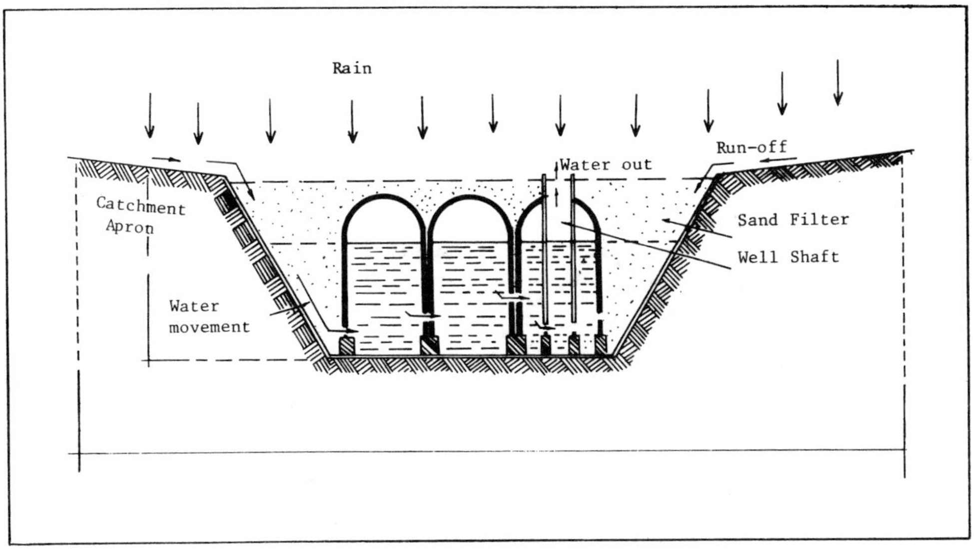

Fig. 9. An experimental catchment system with storage of water for human consumption (Intermediate Technology Development Group, 1969).

(c) Rainwater harvesting

283. *Characteristics:*
Land-use system: Rangeland
Desertification potential: None
Production system: Intensive
Cost: High
Level of technology: Intermediate to complex

Description:

284. Storage bags made of butyl rubber or butyl-coated nylon two-sixteenths of an inch thick confined to a excavated pond are used to store rainfall. The bags are designed with a water inlet, outlet and overflow outlet and in sizes ordinarily up to 200 m^3.

285. Storage tanks used to collect rainwater can be constructed of concrete or metal and may be entirely above ground (see Fig. 10), partially buried or completely buried. Covered or uncovered, they are more economical and practical than storage bags and do not need to bc fenced. Figures 11 and 12 show diagrams of these systems. Tanks may be constructed of wood, cinder block or fibreglass.

286. A promising method for harvesting rainwater is to treat soils with chemicals such as sodium salts that fill the soil or make it water-repellent. Other commonly tried water-repellent chemicals are silicones, latexes, asphalt and wax. Instead of converting the soil itself into the watershedding surface, it might be better in some situations to cover it with a waterproof cover such as plastic sheets, butyl rubber or metal foil.

Evaluation:

287. Some other techniques consists in collecting and storing natural flows in excavated tanks to provide water for livestock. The normal practice for intercepting flows is the contour drain. On a grade of 0.3 per cent to 0.5 per cent it controls run-off and soil erosion at the same time and may be constructed by farmers using their own disk ploughs

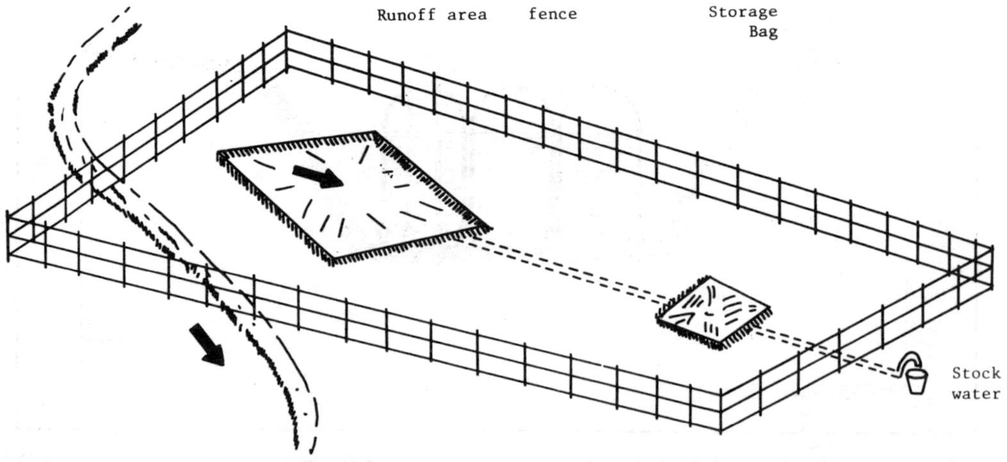

Fig. 10. Diagram of a system for water harvesting with a capacity up to 200 m^3.

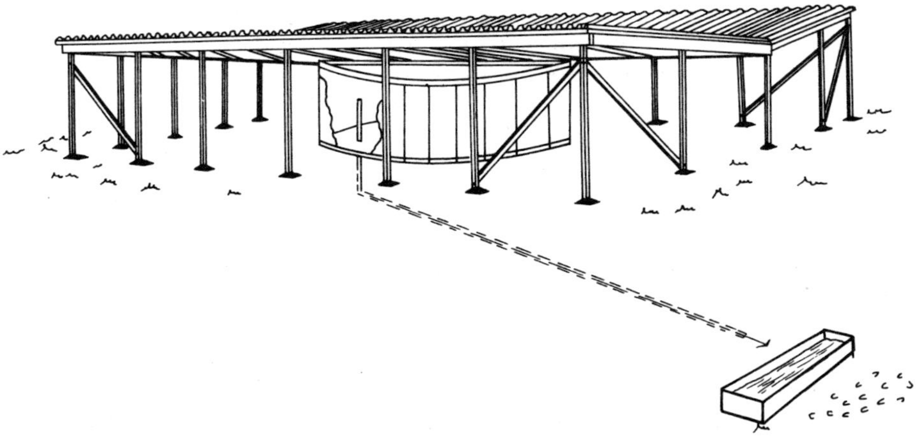

Fig. 11. Roof microwatershed for rainwater collection.

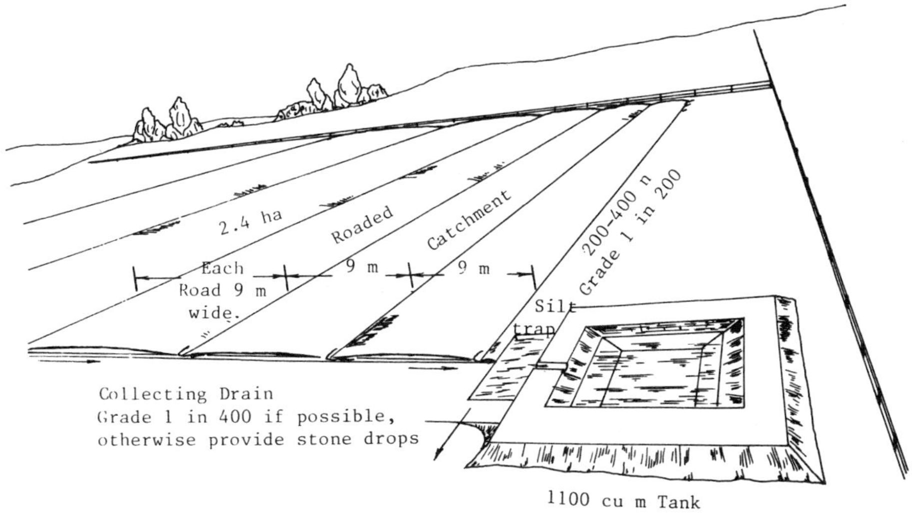

Fig. 12. Rainwater roaded catchment in Western Australia. (Department of Agriculture of Western Australia, 1950).

or by contractors using road graders (Burdass, 1974). The contour drains may also discharge on to a pastured depression or waterway, which takes the water downslope to the tank. Another method of water harvesting is the roaded catchment on the principle of constructing smooth compacted ridges of earth built in parallel, with the furrows between linked to the reservoir directly or by a tail drain (Burdass, 1974, Hollick, 1974). It is extensively used in Australia where it was developed in the years 1949–52. Hollick (1974) recommends that the road widths may vary from 1–6 m, as is shown in Table 7.

288. Procedures must be established to determine the lowest cost combination of catchment area, its efficiency, its storage size, water-demand schedules and rainfall patterns by analyzing the local situation. Assuming 100 per cent run-off, 1 mm of rainfall equals 1 litre per square metre.

TABLE 7. *Recommended road widths in Western Australia (Hollick, 1974)*

Ground slope %	Road width per road length 100–300 m	300–600 m
	m	m
0.5 – 1	3	6
1 – 2	3	5
2 – 5	2	3
Above 5	1	2

289. Catchments are cambered to increase water run-off. Catchments are graded and rolled and shed water with a minimum rainfall of 7.6 mm. They cost US$66–88 per hectare.

290. Storage bags are subject to mechanical damage from vandalism or vermin and their cost is high.

(d) Management and equipment of watering points

291. *Characteristics:*

Land-use system: Rangeland, settlements
Desertification potential: High
Production system: Semi-intensive
Cost: Low to high
Level of technology: Intermediate to complex

Description:

292. Temporary watering points may be excavations which fill up during the rainy season and dry out during the dry season or prolonged drought. Their main value is in enabling livestock to graze larger areas during the rainy season and particularly at the beginning of the dry season. They do not require heavy investments and are interspersed between permanent watering points 3–4 km apart for cattle and 6–8 km for sheep.

293. Permanent watering points are capable of distributing water at all times and in unlimited quantities. The distance between them should be 20–30 km. They are relatively expensive because of their size. The main types of permanent watering points are springs, artificial reservoirs and wells. Their capacity is 5–10 litres per second in discharge or 100,000 m^3 in volume (Baudelaire, 1972).

294. Springs are often located in mountain areas or at the foot of hills. Springs cannot be considered perennial unless they have a steady flow with no danger of drying out during the dry season. It is therefore not recommended that springs with a discharge of less than 5 litres per minute (or 0.3 m^3 per hour) be considered as permanent watering points (Baudelaire, 1972).

295. When larger yields of from 1 to 5 litres per second (or 3–20 m^3 per hour) are found, the water can be impounded without hesitation.

296. If distant, the water has to be delivered to watering points suitably placed for proper use of the pasture. The conduits may be from 5 to 10 km long or even more

when the flow of the spring is plentiful enough for several watering points to be established. To avoid the danger of obstruction, a minimum diameter of 50 mm is recommended. In these circumstances either polyvinylchloride or asbestos-cement conduits are usually the cheapest.

297. Stockwater reservoirs are formed by small dams backing up water. The dams are thrown across narrow draws or valleys or they divert water from large wadis or marigots. Before building, a topographic map should be made of the selected site.

298. Before watering points are established, a chemical analysis of the water must be made to ensure that its total salinity is within tolerable limits (5 g dry residue per litre) and that it contains no toxic salts such as nitrates, no fluorides, and no selenium or molybdenum salts (Baudelaire, 1972).

299. The bacteriological quality of the water provided to animals must always be controlled. The main steps to be followed are: (a) fencing off water sources to protect them against roaming animals; (b) diverting run-off from the impoundment; (c) preventing penetration or infiltration of uncontrolled inflow into dug wells; and (d) preventing animals from standing in the actual water reservoir by building appropriate troughs for them.

300. Animal requirements are shown in Table 8 although they vary with the season. These data may help to rationalize the use of water.

Evaluation:

301. Water is vital to livestock, so pasture zones which do not have satisfactory watering points are abandoned, provoking encroachments upon areas with more water resources. The following studies must be made before establishing new stockwatering points: (a) an inventory of all watering points in the area; (b) quantify the surface and groundwater resources available; and, (c) analyse aerial photographs which show land capability and compare them with points (a) and (b).

Reducing cropland percolation losses in sandy soils

302. *Characteristics:*

Land-use system: Irrigation agriculture
Desertification potential: Slight
Production system: Semi-intensive
Cost: High
Level of technology: Intermediate to complex

TABLE 8. *Average water consumption during the dry season (Baudelaire, 1972)*

Animal	Daily intake in litres	Frequency of watering during dry season
Sheep	4– 5	Once every 2 days
Goats	4– 5	Once a day
Asses and donkeys	10–15	Once a day
Horses	20–30	Once or twice a day
Bovines	30–40	Once a day or once every 2 days
Camels	60–80	Once in 4 or 5 days

Description:

303. Of the many kinds of soil, sand is one of the most challenging. Millions of tillable hectares of sandy soils, widely distributed throughout arid and semi-arid regions, are seldom used for agriculture because of their inability to prevent water from percolating away too rapidly for plant growth and consequent low productivity. Sandy soils thus require frequent irrigation. In many arid regions this is not practical, particularly if rainfall is the only source of water (National Academy of Sciences, 1974).

304. Techniques have been developed to produce artificial underground moisture barriers that keep water and nutrients from percolating below the root zone. Underground moisture barriers are made of continuous films of a water-resistant material placed approximately 60 cm below the soil surface with gaps every 150 m or so for drainage. An asphalt moisture barrier can double the water-holding capacity and prevent excessive deep percolation (Erickson, 1972).

305. Although most barriers have been made of asphalt, any durable material impervious to water can in theory be used. Plastic sheets have been used in East Africa; layers of compost or manure rich in colloids have been used in Hungary (National Academy of Sciences, 1974).

Evaluation:

306. This technique is costly, requires technical assistance and is recommended only for cash crops.

3. Methods of Irrigation

307. Surface methods of irrigation consist in water distribution over the field, with the rate of application exceeding the intake rate of the soil, thereby causing water to flow over the surface.

308. Taking into account the proportion of the soil which is covered with water during the irrigation, the surface methods are divided into two groups: flooding methods which are based on complete soil coverage, and furrow methods which are based on distributing water in furrows methods which are based on distributing water in furrows where the soil is partially immersed. Figure 13 shows several methods of surface irrigation, and Table 9 shows the limitations and advantages of these methods.

Flooding

309. Irrigation by flooding is a method which spreads a thin layer of water over the soil surface to refill the storage capacity of the soil profile. Water conveyance is controlled by the land surface itself.

(a) Wild flooding

310. *Characteristics:*

Land-use system: Irrigation agriculture
Desertification potential: Slight
Production system: Semi-intensive
Cost: Low
Level of technology: Simple

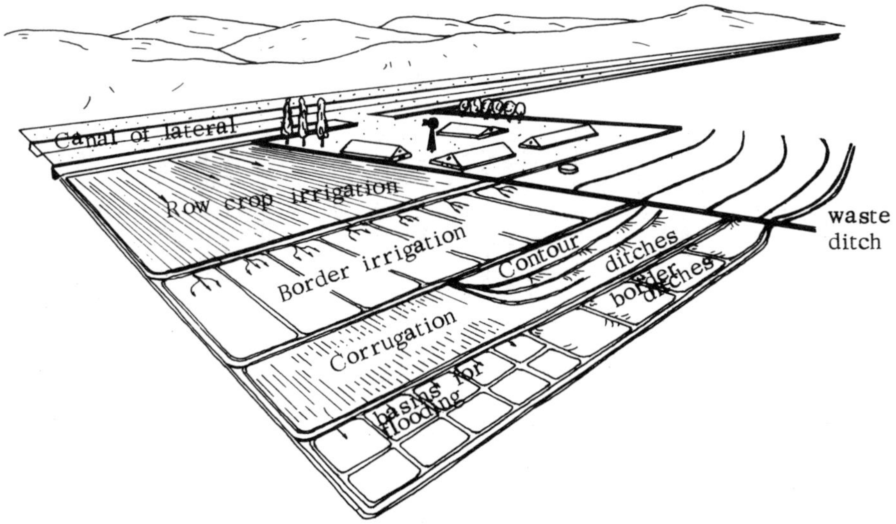

Fig. 13. Various methods of irrigation (Bishop *et al.*, 1967).

Description:

311. The field is divided into strips by irrigation ditches of widths from 20 m to 25 m. This method is used for irrigating perennial dense crops which protect the soil surface against erosion or for irrigating low-income crops on steep land. It is also used for heavy soils with irregular topography.

Evaluation:

312. Wild flooding has a low initial cost in land preparation and consists of spilling water at frequent intervals from a graded ditch. The land must be characterized by a uniformly low slope to prevent soil erosion.

313. The irrigation water is of low cost, requires much manual labour, produces an uneven distribution, and has a low application efficiency (Arnon, 1972).

(b) Flooding from field ditches

314. *Characteristics:*

Land-use system: Irrigation agriculture
Desertification potential: Slight
Production systems: Semi-intensive
Cost: Low
Level of technology: Simple

Description:

315. With this method, water is applied directly to the field from ditches that convey it to various parts of the field without dikes or levees to control the flow. The advancing front of water is controlled by the slope and the rate of water application, and by limited guidance from the irrigator.

TABLE 9. *Adaptations, limitations and advantages of surface irrigation (Bishop et al., 1967)*

Method	Adaptation	Limitations	Advantages
		Flooding	
From field ditches	(1) All irrigable soils	(1) Subdivides fields	(1) Low initial cost
	(2) Close growing crops	(2) High irrigation labour requirements	(2) Adaptable to a wide range of irrigation flows
	(3) Slopes up to 10 per cent	(3) Low water application efficiency	(3) Few permanent structure
	(4) Rolling lands and shallow soils where land grading is not feasible	(4) Uneven water distribution	(4) Run-off from upper areas can be collected and reused
		(5) Possible erosion hazard	
Border strip	(1) All irrigable soils	(1) Extensive land grading required	(1) High water application efficiency possible with good design and operation, regardless of soil type
	(2) Close growing crops	(2) Engineering designs necessary for high efficiencies	(2) Efficient in use of irrigation labour
	(3) Slopes up to 3 per cent for grains and forage crops	(3) Relatively large flows required	(3) Applicable on all soil type
	(4) Slopes up to 7 per cent for pastures	(4) Shallow soils cannot be economically graded	(4) Low maintenance costs
		(5) Dikes hinder cultivation and harvesting	(5) Positive control over irrigation water
Checks or level basins	(1) All irrigable soils	(1) Extensive land grading often required	(1) Good control of irrigation water
	(2) Orchards and close growing crops	(2) Large flows required	(2) High water application efficiency
	(3) Slopes up to 2½ per cent or more when benched or terraced	(3) Initial cost relatively high	(3) Uniform water application and leaching
		(4) Dikes hinder equipment operations	(4) Low maintenance costs
		(5) Maintenance problems on escarpments on steep slopes	(5) Erosion control from irrigation and rainfall
		(6) May effect cropy yields on crops sensitive to inundation	(6) Large streams can be utilized

			Furrow irrigation			
Corrugations	(1)	All irrigable soils	(1)	Moderately high irrigation labour requirements	(1)	Increase efficiency and uniformity over flooding from field ditches on rolling lands
	(2)	Slopes up to 10 per cent	(2)	Short runs required on high intake soils	(2)	Improves border flooding on new lands
	(3)	Close growing crops	(3)	Rough on cultivation and harvesting equipment		
Furrow	(1)	All row crops	(1)	Moderate irrigation labour requirements	(1)	Uniform water applications
	(2)	All irrigation soils	(2)	Engineering design essential for high efficiencies	(2)	High water application efficiency for high efficiencies
	(3)	Slopes up to 5 per cent with row crops and up to 15 per cent for contour furrows in orchards	(3)	Some run-off usually necessary for uniform water application	(3)	Good control of irrigation water
			(4)	Erosion hazard on steep slopes from rainfall	(4)	Control equipment available at low cost such as spiles, siphon tubes and galve

Evaluation:

316. This method requires a uniformly low slope in order to prevent erosion. Because its initial outlays for land preparation are low, it continues to be used. It is not to be recommended, however, because of low efficiency and wastefulness of both soil and water (Taylor and Ashcroft, 1972; Bishop *et al.*, 1967).

(c) Border strip

317. *Characteristics:*

Land-use system: Irrigation agriculture
Desertification potential: Slight
Production system: Semi-intensive
Cost: Low
Level of technology: Simple

Description:

318. This system makes use of parallel earth ridges. The land between two levees is considered a border strip which has a width up to 30 m and a length up to 800 m (Booker, 1974). These measures are functions of the size of the stream and the soil permeability. The field is divided into strips between low parallel levees or borders and so oriented that the water flows slowly as it moves down the slope. The border strips must be leveled between the dikes. This method is used on permeable soils for close-growing crops, orchards and vineyards with slopes lower than 3 per cent or on permanent pastures where slopes of up to 6–7 per cent are permissible. Land grading is generally required. Medium-textured deep permeable soils with deeped-rooted crops are well irrigated with this method.

Evaluation:

319. This system is not recommended for shallow-rooted crops on sandy soils with high water-intake rates where water losses could be produced through deep percolation. The method shows a tendency for excessive water penetration in the middle of the strip leading to salt accumulation.

(d) Basin method

320. *Characteristics:*

Land-use system: Irrigation agriculture
Desertification potential: Slight
Production system: Semi-intensive
Cost: Low
Level of technology: Simple

Description:

321. This is the most widely used irrigation method and consists in applying irrigation water to basins enclosed by levees. Arnon (1972) states that a large stream of water flowing at 150–200 m^3 per hour is used to fill these basins quickly. The method is

suitable for either highly permeable or slightly permeable soils, but it is seldom used for crops sensitive to wet conditions (Booker, 1974). The method is usually adapted to orchards where under favourable conditions it is possible to include several trees in each basin, although sometimes a separate basin is made for each tree.

Evaluation:

322. The main disadvantages of this method are: the levees interfere with the agricultural equipment to cultivate or harvest the crops; it is sometimes difficult to drain excess waters rapidly; considerable land is occupied by the levees, reducing the available area for cultivation. This method is not suited to steeply sloping land with soils which develop deep cracks on drying.

Primitive irrigation

323. *Characteristics:*

Land-use system: Irrigation agriculture
Desertification potential: Slight
Production system: Semi-intensive
Cost: Medium
Level of technology: Simple

Description:

324. In primitive systems, water from surface sources is generally abstracted by means of continuous flow in canals, while water from groundwater sources requires lifting to the field surface.

325. *Kair* farming is a primitive form of irrigation farming. Since the earliest times, people have cultivated grain and vegetable crops on the foot terraces of the large rivers which cross the deserts of Central Asia without any form of irrigation. The narrow strips on nonsaline alluvial soil with fresh groundwater in the vicinity are called kairs.

Evaluation:

326. This traditional method has been used only for small-scale agriculture and under a low level of technology.

Furrow irrigation

327. *Characteristics:*

Land-use system: Irrigation agriculture
Desertification potential: Slight
Production system: Semi-intensive
Cost: Low
Level of technology: Simple

Description:

328. The method consists of running a stream of water in a furrow with a gentle (0.2–0.6 per cent) and uniform slope. The length of the furrows varies from 50 to 300 m (Arnon, 1972) depending on the soil type and size of the stream.

329. Many crops are grown in single rows 75–105 cm apart, the spacing depending on the crop, the machinery to be used in planting, cultivation and harvesting, and the wetting pattern obtained by the lateral movement of water in the soil.

330. Booker (1974) describes three kinds of furrows: (a) contour furrows are used when it is not possible to use straight furrows and are constructed on a predetermined slope; (b) bench furrows are constructed where the slope is too steep (up to 25 per cent) for contour furrows, and land is benched to allow the production of irrigated crops; (c) zigzag furrows are intended to increase the distance that water must travel to reach the end of the irrigation run, in order to reduce the average slope and the velocity of the water flow. There are different patterns in laying out irregular furrows.

Evaluation:

331. According to Booker (1974), the furrow length must be well selected because labour requirements and irrigation costs increase as the length becomes shorter, while the uniformity of water application usually decreases as the furrows become longer. Depending on intake rate, soil texture, slope, erodibility, and size of the stream, furrows which are 15–20 cm deep and 25–30 cm wide at the top normally carry a flow of about 3 litres per second (Booker, 1974). The maximum flow rate allowable is determined by the need to prevent excess run-off, overtopping of the beds, and soil erosion.

Corrugation method

332. *Characteristics:*

Land-use system: Irrigation agriculture
Desertification potential: Slight
Production system: Semi-intensive
Cost: Low
Level of technology: Simple

Description:

333. This method involves the flow of water downslope in small furrows called corrugations or rills. This method minimizes soil surface crusting and is well adapted to close growing crops on sloping lands and on soils with reduced permeability. Corrugations are V-shaped or U-shaped channels about 10 cm deep, spaced 40–75 cm apart, carrying small streams 1–2 m^3 per hour. The length varies from 40 m to 120 m on a slope of less than 10 per cent (Arnon, 1972). On silt loam or clay loam soils, lateral movement of water takes place readily yet the method is not recommended for clay soils due to their low intake rate, nor is it recommended for sandy soils which usually have high intake rates and water losses though deep percolation. Spacing between corrugations will depend on how rapidly the water moves laterally through the soil (Booker, 1974).

Evaluation:

334. Under saline conditions capillary phenomena and subsequent water evaporation will tend to concentrate salts on the soil surface. The same could happen where the irrigation water has a high salt content, so precautions must be taken.

Overhead or sprinkler irrigation

335. *Characteristics:*

Land-use system: Irrigation agriculture
Desertification potential: Slight
Production system: Intensive
Cost: High
Level of technology: Intermediate to complex

Description:

336. This irrigation method uses pipes for water conveyance to various parts of the field to be irrigated where it is then sprayed into the air to fall on the soil surface with uniformity. Because of cost and technical complexities sprinkler irrigation is not practical on very large areas. The principal types of sprinklers are rotating sprinkler heads, fixed jets and perforated pipes. Modern portable systems use light-weight aluminium tubing, and more recently plastic tubing complemented with quick-coupling parts (Arnon, 1972).

337. According to the type of conveyance system, sprinkler systems are classified as follows: Permanent or fixed, semi-permanent or semi-portable, and fully portable. Sprinkler systems are well adapted to soils with high infiltration rates. The rate of water application should not be greater than the infiltration rate of the soil. The application rate is regulated by combinations of pressure, nozzle size, and spacing of nozzles. Sprinkler irrigation systems are also used to apply fertilizer through water, frost protection and temperature control (Christiansen and Davis, 1967).

Evaluation:

338. Sprinkler irrigation is technically and economically feasible wherever the land is too steep or uneven for surface irrigation. Elimination of field ditches increases the net area for plant production and reduces water losses due to seepage and percolation. It also requires less manual labour. Water application is more uniform with sprinklers than with surface irrigation methods. Water application efficiencies should be above 75 per cent.

339. Some of the disadvantages of sprinkler irrigation are: high cost of the equipment and water loss by evaporation greater than with traditional methods. It is not well adapted to windy areas.

Drip or trickle irrigation

340. *Characteristics:*

Land-use system: Irrigation agriculture
Desertification potential: Medium
Production system: Intensive
Cost: High
Level of technology: Complex

Description:

341. Drip or trickle irrigation uses a system of pipes, located on or under the soil, which carries water delivered through outlets located near each plant or fruit tree. The emitters are designed to discharge water at rates of 1–10 litres per hour.

342. The method consists in delivering water to strategically located points near the roots at frequent intervals. According to several authors (Arnon, 1972; Rawitz and Heller, 1969; Booker, 1974), with this method it is possible to adjust water delivery to the absorptive capacity of the soil, to seasonal crop requirements and to climatic conditions. Chemical fertilizers can be applied through the outlets, salt damage is avoided, and the destruction of soil surface structure is eliminated. Under this irrigation method, soil moisture and soil seration are maintained at an optimum for plant growth.

343. The wetting pattern will determine the number of omitters required to irrigate a given area, as well as the rate of flow that should be discharged by each emitter.

344. Standard irrigation methods are particularly wasteful of water in arid lands because extended areas of wetted soil encourage evaporation.

345. In trickle irrigation, small amounts of water are applied at frequent intervals to a specific site near each plant. Under each site a wet area is formed which extends to the plant's roots. The application can be adjusted to replace water continuously as it is used by the plant.

346. Trickle irrigation not only improves the efficiency of water use but also gives much greater control over the replacement and amount of water, which can be adjusted to the soil's absorptive capacity and to the characteristics of the particular crop, its stage of growth, and the climatic conditions.

Evaluation:

347. Evaporation is reduced by this method, increasing water use efficiency. It is applied to trees, vegetables and other cash crops, but because of its high initial cost it is not extensively used for basic crops.

348. This relative new commercial system has several disadvantages: (a) emitters can become clogged; (b) the system requires good technical advice when saline waters are used; (c) it has a high initial cost; (d) it suffers from lack of uniformity in water application; and, (e) in most cases, it needs to be complemented with another irrigation method, usually sprinkler, to leach excess salinity in the root zone between crops.

349. The method is well suited for row crops growing on sandy or gravelly soils in dry regions, where conventional surface methods are not practicable, and where overhead irrigation would lose too much water through evaporation (Arnon, 1972). The main aim of drip irrigation is to produce high yields through its ability to maintain low moisture tensions in the soil, as well as a more or less constant level of nutrients, with high frequency of irrigation. Other advantages of this system are the following: elimination of losses at field margins, the flexible tubes can follow contour lines and curving rows, the small-diameter tubing is easy to handle.

350. Trickle irrigation is potentially important for many irrigated crops in arid lands. It is already extensively used on tree, vine, and row crops, but because of high costs it has not yet been applied to field crops. It has particular promise for slopes or rocky areas where land leveling for conventional irrigation is prohibitively expensive.

Subirrigation

351. *Characteristics:*
Land-use system: Irrigation agriculture
Desertification potential: High
Production system: Semi-intensive
Cost: High
Level of technology: Complex

Description:
352. With this method the root zone is watered by raising the watertable, after which it is lowered again until the next irrigation. It requires a high permanent watertable or a relative impermeable soil layer not too deep. There are two ways to apply the irrigation water:

(a) Through a continuous head of water to maintain a constant supply to replace water loss by consumption.
(b) Through a periodic refill of the ditches holding the water. This method tends to cause salt accumulation in the root zone and can be used only when the soil is regularly leached by rainfall, the irrigation ditches acting as a drainage system (Rawitz and Heller, 1969; Arnon, 1972).

353. With subirrigation, it is necessary to balance the inflow of water, evapotranspiration, and seepage. The main problem is to maintain a balance between inflow and consumption which will produce maximum crop growth (Taylor and Ashcroft, 1972).

Evaluation:
354. Application efficiencies vary from 30–50 per cent and under favourable conditions go up to 70–80 per cent (Kovda *et al.*, 1973).
355. Several advantages of this system are that irrigation can be applied continuously, soil moisture availability can be optimum, thereby increasing yields and saving water up to 30 per cent due to decreased evaporation. However, when subirrigation is used in arid and semi-arid regions, there is a danger of soil salinization if salts occur either in the soil or irrigation water. This method is more practised in humid areas, with very limited application in the arid zones (Criddle and Kalisvaart, 1967).

Pitcher irrigation

356. *Characteristics:*
Land-use system: Irrigation agriculture
Desertification potential: None
Production system: Intensive
Cost: Low
Level of technology: Simple

Description:

357. This method uses unglazed baked earthernware pitchers, which are often available and cheap in developing countries. The pitcher is sunk to its neck in the soil and filled with water; seeds are planted around it. Water passes through the porous wall of the pitcher into the root zone. Experiments in India have demonstrated the growth of melons and pumpkins to maturity with very little water (less than 2 cm/ha for the entire 88-day growing period (National Academy of Sciences, 1974).

Evaluation:

358. This technique has been long used by farmers. It is cheap, simple and allows the irrigation of small fields with high water-use efficiency and relatively high unitary yields. A disadvantage is that the pitchers must be continuously filled.

Innovative irrigation methods

359. *Characteristics:*

Land-use system: Irrigation agriculture
Desertification potential: None
Production system: Semi-intensive
Cost: Medium to High
Level of technology: Intermediate to complex

Description:

360. Innovations in irrigation methods include the basic principles of subsurface, sprinkler and trickle irrigation. Others include the following: (a) plastic or aluminium pipelines convey the irrigation water across uneven land, reducing the cost of land levelling, principally for fruit tree production; (b) one method similar to trickle irrigation uses a fixed-orifice sprayer. It requires less filtering, but evaporation losses are higher and in windy areas it leaves dry pieces of land; (c) in Libya and some other countries, water is mined from an aquifer through a centre-pivot sprinkler. A rigid pipe up to 500 m long is suspended on wheeled carriages, and is rotated around the pivot point (National Academy of Sciences, 1974).

Evaluation:

361. Some innovative irrigation methods present adequate alternatives to traditional techniques. One of the main goals in research and technical assistance has been the generation and diffusion of appropriate methods to achieve the highest water-use efficiency. Unfortunately most of the developing countries persist in low water-use efficiencies and considerable waste of water.

4. Run-off Management

Run-off farming

362. *Characteristics:*

Land-use system: Rainfed agriculture

Desertification potential: Medium
Production system: Semi-intensive
Cost: Low
Level of technology: Simple to complex

Description:

363. In the Negev highlands an area of 300,000 ha of run-off farms once used sophisticated systems by collecting rain contour ditches, by clearing large hillside areas and smoothing the soil. This system was in operation by the time of the Roman occupation. After the Arab conquest (A.D. 630) the ancient desert agriculture practised under an annual precipitation of 100 mm slowly disintegrated. Professor Evenary and his coworkers have reconstructed these techniques in an area of approximate 5 ha in Avdat, Israel, as can be seen in Fig. 14.

364. Run-off farming techniques are applied in Tripolitania (Djebel Nefousa) and in central and Southern Tunisia (Sousse, Matmata, Cafsa) where 10 million olive trees are successfully cultivated under an annual precipitation of 300 mm (Le Houérou, 1975).

Evaluation:

365. These techniques have good potential and could be adopted in many arid and semi-arid zones. By using their basic principles run-off farms, now located in arid and semi-arid lands, could improve considerably their yields of crops and native vegetation.

Chapingo microcatchment system

366. *Characteristics:*

Land-use system: Rainfed agriculture
Desertification potential: Slight
Production system: Semi-intensive
Cost: Medium
Level of technology: Intermediate to complex

Description:

367. The basic principles for microcatchments indicate that the most economical treatment of the run-off collecting area consists of the elimination of weeds and all obstacles which reduce run-off. Other treatments, such as the use of bituminous materials, can be applied if they are not expensive. It is required that the cultivated area be deep enough, have good permeability, adequate fertility and an appropriate water-holding capacity.

368. Anaya *et al.* (1976) have developed a formula to calculate microcatchment size for water harvesting in semi-arid lands where rainfall distribution is adequate. They consider three types of plants: (a) those cultivated in rows (corn, sorghum, soybeans); (b) dense crops such as barley, grasses and wheat and (c) fruit trees. A formula can be applied to the three types of crops as follows:

$$\text{Microcatchment size} = ca + \frac{1}{rc}\,\frac{(CU - P\,ca)}{P}$$

Fig. 14. Run-off farming techniques in the Negev desert for fruit trees and annual crop production.

where *ca* = cultivated area. This is the area horizontally explored by roots. In the case of rows, it is expressed in centimetres, for dense crops in metres, and for fruit trees in square metres and these will define the microcatchment size units (see Figs. 15, 16, 17, and 18)

rc = run-off coefficient

CU = consumptive use during the vegetative cycle (in millimetres)

P = precipitation during the vegetative cycle (50 per cent probability) expressed in millimetres.

Evaluation:

369. The methodology represents an important tool for research in arid and semi-arid areas where consumptive use is larger than the rainfall precipitation during the vegetative cycle.

Watershed management with ridges, furrows and storage tanks

370. *Characteristics:*

Land-use system: Rainfed agriculture
Desertification potential: Medium
Production system: Semi-intensive
Cost: Medium to high
Level of technology: Intermediate

Description:

371. For those semi-arid areas with erratic rainfall distribution and where the soil has a low-holding capacity, water harvesting will consist in allowing the soil to retain as much rainfall as it can, with the excess conducted through grassed ways to surface reservoirs. The following diagram illustrates the principles of this method (Fig. 19).

Evaluation:

372. Storaged water can be utilized for supplemental irrigation during drought periods.

Water-spreading methods

373. *Characteristics:*

Land-use system: Rainfed agriculture, rangeland, forestry
Desertification potential: Slight
Production system: Semi-intensive
Level of technology: Simple–complex

Description:

374. Water-spreading techniques provide simple ways of irrigation and are usually practised in lands adjacent to rivers or small streams which may be permanent or intermittent. Floods are deliberately diverted from their natural courses to spread over adjacent flood plains or be concentrated on valley floors. Portions of land will be flooded occasionally and will produce green feed at times when rain-grown vegetation is

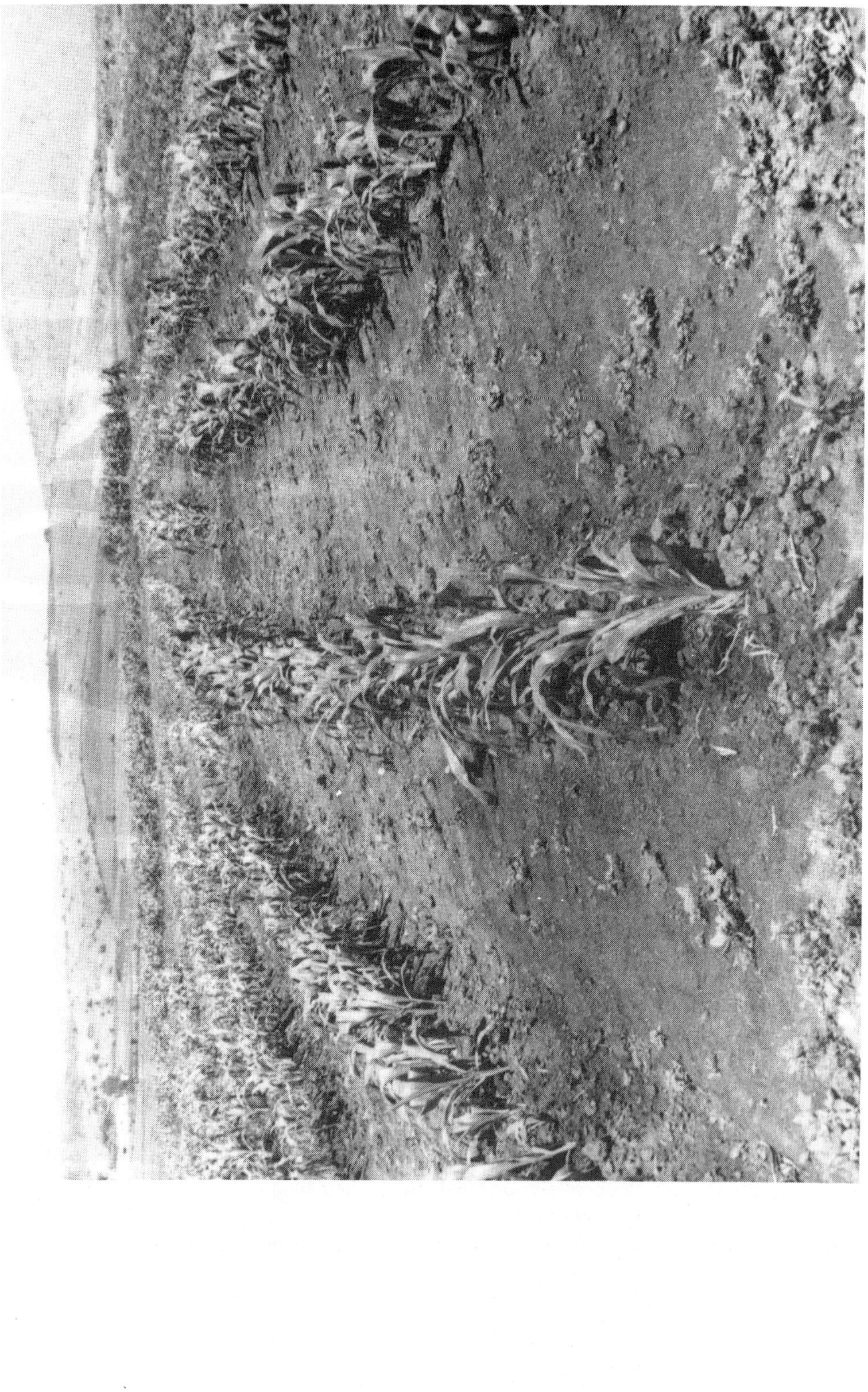

Fig. 15. Experimental site to see the effect of several distances between rows on water availability and plant production.

Fig. 16. Water harvesting method for dense crops.

Fig. 17. Microwatershed for water harvesting in fruit tree production system.

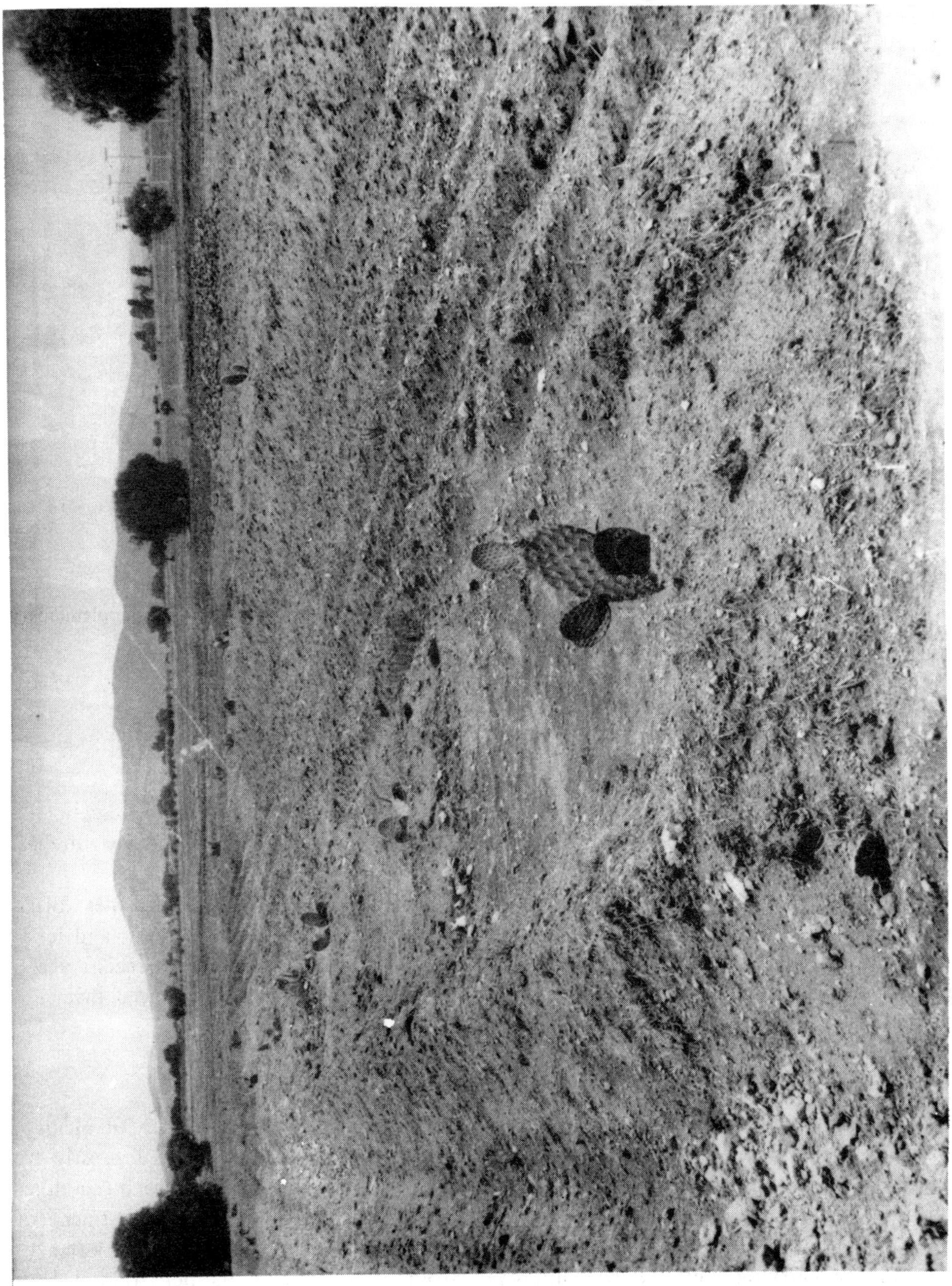

Fig. 18. Partial microwatershed to avoid excess water in the first years. As soon as the tree grows, more run-off area will be dedicated to water harvesting.

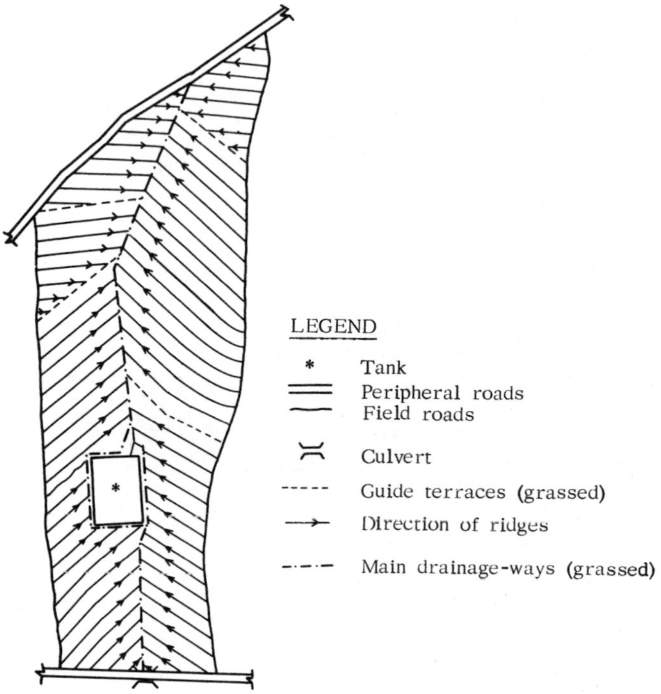

Fig. 19. Diagram of a watershed in ridges and furrows with a run-off storage facility for supplementary irrigation purpose (Kampen *et al.*, 1974).

1. Tank
2. Field roads
3. Guide terraces (grassed)
4. Main drainage ways (grassed)
5. Peripheral roads
6. Culvert
7. Direction of ridges

inadequate. This excess production may extend the grazing season and increase the carrying capacity of the ecosystem. The floodwater is diverted or retarded by ditches, dikes, small ponds or brush fences (see Fig. 20).

375. Semi-nomadic populations build dams of rough mud and skimpy fascines, often brought from long distances, to raise the water of the freshet to bank-top level, and let it off through one or two unlined channels or *sehias* dug in the base soil. In recent years, engineers have developed methods for constructing permanent works with faced diversion dams and lined canals.

Evaluation:

376. An example of the application of this method is found in the north of Mexico, an arid region with a yearly rainfall of 250 mm. The grassland is occupied mostly by *Boutelowa gracilia* covering 35,000 ha; 892,000 m^3 of earth were moved to build dikes across the cheeks and 18,152 km of contour furrows were built with an investment of US$8 per hectare. The production of meat on the hoof was increased from 2.10 kg/ha to 3.63, and the carrying capacity was improved from 34.4 to 22.8 ha per Animal Unit Year in spite of the critical limit for this system of water spreading which is located around 220 mm of annual rainfall (Claveran and Beneitez, 1973).

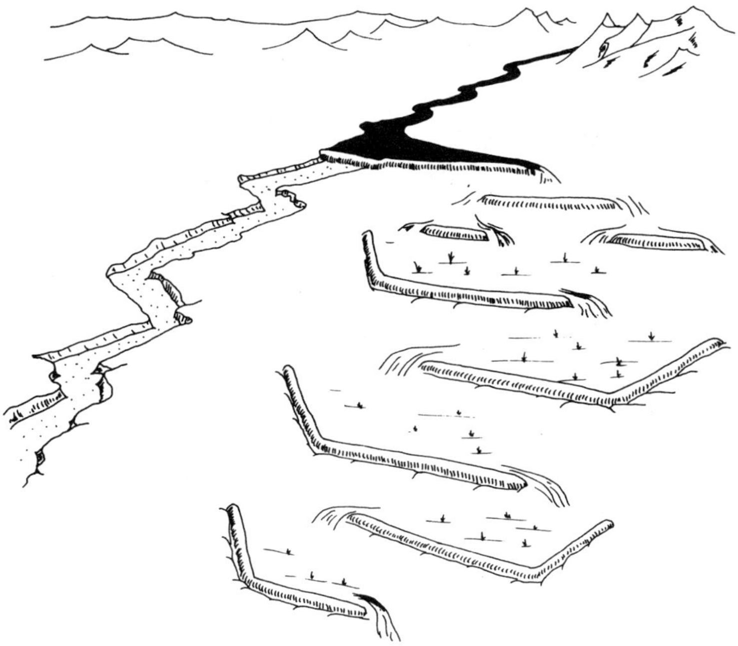

Fig. 20. Sketch of water-spreading ditches. Zigzag pattern slows the torrent of flood water and allows it to penetrate the soil (National Academy of Sciences, 1974).

377. The method is of course an extremely unpredictable agricultural gamble, and its success does not depend solely on the probable frequency of freshets nor on the quality of the diversion works but also, obviously, on the capacity of the soil to retain for long periods water which only reaches it at widely separated intervals.

5. Flood Control

Flood and sedimentation control

378. *Characteristics:*
Land-use system: Settlements forestry, rangeland, rainfed agriculture, irrigation agriculture
Desertification potential: Slight
Production system: Intensive
Cost: Medium
Level of technology: Simple to complex

Description:
379. Channel improvements whose purpose is to increase the conveyance capacity which in turn will decrease the extent and frequency of flooding. Improvements can be accomplished by enlarging the channel cross-section, by realigning the channel, by increasing water velocity and by removing obstructions to flow.

380. Levees (dikes, embankments, bunds) and flood walls represent the commonest method. Levees are suitable for streams having flood plains. A levee must be structurally safe, require minimum maintenance and have provision for drainage facilities, such as flood gates.

381. It is advisable that continuous earthen banks along the natural stream banks be constructed to create an artificial floodway during high flows and thus avoid flooding the agricultural areas.

382. The principal sources of sediment are: (a) sheet and rill erosion in agricultural land rangeland and forests; (b) roads; and, (c) upland gullies, stream channels and valley trenches.

383. Gottschalk (1962) reported that estimates of sediment sources from 157 watersheds indicated that 73 per cent of their sediment yield was due to sheet erosion, 10 per cent from gully erosion, and 17 per cent from other sources such as roadside erosion, stream-bank erosion, and flood plain scour.

384. To reduce flood and sedimentation hazard and to increase soil moisture storage for plant production, various agricultural implements have been used to trench, furrow or pit soil surfaces.

385. Some measures to reduce flood and sedimentation hazards are the following:

(a) Revegetation – vegetation can be restored on depleted and eroded areas by contour trenching and reseeding.
(b) Water spreading consists of a system of dikes constructed to divert flood flows automatically from gullies or arroyos and spread them over adjacent lands.
(c) Diversion and by-passes or emergency floodways – the emergency floodway is operated through a flood gate, or a fuse-plug levee which is broken when the water level in the main channel reaches a predetermined flood state.
(d) Reservoirs, detention and retarding basin – in most countries, reservoirs are used to control flooding. There are two basic types: (i) Detention reservoirs and retarding basins provided with fixed outlets which regulate the outflow in accordance with the stage and volume of water in storage, and (ii) storage/impounding reservoirs which are regulated by gates and valves.
(e) Land treatment measures involve soil and water conservation practices in upstream areas, and include: crop rotation, strip cropping, terracing, contour cultivation, selective planting, reforestation and construction of small ponds.
(f) Non-structural measures include: (i) flood plain management, (ii) flood forecasting and temporary evacuation, (iii) flood insurance and rehabilitation, and (iv) flood damage prevention planning.

Evaluation:

386. It is estimated that the total area protected by drainage and flood control in the world in 1975 was over 119 million ha, the range protected by different countries varying from 20,000 to 47 million ha (Framji and Gara, 1976).

387. India initiated a national programme of flood control in 1954. The overall progress made up to 1971 included: (i) construction of 7063 km of new embankments, (ii) 9377 km of drainage channels, (iii) 191 town protection schemes, and (iv) the raising

of 4585 marooned villages. Through these efforts 6.05 million ha have become reasonably protected. Total investment was Rs 2.283 million (1 US$ = Rs 7.50) with a cost per hectare of Rs 380.

B. SOIL

1. Soil Conservation

Gully Control

388. *Characteristics:*

Land-use system: Irrigation agriculture, rainfed agriculture, rangeland, forestry, transport routes, mining, settlements
Desertification potential: Slight
Production system: Extensive
Cost: Low to high
Level of technology: Simple to complex

Description:

389. Every system of gullies has an independent catchment having a regular flow equivalent to a drainage system. Gullies have been classified by Teswani *et al.* (1973) as shown in Table 10).

TABLE 10. *Classification of gullies (Teswani et al., 1973)*

Description of size	Specifications in metres depth	Bed width	Side slope
Very small	3	18	variable
Small	3	18	variable
Medium	3–9	18	8–15%
Deep and narrow			
(a)	3–9	18	variable
(b)	9	variable	

390. Gullies may be V-shaped or U-shaped. The V-shaped form is typical where the subsoil is resistant to erosion while U-shaped gullies are often found in loess regions and alluvial valleys where both the surface soil and subsoil are easily eroded. Figure 21 shows a U-shaped gully of small size in a semi-arid area of Mexico.

391. Figures 22 and 23 show land degradation when overgrazing reduces the vegetative cover.

392. Control of gullies by vegetation is desirable because it protects the soil against rainfall impact, reduces the velocity of flow and some of the sediment load may be deposited until the gully is filled. Unfortunately, in most areas, the bed of a gully has very low productivity.

393. Planting the sides of gullies is not easy, because they are usually unstable, steep and erodible. If money is available and the cost is justified, the banks can be levelled by hand or by heavy earth-moving machinery, fertilized and then seeded or planted.

Fig. 21. Gully formation in a semi-arid area of Mexico with 500 mm annual precipitation.

Fig. 22. Overgrazed exposed slope with soil and plant cover degradation in a semi-arid area of Mexico (650 mm).

Fig. 23. Grassland deteriorated by soil erosion in a semi-arid area of Mexico (500 mm annual precipitation).

394. Plant choice for gully control should take into account suitable and available species and the use of special planting techniques. Such plants should be characterized by vigorous growth in poor soil conditions and provide good cover for the soil surface. Hudson (1971) mentions the use of cylinders of thin-back polyethylene to form bottomless plant pots filled with fertile soil; by the time the plant has outgrown its reservoir, it has more chance for survival under adverse conditions. This method is suitable for shrubs and trees.

395. A method for establishing grass colonies consists in filling sacks of jute or hessian with fertile soil and laying them in shallow trenches in the bed of the gully after making small cuts in the bag. When the bag rots away the plants are strong enough to survive.

396. According to Hudson (1971), structures for gully control may be classified as temporary (wire bolsters, netting dams, brushwood dams, log dams, and brick weirs) or permanent (silt-trap dams, regulating dams, gully-head dams, drop structures, masonry dams and gabions).

397. Log dams use two rows of vertical posts driven into the bed of the gully and extending up the sides to above flood level, and then logs are packed in between. If the gully has steep sides, it is better to have a rectangular notch in the centre, but the notch must be big enough to pass the whole of the flood. When the logs are packed between the rows of posts, the bottom layer should be sunk below ground surface to avoid seepage and scour underneath. After the top logs have been put in place they are held in position by strong wire tied between the vertical posts (see Fig. 24).

398. A simple but effective method where rocks are abundant and cheap enough for use in gully control is to build a structure of loose rockfill with the stones anchored in place by wire netting (see Fig. 25). More substantial structures can be constructed by forming a structure of rock as can be seen in Fig. 26.

399. Gully control programmes should be established in forestry land. Figure 27 shows a simple and economical method of control.

400. The gabion method was developed in Italy and uses prefabricated rectangular baskets called gabions made of heavy duty wire netting (Fig. 28). The basket is placed in position and filled with stones, then the lid is wired down. The baskets, up to 4 m long by 1 X 1 m, are built up one on top of the other like courses of brickwork and can form large or small structures. The use of galvanized wire ensures a long-lasting resistance to corrosion.

Evaluation:

401. Gullies are important causes of land degradation in ecosystems throughout the world. Their control is expensive and in most cases requires technical assistance. The best way to control gullies is through application of preventive techniques.

Implement for soil and water conservation

402. *Characteristics:*

Land-use system: Irrigation agriculture, rainfed agriculture, rangeland, forestry, transport routes, mining

Desertification potential: Medium

Fig. 24. Log dam for gully control in a semi-arid area of Mexico.

Fig. 25. Temporary structure of wire netting for controlling small gullies in a semi-arid area of Mexico.

Fig. 26. Temporary structure of stones for controlling small gullies in a semi-arid zone of Mexico.

Fig. 27. Gully control in forestry lands of Mexico.

Fig. 28. Permanent structure of gabions for control of a deep and narrow gully in Mexico.

Production system: Extensive
Cost: Low
Level of technology: Simple to complex

Description:

403. Tools, whether used by hand, animal traction or by machines, should be available free if necessary to small farmers who cannot afford to purchase them. Figures 29, 30 and 31 show some of the traditional implements used in developing countries. The proper selection and use of efficient tillage methods is a basic element in preventing land degradation, principally in dryland use systems.

Evaluation:

404. The use of manpower and animal traction represents in many countries more than 80 per cent of the energy used in agriculture production. Agricultural implements adapted to the socio-economic conditions of villagers and small farmers should be studied, designed and produced for free distribution where necessary.

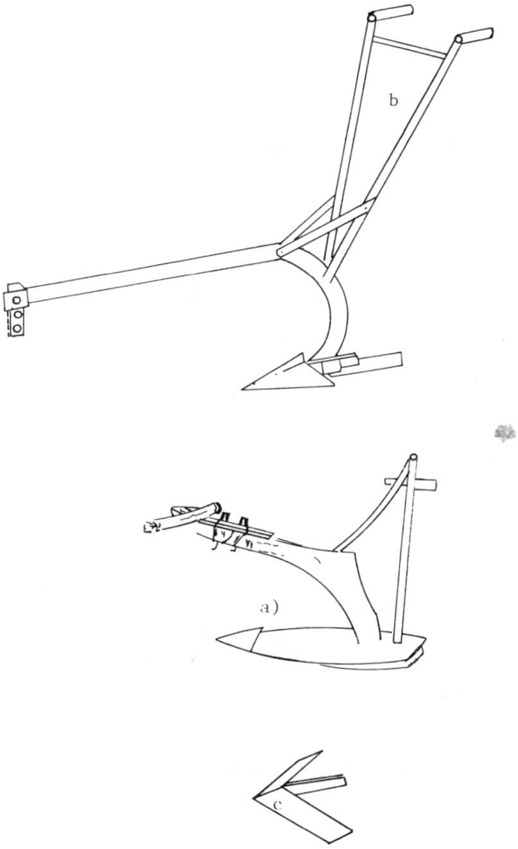

Fig. 29. Improved multi-purpose plough for animal traction: (a) traditional Libyan plough: (b) improved version with exchangeable shares for shallow tillage or intercultivation, beam length approximately 75 cm; ground clearance 35 cm; (c) weeding sweep or mahacha (FAO, 1969).

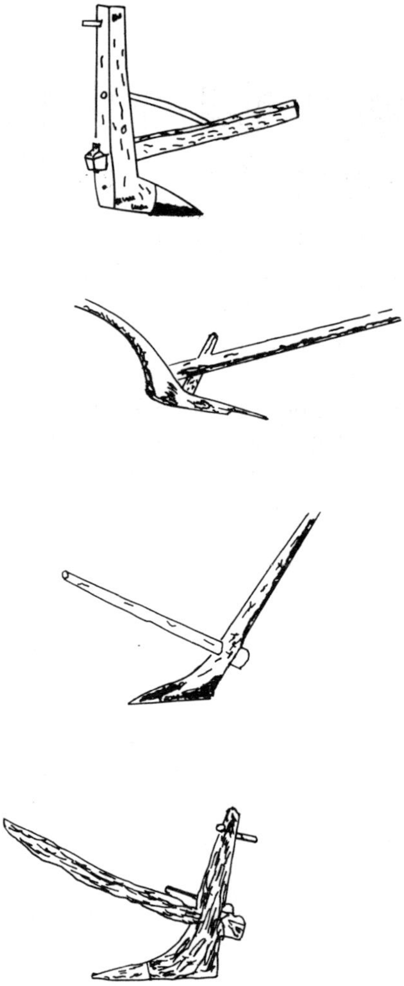

Fig. 30. Traditional ploughs from: (a) Afghanistan, (b) Syria and Central Iraq, (c) India, (d) Turkey (FAO, 1969).

Mechanical practices

405. *Characteristics:*

Land-use system: Irrigation agriculture, rainfed agriculture, rangeland
Desertification potential: High
Production system: Semi-intensive
Cost: Medium to high
Level of technology: Intermediate–complex

Description:

406. Channel terraces or contour ridges: There are several types of channel terraces (see Fig. 32). Such terraces involve some form of earth-moving work at right angles to the steepest slope to intercept and control run-off. They have been given a variety of

Fig. 31. Land preparation by using animal traction and a rudimentary plough in an arid region of Mexico.

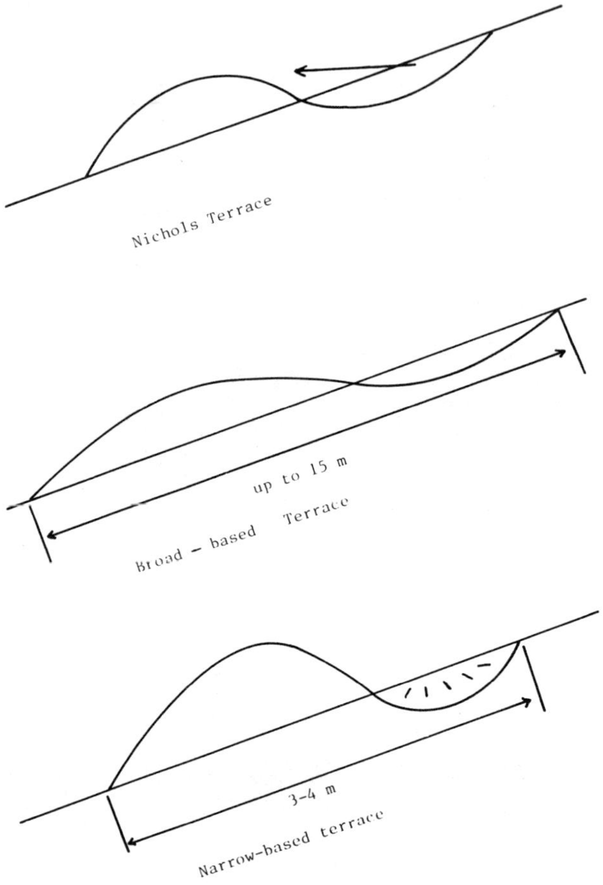

Fig. 32. Types of channel terraces (Hudson, 1971).

names: in Spanish *terraza*, in the United States channel terraces, in Commonwealth countries ridges or bunds. Channel terraces include an excavated channel and a bank formed on the downhill side with the earth from the excavation. A slight gradient at non-erosive velocities is used in arable lands for the gentle drainage of excess water.

407. Contour ridges or absorption terraces are constructed on the contour to hold the water and promote infiltration into the soil. They can be open-end and close-end terraces.

408. In Rhodesia (Hudson, 1971), the contour ridges mean graded open-end channels with raised banks and broad bases which are shallow and gently sloping and can easily be crossed by tractors, so that row crops continue over them. The narrow-based ridge is also a graded channel but with a steep-sided bank; it cannot be crossed by tractors.

409. The bench-terracing method consists in converting a steep slope into a series of steps. There are several types of benches, which may have horizontal or nearly horizontal ledges and almost vertical walls between the ledges (see Fig. 32). These terraces are typical on steep slopes and require a lot of labour. Vertical walls require soil stabilization with stone, brick or timber, or with vegetation in very stable soils. Under irrigation, the bench terrace requires a raised lip at the outer edge to retain water.

410. A technique which both controls soil erosion and conserves run-off more efficiently than conventional terrace systems has been designed by Zingg and Hauser (1959) and is called the "Zingg conservation bench terrace". It has been extensively applied in the Great Plains. Two terrace systems mentioned above can be seen in Fig. 33.

411. With slopes greater than 10 per cent, the method to adopt is level terraces, ridges or bunds laid out on the contour. These terraces are generally low and wide, so machinery can operate on them easily; their construction may be accomplished with a road grader, disk or mouldboard plough, or a one-way disk. Several considerations must be taken into account: (a) Their use may not be advisable where topsoil is shallow, as is common in semi-arid areas. (b) To eliminate ponding, ridges are placed on a grade of 0.2–0.5 per cent. (c) Proper and continuous maintenance of the terraces is required.

412. Rege (1964) found that contour farming may reduce soil loss by 50 per cent and water loss by 15–20 per cent on land planted with furrow crops such as corn or cotton, and the increase in yield can be 5–10 per cent. The effect is better in low rainfall vegetative cycles.

Contouring

413. As an alternative to terracing, the farmer may use contour strips, that is, alternate strips of protective crops. Contouring is used in control water erosion. Tillage operations are carried out as nearly as possible on the contour. A guideline is laid out on each ploughland, and the backfurrows or deadfurrows are ploughed on these lines. On small fields of uniform slope, one guideline may be sufficient.

414. On gently sloping land (<10 per cent), contouring will reduce the velocity of overland flow. If ridge or lister cultivation is practised, the storage capacity of the furrows is increased. When contouring is used alone on steeper slopes or under conditions of high rainfall intensity and soil erodibility, there is an increased hazard of gullying because row breaks may release stored water.

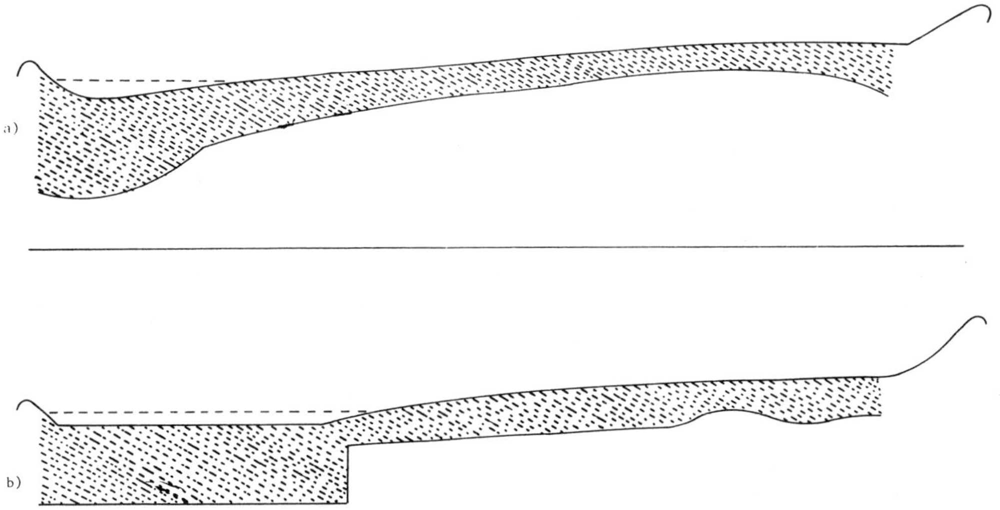

Fig. 33. Cross-section of (a) Conventional level terrace and (b) Zingg conservation bench terrace, showing soil moisture distribution (Zingg and Hauser, 1959).

Evaluation:

415. The rate of terrace construction is affected by several factors: soil moisture, uniformity of land slope, type of crop, organic residues, soil tilth, and local obstructions. The cost of construction will be lowest with the mouldboard plough if channel capacity is not taken into account. Terraces once constructed will require continuous care to keep them productive.

Soil conservation on roads

416. *Characteristics:*

Land-use system: Irrigation agriculture, rainfed agriculture, rangeland, forestry, transport routes, mining, settlements
Desertification potential: High
Production system: Intensive
Cost: High
Level of technology: Intermediate to complex

Description:

417. In many developing countries there exist national programmes for hand-graded roads. In some of them, the roadside soil conservation techniques are not adequate and problems of land degradation are increasing (see Fig. 34).

418. The successful control of roadside erosion requires soil conservation technologies. Some of the conditions responsible for land degradation along roads are the following: steep slopes, excessive surface run-off, highly erodible soil types and deficient vegetative cover (Skurlow, 1959). Prevention and stabilization measures to control land disturbance along roadsides must be part of plans for road construction. Road erosion can be classified as mass erosion (landslides, dry crop) or surface erosion.

419. Locating new roads is a difficult task. New methods, however, simplify the analysis of soil type, landscape, vegetative cover and topography by using aerial photographs. According to Hudson (1971), the first rule of road location is to put them on crests wherever possible. This allows run-off to be easily discharged on both sides of the road. When this first rule is not possible, the alternative is to put the road on a gentle grade (0.01 to 0.02) fairly close to the true contour. When roads go diagonally down the side of a hill on grades steeper than 0.05 the drainage system must be lined and that increases the cost (see Fig. 35). It is usually better to use a zigzag layout or the combination of some lengths on gentle grades and some segments of the road straight down the slope.

420. In installing drainage systems during road construction, swamps and permanent wet areas should be avoided. On earth or gravel roads, wheel ruts can reduce the amount of water reaching the side of the road, and with a certain slope each rut will become a scouring water course. This problem can be avoided by diverting the water sideways with very gentle depressions in the road surface. Road erosion inhibits road use or even makes roads impassable when they are not paved. Soil erosion is sediment transport creating damage to downstream cultural and ecological settings.

Fig. 34. Gully formation along a roadside in a rural semi-arid area of Mexico (500 mm annual precipitation).

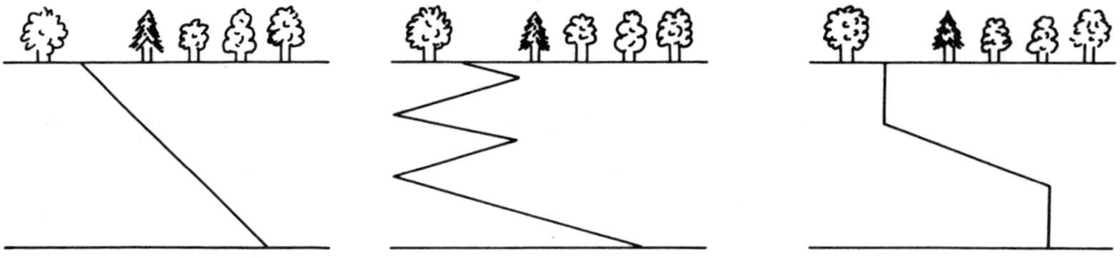

Fig. 35. Road alignment down steep hills. A is undesirable because drains on so steep a grade must be lined and are expensive; B or C, using grades suitable for grass-lined drain channels, are preferable (Hudson, 1971).

421. Megahan (1976) lists the following guidelines to reduce land degradation during road location: (a) select locations where present erosion can prevent greater erosion problems; (b) take advantage of natural features (natural benches, low gradient slopes, ridge tops); (c) if necessary, combine short road segments with steeper gradients to avoid problem areas or to take advantage of terrain features; (d) avoid midslope locations on long, steep, unstable slopes, especially where bedrock is highly weathered or soils are plastic; (e) avoid slide-prone areas (seeps, clay, concave slopes, hummocky topography, and rock layers that tend to dip parallel to the slope); (f) vary road grades where possible to reduce concentrated flow in road drainage ditches and culverts; (g) select drainage crossings to minimize channel disturbance during construction and to minimize approach cuts and fills; (h) locate roads far enough above streams to provide an adequate buffer or be prepared to catch sediment moving downslope below the road.

422. Road design criteria must be flexible to reduce erosion hazards under diverse conditions. According to Megahan (1976) and other authors (Ramos *et al.*, 1974; Hudson, 1971), revegetation and associated practices are important elements of the design process. Future maintenance needs must be assessed to assure the stability of this important land use. Megahan (1976) and Ramos *et al.* (1974) give complete descriptions of erosion control practices and costs that should be considered in the road-design process.

Evaluation:

423. There are many examples of dramatic erosion effects caused by road construction. Prevention rather than control of land deterioration is the most effective way to reduce erosional effects; prevention may also avoid irreparable damage or repairs which can exceed original constructions costs.

Erosion control in rangelands

424. *Characteristics:*

Land-use system: Rangeland
Desertification potential: High
Production system: Semi-intensive
Cost: Medium
Level of technology: Simple to complex

Description:

425. The basic control mechanism on grazing lands is obviously the maintenance of vegetative cover. Overgrazing is the main cause of soil and plant degradation on rangelands. The provision of harvested forage to feed cattle during periods of poor vegetative growth is one of the keys to control wind and water erosion.

426. According to FAO (1960), problems of erosion on rangelands can be overcome by: (a) supplying adequate watering sites, and if possible, moving them from time to time; (b) fencing or otherwise excluding animals from very erosive portions of the pasture; (c) avoiding placement of gates or lanes on erosive sites and moving them if an erosion problem occurs; (d) providing wind barriers to protect permanent lanes and watering sites; (e) applying corrective measures promptly to potentially erosive areas; (f) application of manure or commercial fertilizers where economics permit; (g) contour furrowing to reduce run-off and increase rainfall efficiency; and (h) rotational grazing to maintain desirable species and increase productivity.

Evaluation:

427. These principles of control apply to all rangelands, but the relative utility of each varies with the climate, soil, plant and land-use conditions. Usually not one but a combination of methods will be most effective and dependable.

Erosion control in coal mines

428. *Characteristics:*

Land-use system: Mining
Desertification potential: Slight
Production system: Semi-intensive
Cost: Medium
Level of technology: Simple to complex

Description:

429. One way of obtaining coal is the open-cut method (Fig. 36).

430. Reclamation of coal mine spoils in the United States by planting orchard and forest trees was started in 1918. Besides many timber species, fruit trees, shrubs and grasses have been planted with varying success. Planting trees on spoils has cost around US$55 per hectare. This cost is much greater than the cost of seeding grass and legumes. Returns from the trees begin only 10 or 12 years later when fence posts can be cut from black locusts. The main income will not materialize until pulpwood or saw lumber can be harvested, usually after 30–40 years. Where calcareous spoils are seeded to grass and legume mixtures, grazing will bring returns within a very few years.

431. In addition to the cost of grading are expenses for liming, fertilizing, seeding and fencing which may be considerable, especially on more acid spoils.

Evaluation:

432. Less labour is needed to produce a ton of coal by stripping than by underground mining. As a rule, strip mining recovers over 95 per cent of the coal in place while underground mining recovers only 50–60 per cent. Strip mining leaves the overburden in

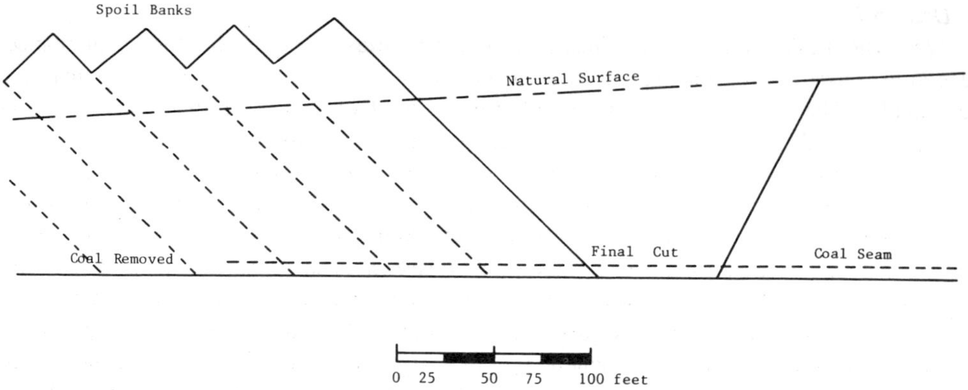

Fig. 36. Schematic cross-section of abandoned strip coal mine (Kohnke, 1950).

steep and jagged ridges that make the landscapes appear desolate and unproductive. The spoil banks affect plant growth and land use, and these areas can be reclaimed only by considering topography, erosion, soil, water, vegetation and wildlife. Erosion is intense on fresh spoils, as long as they are not covered by a mantle of vegetation. In the case of the United States, strip mining has added a great number of ponds and lakes to regions that previously had little open water. Most mine waters are high in sulphate. Very soon after land has been stripped for coal, vegetation of some sort appears on the spoil banks.

433. In view of the lower cost of producing coal by the strip method compared to underground mining, it seems possible that reclamation costs could be included in the price of strip-mined coal.

Soil stabilizers

434. *Characteristics.*

Land-use system: Settlements, transport routes, irrigation agriculture, mining, rainfed agriculture, rangeland, forestry.
Desertification potential: None
Production system: Intensive
Cost: High
Level of technology: Complex

Description:

435. To use oil, bitumen and latex to stick soil particles together and improve germination, a seed bed is prepared, seed and fertilizers are applied, and then the mixture is sprayed on top. In some cases, the seed and fertilizer are added to the sprayed mixture. Another technique is to combine the bulk of a straw mulch with the fixing power of bitumen using two nozzles, one spraying chopped straw blown by air, the other spraying a bitumen emulsion in water. Still another method uses glass fibre sprayed out as a continuous thread through a special high-pressure air nozzle, forming a light foam-like mulch over the soil surface. Glass fibre is more expensive than organic mulches but is not subject to decomposition. On steep slopes, mulching materials can be anchored by spot applications of bitumen. In Israel, Morin and Agassi (1973) have used the hydroseeding

method on slopes up to 45 per cent by applying as mulch material Silva fibre at 1000 kg/ha combined with a 400 kg/ha 1:20 dilution of Curasol and seed mixture, 30 kg/ha of Atriplex, 20 kg/ha of Polygonum and 30 kg/ha of Medicage sativa.

Evaluation:

436. The above spray-on methods are expensive and require a lot of labour.

Sand-dune stabilization

437. *Characteristics:*

Land-use system: Irrigation agriculture, settlement, transport routes, rainfed agriculture, rangeland, forestry, mining

Desertification potential: Medium

Production system: Extensive

Cost: Medium

Level of technology: Intermediate to complex

Description:

438. For afforestation purposes, it is necessary to consider the soil moisture régime through the year at different depths. Planting should be made as deep as possible according to tree tolerances. In general, the depth of planting is shallower in depressions (50–70 cm) and deeper on the high sites (dune slopes of 80–120 cm and high dunes of 150 cm). Cuttings or transplants are carefully covered with moist sand and watered with 5–10 litres. Seedlings should be cut 10 cm above soil level. This method has an efficiency of 65 per cent on top of dunes, and of 90 per cent in depressions (Costin *et al.*, 1974).

439. Soil moisture content must be taken into account for the type of cultivation (mixture, spacing) and method of planting. In general, soil moisture increases with depth, and sand dunes have the characteristic of preserving moisture because of low capillarity.

440. Fencing built with dead materials such as reeds or branches of shrubs and trees are cut into bundles about 70 cm long and buried 15–20 cm deep in ditches. Reed fences should be erected towards the prevailing wind as they represent a mechanical device for stopping sand movement and they help the natural regeneration of some grasses.

441. *Tamarix aphylla* is well adapted for afforestation. It is drought resistant, wind and salt tolerant and it grows rapidly in its early stages (in 100 days it can reach a height of 70–105 cm). It acts as a barrier against wind. According to Costin *et al.* (1974), afforestation requires well-developed stocks for transplanting of fresh cuttings of trees (no more than 2 days after collection and with a diameter of 1.5–2 cm. Other very good species which spread their crowns and enrich the soil with organic matter are: *Prosonis juliflora, Acacia tortilia, A. cyanophylla, Parkinsonia aculeata* (it can reach 4 m in height within 2 years after planting), *Calligonum comosum* and *Vernonia spp.* A mixture of Tamarix and these broad-leaved species in alternate rows is recommendable. Under prolonged drought periods, irrigation may be necessary.

442. As a rough approximation, 1 ha of sand dune fixation and afforestation with these intermediate technologies costs about 55 S.Y. Dinars. With mechanization for

digging the holes, cutting the reeds and sprinkler irrigation in the nursery, the cost could be reduced by 50 per cent.

443. In Tunisia, coastal sand dune fixation totals 30,000 ha, now completely stabilized and supporting a healthy coniferous forest (*Pinus pinea* and *P. pinaster).*

444. The cost of sand-dune fixation in India has been calculated as follows:

(a) Protection against biotic damage through fencing with angle iron posts and barbed wire. Rs 6 per running metre.
(b) Sand-dune fixation by establishing barriers from the crest down the slope of the dune, collecting local brushwood and burying it in lines to 2–5 m across the wind direction has a cost of Rs 100 per hectare.
(c) Establishment of vegetation to stabilize the sandy surface in the interspace between mulch barriers with tree and/or grasses. The brick planting technique uses seedlings raised on bricks (a mixture of clay, sand and manure and then sundried), and planting does not need watering. Seedlings of 0.8 to 1.2 m should be planted 50–80 cm deep and about 0.40 cm above the ground so that the root zone comes in direct with the moist sand. The cost of planting is Rs 300 per hectare.

445. Fixed sand dunes should be closed to cropping, grazing and other influences for at least 10 years. From the tenth year onwards, the trees should be lopped for top feeds and felling rotation may be commenced. On a block of 200 ha, the cost of replacement of casualties is Rs 80 to 100 per year.

2. Soil Moisture Conservation

Tied ridge cultivation

446. *Characteristics:*
Land-use system: Rainfed agriculture
Desertification potential: Slight
Production system: Extensive
Cost: Low
Level of technology: Simple

Description:

447. With this method, land is always maintained in ridges laid out across the slope. The furrows are dammed into a series of rectangular depressions. The ridges and the cross-ties can be made using handpower, animal traction or machinery. Several advantages of this method are: (a) it controls soil erosion; (b) it increases crop yield in most cases; (c) it improves water-use efficiency under rainfed agricultural conditions; (d) it maintains soil structure, and (e) controls weeds (FAO, 1966).

448. Hudson (1971) recommends a back-up system of conventional channel terraces to cope with flood run-off when exceptional storms cannot be contained. He adds that the method will work well on permeable soils but will be unsuccessful on grade with the ties lower than the ridges where failure and sudden run-off will be along each ridge and not down the slope.

Evaluation:

449. This is a very simple method to reduce water erosion and improve rainfall-use efficiency. The stability of agricultural systems depends on the bulk of the soil being held in place at all times. If rainfall infiltrates into the root zone, there will be no surface run-off and no water erosion. This method could be used by farmers in many developing countries employing no more than a shovel for tied ridges on row crops.

450. Disadvantages are if the soil becomes too wet and the depressions fill up and then overflow, the ridges may be broken, also the method is not well adapted to those areas with poorly distributed rainfall.

Range pitting, or interrupted contour furrows

451. *Characteristics:*

Land-use system: Rangeland
Desertification potential: Slight
Production system: Extensive
Cost: Low
Level of technology: Intermediate

Description:

452. This method consists in creating small depressions or pits in rangeland which catch and hold rainfall and run-off. Moisture held by the pits becomes available for seed germination and plant production. The water-holding capacity of the pits can be evaluated by placing a sheet of plastic material over them and measuring the amount of water held. Whitney *et al.* (1967) designed, constructed and evaluated several machines for digging pits and found that those made by the disk plough and the lister provided better conditions for introducing seed.

453. Another way of holding run-off and increasing water infiltration into the soil is by constructing pasture furrows with one pass of a single furrow plough. These will act as small open drains on level grade, following the contour and fairly close together (Hudson, 1971).

454. Fitzgerald, mentioned by Heady (1975), describes a method which uses an opposed-disk plough with a central deep reaper and attached seeder box which makes interrupted contour furrows that do not channel water. Pits and ploughed strips have retained seed and water-holding characteristics for 5 years and have helped to rehabilitate bare and badly eroded flatlands in the Ord River catchment in Australia. According to Barnes (1952), pitting can be done at a rate of approximately 2 ha per hour, using a small tractor at speeds of 8–10 km per hour.

Evaluation:

455. This method, extensively used in the Great Plains of the United States, was developed in the 1930s and is restricted to areas with less than 20 per cent slope. As pits fill in after a few years, their effect on increasing productivity should be evaluated under different conditions before they are recommended. Yet this is an established method to reduce excessive run-off, to encourage water infiltration and to increase plant production on native range.

The fallow system

456. *Characteristics:*

Land-use system: Irrigation agriculture, rainfed agriculture
Desertification potential: Slight
Production system: Semi-intensive
Cost: Medium
Level of technology: Intermediate

Description:

457. The amount of moisture stored during fallowing between crops is often not more than 20 per cent of the rain received, which together with that falling during the crop season, usually assures a satisfactory harvest. In addition, fallowing enhances the nutrient content of the soil. This method may be applied in areas with less than 400 mm of rain per year.

458. There are three classes of fallow: (a) bar fallow, where the soil is inverted with a disk or mouldboard plough and very little residue is left on the surface; (b) mixed fallow, prepared with a one-way disk plough or disk tiller, in which the soil is partially inverted and residue is incorporated into the soil and left on the surface; and, (c) stubble mulch fallow, where a maximum amount of residue is left on the surface, is prepared with surface tillage implements such as chisel ploughs, sweep implements and rod weeders.

Evaluation:

459. In North America it is common to find a strip-cropping system which consists of alternate strips of crop and fallow, usually 15–75 m wide, running either on the contour or perpendicular to the prevailing wind for wind erosion control. In Australia a cover crop is planted after the harvest of cereals, which is either grazed or harvested for hay or is left on the land to provide protection or to improve productivity. In developing countries this system is not extensively used in spite of its advantages.

3. Salinity Control

Soil salinity control

460. *Characteristics:*

Land-use system: Irrigation agriculture, rainfed agriculture
Desertification potential: Slight
Production system: Semi-intensive
Cost: High
Level of technology: Complex

Description:

461. In order to control salinity, it is very important to know: (a) salt balances and leaching requirements of the area in relation to the crops to be grown; (b) amount and

quality of available water; (c) soil permeability; (d) drainage characteristics; (e) cropping pattern; and, (f) design of irrigation projects (Paliwal, 1972). Salinity can be prevented by the maintenance of a satisfactory salt balance.

462. A low soil permeability is the greatest handicap to reducing soil salinity. When the rate of evaporation at the soil surface is greater than the soil permeability, salt content will increase with time. If low soil permeability is due to compacted layers, they must be broken by ploughing. If low permeability is due to high exchangeable sodium content, it can be improved by adding gypsum or any other cheap source of soluble calcium.

463. Planting technique is also very important, because if moisture content is reduced, salt concentration becomes high near the seed and may be detrimental to its germination and further development. In flooded or basin irrigated soils, this phenomenon is not as harmful as it is when crops are grown in furrows. It is well known that salt content on ridges is high as compared to slopes, so the adverse effect of salinity can be alleviated to some extent by planting on sloping beds.

464. The control of soil salinity is achieved by leaching the salts below the root zone in the soil profile which varies from 50 to 250 cm deep depending on crop type. Sodium soils require additives such as gypsum which is extensively applied in Colombia, Mexico, India, the USSR, Hungary, Israel, Turkey, and Australia in rates which vary from 2 to 50 tonnes per hectare. Where a natural source of soluble calcium is available in the soil profile, no additives are required.

465. Some countries use grasses and legumes (Australia, Iraq, Israel) and cereals (rice in China, Turkey and Pakistan) to enhance the reclamation process. If the initial level of salinization is high, it will be necessary to pre-leach the topsoil before tolerant plants can be established. In Pakistan, where heavy soil textures, high sodium content and low infiltration velocity were present, the use of a legume (*Sesbania aculeata*) helped in favouring infiltration and the leaching operations. Greece and the Republic of China also use this plant. In Australia (Swan Hill Irrigation Research Farm) cultivation, cropping and mulch helped to hasten soil salinity control. Gypsum applications may be combined with manure and compost.

466. Leaching displaces salts from the root zone with water of lower salt concentration and represents the only way of reclaiming soils which have been under improper management in terms of irrigation and drainage. Methods of leaching are flooding (continuous or intermittent) and sprinkling. The Tolima project in Colombia had 7000 ha of sodium soils (ESP = 34–77). By adding 2 tonnes per hectare of sulphur in two stages and leaching, rice yields were doubled to tripled at a cost of US$110 per hectare and the ESP was reduced to 6–10.

467. In order to obtain a good salt balance near the root zone, an efficient drainage system is essential, consisting of either open drains or tile lines, according to the local conditions, to drain out water until the watertable is below the root zone. In areas where saline water or saline soil are used for agriculture, more salt-tolerant crop varieties must be selected because they have low leaching requirements. By this technique, the amount of water required for irrigation is reduced. When a crop with low evapotranspiration is planted, this also reduces leaching requirements.

468. The objectives of reclamation are to reduce the salt concentration at the roots to a level compatible with crop growth and to reduce the exchangeable sodium level where necessary. Such reclamation can be achieved by physical amelioration, biological

amelioration, chemical amelioration, hydrotechnical amelioration, and the synergic effect of combinations of reclamation methods (Hutchinson, 1973). These methods include:

(a) Deep ploughing is one of the physical amelioration methods which is more effective on stratified soils having impermeable layers. Subsoiling is another physical method, and it consists of pulling vertical strips of steel or iron through the soil to open channels. Sanding, which means to incorporate sand into fine-textured soil, improves its permeability which in turn facilitates the leaching of salts. Also profile inversion can be suited to some cases, and this is accomplished by removing the surface soil, deep-ploughing the subsoil and substratum and finally replacing the surface soil.
(b) Biological amelioration acts by improving soil permeability and by releasing carbon dioxide during respiration and decomposition. Reclamation of soils can be enhanced by the presence of plants. The beneficial effects of plants may be due to the rapid removal of entrapped air by the roots, thus improving soil permeability, or to the increased dissolution of calcium carbonate in the presence of CO_2 evolved from plant roots, and the addition of organic matter.
(c) Chemical methods use chemicals to neutralize soil reactions with gypsum and sulphur the most common additives for the reclamation of alkaline soils.
(d) Hydrotechnical amelioration consists of leaching and drainage which are basic requirements for the successful reclamation of saline and alkaline soils. A drainage system installed in saline soils must desalinize the topsoil layer and the upper subsoil and water-bearing horizons.

469. Soil reclamation can be speeded up by a combination of various methods, the choice depending on basic principles. Saline soils require drainage and leaching with a sufficient amount of water. The quality of water required for leaching and the degree to which the soil will be leached depend on the initial soil salinity, the quality of the leaching water, the method of leaching, the soil depth to be leached, and the soil moisture and physical characteristics.

470. Alkaline soils require neutralization of free sodium and replacement of exchangeable sodium with calcium (gypsum or acidifying agents), then the removal of sodium salt which results from the reaction. Another method of reclaiming alkaline soils is the successive dilutation of initially high-salt water (seawater). By starting with water of high electrolytic concentration, good soil permeability can be maintained. An equivalent dilution method is based on the principle that divalent cations tend to replace monovalent cations as a result of dilution of the soil solution (Reeve and Bower, 1960).

471. In the reclamation of saline-alkaline soils, calcium salts or substances that promote the solution of calcium carbonate are added prior to leaching or along with irrigation water to promote removal of exchangeable sodium. The soluble calcium prevents the soil from becoming non-saline alkaline in the first place, and therefore deterioration of the structure is prevented.

Evaluation:

472. In order to keep salinity under a safe limit in relation to a particular crop variety, it is essential to irrigate more frequently and with more water to satisfy leaching requirements and evapotranspiration needs. For success these factors must be also taken

into account: soil permeability, drainage conditions and salt balance behaviour of the crop, and the socio-economic condition of the villagers.

473. Prevention techniques are much more practical and economical than reclamation measures in controlling soil salinity.

4. Drainage

Drainage methods

474. *Characteristics:*

Land-use system: Irrigation agriculture, rainfed agriculture, rangeland
Desertification potential: Slight
Production system: Semi-intensive
Cost: High
Level of technology: Simple to complex

Description:

475. Under irrigation, drainage is required to prevent waterlogging and the concentration of injurious salts in the soil. Seepage from irrigation canals and percolation of irrigation water from the field contribute large quantities of water to the subsoil. The United States Bureau of Reclamation (1952) reported that approximately 25 per cent of the water entering unlined canals and laterals is lost by seepage before it reaches the farmer's field. Drainage as a complement of irrigation has been understood only in the present century.

476. Percolation losses include canal seepage and infiltration losses in the field. Field percolation losses occur in moderately permeable soils. With gravity irrigation, they will come to 20 per cent to 30 per cent of the field supply as an average over the year, 30 per cent to 50 per cent during the wet season, and 10 per cent to 20 per cent in the dry season. Under sprinkler irrigation, percolation losses amount to 10 per cent to 15 per cent of the water supplied (Hutchinson, 1973).

477. The drainage needed for salinity control is defined by the concept of salt and water balance. The salt content of the soil moisture in the unsaturated root zone, the leaching coefficient and the fluctuations of monthly and yearly soil moisture are decisive for the design of the drainage system.

478. Drainage can be divided into two types: (a) surface drainage which helps to remove excess water from irrigation and precipitation, and (b) subsurface drainage which helps to control the groundwater level and salts (Hutchinson, 1973). Surface drainage is accomplished by open ditches and lateral drains. Subsurface drainage is accomplished by a system of open ditches and buried tube drains into which water seeps by gravity, and also by pumping from wells to lower the watertable. Channels can be made of pipe or tile, and channels molded into the soil without lining or other support are known as mole drains. The main advantage of subsurface drainage is that it does not occupy any land surface.

479. Open channels or ditches are the oldest method of drainage. They can be recommended for the quick removal of large quantities of water and are well suited to

intercepting runoff from high areas. Their main disadvantage is that they occupy land that could otherwise be cropped.

480. Open ditches can carry water from the surface and subsurface drainage of land areas. They usually present maintenance problems that interfere with agricultural operations. Their advantages are low cost and in many areas they represent the only possibility.

481. Surface field drains are well suited to humid areas mainly in flat land with impermeable subsoil and where tube drainage would be uneconomical. There are three types: graded depressions, smoothed depressions and parallel field drains recommended for flat fields. Where dead furrows are left by ploughing lands at the same location for several years, the drainage system is called bedding.

482. Subsurface types are: pipe or tube drainage, mole drainage, combined mole-pipe drainage, and subsoiling.

483. Pipe drainage has a high installation cost, but the amount of work involved in keeping the system in good condition is less than with any other form of drainage, and the system can be used in any type of soil. The following types of tube drainage are common: the random system, used where the topography is undulating or where the field exhibits isolated wet areas, the herringbone system which is used where the main is in the direction of greatest slope with good grades for laterals; the gridiron system which is suited to flat fields; and, interception drains, which collect seepage moving down a slope.

484. The most common materials used for the pipe lines are clay, concrete and plastic. Clay pipes are available in lengths of about 30 cm and have a range of internal diameters. Concrete pipes can be used if clay pipes are not available or if greater diameters are required (more than 15 or 20 cm). A wide variety of plastic pipes have been manufactured, but only a limited number have found application in practice (Cavelaars, 1974).

485. The mole drainage method consists of forming the drain without artificial lining and creating cracks or fissures in the soil which form secondary channels through which the water travels to the mole channels. Its use is restricted to soil with enough clay to retain the shape of the channel. This method is cheap, simple and particularly useful for soils with low permeability. The primary aim of such a system is to remove excess water from the surface or topsoil.

486. With a combination mole-pipe drainage it is possible to obtain the advantages of both systems. Improvement of water percolation through the soil resulting from the mole drainage lowers the installation costs of the pipe drains because the pipes can be laid farther apart.

487. If thin impermeable layers occur within the drainage depth, either due to natural phenomenon or to artificial causes such as continued ploughing to the same depth, they can be broken up by some form of subsoiling machine.

488. Other drainage systems which are useful only under certain conditions are the interceptor systems and pumping for drainage. In the first, an interceptor drain is installed to collect lateral or horizontal flow which comes from known upslope, thereby preventing water from reaching an area to be protected. Pumping groundwater is an effective method for lowering a high groundwater table, in order to reduce the salinity hazard in irrigated areas. Pumping vertical wells can be divided into: (a) pump wells (discharge wells) from which water is pumped, often from great depths; (b) relief wells (flowing wells) which penetrate into artesian beds and remove water by free flow, and (c) inverted wells

(recharge wells) which consist of a vertical hole penetrating the underlying porous rock or other permeable material into which water is drained away (Hutchinson, 1973).

489. Chemical methods of drainage are applied mainly to heavy soils by using polystyrene, peat, or other bulky materials, and their problems are largely those of cost (Trafford, 1971).

490. Surface drainage must be used on poor-quality land with low agricultural production. Subsurface drainage must be used in areas where excessive water in the soil is the main problem (Theobald, 1968).

Evaluation:

491. It is estimated (Israelsen and Hansen, 1965) that from 20 per cent to 30 per cent of the irrigated lands in arid and semi-arid regions need improved drainage for permanent and productive agriculture. Estimates indicate that about 80 million ha of land are at present protected by drainage in different parts of the world. According to Theobald (1968), zones suffering from soil damage through waterlogging and salinization represent 50 per cent of the total area under irrigation in Iraq, 23 per cent in Pakistan, 50 per cent in the valley of the Euphrates in Syria, and 30 per cent in Egypt.

492. A number of factors must be considered in selecting a drainage system, some of which are: the nature and size of the drainage materials, the physical condition of the soil, particular cause of the drainage problem, topography, costs, availability of equipment and technical advice, quantity and pattern of rainfall, continuity of strata, hydraulic conductivity and porosity of the various strata.

C. PLANTS

Multiple cropping

493. *Characteristics:*

Land-use system: Irrigation agriculture, rainfed agriculture
Desertification potential: Medium
Production system: Semi-intensive
Cost: Medium
Level of technology: Simple to complex

Description:

494. Multiple cropping. This will generally refer to a regular sequence of more than one annual food, fodder or industrial crop planted and harvested in the course of a 12-month period and grown in basically pure stands on the same piece land under a system of permanent agriculture. There are several types:

495. Green manure crops. These crops are legumes which are grown for their soil-improving qualities or as a cover crop and which are ploughed under at the end of the growing season. They are not harvested and do not by themselves produce a food or feed crop.

496. Mixed farming. This practice can involve: (a) the growing of annual crops under or with perennial crops (shrubs or trees); and (b) the growing of crops together with the

raising of livestock. Perennial crops are excluded because they are not planted and harvested in the course of one year.

497. Intercropping. In many tropical areas, more than one crop – in some cases dozens – are planted on a given piece of land. The various crops may be planted at the same or at different times. Usually, however, they are not planted in sequence. Where intercropping involves only few crops, several variants arise which could be counted as forms of multiple cropping.

Evaluation:

498. Some of the disadvantages of multiple cropping are: (a) it is difficult to determine and apply fertilization formulae for each individual crop; (b) it is not easy to eliminate weeds by hand cultivation when there are several mixed crops; (c) control of pests is more difficult; and (d) this is a more intensive system of agriculture and requires more inputs and better training of farmers.

499. Some of the advantages are that under rainfed agriculture the risks are lower, and that this method provides an opportunity to produce at least one or two basic crops.

Vertical nomadism

500. *Characteristics:*

Land-use system: Rangeland
Desertification potential: High
Production system: Extensive
Cost: Low
Level of technology: Simple to complex

Description:

501. In general, densities of population are very low in nomadic herding regions (0.5 per km^2 in the Mongolian steppe, 0.5 per km^2 in the Sahara, 3 per km^2 in the Sahel and 17 per km^2 in East Africa).

502. Semi-nomads are more common than true nomads. They sometimes exchange animal products for grain with sedentary oasis dwellers.

503. Grazing lands in nomadic herding regions usually have low productivity and a very large area is needed to sustain the animals, but this varies with the precipitation available as can be seen in Table 11.

504. In the Middle East, a migratory unit consists of five or six families. For minimum subsistence, each family will require 25–60 goats and sheep or 20–25 camels. Cattle need

TABLE 11. *Area required to feed one livestock unit (1 livestock unit = 1 cow or 7 sheep)*

Annual rainfall (mm)	No. of hectares per livestock unit
50–100	50 or more
200–400	10–15
400–600	6–12

better pastures and more water to drink, and they are dominant in the areas of the Sahel where annual rainfall is at least 800 mm.

Evaluation:

505. As long as an equilibrium is established between plant productivity and animal pressure the system of production or the land-use system will not deteriorate.

Revegetation for fodder production

506. *Characteristics:*

Land-use system: Rangeland
Desertification potential: None
Production system: Extensive
Cost: Medium
Level of technology: Simple to complex

Description:

507. Species selection and seeding operations in pastoral lands should be conducted in accordance with: adaptability to soil and climate; vegetative cycle; expected yields; physiological response to grazing; nutritive value; palatability; compatibility in mixture of species; and forage production to meet seasonal deficiencies.

508. Le Houérou (1975) suggests that between the 350–400 mm isohyets reseeding practices for the improvement of rangelands can be successful with these species: *Medicago sativa, Oryzopsis holciformis, O. miliacea, Cenchrus ciliaris, Lotus creaticus, Sangrisorba minor, Hudysarum carnosum, Lolium rigidum, Phalaris truncata, Agropyron elongatum, Ehrharta calcycina,* and *Trifolium subterraneum*. In places with less than 350 mm it is necessary to explore the possibilities for reseeding.

Evaluation:

509. A properly balanced relationship introducing shrubs-natural vegetation-grazing animals (domesticated or wildlife) should be reached and maintained in order to avoid land degradation.

510. Where there is no possibility of irrigation, improved land use is necessary for more rational exploitation of the local potential under the given environmental conditions. The natural vegetation of arid and semi-arid zones consists mainly of annual species, perennial and drought-resistant shrubs which could mean a stable element and a supply source of forage both in critical periods of the annual grazing cycle and during drought periods.

511. Experimental results in an area of 160–200 mm of annual rainfall showed the good palatability of *Atriplex numularia, A. breweri, Cassia aturrii* and *Acacia victoriae* during the winter months. Palatability must be evaluated as a function of the chronological priority of species in being eaten, or the time spent by animals in grazing them and of the quantity of matter consumed. Generally the degree of palatability varies during the year (Forti, 1971).

Grazing systems

512. *Characteristics:*

Land-use system: Rangeland
Desertification potential: High
Production system: Extensive
Cost: Low to medium
Level of technology: Simple to complex

Description:

513. The most effective organizations for grazing systems are those with defined land rights and which constitute discrete management units. The greatest concentration to date is found in East Africa where there is a variety of ranching companies and cooperatives, group ranches, grazing associations and individual ranches, each designed for different conditions.

514. South Australians have developed advanced systems of agriculture adapted to the climate and soils of that Mediterranean zone.

515. Low intensity grazing system. The pastoral zone receives less than 150 mm of rain per year. It covers 823,200,000 ha and constitutes 83 per cent of South Australia. Sheep are protected from wild dogs (dingo) with 8000 km of fencing run across the state in an unbroken barrier of wire netting 1.8 m high. Stocking rates vary from 5 to 25 animals per 100 ha.

516. Most of the pastoral zone is administered by the Pastoral Board with the state owning the land under lease. The lessee must improve the land, must prevent overgrazing and must provide fencing. With this legislation and organization, land deterioration is prevented (Webber *et al.*, 1976).

517. An example of a typical station in the South Australian pastoral zone:

Rainfall:	210 mm
Area:	13,300 ha
Tenure:	Perpetual lease (99 years)
Subdivisions:	15 paddocks
Water:	3 bores, 10 earth tanks
Livestock:	4500 sheep
Improvements:	Owner's house, shearing shed, yards
Machinery and plant:	One truck, one tractor, one utility van, two motor cycles, earth-moving equipment, shearing plant and crutching plant.
Labour:	Owner-operator and son
Prevent stocking:	One sheep per 3 ha

518. Provision of enough watering points reduces two problems – overgrazing around the water supply and undergrazing in large areas away from the water.

519. Other complementary techniques which control the rate of stocking are: the control of rabbits by the virus disease myxomitosis, an operation to control blow-fly damage, using motor cycles for mustering and moving the sheep, and soil and water conservation practices.

520. Among the disadvantages of this system, the farmer is dependent on one product, wool.

521. The ley farming system in the cereal zone (250–500 mm). Ley farming is an integrated system of cereal (wheat, barley and oats) and livestock production (sheep and cattle). The legume pastures integrate the system by providing nitrogen (60–70 kg/ha in one season) for the crops and high-quality feed for livestock.

522. Several factors out of which the ley farming system developed were the availability of tractor power to plough and sow in the autumn, the use of seed of pasture legumes, application of superphosphate, contour banks to control soil erosion to reduce run-off and make maximum use of rainfall. Ley farming is an advanced, integrated system. It was introduced in the 1930s and by 1960 it was an established practice in South Australia.

523. Example of a typical farm in the cereal zone of Australia (375–400 mm):

Area	600 ha
Cropping:	Wheat 160 ha, 100 ha barley, and 20 ha oats
Pasture:	280 ha
Subdivision:	15 main paddocks, 3 small paddocks, 40 km of fencing
Water:	1 pint point per paddock, 5 km of piping
Livestock:	1000 sheep comprised of 500 Merino ewes, 300 wethers, 200 young sheep, 25 beef cows
Improvements:	Owner's house, shearing shed, implements shed, grain silos
Machinery and Plant:	2 tractors (50 and 80 hp), 1 header/harvester, 1 cultivator (29 hoes), 1 disk plough (18 disk), 1 combine-seeder (24 row), harrows, wideline scarifer, slasher and hay rake, bulk grain equipment, truck, utility and workshop equipment.
Labour:	1 owner/operator, 3–4 months casual labour for baling, grain carting and contract shearing.

524. The average production of this farm runs at 6000 kg of wool, 6000 kg of meat and 420 tonnes of cereal grains, with emphasis on the use of medio pasture (*Medicago littoralis*) and the addition of super-phosphate. Due to fluctuations and low prices of wool, farmers buy young heifer calves, often less than 1 week old. In this way they build up their herds of beef cattle slowly but inexpensively. Fallows are prepared in August–September when grassy pastures are flowering. These areas are sown to crop in the following May. Unfortunately, this modern technology cannot be applied in those countries with high animal population densities.

525. Other biological recovery techniques include:

526. Fencing. In South Australia, the commonest method of fencing consists of post and wire construction for sheep. Cattle require better and stronger fences incorporating barbed wire.

527. Management of pastures. The technical approach to management of pastures includes rotational grazing, deferred grazing and combinations of these systems. Other techniques include control of weeds and posts, livestock husbandry, use of fertilizers.

528. Legume pastures. The ley farming system was designed to obtain maximum grazing for livestock throughout the year and to ensure enough seed production so that the pasture regenerates well in the following year. The sowing rates are 10–12 kg/ha, at about 1 cm in heavy-textured soils and 4 cm deep on light-textured soils with adequate control of weeds and avoidance of heavy grazing soon after germination, allowing the vegetation of flower. Hay made from legume pasture helps the farmer to feed his animals over the dry summer and fall months.

Evaluation:

529. Legislation is needed to prevent overgrazing. Education and training through extension methods are also basic to assist graziers, to apply new technologies and adopt better forms of organization. Training should be based on existing facilities.

530. The question of which combination of species is likely to make the most efficient use of pasture and other resources should always receive the most careful consideration. Farmer pastoralists whose diet is predominantly milk can substitute grain foods for a proportion of the milk diet, which offers the prospect of reducing overgrazing without reducing the number of people (FAO, 1974).

531. The system used by Australians is stable, is increasing the level of soil fertility, and has controlled soil erosion. The policies that have been developed for these regions lead the world in the management of stable livestock production in arid area through integrated farming systems (Webber *et al.*, 1976).

Afforestation:

532. *Characteristics:*

Land-use: Irrigation agriculture, rainfed agriculture, rangeland, forestry, settlements, transport routes, mining
Desertification potential: None
Production system: Extensive
Cost: Low to high
Level of technology: Intermediate

Description:

533. Afforestation programme with the planting of shrubs and trees tend to stabilize the land, but must include a careful appraisal of environmental conditions.

534. Afforestation for fuel in rural areas is needed for the cutting and gathering of wood. Several governments are providing small cookers which use petroleum in order to reduce this problem. Another approach is rural afforestation which consists in planting fast-growing trees.

535. Le Houérou (1975) describes the following species usable for firewood:

536. In the Mediterranean arid zone:

Tamarisks: *Tamarix aphylla, T. stricta*, *T. nilotica.*

Eucalyptus or Gumtrees: *Eucalyptus microtheca, E. oleosa, E. brockwayi, E. torquata, E. salubris, E. dunsasii, E. sargentii, E. gongylocarpa, E. woodwardi, E. interexta, E. sideroxylon, E. gracilis, E. occidentalis.*

Acacias: *Acacia farnesiana, A. senegal, A. vitorias, A. peuce, A. ligulata, A. cyanophylla, A. tortillis, A. raddiana.*

Organtrees: *Argania sideroxylon.*

Pinetrees: *Pinus halepensis.*

Cypresses: *Cupressus sempervirens, C. arizonica.*

Casuarinas: *Casuarina equisetifolia, C. tenuissima, C. cunning-hamiana, C. stricta.*

Russian olives: *Eleagnus angustifolius.*

Sohinus: *Schinus moole, Sch. terebinthifolia.*

Parkinsonias: *Parkinsonia aculeata.*

537. In the Tropical arid zone:

Tamarisks: *Tamarix aphylla, T. molotica, T. stricta.*

Gumtrees: *Eucalyptus microtheca, E. crebra, E. rudis, E. camaldulensis, E. tereticornis.*

Acacias: *Acacia albida, A. senegal.*

Mosquite: *Prosopis juliflora, P. africana, P. chilensis, P. cineraria.*

The Neem: *Azadirachta indica.*

538. These species have a potential wood production of 10–15 m^3/ha/yr, and can be used as windbreaks and shelterbelts. Kassas (1966) mentions that *Prosopis chilensis* was introduced into the Sudan in 1928. It is capable of surviving in low-rainfall environments, produces valuable materials, provides effective protection against soil erosion and is also a good shelterbelt. This plant (bush or tree), once established, produces good fodder and it can withstand repeated cutting.

539. In the surroundings of Beersheba, Israel, with an annual rainfall of about 200 mm, the afforestation method consists of planting trees in small patches, locally called *liman*, located in lowlands where run-off concentrates and is storaged for tree development. The advantages of these limans is that they utilize excessive run-off and provide shade for animal life and recreational sites for people. The difficulty is to establish the correct relationship between the run-off areas and the planted areas. It is not easy to establish the correct tree density and some of the trees may die because of scarcity of water (see Fig. 37).

Shelterbelts and windbreaks

540. *Characteristics:*

Land-use system: Irrigation agriculture, rainfed agriculture, rangeland, settlements, transport routes, mining
Desertification potential: None
Production system: Extensive
Cost: Low to High
Level of technology: Intermediate to complex

Description:

541. There is a great diversity of methods for planting shelterbelts under different ecological conditions. In general these windbelts have these purposes: (a) to protect fields,

Fig. 37. Eucalyptus plantation 8 years old in small patches (usually 1500 m²) near Beersheba, Israel.

rangelands, highways and settlements against high winds, dust storms and sand-dune encroachment; (b) to produce fuel wood and small-sized timber for local consumption; (c) to increase production by providing a shelter, food supply and shade for domestic animals and wildlife, (d) to beautify the landscape.

542. For those areas with high wind-speed and varied directions as in the People's Democratic Republic of Yemen, a chessboard pattern of windbelts is recommended. According to Armand *et al.* cited by Coetin (1974), the distance of highest effectiveness of shelterbelts is about 20 times the tree height; on the leeward side and 5 times on the windward side. For the People's Democratic Republic of Yemen, the calculated distance for trees 15 m in height would be 375 m, but because of strong winds and light soils, distances should be reduced to 200–300 m for cultivated crops and 500–600 m between secondary shelterbelts.

543. The choice of species is a function of the soil and climatic conditions. *Casuarina equisetifolia* is an evergreen tree, tall, with a very straight stem and conical crown. It can grow 3 m in the first year and 7.5 m after 3 years. *Azadirachta indica* is tall and fast growing with a height of more than 15 m. It is a drought-resistant tree and grows under dry-farming conditions where the groundwater table is not too deep. *Parkinsonia aculeata* is adapted to irrigated and non-irrigated areas. It is an evergreen decorative and very fast-growing in the first years. *Zizyphus jujuba* and *Acacia nilotica var indica* are palatable trees and appreciated as fodder sources.

544. The planting of shelterbelts should be made by transplants grown in the polythene bags for 6 to 8 months. For *Parkinsonia*, 3 months may be enough. Tamarisk can be planted by using cuttings. Planting is usually made during the winter or early in the spring in pits of 40 × 40 cm on channels. Irrigation can be required in the first years. Once the shelterbelts are established, proper management is mandatory, and pruning of the lower branches should start in the early stages. To ensure permanent protection, an average of 20 years rotation of cutting should be considered. One of the more economic and efficient forms of shelterbelts in arid areas is a narrow form of two rows in fertile soils or of three to five rows on sandy soils (see Fig. 38).

Evaluation:

545. Shelterbelts usually improve environmental conditions through wind-speed reduction, increasing the relative humidity and reducing crop transpiration and evaporation. Cosin *et al.* (1974) report increases in yields of 10–20 per cent for cotton, 15 to 50 per cent for cereals, 160 per cent for apples, 60 per cent for tomatoes, 115 per cent for carrots, and 60–70 per cent for pastures. Animal production is also increased and wool production per head may be greater by 20 per cent.

546. One of the disadvantages of shelterbelts in crop areas is the competition of the roots of the trees for water and light. Sometimes they act as hosts for insects, but in general they produce more benefits than losses.

547. Tree planting programmes and the subsequent conservation and management of trees require a permanent increase in technical staff and training at various levels.

Reforestation:

548. *Characteristics:*

Land-use system: Forestry

Fig. 38. Shelterbelt of tamarisks in Dakar, Senegal (750 mm annual rainfall).

Hall, A.E. / Cannell, G.H., Lawton, H.W.

Agriculture in Semi-arid Environments

Ecological Studies 34

Berlin, Heidelberg, New York 1979

Desertification potential: Slight
Production system: Extensive
Cost: Medium
Level of technology: Simple to complex

Description:

549. Reforestation practices are important in providing a more healthy and productive biosphere. According to Kaul (1968), in Jodhpur, India, with a mean annual rainfall of 366 mm and a deep sandy soil, *Eucalyptus melanophyloia* has shown the greatest promise. *Acacia tortilis* proved to be the most promising introduction. Eight-year-old trees at Jodhpur and 4-year-old trees at Pali recorded 100 per cent survival, and their performance on shifting dunes as well as on skeletal soils has been very good. In Pakistan, the exotic species Australian *Acacia* spp, and *Eucalyptus* spp, *Prospis juliflora* (Mexico) and *P. glandulosa* have done best compared to the indigenous species.

550. A reforestation programme should include: (a) selection of proper species; (b) seedling production; (c) soil and water conservation techniques; (d) protection against disease and fire and massive participation of citizens and governments in the planting and maintenance of trees and shrubs (see Figs. 39 and 40).

Evaluation:

551. The best protection against land degradation and the best stabilization of the soil is provided by applying simple and appropriate technologies of soil and water conservation complemented with afforestation and reforestation, which also help to control sedimentation and avoid watershed deterioration.

552. Reforestation practices are very important as a source of wood for multiple purposes, as effective measures to control sediments and as an important component of a healthy biosphere.

Fencing

553. *Characteristics:*

Land-use system: Rangeland, forestry, rainfed agriculture, irrigation agriculture, transport routes, mining, settlements
Desertification potential: Slight
Production system: Extensive
Cost: High
Level of technology: Simple to complex

Description:

554. *Acacia albida* has been used as a tree cover and has great value as a soil stabilizer against erosion. It provides considerable quantities of organic matter, nitrogen and phosphorus. Its pods and leaves provide cattle fodder during periods of scarcity.

555. As an example of biological recovery on deep sandy soils (in Mauritania north of Adrar) where the average annual precipitation is about 80 mm, some enclosures were set up in the 1950s and are still in excellent condition (Monod, 1951; Naegele, 1959; Adam, 1957, cited by Le Houéron, 1975). Another example cited by Le Houérou (1975)

Fig. 39. Rural reforestation programme with the enthusiastic participation of villagers (men, women, children) in San Bartolo Ozacalpan, Mexico, in a semi-arid area with 500 mm of annual precipitation.

Fig. 40. Reforestation programme of the Mexican government to reclaim deteriorated areas and control sediments in an upland watershed (650 mm of annual precipitation).

is to be found in Southern Tunisia (Nefta) where a site with 80 mm rainfall has been protected for 79 years with shrubby regetative cover 2–3 m high.

Evaluation:

556. Many enclosures of several hundreds to several thousands of hectares have been established in different arid and semi-arid areas to test their effect on biological recovery.

557. Bhimaya *et al.* (1966) evaluated the costs of a different method of fencing, as shown in Table 12. They conclude that although the initial cost of erection of angle iron posts with barbed wire fencing is rather high, it was the most efficient.

TABLE 12. *Cost of erection and maintenance of different types of fencing on the range (Bhimaya et al., 1966), India*

Type of fencing	Number of areas studied	Erection cost, Rs. per running metre	Maintenance cost, Rs, per metre,
Angle iron posts with barbed wire	12	1.64	0.03
Angle iron posts with woven wire	3	2.18	0.04
Stone posts with barbed wire	5	2.22	0.05
Wooden posts with barbed wire	2	1.70	0.06
Ditch and mound	18	1.00	0.24
Corewall and thorns	10	0.58	0.15
Stone wall	1	0.91	0.41
Live hedge with cactus	1	0.12	0.12

Fire control

558. *Characteristics:*

Land-use system: Rangeland, forestry, rainfed agriculture, irrigation agriculture, settlements

Desertification potential: Slight

Production system: Semi-intensive

Cost: Low to high

Level of technology: Simple to complex

Description:

559. According to FAO (1968), there are three methods of fire control: (a) by using water, sand or soil; (b) by constructing firebreaks and (c) by the use of backfires. In those places where there are flash fuels such as dry grasses, the three methods should be combined. If fire occurs on the soil surface and on a small scale, it can be controlled by using water and a firebreak parallel to the fireline.

560. The following steps may be useful for a fire-control plan: (a) estimation of the magnitude of the problem and its spreading potential; (b) selection of the place to start the action; (c) location of the control line; (d) construction of the fireline; (e) using backfire if necessary; (f) mopping up to extinguish residual fires and (g) a patrol service.

Evaluation:

561. Fire control needs the active participation of citizens, which may be achieved through an alarm system, with weather broadcasting stations located in the forests, with lookout towers, telephone networks and air patrols. Economic losses and environmental deterioration may be of immense magnitude if prevention measures are not taken.

BIBLIOGRAPHY

Anaya, M. G., Tovar, S. J. and Macias, Y. A. (1976) *Métodos de captación de lluvia para zonas agrícolas con temporal deficiente*, Colegio de Postgraduados, Chapingo, Mexico.

Arnon, I. (1972) *Crop Production in Dry Regions*, Vol. 1, *Background and Principles*, Leonard Hill, London, pp. 217–224.

Baudelaire, J. P. (1972) Water for livestock in semi-arid zones, *World Animal Review* No. 3, FAO, pp. 1–9.

Barnes, O. K. (1952) *Pitting and Other Treatments on Native Range*, Wyoming Agricultural Experimental Station.

Barry, R. G. (1969) The world hydrological cycle, in *Water, Earth, and Man*. R. J. Chorley ed., Methuen, pp. 11–31.

Bhimaya, C. P. *et al.* (1966) *The Economics and Efficiency of Different Types of Fencing for Soil Conservation in Rajasthan*, Central-Arid Zone Research Institute, Jodhpur, India.

Bishop, A. A., Jensen, M. E. and Hall, W. A. (1967) Surface irrigation systems, in R. M. Hagan, H. R. Haise and T. W. Edminster, eds.; *Irrigation of Agricultural Lands, Agronomy*, No. 11, Madison, Wisconsin, pp. 865–884.

Booker, L. J. (1974) *Surface Irrigation*, FAO Agricultural Development Paper No. 95, Rome.

Borgstrom, G. (1973) *World Food Resources*, Kingswood House-Heath & Reach, Guildford and London.

Burdass, W. J. (1974) Water harvesting for livestock in Western Australia, in *Proceedings of the Water Harvesting Symposium*, Phoenix, Arizona.

Cavelaars, J. C. (1974) Subsurface field drainage systems, in *Drainage Principles and Applications: Design and Management of Drainage Systems*, Publication 16, Vol. IV.

Chatel, B. (1976) *Solar Energy Applications for the Arid Areas*, Office of Science and Technology, United Nations, New York.

Christiansen, J. E. and Davis, J. R. (1967) Sprinkler irrigation systems, in R. M. Hagan, H. R. Haise and T. W. Edminster, eds., *Irrigation of Agricultural Lands, Agronomy* No. 11, Madison, Wisconsin, pp. 885–904.

Claveran, R. and Beneitez, A. (1973) *Rentabilidad financiera de una inversión de agua en pastizales áridos*, Banco de México y A.N.D.S.A, México City.

Cornejo, A. T. (1970) Resources of arid South America, in *Arid Lands in Transition*, AAAS, Washington, D.C.

Costin, E. *et al.* (1974) *Sand Dune Fixation and Afforestation in Some Semi-Desert Countries of the Middle East*, Inter-Regional Training Centre on Heath Land and Sand Dune Afforestation, FAO, Rome.

Criddle, W. D. and Kalisvaart, C. (1967) Subirrigation systems, in R. M. Hagan, H. R. Haise and T. W. Edminster, eds., *Irrigation of Agricultural Lands, Agronomy*, No. 11, Madison, Wisconsin.

Dakshinamurti, C. A., Michael, M. and Mohan, S. (1973) *Water Resources of India*, Water Technology Centre, Indian Agricultural Research Institute, New Delhi.

Dalrymple, D. G. (1973) *Controlled Environment Agricultural, A Global Review of Greenhouse Food Production*, U.S. Dept. of Agriculture, Washington, D.C.

Dept. of Agriculture of Western Australia (1950) *Roaded Catchments of Farm Supplies*, Bulletin 2393, Perth, Western Australia.

Despois, J. (1961) Development of land use in Northern Africa, in *A History of Land Use in Arid Regions*, UNESCO, Paris.

Donnan, W. and Schwab, G. O. (1974) Current drainage methods in the U.S.A., in J. Van Schilfgaarde, ed., *Drainage for Agriculture, Agronomy* No. 17, Madison, Wisconsin.

Dufflie, J. A. and Beckman, W. A. (1976) Solar heating and cooling, *Science*, **191**: No. 4223.

Evenary, M., Shanan, L. and Tadmor, N. (1971) *The Negev, The Challenge of a Desert*, Harvard University Press, Cambridge, Mass.

Erikson, A. E. (1972) Improving the water properties of a sand soil, in D. Hillel, ed., *Optimizing the Soil Physical Environment Toward Greater Crop Yields*, Academic Press, New York and London.

FAO (1960) *Soil Erosion by Wind and Measures for its Control on Agricultural Lands*, FAO Agric. Development Paper No. 71, Rome.

FAO (1966) *Equipment and Methods for Tied Ridge Cultivation*, prepared by the Cultivation Department of the National Institute of Agricultural Engineering (U.K.).

FAO (1968) *Métodos de lucha contra los incendios forestales*, Barcelona, Spain.

FAO (1969) Hopfen, H. J. *Farm Implements for Arid and Tropical Regions*, Rome.

FAO *Tillage and Seeding Practices and Machines for Crop Production in Semi-arid Areas*, Agricultural Development Paper No. 92.

FAO (1974) *The Ecological Management of Arid and Semi-Arid Rangelands in Africa and the Near East*, Report of an Expert Consultation held in Rome with the support of UNEP, Rome.

Figueroa, S. B. (1975) Pérdidas de suelo y nutrimentos y su relación con el uso del suelo en la cuenca del Rio Tezcoco, Tésis de Maestro en Ciéncias, Colegio de Postgraduados, Chapingo, México.

Forti, M. (1971) *Introduction of Fodder Shrubs and Their Evaluation for Use in Semi-arid Areas in the North Western Negev*, The Negev Institute for Arid Zone Research, Beersheba, Israel.

Framji, K. K. and Gara, B. C. (1976) *Flood Control in the World, A Global Review*, International Commission on Irrigation and Drainage, New Delhi.

Gottschalk, L. C. (1962) *Effects of Watershed Protection Measures on Reduction of Erosion and Sediment Damage in the United States*, Intern. Ass. Sci. Hydrol. Publ. 59.

Grigg, D. B. (1974) *The Agricultural Systems of the World, An Evolutionary Approach*, Cambridge University Press.

Hare, F. K. (1976) *Climate and Desertification*, a review prepared for the United Nations Conference on Desertification.

Heady, M. F. (1975) *Rangeland Management*, McGraw-Hill, New York.

Hill, E. S. (1966) *Arid Lands, A Geographical Appraisal*, Methuen, London, UNESCO, Paris.

Hollick, M. A. (1974) *The Design of Roaded Catchments for Farm Water Supplies*, Part I, *Design to Minimise Erosion*, Department of Civil Engineering, University of Western Australia.

Hudson, N. (1971) *Soil Conservation*, Cornell University Press, Ithaca, New York.

Hutchinson (Publishers) (1973) *Irrigation, Drainage and Salinity*, London.

Intermediate Technology Development Group (1969) *The Introduction of Rainwater Catchment Tanks and Micro-Irrigation to Botswana*, Parnell House, London.

Israelsen, O. W. and Hansen, V. E. (1965) *Irrigation Principles and Practices*, John Wiley and Sons, New York.

Jackson, Ray D. and van Bavel, C. H. M. (1965) Solar distillation of water from soil and plant materials: a simple desert survival technique, *Science*, 149 (3690).

Kaczmarek, Z. *The Promise of Technology, Scale and Ingenuity*, Topic II of Draft Agenda, United Nations Water Conference.

Kampen, J. (1974) Soil and water conservation and management in farming systems research for the semi-arid tropics, in *International Workshop on Farming Systems*, ICRISAT, Hyderabad, India.

Kassas, M. (1966) *Plant Life in Deserts, in Arid Lands: A Geographical Appraisal*, E. S. Hills, ed., Methuen, London, UNESCO, Paris.

Kassas, M. (1970) Desertification versus potential for recovery in circum-Saharan territories, in *Arid Lands in Transition*, H. Dregne, ed., American Association for the Advancement of Science, Washington, D.C.

Kates, R. W. and Johnson, D. R. (1976) *Population, Society and Desertification*, a review prepared for the United Nations Conference on Desertification.

Kaul, R. M. (1968) *The Indian Sub-Continent: Indo-Pakistan, Afforestation in Arid Zones*, Junk, The Hague, FAO.

Kohnke, H. (1950) *The Reclamation of Coal Mine Soils*, Purdue University Agricultural Experiment Station, Lafayette, Indiana.

Kovda, V. A. (1974) Problems of land degradation of arid soils, in *A World Assessment of Soil Degradation*, FAO, Rome.

Kovda, V. A., van den Berg, C. and Hagan R. M. eds. (1973) *Irrigation, Drainage and Salinity, An International Source Book*, FAO/UNESCO.

Le Houérou, H. N. (1975) *Science, Power and Desertification*, International Geographical Union Working Group on Desertification, Dept. of Geography, University of Cambridge.

Lustig, H. (1974) *Solar Energy; The State of the Art and the Art of the State*, Division of Technology, Higher Education and Research, UNESCO, Paris.

Makhijani, A. and Poole, A. (1975) *Energy and Agriculture in the Third World*, Ballinger Publishing Co., Cambridge, Mass.

Megahan, W. (1976) Reducing erosional impacts of roads, in *Watershed Management Guidelines and Examples*, FAO Forestry Department, Rome.

Morin, J. and Agassi, M. (1973) *Methods and Materials for Roadside Stabilization by Vegetation*, Soil Conservation and Drainage Division, Ministry of Agriculture, Tel Aviv, Israel.

National Academy of Sciences (1974) *More Water for Arid Lands, Promising Technologies and Research Opportunities*, Washington, D.C.

Paliwal, K. V. (1972) *Irrigation with Saline Water*, Water Technology Centre, Indian Agricultural Research Institute, New Delhi, Monograph No. 2.

Pimentel, D. (1976) *Energy in Food (and Feed)*, American Association for the Advancement of Science, Boston.

Ramos, F., Dominguez, L., Jiménez, F. and Soriano, C. (1974) *Tratamiento funcional y paisajistico de taludes artificiales*, Ministerio de Agricultura, Instituto Nacional para la Conservación de la Naturaleza, Madrid.

Rawitz, E. and Heller J. (1969) Irrigation methods and equipment, in B. Yaron, R. Danfors and Y. Vaadia, eds., *Irrigation in Arid Zones*, draft edition, Beth Dagan, Israel.

Rege, N. D. (1964) Managing arable lands in arid and semi-arid regions, in *Proceedings of the Symposium on Problems of the Indian Arid Zone*, Jodhpur, New Delhi.

Reeve, R. C. and Bower, C. A. (1960) Use of high-salt waters as a flocculant and a source of divalent cations for reclaiming sodic soils, *Soil Science*, **90**: 139–144.

Schumacher, E. F. (1973) *Small Is Beautiful*, Harper & Row, New York.

Skurlow, J. U. E. (1959) *Control of Roadside Erosion by Soil Conservation Methods.*

Stout, B. A. and Downing, C. M. (1975) Agricultural mechanization policy, Expert Meeting of the Effects of Farm Mechanization on Production and Employment, Rome, Italy.

Taylor, S. A. and Ashcroft, G. L. (1972) *Physical Edaphology, The Physics of Irrigated and Non-irrigated Soils.*

Tejwani, K. G., Gupta, S. K. and Mathur, H. N. (1975) *Soil and Water Conservation Research, 1956–71*, Indian Council of Agricultural Research, New Delhi.

Tewari, A. K. (1976) *Rough Pastures and their Utilization in Jaisalmer Desert*, University of Jodhpur, India.

Theobald, G. A. (1968) *Methods and Machines for Tile and Other Tube Drainage*, FAO Agricultural Development Paper No. 78, Rome.

Trafford, B. D. (1971) Drainage of heavy soils, in *Drainage of Heavy Soils, Irrigation and Drainage*, Paper No. 6, FAO, Rome.

UNESCO (1975) *Obstacles to Development of Arid and Semi-arid Zones*, Committee on Science and Technology Development.

U.S. Department of the Interior, Bureau of Reclamation (1952) *Canal Linings and Methods of Reducing Costs.*

United Nations Economic and Social Council (1975) *Obstacles to Development of Arid and Semi-arid Zones*, Report of the Ad-hoc Interagency Task Force, World Programme of Development Research and Application of Science and Technology to Solve the Special Problems of the Arid Areas.

U.S. Public Health Service (1963) *Manual of Individual Water Supply Systems*, Publication No. 24, Washington, D.C.

Wagner, E. G. and Lanoix, J. N. (1959) *Water Supply for Rural Areas and Small Communities*, WHO, Geneva.

Warren, A. (1976) *Ecological Change and Desertification*, a review prepared for the United Nations Conference on Desertification.

Watt, S. B. (1975) Approaches to water pumping in West Africa, Seminar on Small Scale Water Resources Development in West Africa, Ouagadougou, Upper Volta.

Webber, G. D., Cocks, P. S. and Jefferies, B. C. (1976) *Farming Systems in South Australia*, Department of Agriculture and Fisheries, Adelaide, South Australia.

White, G. W. (1962) Alternative uses of limited water supplies, in *The Problems of the Arid Zone.*

White, G. W. (1976) Resources and needs: assessment of the world water situation, United Nations Water Conference.

Whitney, R. W., *et al.* (1967) *Pasture Pitting Machines*, Oklahoma Agricultural Experimental Station, Bulletin No. 657.

Zingg, A. W. and Hauser, V. L. (1959) Terrace benching to save potential runoff for semi-arid land, *Agronomy Journal*, **51**: 289–292.